UNDERSTANDING ORGANIC CHEMISTRY

UNDERSTANDING ORGANIC CHEMISTRY

Bridging Fundamental and Advanced Courses

Fourth Edition

ANN M. FABIRKIEWICZ
Department of Chemistry
Randolph College
Lynchburg, Virginia

JOHN C. STOWELL[†]
Department of Chemistry
University of New Orleans
New Orleans, Louisiana

WILEY

Published by John Wiley & Sons, Inc., Hoboken, New Jersey.
Published simultaneously in Canada.

Library of Congress Cataloging-in-Publication Data applied for

Hardback ISBN: 9781394179831
ePDF: 9781394179848
epub: 9781394179855

Cover Design: Wiley
Cover Image: Courtesy of Ann Fabirkiewicz

Set in 11/14pt TimesLTStd by Straive, Chennai, India

CONTENTS

PREFACE TO THE FOURTH EDITION

In this new edition, I've tried to hold true to the features that I appreciated in this text—size, readability, references to the primary literature, and particularly the open-access literature. Organic chemistry owns an ever-expanding knowledge base, and new methods are always under investigation. Environmental concerns are increasingly important, and green chemistry is integrated throughout the text. The organization of the chapters remains the same, with foundational topics like nomenclature and stereochemistry presented early, followed by mechanistic topics as a foundation for the study of reactions, but the book is designed to allow chapters to be introduced independently. Readers will find new sections in several chapters and updated examples in many places. Additional problems have been added, and a partial answer key included as an appendix. Because of its dependence on internet resources, Chapter 2 has been completely updated. Students are encouraged to consult the cited literature frequently to engage with the content in greater depth, and to gain understanding of the motivation for chemical studies and an appreciation of the details of the experiments. Reading the literature of organic chemistry will help you see and recognize patterns and become familiar with the discipline and the challenges and curiosities it holds.

My thanks to those who reached out with suggestions from the previous edition. Comments and corrections are always welcome. Please email me at afab@randolphcollege.edu and let me know what you think.

A project like this doesn't occur in isolation and I'm grateful for the support of so many people. To the colleagues who helped me through some long days, to my husband who always has dinner on the table, to my son who is an invaluable source of support and honest feedback, to my brother who always texts me back—thank you. Tama Paul and the team at Wiley have helped immeasurably with this work and I appreciate all their help.

Ann M. Fabirkiewicz
Lynchburg, Virginia
March 2025

PREFACE TO THE THIRD EDITION

When I read the first edition of Stowell's Intermediate Organic Chemistry, I knew I had found an ideal text for a one-semester advanced organic chemistry course. The length of the text was manageable in one semester, and the extensive references into the primary literature allowed that introduction for students interested in pursuing organic chemistry in greater depth and made the text a useful reference beyond the scope of the course.

In this revision, I tried to stay true to the features that attracted me to the text initially, well described in Professor Stowell's preface to the second edition. The organization of the book remains much the same. The chapter on mechanisms has moved earlier in the book, as has the chapter on pericyclic reactions. My inclination is to teach mechanisms early as a foundation for the study of reactions, but all the chapters have been written with the intent that they can be introduced to the students in any order. As much as possible, open-source materials are referenced allowing students ready access to the original literature, but at the same time, the early work on a topic is found in older literature, and that historical value is not ignored. The second chapter has been completely rewritten to focus on internet accessible resources. The last chapter has also been completely rewritten as a survey of organic spectroscopy.

I have tried to make this text as error free as possible, knowing the frustration of a missing or mistaken reference, or a bond misplaced in a chemical structure. I welcome input from readers of this text. Please contact me directly at afab@randolphcollege.edu.

I want to thank my husband Steve for his patience and support during the preparation of this manuscript. My son Adam is a source of support and insight. My mom is an inspiration. My colleague Abraham Yousef at Sweet Briar College was helpful in obtaining Figures 10.6 and 10.14. Jonathan Rose and the team at Wiley have been responsive to all my questions and their assistance is invaluable. The support of family and colleagues has made this work possible.

ANN M. FABIRKIEWICZ
Lynchburg, Virginia
November 2014

PREFACE TO THE SECOND EDITION

Consider the typical student who, having finished the two-semester introductory course in organic chemistry and then picking up an issue of the *Journal of Organic Chemistry*, finds the real world of the practicing chemist to be mostly out of reach, requiring a higher level of understanding. This text is intended to bridge that gap and equip a student to delve into new material.

There are two things to learn in studying organic chemistry. One is the actual chemistry, that is, the behavior of compounds of carbon in various circumstances. The other is the edifice of theory, vocabulary, and symbolism that has been erected to organize the facts. In first-year texts, there is an emphasis on the latter with little connection to the actual observables. Thus students are able to answer subtle questions about reactions without knowing quite how the information is obtained. Chemistry is anchored in observations of specific cases, which can be obscured by the abstractions. For this reason this text includes specific cases with more details and literature references to illustrate the general principles. Understanding these cases is an exercise to ensure the understanding of those general principles in a concrete way.

Many of the problems begin with raw data and require multistep thinking. Therefore the student must solve a problem from the beginning rather than from a half-finished setup, and more thought and puzzling is necessary.

This book is a selection. Out of the endless possibilities of reactions, a specific limited set was chosen and the chapters are consistent within that set. Likewise, the problems in the chapters were chosen for requiring only that set plus reactions from the introductory organic courses. These selections were made on the basis of their appearance in many current journal articles. Review references are given to aid a student who wishes to go further with these topics. Subjects that are generally well covered in introductory courses are omitted or briefly reviewed here. Advanced topics are treated to a functional level but not exhaustively.

For those familiar with the first edition, you will find that major changes have been made in all the chapters. New examples have been added, and many others have been replaced with better ones. The explanations have been elaborated and illuminated with more detail. The same 10 chapters have been retained, but new sections have been added, and material reorganized. Substantial updating has been done including stereochemical terminology, and the NMR chapter has been reframed in terms of pulsed high-field spectrometers. About twice as many exercise problems are available at the end of most chapters.

John C. Stowell
New Orleans, Louisiana
October 1993

PREFACE TO THE FIRST EDITION

Consider the typical student who has finished the two-semester introductory course in organic chemistry and then picks up an issue of the *Journal of Organic Chemistry*. He or she finds the real world of the practicing organic chemist to be mostly out of reach on a different level of understanding. This text is intended to bridge that gap and equip a student to delve into new material.

There are two things to learn in studying organic chemistry. One is the actual chemistry, that is, the behavior of compounds of carbon in various circumstances. The other is the edifice of theory, vocabulary, and symbolism that has been erected to organize the facts. In first-year texts, there is an emphasis on the latter with little connection to the actual observables. Thus students are able to answer subtle questions about reactions without knowing quite how the information is obtained. Chemistry is anchored in observations of specific cases, which can then be obscured by the abstractions. For this reason this text includes specific cases with more details and literature references to illustrate the general principles. Understanding these cases is also an exercise to ensure the understanding of those principles in a concrete way.

This book is by necessity a selection. Subjects that are generally well covered in introductory texts are omitted or briefly reviewed here. Advanced topics are treated to a functional level, but not exhaustively.

Specifically, the subjects are those necessary for understanding and searching the literature and some topics that are the elements of many current journal articles. The outcome is a selected sampling on a scale with latitude for creative lecturers to amplify with their own selections. Advanced texts give much attention to classical work that led to modern understanding. This text, with all due respect to the originators, does not cover that familiar ground but again covers current examples grounded in those classical ideas with modern interpretation.

There is a modicum of arbitrariness in the selections, and, considering this text as a new experiment, the author would welcome suggestions for substitutions and improvements.

John C. Stowell
New Orleans, Louisiana
November 1987

ABOUT THE COMPANION WEBSITE

This book is accompanied by instructor website. The URL for the site is

www.wiley.com/go/fabirkiewiczorganic4e

The instructor's website includes:

- PowerPoint Slides
- Solutions to Chapter Problems

1

READING NOMENCLATURE

Organic chemistry is understood in terms of molecular structures as represented pictorially. Cataloging, writing, and speaking about these structures requires a nomenclature system, the basics of which you have studied in your introductory course. To go further with the subject, you must begin reading journals, and this requires an understanding of the nomenclature of complex molecules. This chapter presents a selection of compounds to illustrate the translation of names to structural representations. The more difficult task of naming complex structures is not covered here because each person's needs will be specialized and can be found in nomenclature guides [1–5]. Most of the nomenclature rules are used to eliminate alternative names and arrive at a unique (or nearly so) name for a particular structure; thus, when beginning with names, you will need to know only a small selection of the rules in order to simply read the names and provide a structure. Although the subject of nomenclature is vast, these selections will enable you to understand many names in current journals.

Understanding Organic Chemistry: Bridging Fundamental and Advanced Courses,
Fourth Edition. Ann M. Fabirkiewicz and John C. Stowell.

Companion website: www.wiley.com/go/fabirkiewiczorganic4e

1.1 ACYCLIC POLYFUNCTIONAL MOLECULES

(2*E*,6*E*)-2,6-Dimethylocta-2,6-diene-1,8-diol

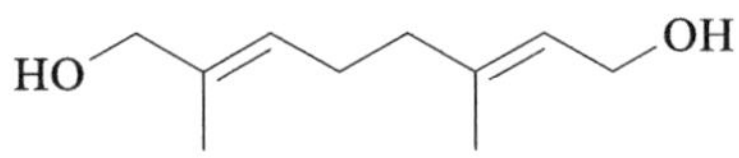

The alcohol groups are the highest in priority, so the molecule is named as an alcohol. Two alcohol groups, located at the ends of the molecule, make it a diol. The molecule is numbered so as to give the remaining groups, the alkenes and the methyls, the lowest possible numbers. Both double bonds are in the *E* or trans configuration.

3-(*S*)-*trans*-1-Iodo-1-octen-3-ol methoxyisopropyl ether

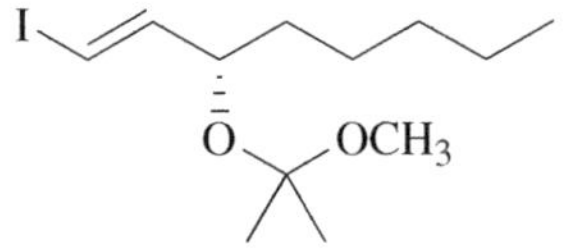

This is an example of a derivative name, that is, the first word is the complete name of an alcohol, and the other two words describe a derivatization in which the alcohol is converted to an ether (ketal). Such a name would be useful in discussing a compound that has the ketal present as a temporary entity, for example, as a protecting group.

Ethyl (*E*,3*R**,6*R**)-3,6,8-trimethyl-8-[(trimethylsilyl)oxy]-7-oxo-4-nonenoate

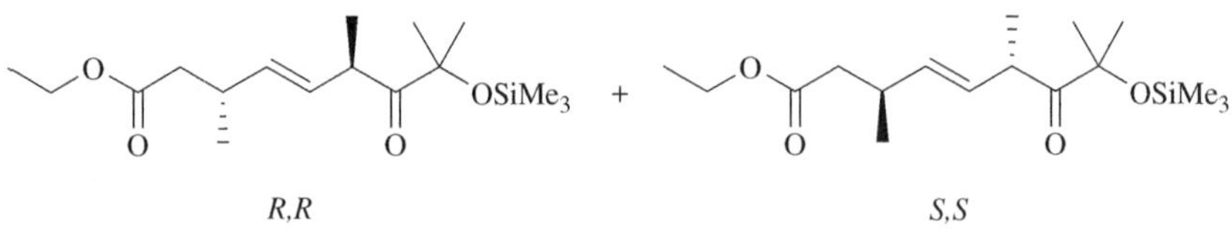

This is the ethyl ester of a nine-carbon unsaturated acid with substituents. The *oxo* indicates that there is a keto function on carbon 7. Be careful to distinguish this from the prefix *oxa*-, which has a different meaning; see Section 1.6. The asterisks indicate that the configuration

designation is not absolute but rather represents the 3*R*,6*R* stereoisomer and/or its enantiomer. Thus, this name represents the *R*,*R* and/or the *S*,*S* isomers, but not *R*,*S* or *S*,*R*. This designation excludes diastereomers and is a common way to indicate a racemate.

1.2 MONOCYCLIC ALIPHATIC COMPOUNDS

2-(4-Acetylphenyl)cyclopentan-1-one

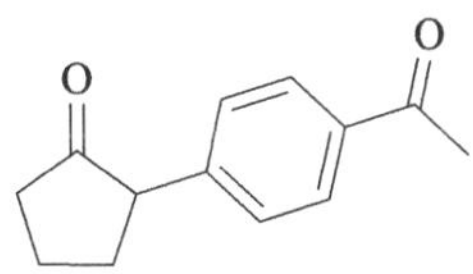

The benzene ring is named as a substituent with an acetyl group in the fourth position. This substituent is located on C-2 of the cyclopentanone ring, with C-1 bearing the carbonyl group.

[2*S*-(1*E*,2α,3α,5α)]–[3-(Acetyloxy)-2-hydroxy-2-methyl-5-(methylethenyl) cyclohexylidene]acetic acid ethyl ester

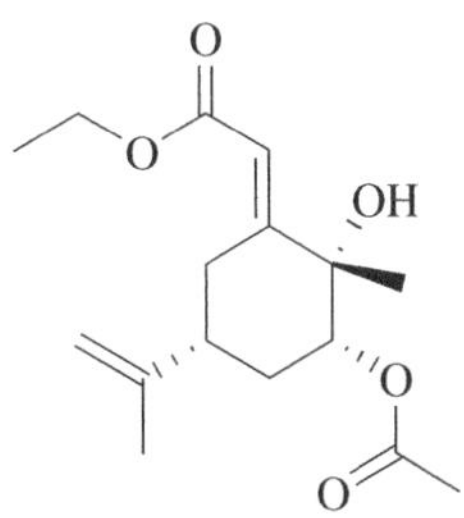

The *ylidene* indicates that the cyclohexyl is attached to the acetic acid by a double bond, and the ethyl ester is indicated at the end for simplicity. The double-bonded ring atom is carbon 1, and the substituents on the ring are placed on the ring according to their locant numbers. The *E* indicates the geometry of the double bond. All the α substituents reside on one face of the ring, cis to each other. Any β substituents would reside on the opposite face of the ring, trans to the α substituents. Where two substituents are on the same ring atom, as on carbon 2 in this case, the Greek letter indicates the position of the higher-priority substituent. Here, the hydroxy, acetyloxy, and methylethenyl groups are all cis to each other on the ring.

1.3 BRIDGED POLYCYCLIC STRUCTURES

The nomenclature of bridged polycyclic systems requires additional specifications. A bicyclic system would require two bond breakings to open all the rings, a tricyclic system, three, and so on. Rather than viewing these compounds as rings, certain carbons are designated as bridgeheads from which the bridges branch and recombine. In the system below, the first bridgehead is designated as carbon 1, and the system is numbered around the largest bridge to the second bridgehead, carbon 5. Numbering continues around the medium bridge, then the smallest bridge, as shown. The compound is named bicyclo[3.2.1]octane.

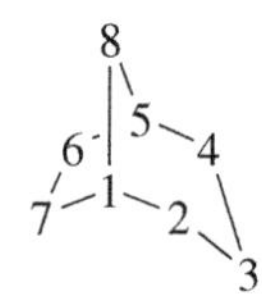

All bicyclo compounds require three numbers in brackets, tricyclo require four, and so on, and these numbers indicate the number of carbons in the bridges and are used to locate substituents, heteroatoms, and unsaturation. The name of the parent alkane includes the total number of atoms in the bridges and bridgeheads (excluding substituents) and is given after the brackets. The use of prefixes *exo*, *endo*, *syn*, and *anti* to indicate stereochemical choices is demonstrated generally as shown below.

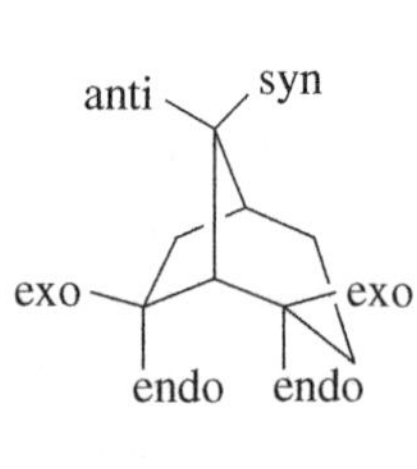

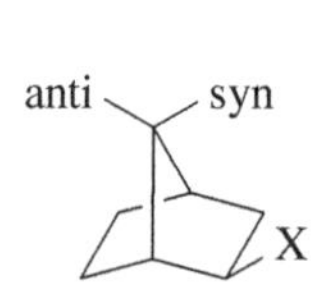

In tricyclic compounds, the relative stereochemistry among the four bridgeheads requires designation. Look at the largest possible ring in the molecule and consider the two faces of it. If there are no higher-priority substituents on the primary bridgehead atoms, the smallest bridge (but not a zero bridge) defines the α face. If the smallest non-zero bridge at the secondary bridgeheads is on the same face of the large ring as the α-defining one, it is also designated as α; that is, the two are cis to each other.

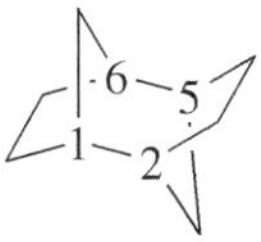

1α,2β,5β,6α

1α,2α,5α,6α

If they are trans, there will be two αs and two βs as illustrated. If there is a zero bridge, the position of the bridgehead hydrogens is indicated with Greek letters, as in the dione example below.

(1α,2β,5β,6α)-Tricyclo[4.2.1.0^{2,5}]non-7-ene-3,4-dione

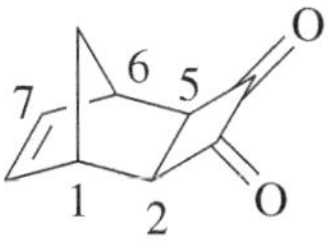

Starting with a pair of bridgeheads, draw the four-, two-, and one-carbon bridges. The zero bridge then connects carbons 2 and 5 as indicated by the superscripts, thus making them bridgeheads also. At bridgeheads 1 and 6, the smallest bridge is considered a substituent and given the α designation at both ends. At bridgeheads 2 and 5, the βs indicate that the hydrogens are trans to the α bridge.

[1*S*-(2-*exo*,3-*endo*,7-*exo*)]-7-(1,1-Dimethylethyl)-3-nitro-2-phenylbicyclo[3.3.1]nonan-9-one

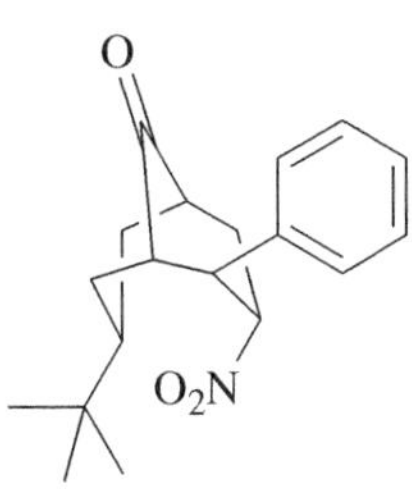

This bicyclo system has bridges with three, three, and one carbons each, indicated by the bracketed numbers separated by periods. Carbon 2 carries a phenyl ring that projects toward the smaller neighboring one-carbon bridge rather than the larger three-carbon bridge, as indicated by 2-*exo*. The 1,1-dimethylethyl group is also *exo*. This group is commonly called *tert-butyl. Chemical Abstracts* names substituents based on the linear group, thus naming ethyl as the longest chain, and the two methyl groups as substituents. The prefixes *exo* and *endo* indicate the stereochemistry.

Sometimes, a bridgehead substituent will have a higher priority than the smallest bridge thereon. The designation for that bridgehead will indicate the position, α or β, of that higher-priority substituent rather than the bridge, as illustrated in the next example.

(1α,2β,4β,5β)-5-Hydroxytricyclo[3.3.2.0^{2,4}]deca-7,9-dien-6-one

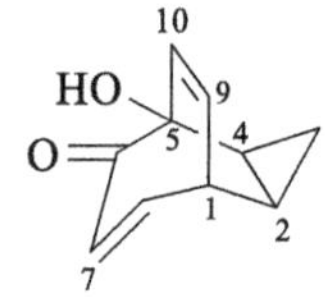

At bridgehead 1, the smallest bridge, between carbons 9 and 10, is considered a substituent on the largest ring and is designated α. The hydrogens at carbons 2 and 4 are trans to it and are designated β. The OH group on carbon 5 is a higher-priority substituent than the C9–C10 bridge and is trans to the bridge; thus, it is labeled β.

1.4 FUSED POLYCYCLIC COMPOUNDS

Fused-ring compounds have a pair or pairs of adjacent carbon atoms common to two rings. Over 35 carbocyclic examples have trivial names, some of which need to be memorized as building blocks for names of more complex examples. The names end with *-ene*, indicating a maximum number of alternating double bonds. A selection is illustrated in Table 1.1, showing one resonance form for each. Others can be found online [6].

Fusing more rings onto one of these basic systems may give another one with a trivial name. If not, a name including the two rings or ring systems with bracketed locants is used, as in the following example.

TABLE 1.1 Trivial Names of Some Fused Polycyclic Hydrocarbons

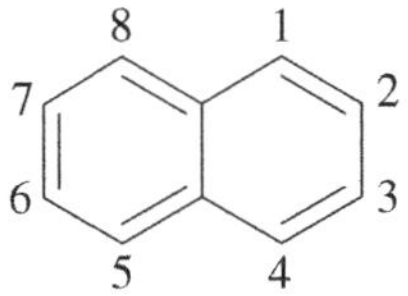

Naphthalene

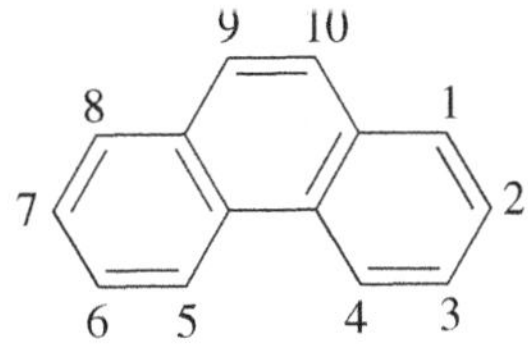

Phenanthrene[a]

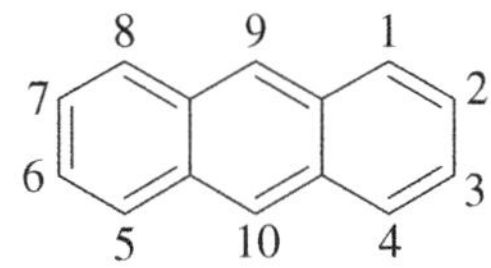

Anthracene[a]

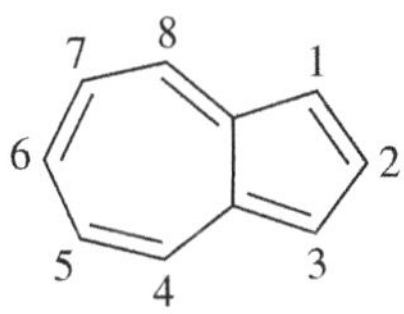

Azulene

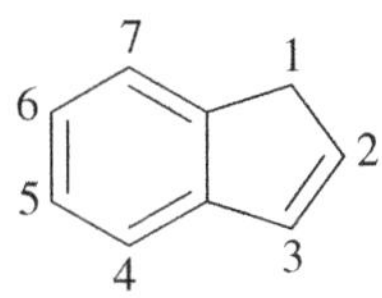

Indene

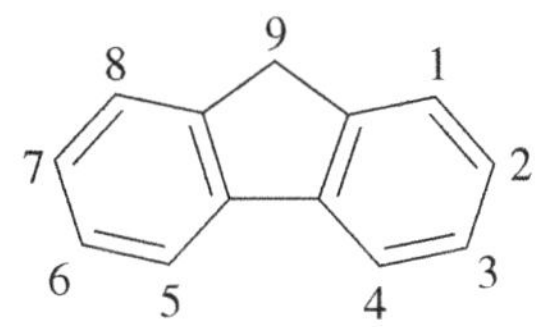

Fluorene

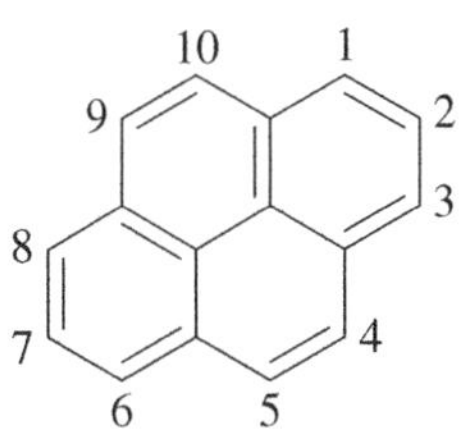

Pyrene

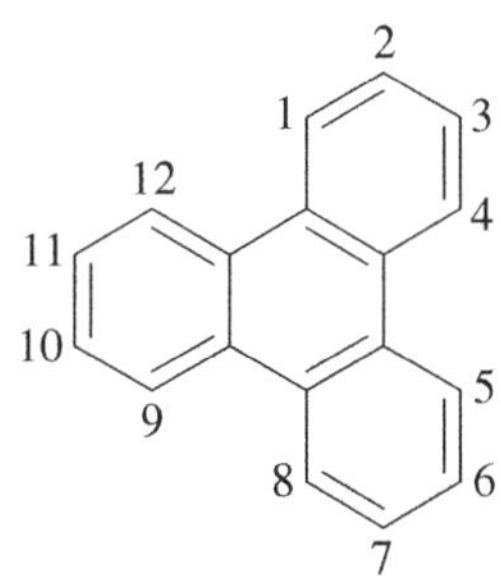

Triphenylene

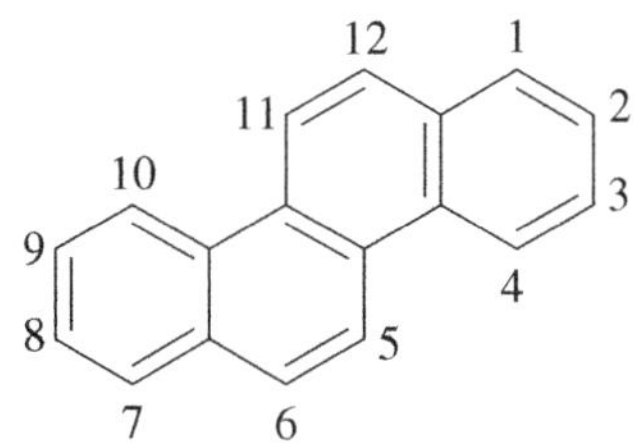

Chrysene

[a]Exceptions to systematic numbering.

Benz[*a*]anthracene-7,12-dione

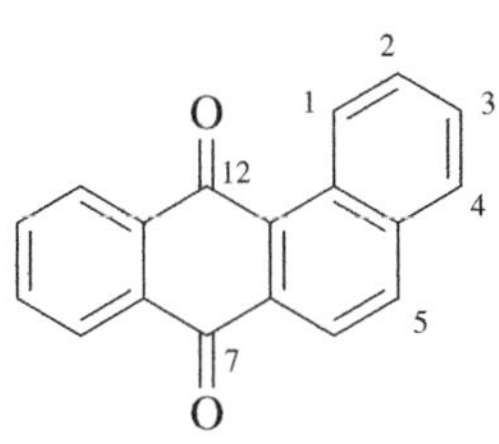

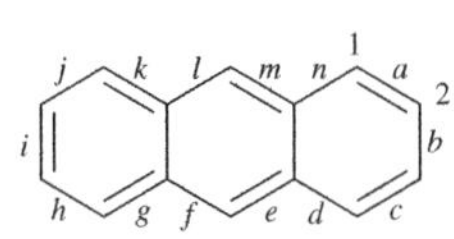

Anthracene

A benzene ring is fused to one side of anthracene, so the sides of the anthracene are labeled *a*, *b*, *c*, and so on, where carbons numbered 1 and 2 constitute side *a* and 2 and 3 constitute side *b*, continuing in order for all sides. The earliest letter possible on the anthracene is used to indicate the position of fusion, and the ring fused to it appears first in the name. The "o" ending of benzo is deleted here because it would be followed by a vowel.

The final ring system is then renumbered to locate substituents, or sites of reduced unsaturation. To renumber, first orient the system so that a maximum number of fused rings are in a horizontal row. If there are two or more choices, place a maximum number of rings to the upper right. Then number clockwise starting with the carbon not involved in fusion in the most counterclockwise position in the uppermost, farthest-right ring. See the numbering in the systems with trivial names above. Atoms at the fusion sites, which could carry a substituent only if they were not π-bonded, are given the number of the previous position with an *a*, *b*, *c*, and so on. Where there is a choice, the numbers of the fusion carbons are minimized too; for example, 4b < 5a:

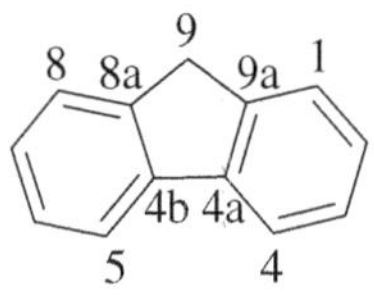

Correct

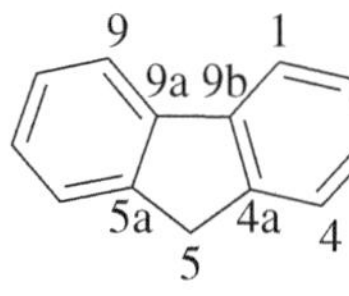

Incorrect

1*H*-Benz[*cd*]azulene

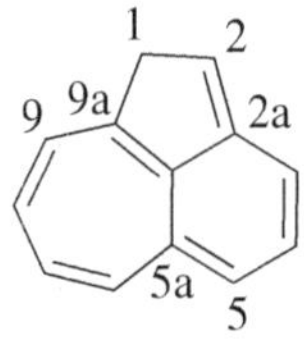

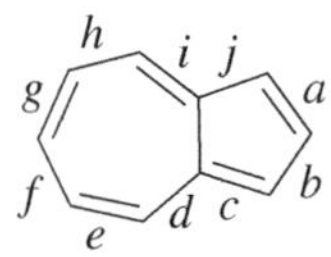

azulene

First, letter the sides of azulene. The benzene ring is fused to both the *c* and *d* faces as indicated by the brackets. Now, reorient the system for numbering. The choice of which two rings go on the horizontal axis and which one is in the upper position is determined by which orientation gives the smallest number to the first fusion atom—in this case, 2a instead of 3a, or 4a. In molecules where one carbon remains without a double-bonding partner, it is denoted by *H*. This is called *indicated hydrogen* and is used even when an atom other than hydrogen is actually on that carbon in the molecule of interest.

trans-1,2,3,4-Tetrahydrobenzo[*c*]phenanthrene-3,4-diol diacetate

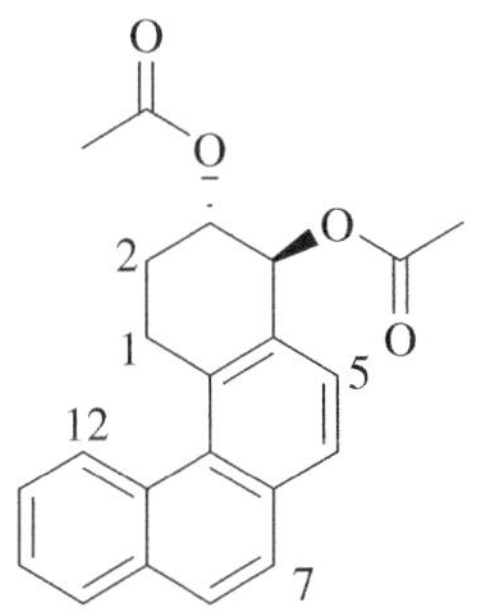

Phenanthrene is numbered as shown in Table 1.1. The *c* side falls between carbons 3 and 4. A benzene ring is fused to the *c* side, and a new systematic numbering is made. Carbons 1–4 have hydrogens added to saturate two double bonds, and then carbons 3 and 4 have hydroxy groups substituted for hydrogens in a trans arrangement. Finally, the compound is named as the acetate ester at both alcohol sites.

6,13-Dihydro-5*H*-indeno[2,1-*a*]anthracene-7,12-diol

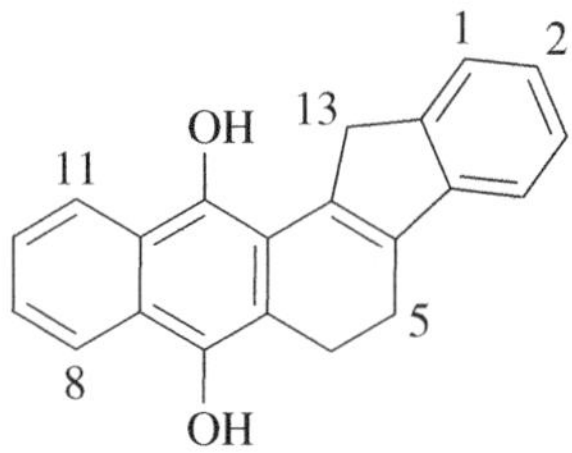

The fusion of this ring system is given in the brackets. The bracketed *a* precedes anthracene, indicating that the sides of anthracene will be lettered, and the bracketed 2,1 follows the indene, thus the indene numbering indicated in Table 1.1 is used to give the point of fusion. The order of the numbers indicates the direction of the fusion; thus, the carbons of indene are fused in order 2, then 1 to the *a* side of anthracene, with the number 1 carbon of anthracene constituting the number 2 carbon of indene, and the number 2 carbon of anthracene, the number 1 carbon of indene. The united system is then renumbered according to the rules, and the substituents, added hydrogen, and indicated hydrogen are placed accordingly. The indicated hydrogen is assigned to the lowest-numbered atom not involved in double bonding.

The other direction of fusion is seen in 1*H*-indeno[1,2-*a*]anthracene:

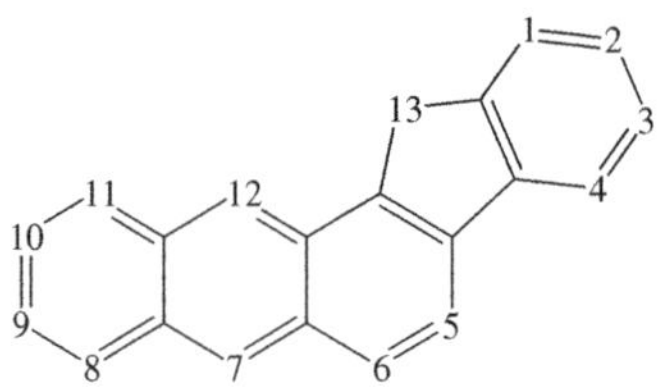

[2,1-*a*] fusion

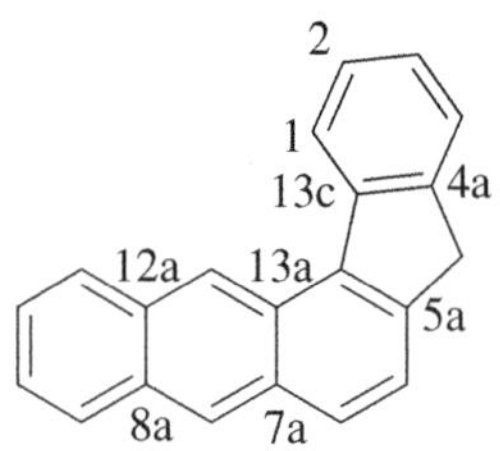

[1,2-*a*] fusion

Note that here a different orientation is used because it gives lower fusion numbers.

1.5 SPIRO COMPOUNDS

In spiro compounds, a single atom is common to two rings. There are two kinds of nomenclature for these. Where there are no fused rings present, the carbons of both rings are counted in one series as in the bicyclic nomenclature, and the hydrocarbon name includes the carbons of both rings, as in the following example.

1,5-Dioxaspiro[5,5]undecane-3-one

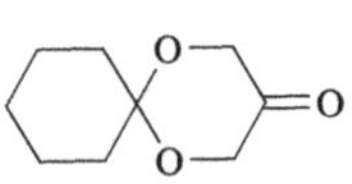

Of the eleven atoms, one is the spiro atom and two are oxygens. Locant numbers begin at an atom adjacent to the spiro one, an oxygen in this case, then continue around the ring to the carbonyl group, then to the second oxygen. The numbers in brackets indicate the number of atoms attached to the spiro one.

When ring fusion is also present, the two ring systems that share the spiro atom are given in brackets with splicing locants as shown in the following example.

(*S*)-4′-Chloro-3,4-dihydro-1′H, 5*H*-spiro[furan-2,2′-naphthlene]-1′,5-dione

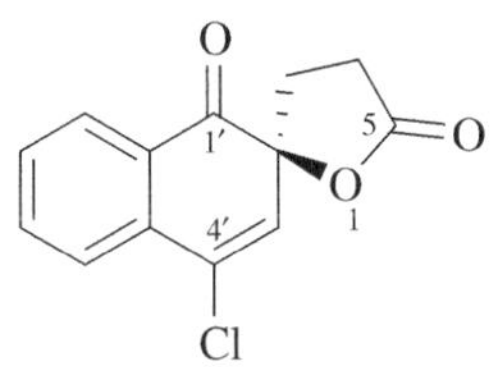

Two separate numbering systems are used for the two rings. The furan ring is given the unprimed numbers, and numbering begins at the oxygen and moves to the fusion site. The naphthalene ring is numbered as usual (Table 1.1) and identified with primes. Within the brackets, 2,2′ indicates the position of the spiro attachment, with the number referring to the ring it is closest to (so 2 indicates the position in the furan ring and 2′ the position in the naphthalene ring). The spiro linkage requires indicated hydrogens in both rings, assigned to positions 1′ and 5.

1.6 MONOCYCLIC HETEROCYCLIC COMPOUNDS

Systematic and trivial names are both commonly in use for heterocyclic compounds. The systematic names consist of one or more prefixes from Table 1.2 with multipliers where needed, designating the heteroatoms, followed by a suffix from Table 1.3 to give the ring size with an indication of the unsaturation. This is preceded by substituents. Thus, oxepin is a seven-membered ring including one oxygen and three double bonds. The ring size is explicit in some of the suffixes as *ep*, *oc*, *on*, and *ec*, which are derived from h*ep*tane, *oc*tane, n*on*ane, and d*ec*ane, respectively. Numbering begins with the element highest in Table 1.2 and continues in the direction that gives the lowest locant to the next heteroatom.

TABLE 1.2 Prefixes in Order of Decreasing Priority[a,b]

Oxygen	ox-
Sulfur	thi-
Selenium	selen-
Nitrogen	az-
Phosphorus	phosph- (or phosphor- before -in or -ine)
Silicon	sil-
Boron	bor-

[a]Adapted from Ref. [7].
[b]An "a" is added after each prefix if followed by a consonant.

TABLE 1.3 Suffixes Indicating Ring Size[a]

	Containing Nitrogen		Containing no Nitrogen	
Atoms in the Ring	Maximally Unsaturated	Saturated	Maximally Unsaturated	Saturated
3	-irine	-iridine	-irine	-irane
4	-ete	-etidine	-ete	-etane
5	-ole	-olidine	-ole	-olane
6	-ine	—[b]	-in	-ane
7	-epine	—[b]	-epin	-epane
8	-ocine	—[b]	-ocin	-ocane
9	-onine	—[b]	-onin	-onane
10	-ecin	—[b]	-ecin	-ecane
>10[c]	—	—	—	—

[a]From Ref. [7]/IUPAC.
[b]Use the unsaturated name preceded by "perhydro."
[c]Use the carbocyclic ring name with heteroatom replacement prefixes: oxa-, thia-, and so forth.

Methyl 4-(5,5-dimethyl-1,3,2-dioxaborinan-2-yl)benzoate

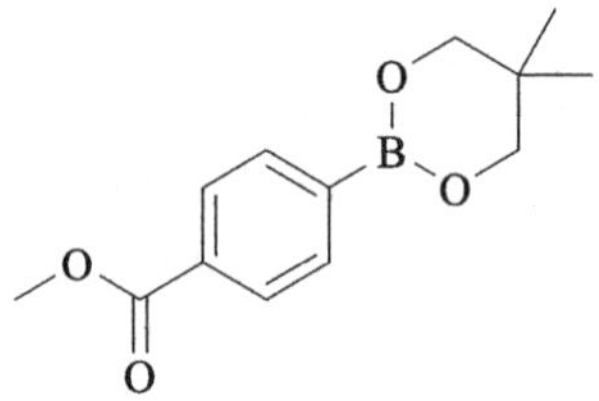

The space after methyl and the “ate” ending tell you this is a methyl ester. The number 4 in front of the parentheses tells you the location of the substituent group, which is named in parentheses, on the benzoate ester. Numbering in the substituent ring begins at one oxygen and proceeds around the ring to give the heteroatoms the lowest possible numbers. Without the boron, this would be a 1,3-dioxanyl ring. Note the change that occurs to indicate the boron at position 2.

(*R*)-4-Phenyl-3-(1,2-propadienyl)oxazolidin-2-one

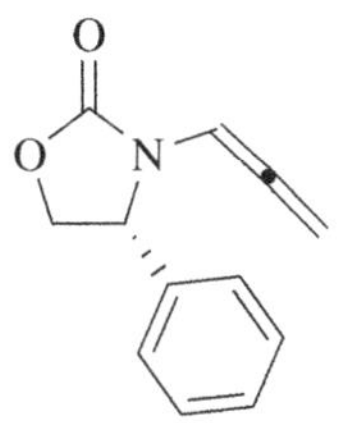

The numbering begins with oxygen for priority. The 1,2-propadienyl group is thus located at position 3, which is the nitrogen. The *–olidin–* ending specifies a saturated five-membered ring, with *–one* indicating the presence of the ketone.

4,7-Dihydro-2-methoxy-1-methyl-1*H*-azepine

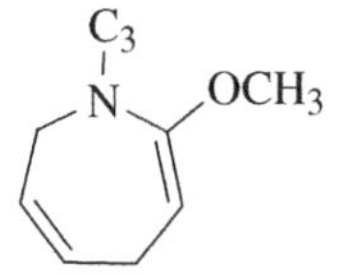

Azepine indicates an unsaturated seven-membered ring containing one nitrogen. Because of the odd number of atoms, at least one of the seven must have an indicated *H*. The dihydro indicates additional hydrogens on two other carbons; therefore, there is one π bond less than maximally unsaturated. Notice that the indicated hydrogen is assigned the lowest-numbered atom not in double bonding (the nitrogen) and *then* replaced by the substituent.

TABLE 1.4 Trivial Names of Some Heterocycles

Saturated				
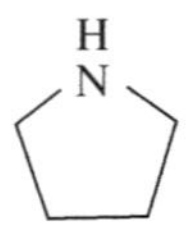	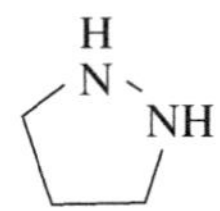	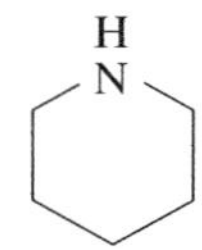	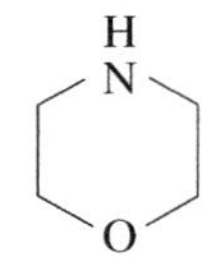	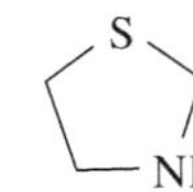
Pyrrolidine	Pyrazolidine	Piperidine	Morpholine	Thiazolidine
Unsaturated				
	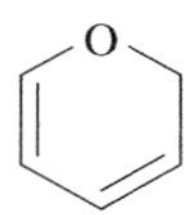	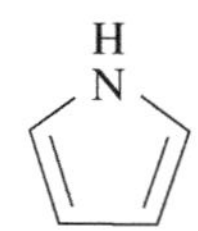	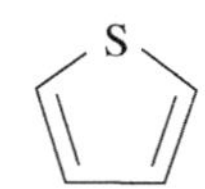	
Furan	2*H*-pyran	Pyrrole	Thiophene	
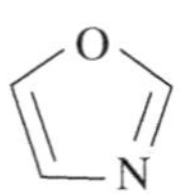	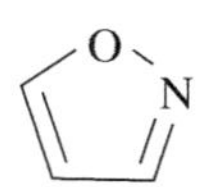	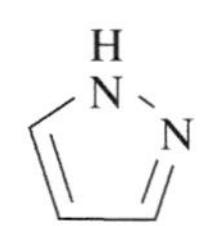	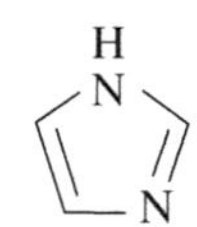	
Oxazole	Isoxazole	Pyrazole	Imidazole	

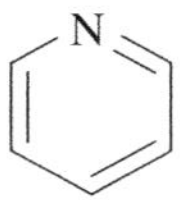
Pyridine

Pyridazine

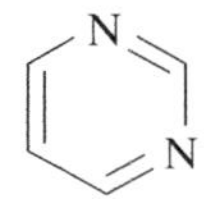
Pyrimidine

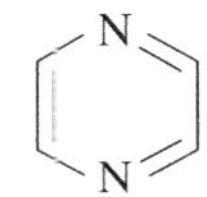
Pyrazine

Fused

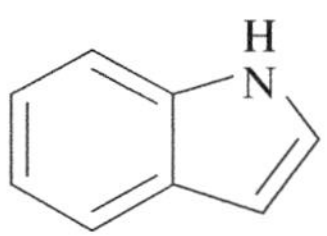
Indole

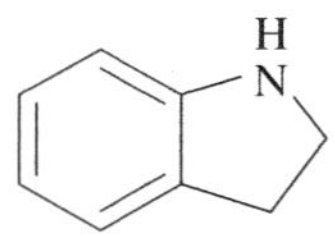
Indoline

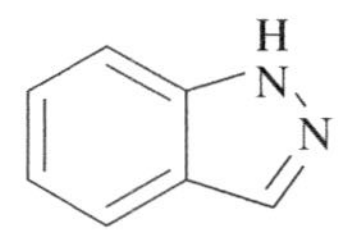
Indazole

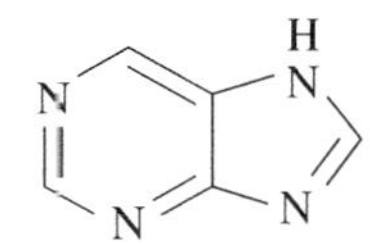
Purine

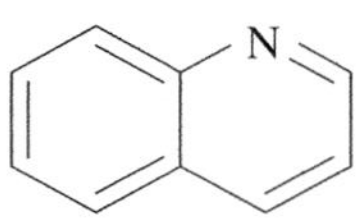
Quinoline

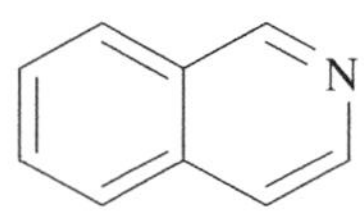
Isoquinoline

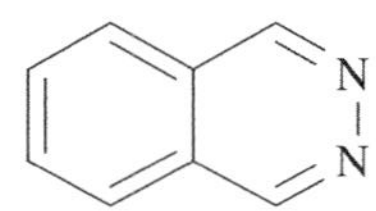
Phthalazine

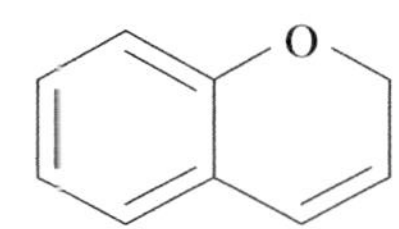
Chromene

(6*R*,14*R*)-6,14-Dimethyl-1,7-dioxa-4-(1-propylthio)cyclotetradec-11-yne-2,8-dione

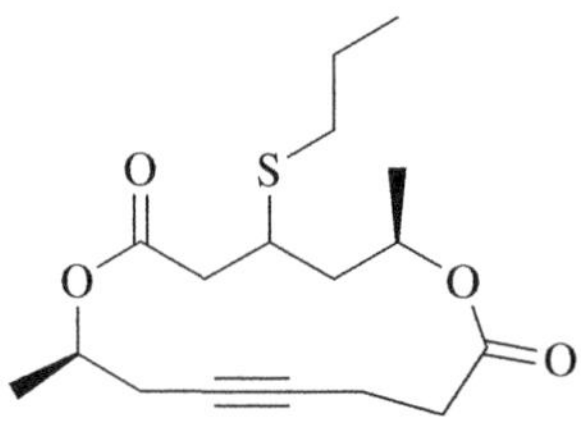

The ring is larger than 10 members; therefore, the hydrocarbon ring name cyclotetradecyne was used, modified by 1,7-*dioxa*, which indicates a replacement of carbons 1 and 7 with oxygens. The *a* ending on *oxa* indicates replacement. The numbering begins at a heteroatom and proceeds to the other heteroatom by the shortest path. The stereochemistry at position 4 is unspecified.

Many five- and six-membered heterocycles and fused ring systems have trivial names that are preferred over the systematic names. Table 1.4 provides a selection of the more common ones. Additional names can be found online [8].

1.7 FUSED-RING HETEROCYCLIC COMPOUNDS

The names of the heterocycles in the previous section, along with the rules for fusion in Section 1.4, are the basis for the names below. As in Section 1.4, numbering always begins at a nonfused atom adjacent to a fused atom, but since there are several possible orientations for the molecule, a hierarchy of rules is followed.

1. First, give the lowest possible number to the first heteroatom regardless of the priority of the atom.
2. Then, if this allows more than one choice, choose the one that gives the second heteroatom the lowest number. If there are still multiple choices, minimize the number of the third, and so on.
3. Then, if there are still multiple choices, give the lower number to the higher-priority heteroatom.
4. Then give the lowest number to the first fusion atom.
5. Finally, give the lowest numbers to the substituents.

3-Bromo-2-(2-chloroethenyl)benzo[*b*]thiophene

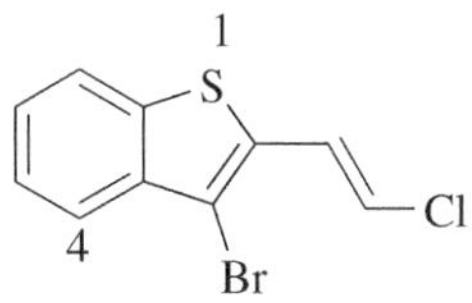

The parent name, thiophene is found in Table 1.4. Benzene is fused to the *b* side. The final numbering begins at the most counterclockwise atom not involved in fusion, arranged to give the heteroatom the smallest possible number, in this case 1.

7-Bromo-1*H*-2,3-benzothiazine

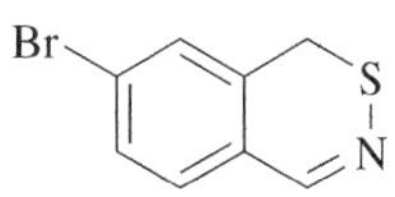

The parent name thiazine is built from Tables 1.2 and 1.3. Benzene is fused to the thiazine ring. Applying the rules for numbering fused systems, sulfur is higher-priority than nitrogen, and the lowest number for it is 2. The 3 locates the nitrogen. In this case, no bracketed site of fusion is specified because the fusion must precede the atom numbered 1. This is usual when there is more than one heteroatom and the fusion is simply benzo. The presence of one divalent atom in a six-membered ring excludes another atom from double bonding, thus the indicated hydrogen.

1,2-Dihydro-3-methylbenzo[*f*]quinoline

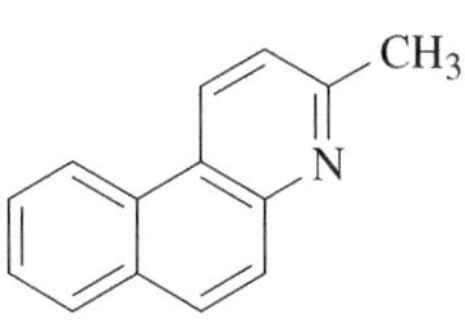

The sides of quinoline are lettered following the numbering system, and the benzene is fused to side *f*. The whole system is renumbered, orienting as directed in Section 1.4, that is, the maximum number of rings in a horizontal row, maximum in the upper right, giving the heteroatom the lowest possible number, and numbering from the most counterclockwise nonfused atom in the upper right ring.

When two heterocyclic rings are fused, the sides and direction of fusion are indicated in brackets as in the carbocyclic cases in Section 1.4. Examples of the six possible fusions between pyridine and furan are shown. Furo[3,2-*b*] indicates that carbons 3 and 2 of the furan are the 2 and 3 carbons of the pyridine, which make up the *b* side of the pyridine, respectively. The two-ring system is renumbered following the hierarchy of rules given at the beginning of this section.

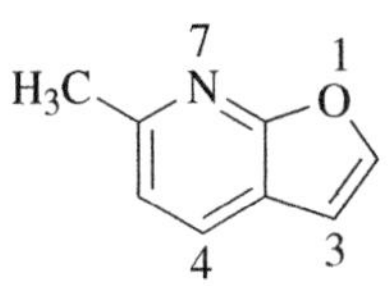

6-Methylfuro[2,3-*b*]pyridine

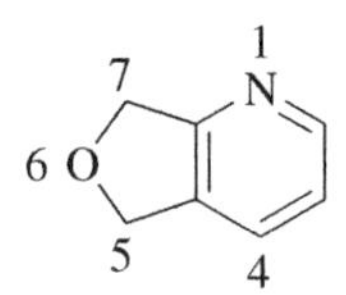

5,7-Dihydrofuro[3,4-*b*]pyridine

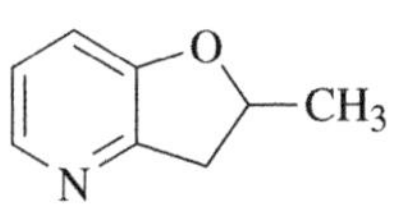

2,3-Dihydro-2-methylfuro [3,2-*b*]pyridine

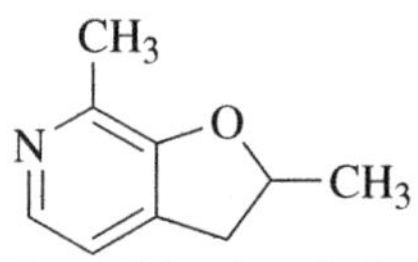

2,3-Dihydro-2,7-dimethylfuro[2,3-*c*]pyridine

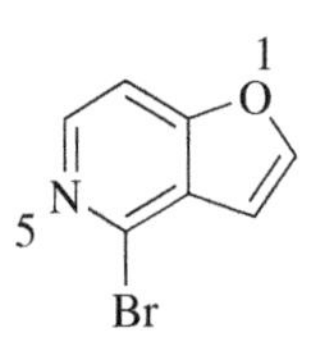

4-Bromofuro[3,2-*c*]pyridine

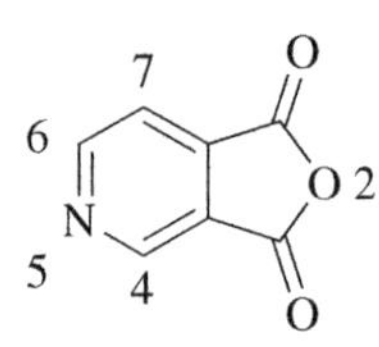

Furo[3,4-*c*]pyridine-1,3-dione

Thieno[3,4-*c*]pyridine

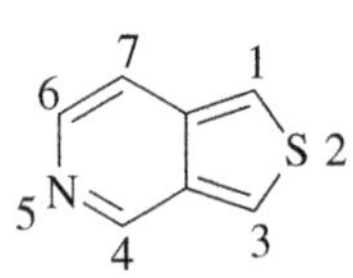

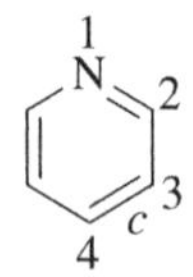

Pyridine

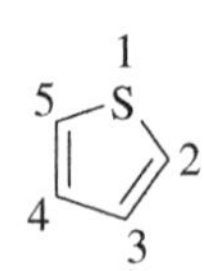

Thiophene

Thiophene, as the first ring named in a fused system, is shortened to thieno. The thieno ring is fused to the *c* side of pyridine, and the new system is renumbered to follow the rules from Section 1.4.

2-Methyl-2*H*-thiazolo[4,5-*e*]-1,2-oxazine

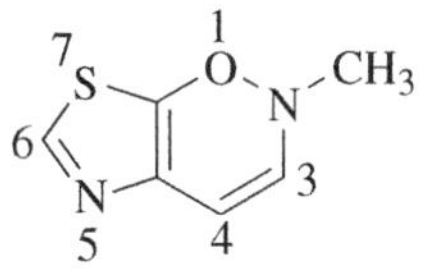

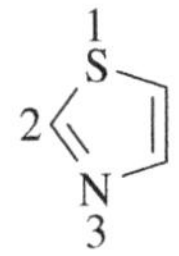

1,3-thiazole

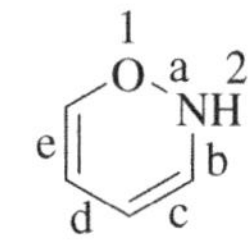

1,2-oxazine

The name *thiazole*, like *oxazole*, means 1,3-thiazole. The atoms of thiazole are numbered, and the sides of 1,2-oxazine are lettered. The fusion is drawn with atoms 4 and 5 of thiazole as atoms 5 and 6 of the oxazine. The system is then renumbered using the hierarchy of rules. You should flip your initial drawing about both the *x* and *y* axes to consider all four orientations to find the correct numbering.

2-(4-Cyanophenyl)-1-[2-(3,4-dimethoxyphenyl)-ethyl]-1*H*-benz[*d*]imidazole-5-carboxylic acid ethyl ester

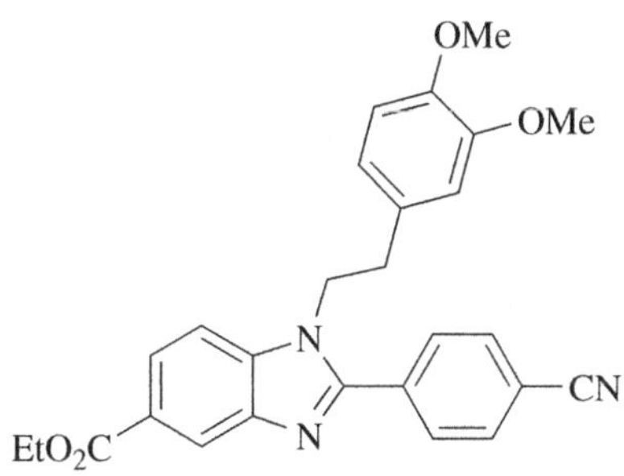

The fusion in this ring places the benzene on the *d* side of the imidazole ring. Renumbering the new ring following earlier rules gives the locants as shown.

1.8 BRIDGED AND SPIRO HETEROCYCLIC COMPOUNDS

Bridged and spiro heterocyclic compounds are named using the replacement nomenclature; that is, the hydrocarbon name is used with oxa, aza, and so on to substitute heteroatoms for carbons as was seen in large ring monocyclic compounds in Section 1.6.

3-(2-Methylphenyl)-3,8-diazabicyclo[3.2.1]octane

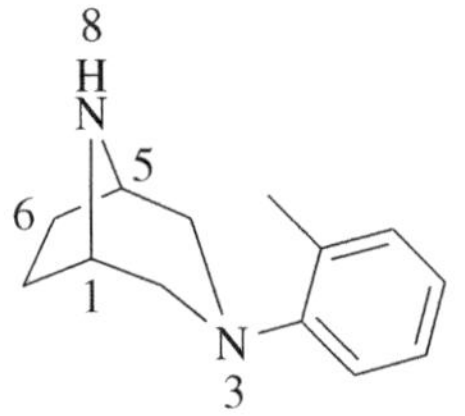

3-Oxa-1-azaspiro[4.5]decan-2-one

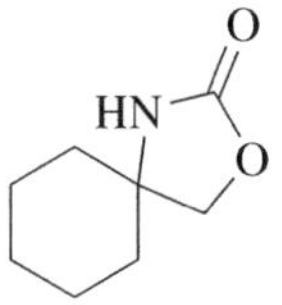

2-(2′,6′-Dimethoxyphenyl)-1,3-oxazaspiro[5.5]undeca-2,7,10-trien-9-one

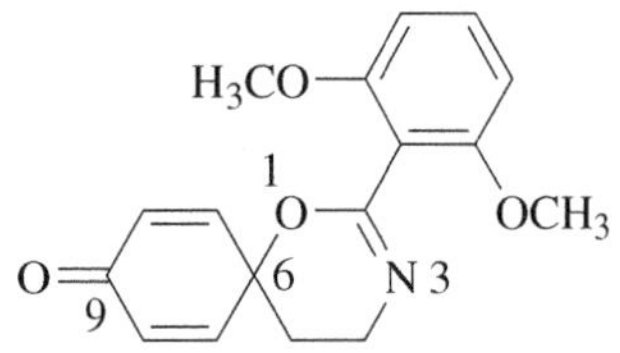

Many of the names that you will see in journals should be understandable by analogy from the examples studied here, but certainly, there are many more complicated cases beyond the scope of this chapter. The chapter references should be consulted for them.

RESOURCES

1. Many of the references in this chapter and the next can be helpful in finding the names of organic compounds.
2. The website maintained by ACD Labs with the contents of the IUPAC Blue Book, https://www.acdlabs.com/resources/iupac-blue-book/, is very helpful.
3. The website maintained by Chemical Abstracts Service, https://commonchemistry.cas.org/, can identify synonyms from CAS Registry numbers and names.

PROBLEMS

Draw complete structures for each of the following compounds. Questions marked with an asterisk are answered in Appendix C.

1.1* 2,4-Dimethylbenzo[*g*]quinoline

1.2* Spiro[5.7]trideca-1,4-dien-3-one

1.3* (1α,2β,5β,6α)-Tricyclo[4.3.1.1^{2,5}]undecane-11-one

1.4* 2*H*-3,1-Benzothiazine

1.5* (*exo,syn*)-2-(1-Pyrrolidinyl)bicyclo[3.3.1]nonan-9-ol

1.6* (±)-Dimethyl 2-((5-hydroxy-4-oxo-3,5-diphenylcyclopent-2-en-1-yl)methyl) malonate

1.7* (2*S*,3*S*)-3-Acetyl-8-carboethoxy-2,3-dimethyl-1-oxa-8-azaspiro [4.5]decane

1.8* 2-[(1*S*)-3-oxocyclopentyl]propanedioic acid dimethyl ester

1.9* Dibenzo[*a*,*e*]cyclooctadiene

1.10* 6-Methylbenzo[*d*]thiazole-2-carbonitrile

1.11 2-(Bromomethyl)-4,7-dimethoxyfuro[2,3-*d*]pyridazine

1.12 Spiro[cyclopentane-1,3′-bicyclo[4.1.0]heptane]

1.13 (1*R*,3*R*,5*S*)-*endo*-1,3-Dimethyl-2,9-dioxabicyclo[3.3.1]nonane

1.14 1-Benzoyl-2-phenylaziridine

1.15 3-Butyryl-2-(3-chloropropyl)-1-(methoxycarbonyl)-1,2-dihydro-pyridine

1.16 7-Methyl-7*H*-benzo[*c*]fluorene-7-carboxylic acid

1.17 (1*R*,6*S*,7*S*)-4-(*t*-Butyldimethylsiloxy)-6-(trimethylsilyl)bicyclo [5.4.0]undec-4-en-2-one

1.18 (1α,2β,3α,5α)-6,6-Dimethylbicyclo[3.1.1]heptane-2,3-diol

1.19 [1*R*-(1α,2β,6α)]-4,7,7-Trimethylbicyclo[4.1.0]hept-3-en-2-ol

1.20 6-(Benzyloxycarbonyl)-3-cyano-4-chloro-6-azabicyclo[3.2.1] oct-3-ene

1.21 Thieno[3,4-*b*]pyridine

1.22 1-Phenyl-1,4,5,6,7,8,9,10-octahydrocyclonona[*d*][1,2,3]triazole

1.23 5′-Acetyl-4′-amino-1,3-dihydro-6′-methyl-1,3-dioxospiro[2*H*-indene-2,2′-2*H*-pyran]

1.24 5-Methylbenzo[*b*]chrysene

1.25 Methyl 7-methylbicyclo[4.2.0]octa-1,3,5-triene-7-carboxylate

1.26 3-Amino-5,6,8,9-tetrahydro-7*H*-pyrazino[2,3-*d*]azepine-2,7-dicarboxylic acid diethyl ester

1.27 2-Benzoyl-1,6,7,11b-tetrahydro-2*H*-pyrazino[2,1-*a*]isoquinoline-3,4-dione

1.28 [1*R*-(1α,2β,4α)]-4-chloro-2-methylcyclohexanecarboxylic acid

1.29 8-Bromo-7-chloro-2-(ethylthio)-4*H*,5*H*-pyrano[3,4-*e*]-1,3-oxazine-4,5-dione

1.30 1′,2′-Dihydro-4-methyl-2′-oxospiro[4-cyclohexene-1,3′-[3*H*] indole]-2,2-dicarboxylic acid diethyl ester

1.31 2,4,6-Trichlorophenyl 3,4-dihydronaphthalene-2-carboxylate

1.32 7-Chloro-2-methyl-1,4-dihydro-2*H*-isoquinolin-3-one

1.33 4-(Methylthio)-2-phenyl-3-quinazoline

1.34 [(1*S*,2*R*)-2-methoxy-1-propylpentyl]thiobenzene

1.35 (2*R*,3*S*)-3-hydroxy-2-methyl-1-morpholino-5-phenylpentan-1-one

1.36 (*E*)-1,3-diphenylhexa-3,5-diene-1,2-dione

1.37 2-Chloro-6,7-dimethoxy-1,2,3,4-tetrahydroisoquinoline

REFERENCES

1. Panico, R.; Powell, W. H.; Richer, J.-C. (preparers). *A Guide to IUPAC Nomenclature of Organic Compounds (Recommendations 1993)*; Blackwell Scientific Publications: Oxford, UK, 1993.
2. Rigaudy, J.; Klesney, S. P. (preparers). *Nomenclature of Organic Chemistry*; International Union of Pure and Applied Chemistry/Pergamon Press: Oxford, 1979.

3. Advanced Chemistry Development Labs. IUPAC Nomenclature of Organic Chemistry Home Page. https://www.acdlabs.com/resources/iupac-blue-book/. (This website includes the contents of references 2 and 3 in html format.) (accessed January 31, 2024).
4. American Chemical Society. *Naming and Indexing of Chemical Substances for Chemical Abstracts*; American Chemical Society: Columbus, OH, 2008. https://www.cas.org/sites/default/files/documents/indexguideapp.pdf (accessed January 31, 2024)
5. Gupta, R. R.; Kumar, M.; Gupta, S. S. *Heterocyclic Chemistry*, Volume I; Springer: Berlin, 1998, pp. 3–38.
6. Advanced Chemistry Development Labs. Fused Polycyclic Hydrocarbons. http://www.acdlabs.com/iupac/nomenclature/79/r79_63.htm (accessed January 31, 2024).
7. Advanced Chemistry Development Labs. Heterocyclic Systems. Rule B-1. Extension of the Hantzsch-Widman System. http://www.acdlabs.com/iupac/nomenclature/79/r79_702.htm (accessed January 31, 2024).
8. Advanced Chemistry Development Labs. Heterocyclic Systems. Rule B-2. Trivial and Semi Trivial Names. http://www.acdlabs.com/iupac/nomenclature/79/r79_951.htm (accessed January 31, 2024).

2

ACCESSING CHEMICAL INFORMATION

A truly vast and ever-growing body of organic chemical information is recorded in the chemical literature and in online databases. In a research library practically all of it can be searched quickly through these databases. Because print resources are still available, they will be discussed briefly, but online sources will be the focus of this chapter.

2.1 DATABASES

Databases gather information about compounds that can include physical constants, spectral data, and toxicity and hazard information. Much of this sort of information can be accessed through a number of familiar publications such as the *CRC Handbook* [1] or *The Merck Index* [2]. *The Merck Index* is also available online, but full access requires a subscription.

Online databases are able to compile a great deal of information in a format that is readily searchable. PubChem [3] is maintained by the National

Understanding Organic Chemistry: Bridging Fundamental and Advanced Courses, Fourth Edition. Ann M. Fabirkiewicz and John C. Stowell.

Companion website: www.wiley.com/go/fabirkiewiczorganic4e

Center for Biotechnology Information (NCBI) and provides information on the biological activity of small molecules in three databases: Substances, Compounds, and BioAssays. As of this writing, there are about 318 million substances and 118 million compounds indexed at the site, as well as over 40 million publications for the compounds in the compound database. "Compounds" are identified by name or structure, while "substances" may include mixtures of compounds, such as plant extracts. It is generally more useful to search the compound index by name or structure.

ChemSpider [4] is maintained by the Royal Society of Chemistry and allows text and structure searching for over 100 million compounds. Spectra, properties, and a wealth of other information are available.

CAS Common Chemistry [5] is provided by Chemical Abstracts Service (CAS) and links chemical names and CAS registry numbers for about 500,000 chemicals. This free database is particularly useful when the name of a chemical is confusing or uncertain as it will allow searching with the CAS number. It includes some physical properties as well.

The Millipore Sigma catalog is available online [6] and provides CAS registry numbers, Beilstein and PubChem ID numbers, in addition to physical constants and Safety Data Sheets (SDS). The catalog allows searching by name and structure.

The National Institutes of Science and Technology (NIST) maintains the NIST Chemistry WebBook, which is an extensive database of chemical information [7]. Included are thermochemical data, spectral data, and kinetic data, which are searchable by name, structure, molecular weight, and other options that are compound-specific.

2.2 CHEMICAL LITERATURE

The chemical literature can be organized into several categories: primary literature, review literature, and abstract literature. Abstract journals such as *Chemical Abstracts*, as well as websites such as the one maintained by the American Chemical Society for accessing their journals [8], allow ready searching of the chemical literature.

The primary literature includes journals that publish original research findings. Lavoisier's *Annales de Chimie* began publication in 1789 and is one of the oldest chemical journals. Examples of current publications include the *Journal of the American Chemical Society*, the *Journal of Organic Chemistry*, and *Tetrahedron*, to name a very few. Most journals require a subscription, and a few offer open access to their materials. Most chemical societies publish a range of journals specific to individual subfields of chemistry, as well as journals covering a broad range of chemical topics.

The review literature includes journals that publish summaries of the current knowledge on a particular topic. Examples include *Accounts of Chemical Research*, *Chemical Reviews*, and *Chemical Society Reviews*. These publications can be particularly helpful as you begin a project because they include substantial lists of references in the primary literature.

An abstract is a summary of the important discoveries and conclusions in a paper and provides a way into the content of research publications. Abstract literature allows access to this information in a searchable format. Many of the early journals published annual author and subject indexes. *Chemisches Zentralblatt*, which began in 1830 and ended in 1969, is one of the most reliable means into the very early chemical literature. A searchable database is available by subscription through SciFinder [9].

Reaxys, a subscription database maintained by Elsevier, contains information dating back to 1771. It includes content from *Beilstein's Handbuch der Organischen Chemie*, 16,000 journals, and 35 million patents. The database contains preparations, properties, and spectral data for millions of chemicals [10]. Reaxys includes an artificial intelligence (AI)-assisted retrosynthetic tool as well [11].

The print edition of *Chemical Abstracts* was published from 1907 to 2010. Abstracts were published in weekly issues, and 6 months of these issues constituted a volume that was accompanied by six indexes: *General Subject Index*, *Chemical Substance Index*, *Formula Index*, *Index of Ring Systems*, *Author Index*, and *Patent Index* [12]. Ten volumes (5 years of indexes) were combined in the *Collective Indexes*; the 11th collective index covered the years from 1982 to 1986 and comprised 93 bound books occupying 17 feet of shelf space, and this was just the index! The chemical literature continues to expand dramatically each year. It is easy to see why the print version is no longer published.

Chemical Abstracts Service, a division of the American Chemical Society, currently covers over 50,000 periodicals and translates and abstracts publications in more than 50 languages. Coverage in this database includes patents extending back to 1808, *Chemisches Zentrallblatt* from 1897 to 1906, and the Royal Society of Chemistry journals from 1841 to 1906, among others. This database is accessed through SciFinder.

SciFinder [9] is a powerful search engine and requires a subscription, although a limited number of free searches are available as a benefit of membership in the American Chemical Society. Searches can be conducted under search headings that include References, Substances, and Reactions. A series of excellent tutorials and videos are available for training purposes [13]. Under References, searches can be conducted by author, company, or journal name, or by document identifier, patent number, or research topic. Under Substances, options include chemical structures, molecular formulas, and properties. Markush searching is also available. Markush structures are found in the patent literature and allow chemically similar structures to be searched at the same time, using –R to avoid specification of individual groups. Under Reactions, a search of structures of either reactants or products, or both, is permitted, and advanced search options allow for specification of solvents, other functionality and reaction details, and dates. Results of a reaction search can be further narrowed by yield or number of steps, for example. A retrosynthesis planner is also available.

2.3 SYNTHETIC PROCEDURES

Synthetic procedures for many organic molecules are compiled in both print and online sources. Due to the massive amount of information, print sources are gradually becoming online resources, often requiring a subscription for access. A few are discussed here in detail, and librarians and search engines will readily find more sources of such information.

Science of Synthesis is published by Thieme, and is available by subscription, though some of the content is open access. This is an organized, completely referenced, very detailed collection of methods of preparing essentially all classes of organic compounds plus their reactions. It includes selected experimental details and extensive tables of examples. The Original Edition was published between 2000 and 2010 and includes 48 volumes. *Knowledge Updates* are published annually since 2010 and include new and updated content. The Reference Library includes content on techniques, applications, and developments in organic chemistry [14]. *Methoden der Organischen Chemie* is commonly known as "Houben-Weyl" for the editors of its first edition. Begun in 1909 with references back into the 1800s, this content is available through *Science of Synthesis*.

Organic Syntheses began as a result of the need for research-grade chemicals in the United States during World War I. The popularity of the initial set of pamphlets describing the preparation of 111 compounds inspired the first annual volume of 84 pages in 1921 [15]. In 1998, the editorial board decided to place all past and future volumes online with open access [16]. These are searched by structure or keyword. All procedures have been checked and carefully annotated with experimental details and hazards. Citations to the original publications are often available and addenda are linked to the initial article so that changes and additions to the initial procedure are readily found.

Organic Reactions is a series of 114 volumes, as of this printing, that began publication in 1942 under the direction of Professor Roger Adams, who also initiated *Organic Syntheses*. Each volume consists of one or more chapters, with each chapter covering a reaction of particular importance to organic chemistry. Over 300 reactions are covered in the complete work. The printed books are published by Wiley [17] and the individual chapters are listed on a searchable Wikipedia page [18], with links to partial contents. Full access requires a subscription.

ChemSpider SyntheticPages are maintained by the Royal Society of Chemistry. This resource is open access and reviewed by the community. While it does not allow literature searching, it does allow access to a collection of current postings by the user base that is validated by user comments. Posts include experimental details and helpful tips [19].

2.4 HEALTH AND SAFETY INFORMATION

Health and safety information for commercially available chemicals is found on SDS, which are available online and from manufacturers. The

TABLE 2.1 Safety Data Sheet Subsections

Section	Title	Section	Title
1	Identification	9	Physical and chemical properties
2	Hazard identification	10	Stability and reactivity
3	Composition	11	Toxicological information
4	First-aid measures	12	Ecological information
5	Fire-fighting measures	13	Disposal consideration
6	Accidental release measures	14	Transport information
7	Handling and storage	15	Regulatory information
8	Exposure controls/personal protection	16	Other information

United Nations adopted the Globally Harmonized System of Classification and Labelling of Chemicals (GHS), which imposes worldwide consistency on the classification of hazards such as toxicity and flammability [20]. SDS are standardized in the GHS and include the sections shown in Table 2.1 [21]. More detailed descriptions of the content of each section can be found online [21]. The GHS requires pictograms on the labels of all chemicals, and those pictograms and a brief description are provided in Table 2.2.

The Millipore Sigma catalog, available online [6], offers SDS for all chemicals sold. *Organic Syntheses* [16] reviews each of its published procedures for safety. Beyond Benign has assembled a Toxicology for Chemists curriculum that is open access and available online [22].

The Hazardous Substances Data Bank (HSDB) has been incorporated into PubChem [3] and no longer exists as a separate entity. To find HSDB information specifically, within a PubChem search, scroll to Information Sources and choose HSDB to filter the data.

The Chemical Safety Library is an open-source, searchable database of hazardous chemical incidents reported by individuals [23].

The Consumer Product Information Database (CPID), formerly part of the TOXNET database hosted by the National Library of Medicine, is a repository of information on commercially available products separated by categories such as Personal Care or Home Office. Foods and pharmaceuticals are not included. Searches can be conducted by brand, product type,

TABLE 2.2 Chemical Label Pictograms

Health hazard	Flame	Exclamation mark

Carcinogen, mutagenicity, reproductive toxicity, respiratory sensitizer, target organ toxicity, aspiration toxicity	Flammables, pyrophorics, self-heating, emits flammable gas, self-reactives, organic peroxides	Irritant (skin and eye), skin sensitizer, acute toxicity (harmful), narcotoxic effects, respiratory tract irritant, hazardous to ozone layer (non-mandatory)
Gas cylinder	Corrosion	Exploding bomb

Gases under pressure	Skin corrosion/burns, eye damage, corrosive to metals	Explosives, self-reactives, organic peroxides
Flame over circle	Environment (non-mandatory)	Skull and crossbones

Oxidizers	Aquatic toxicity	Acute toxicity (fatal or toxic)

Reprinted from https://www.osha.gov/sites/default/files/publications/OSHA3491QuickCardPictogram.pdf/OSHA/Public Domain.

manufacturers, and ingredients, and results will include links to similar products, other products that contain the same ingredients, and SDS for consumer products. Data for 25,000 products are available [24].

PROBLEMS

Questions marked with an asterisk are answered in Appendix C.

2.1* Find a synthesis for the compound below. Give two names by which this compound is known. What is this compound used for? Provide a reference for a procedure using this compound as a reagent.

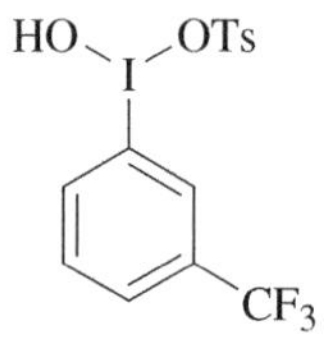

2.2 Find a preparation for the following compound. Give the reference, the melting point, and a published name for the compound. What is the frequency of the carbonyl stretch?

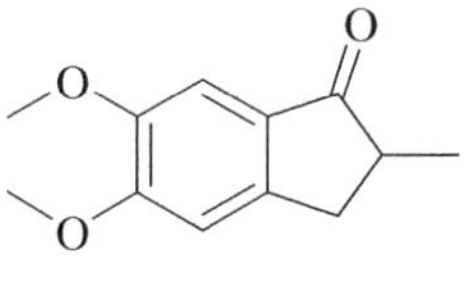

2.3* What is the structure of crizotinib, and what is it used for? What is the International Union of Pure and Applied Chemistry (IUPAC) name of this compound?

2.4 Give the boiling point, water solubility, and octanol/water partition coefficient ($K_{o/w}$) for 1-heptanol.

2.5* Locate a compound of commercial significance that is an example of the following ring system. Give the chemical structure, and IUPAC and commercial names for that compound. Cite the patent owner and at least one patent number.

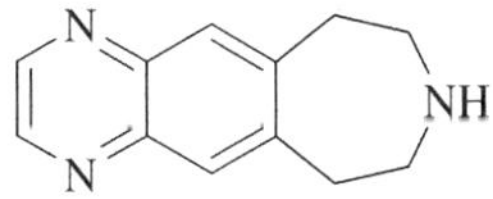

2.6* Locate compounds where *R* is a substituent. To what class of drugs do these compounds belong? Find a reference that describes the compounds and their effectiveness.

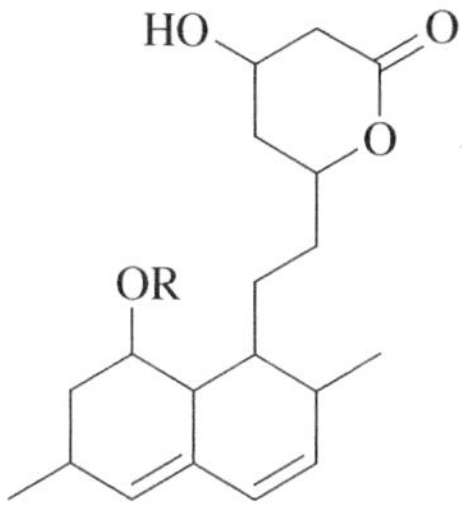

2.7 What is the melting point for gentisic acid? Give the structure of the molecule and a systematic name.

2.8 Find the ultraviolet spectrum of 4-aminobenzophenone. List two wavelengths of maximum absorbance.

2.9 What is the National Fire Protection Association (NFPA) hazard classification for biphenyl? What is the threshold limit value (TLV) for this compound?

2.10 Ciguatoxin is the causative agent of ciguatera, a form of food poisoning associated with contaminated fish. What are the symptoms of ciguatera? Provide a reference for the synthesis of ciguatoxin. What is its molecular weight?

2.11* What is the IUPAC name for (+)-β-conhydrine? What is its CAS number? Where is it found in nature?

2.12* What is BDQ and what is it used for? What is the specific rotation of the optically pure compound? What GHS symbols would be found on its label?

2.13* You will find binol derivatives throughout the pages of this text. Why is it so popular? Give the structure of (*S*)-binol. Cite one synthesis of (*S*)-binol and at least one example of its use in synthetic organic chemistry.

REFERENCES

1. Rumble, J. R., Ed. *CRC Handbook of Chemistry and Physics*, 104th ed.; CRC Press, Taylor and Francis Group: Boca Raton, FL, 2023. Also available by subscription at https://hbcp.chemnetbase.com/ (accessed May 20, 2024)
2. O'Neil, M. J., Ed. *Merck Index*, 15th ed.; Merck: Whitehouse Station, NJ, 2013. Also updated regularly and available by subscription at https://merckindex.rsc.org/ (accessed May 20, 2024)
3. National Center for Biotechnology Information. PubChem. http://pubchem.ncbi.nlm.nih.gov (accessed May 20, 2024).
4. Royal Society of Chemistry. ChemSpider. http://www.chemspider.com/ (accessed May 20, 2024).
5. Chemical Abstracts Service. Common Chemistry. https://commonchemistry.cas.org/ (accessed May 20, 2024).
6. Millipore Sigma. https://www.sigmaaldrich.com/US/en/products/chemistry-and-biochemicals (accessed May 20, 2024).
7. National Institute of Standards and Technology. NIST Chemistry WebBook. http://webbook.nist.gov/chemistry (accessed May 20, 2024).
8. American Chemical Society. ACS Publications Division Home Page. http://pubs.acs.org (accessed May 20, 2024).
9. Chemical Abstracts Service. SciFinder. https://www.cas.org/cas-data (accessed May 20, 2024).
10. Elsevier. Reaxys. https://www.elsevier.com/products/reaxys (accessed May 20, 2024).
11. Elsevier. Reaxys. https://www.elsevier.com/products/reaxys/predictive-retrosynthesis (accessed May 20, 2024).
12. Schulz, H.; Georgy, U. *From CA to CAS Online: Databases in Chemistry*, 2nd ed.; Springer-Verlag: Berlin, 1994.
13. CAS. CAS SciFinder. https://www.cas.org/cas-scifinder-discovery-platform/cas-scifindern (accessed May 20, 2024).
14. Thieme Chemistry. Science of Synthesis. https://www.thieme.de/en/thieme-chemistry/sos-content-54782.htm (accessed May 20, 2024).
15. Organic Syntheses, Inc. History of Organic Syntheses. http://www.orgsyn.org/history.aspx (accessed May 20, 2024).
16. Organic Syntheses, Inc. Organic Syntheses. http://www.orgsyn.org/Default.aspx (accessed May 20, 2024).

17. Evans, P. A.; Denmark, S. E.; Overman, L. E.; Paquette, L. A.; Kende, A. S.; Dauben, W.; Cope, A. C.; Adams, R., Eds. *Organic Reactions*, Volume 1–114; Wiley: New York, 1942–2023.
18. Organic Reactions, Inc. Published Organic Reactions Chapters. https://www.organicreactions.org/published-chapters/ (accessed May 20, 2024).
19. Royal Society of Chemistry. ChemSpider SyntheticPages. https://cssp.chemspider.com/About.aspx (accessed May 20, 2024).
20. *United Nations Economic Commission for Europe. Globally Harmonized System of Classification and Labelling of Chemicals (GHS)*, United Nations; 10th rev. ed., 2023. https://unece.org/sites/default/files/2023-07/GHS%20Rev10e.pdf (accessed May 20, 2024).
21. United States Department of Labor. OSHA Brief: Hazard Communication Standard: Safety Data Sheets. https://www.osha.gov/sites/default/files/publications/OSHA3514.pdf (accessed May 20, 2024).
22. Beyond Benign. The Toxicology for Chemists Curriculum. https://www.beyondbenign.org/curr-toxicology-course-modules/ (accessed May 20, 2024).
23. CAS. Chemical Safety Library. https://safescience.cas.org/ (accessed May 20, 2024).
24. Consumer Product Information Database. https://www.whatsinproducts.com/ (accessed May 20, 2024).

3

STEREOCHEMISTRY

The shapes and properties of molecules can depend not only on the order of connection of atoms but also on their arrangement in a three-dimensional space. Molecules differing only in configuration are called *stereoisomers* and are the principal subject of this chapter [1].

3.1 REPRESENTATIONS

Some organic molecules such as benzene are planar as defined by the point locations of all nuclei present. These are easily represented on the planar printed page.

Most organic molecules are three-dimensional structures, best viewed and represented in solid molecular models. The necessity of using paper requires pictures that show depth, as perspective does in artwork and photography. The mere projection onto the plane of the paper, as in the shadow of a molecular model, loses the real difference between left- and

Understanding Organic Chemistry: Bridging Fundamental and Advanced Courses, Fourth Edition. Ann M. Fabirkiewicz and John C. Stowell.

Companion website: www.wiley.com/go/fabirkiewiczorganic4e

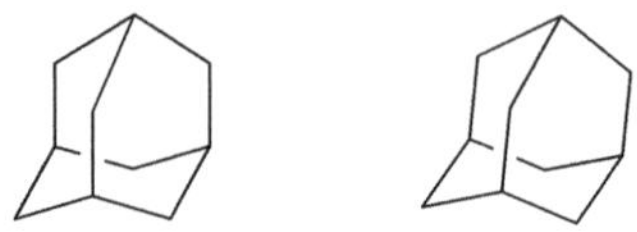

FIGURE 3.1 A stereo pair of images of adamantane.

right-handed structures. The best alternative on paper is a stereo pair of pictures as exemplified in Figure 3.1. The image on the left is for your left eye and that on the right is for your right eye, a pair of views representing a 6° rotation of the molecule. These are precisely scaled and oriented no farther apart than the separation between your eyes. You can view them through a commercially available stereopticon viewer, but with some practice, it is possible to view them without the aid of devices. Stare at the two images with the page directly in front of you and gradually allow your eyes to cross. The image that appears between the two images on the page will appear to be three-dimensional. There are other suggestions for viewing such images [2]. Most commonly in journals and handwritten materials, we use representations where depth is portrayed via conventions. These are exemplified in Figure 3.2 for 2-butanol.

In the dot-and-wedge convention, the group on the broad end of the wedge is defined as being above the plane of the paper, the dotted bond extends below the plane of the paper, and the line bonds are in the plane of the paper. In the line drawing, the hydrogens on carbon are not shown but are defined as completing the tetravalency of carbon. In the Fischer projection, the center of the crossed lines is a carbon atom, and those bonds emanating from it to the side are defined as extending above the plane of the paper toward the viewer and those extending toward the top and bottom of the page are defined as extending below the plane of the paper, away from the viewer.

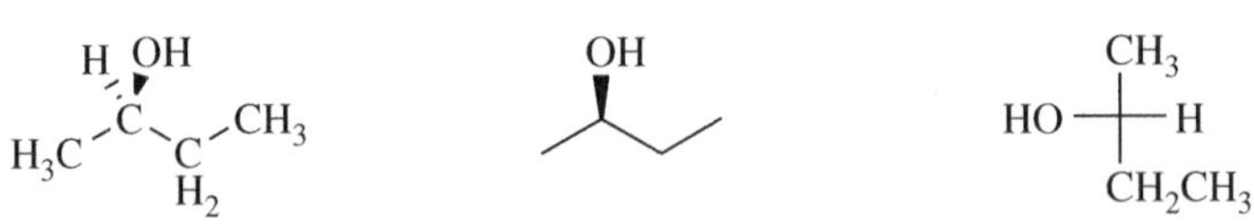

Dot and wedge Line drawing Fischer projection

FIGURE 3.2 (*R*)-2-Butanol.

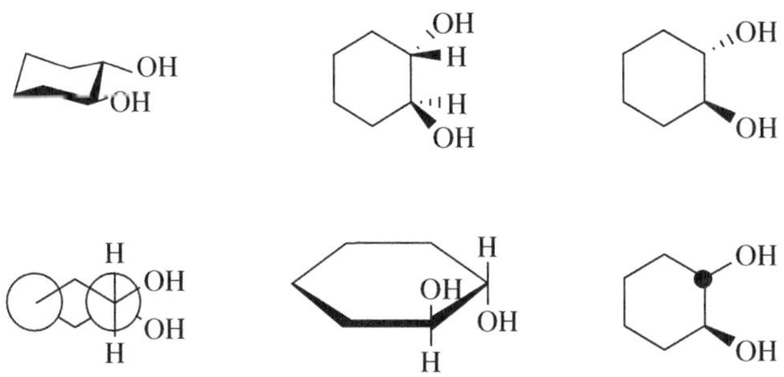

FIGURE 3.3 (*S*,*S*)-*trans*-1,2-Cyclohexanediol.

Related conventions are used for portraying ring compounds as exemplified in Figure 3.3 for (*S*,*S*)-1,2-cyclohexanediol. Many authors will draw one enantiomer of each molecule in a reaction scheme when they are actually using racemic materials. Their text should indicate this meaning.

3.2 VOCABULARY

Much of the terminology used to define the relationships between three-dimensional molecules is covered in introductory texts, and summaries are available [1, 3]. A molecule or other object that is different from its own mirror image (e.g., a glove) is *chiral*. Molecules that are identical to their mirror image are *achiral*. The conformational flexibility of molecules allows for many different representations; therefore, when testing a pair of structures for identity or a mirror-image relationship, the models should be flexed or the drawings redrawn to attempt a match. A left-hand fist is not the mirror image of an open right hand, yet we refer to left and right hands as mirror images generally.

A chiral molecule and its mirror-image molecule are *enantiomers*, that is, the two molecules are different. A pair of gloves is an enantiomeric pair. A *racemic mixture* or *racemate* is a combination of equal amounts of enantiomers, while a combination of unequal amounts of enantiomers is called *aracemic* or *scalemic*. A sample of a single enantiomer is termed *enantiopure*.

Chiral molecules contain one or more stereogenic atoms. A *stereogenic atom* is an atom bearing several groups where an interchange of two groups will produce a stereoisomer (enantiomer or diastereomer) of the original. Carbon 2 in 2-butanol and carbons 1 and 2 in 1,2-cyclohexanediol are stereogenic atoms. The two possible spatial arrangements about a stereogenic atom are called *configurations*, and each one is designated (*R*) or (*S*) according to the Cahn–Ingold–Prelog system [4].

If a molecule contains more than one stereogenic atom, there will usually be *diastereomeric* pairs. Diastereomers have the same order of connection of atoms in their structures, but one differs in spatial arrangement from the other and from the mirror image of the other in all reasonable conformations. Diastereomeric substances must, therefore, differ in many physical properties. The term *diastereomer* is also used to relate *cis* and *trans* alkenes and *cis* and *trans* ring compounds even if they are achiral. The term *stereoisomer* includes enantiomers and diastereomers. All three stereoisomers of 1,2-cyclohexanediol are shown in Figure 3.4. The (*S*,*S*) and (*R*,*R*) isomers are mirror images and are therefore enantiomers. The (*R*,*S*) isomer is different from either (*S*,*S*) or (*R*,*R*) and is a diastereomer of each. Note that the (*R*,*S*) isomer is achiral despite the presence of stereogenic atoms. Achiral molecules containing tetrahedral stereogenic atoms are termed *meso*. Meso structures may be recognized by the presence of a mirror plane within the molecule in certain conformations, such as a boat conformation in this case.

You can expect more stereoisomers for structures that contain more stereogenic atoms, and the maximum number of possible stereoisomers is 2^n, where n is the number of chiral carbons. Each additional stereogenic atom doubles the number of stereoisomers except where meso compounds occur, or where a polycyclic ring system prohibits some configurations. Compound **3.1**, tricyclo[4.4.0.0^{2,8}]decane, contains four stereogenic atoms, but there are only two stereoisomers.

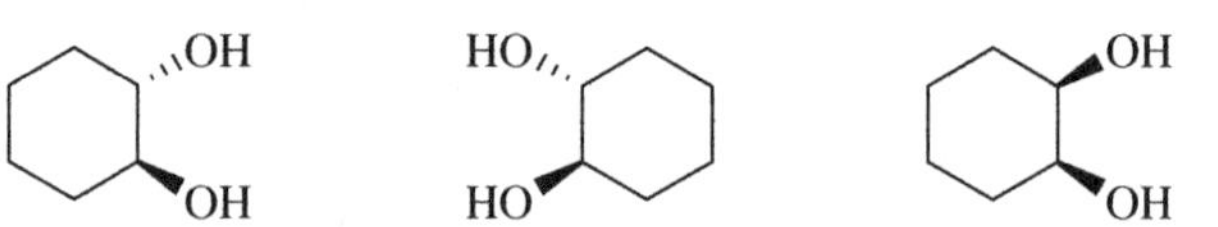

(*S*,*S*) (*R*,*R*) (*R*,*S*)

FIGURE 3.4 1,2-Cyclohexanediol stereoisomers with configurational designation.

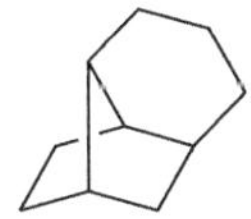

3.1

Certain diastereomeric relationships are designated by prefixes derived from two carbohydrates. D-Threose (**3.2**) and D-erythrose (**3.3**) have two stereogenic atoms and both bear an –H and an –OH group. Other molecules that differ in the analogous fashion are prefixed *threo-* and *erythro-*, as generally represented in Figure 3.5.

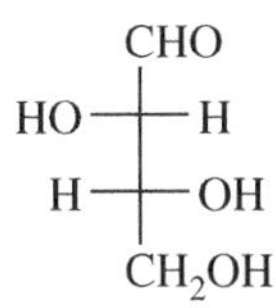

3.2

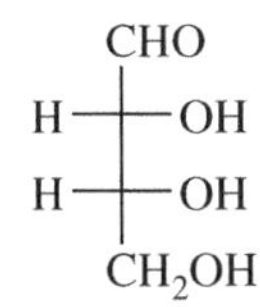

3.3

Chirality is a property of the whole molecule, as concluded by Pasteur even before the structural theory was established, and does not require the presence of a stereogenic atom. Molecules with a twist along an axis such as allenes like 1,3-diphenyl-1,3-di-α-naphthylallene (**3.4**) [5], spiro compounds like 2,6-dichlorospiro[3,3]heptane (**3.5**), and exocyclic double-bonded compounds like 1-methyl-4-methylenecyclohexane (**3.6**) can be chiral. Crowding of groups in a molecule may restrict rotation about single bonds or prevent planarity, again generating a twist, giving conformational enantiomerism ((*S*)-(−)-1,1′-binapthyl-2,2′-diol, (*S*)-(−)-BINOL), **3.7** and 4,5-diphenylphenanthrene, **3.8** [6]).

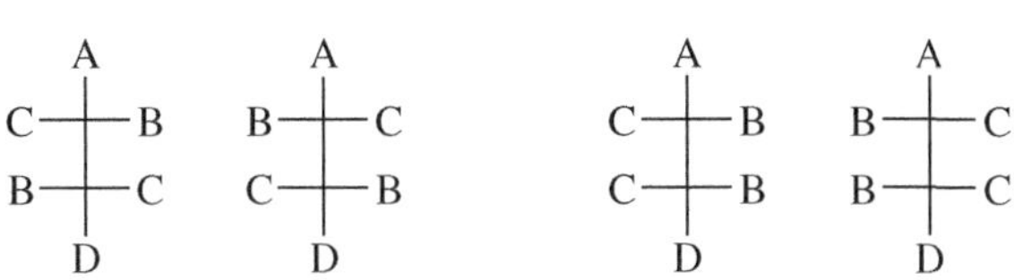

threo- *erythro-*

FIGURE 3.5 Fischer projections of threo and erythro stereoisomers.

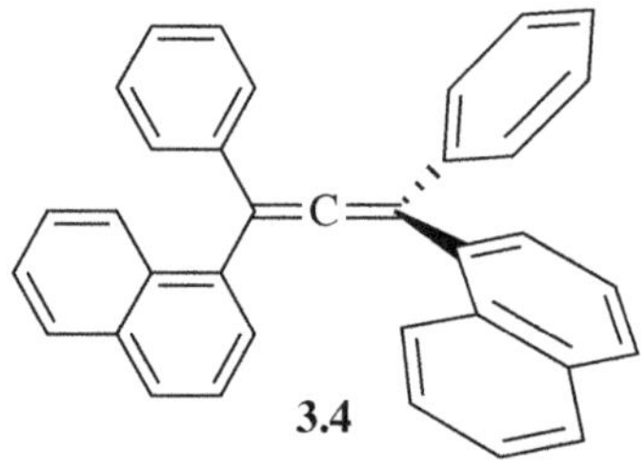

3.4

3.5

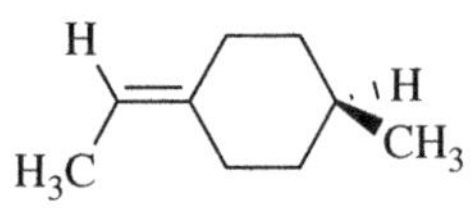

3.6

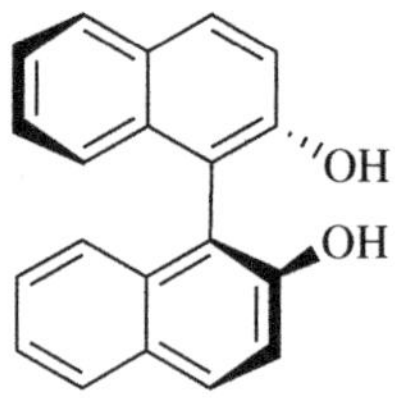

3.7

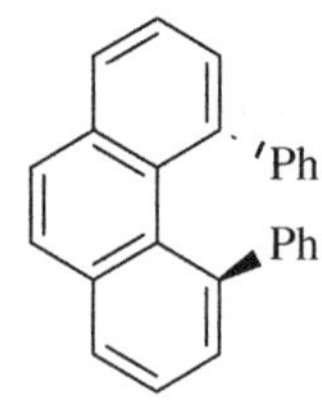

3.8

Finally, some molecules having a planar portion with one face distinguished from the other and lacking a plane of symmetry are chiral. Examples include paracyclophane-1-carboxylic acid (**3.9**) [7] and *trans*-cyclooctene (**3.10**) [8].

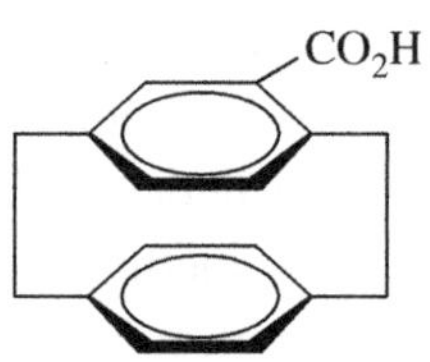

3.9

3.10

3.3 PROPERTY DIFFERENCES AMONG STEREOISOMERS

Thus far, we have considered structures. Structural differences should be manifest in measurable property differences in actual substances.

Diastereomers have different shapes, and the different manner in which each assembles with its own kind in the liquid or solid state will lead to different physical properties such as density, refractive index, melting point, and boiling point. In the gaseous state, shape is of little consequence because of the lack of assembly. Diastereomers will also have dipolar differences, which will influence interaction with radiation, leading to different spectroscopic properties.

The difference between enantiomers is very subtle. A homogeneous sample of a pure enantiomer will have properties dependent on the intermolecular attractions in the sample, and the mirror image will be no different in that regard. Thus, the melting point, boiling point, refractive index, density, and spectra of each will be identical. Enantiomers will interact identically with achiral materials, but if each is allowed to interact with one enantiomer of a chiral material, the combinations will be diastereomeric, and therefore different. That chiral material can be a solvent, an adsorbant, a complexing agent, or a reactant.

Ratios of enantiomers in scalemic materials can be measured with nuclear magnetic resonance (NMR) spectroscopy, using either chiral derivatizing agents or chiral solvating agents which make the enantiomers into diastereomeric molecules or complexes [9]. Mosher's acid, α-methoxy-α-trifluoromethylphenylacetic acid (MTPA) (**3.11**), is a chiral derivatizing agent that allows for determination of absolute configuration. The protocol can even be performed within an NMR tube, allowing analysis of very small samples [10]. Enantiopure (*S*)-(+)-1-methoxy-2-propylamine (**3.12**) reacts with chiral α-substituted aldehydes to give the diastereomeric imines. These show separate signals in ^{1}H NMR which can be measured to determine the ratio of enantiomers [11]. A chiral solvating agent such as enantiopure 2,2,2-trifluoro-1-(9-anthryl)ethanol (TFAE, Pirkle's alcohol, **3.13**) combined in a nonpolar solvent with a chiral solute causes nonequivalence in the ^{1}H NMR spectrum of the enantiomers of solute [12]. The solute must be capable of hydrogen bonding with the solvating agent in order to form diastereomeric complexes.

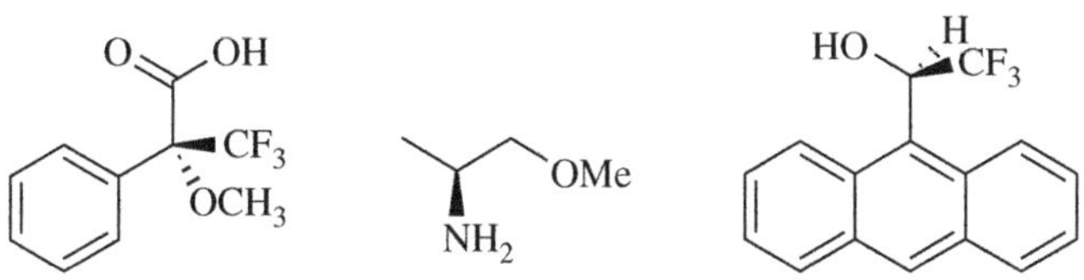

3.11 **3.12** **3.13**

Much larger differentiations between enantiomers may be brought about with lanthanide shift reagents such as tris[3-((heptafluoropropyl) hydroxymethylene)-*d*-camphorate]europium, [Eu(hfc)$_3$] (**3.14**) [13], or samarium(III)-(*R*) or (*S*)-propylenediaminetetraacetate, Sm-pdta (**3.15**) [14]. These paramagnetic complexes associate with polar sites in molecules and cause large changes in ^{1}H and ^{13}C NMR chemical-shift values for nearby atoms. The chiral ligands in the complexes cause different shifts in enantiomeric molecules [15]. Although the shift differences are larger than those found with solvating agents, shift reagents cause line broadening and require optimized mole ratios between the reagent and the chiral substance under analysis.

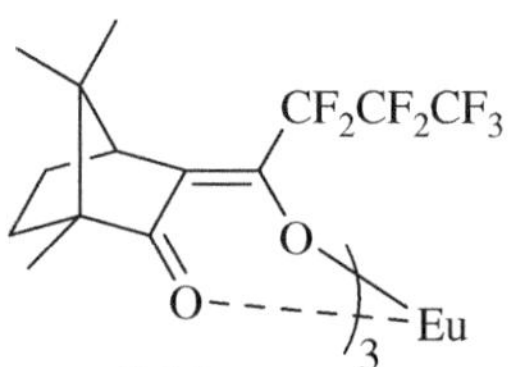

3.14

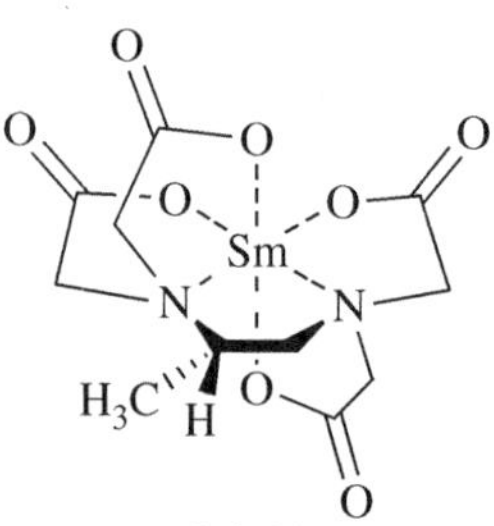

S-**3.15**

It is common to express the relative amounts of enantiomers as percent enantiomeric excess (ee). A similar diastereomeric excess (de) or diastereomeric ratio (dr) can be calculated for those compounds. In the NMR resolved examples given earlier, enantiomeric excess is calculated from the areas of the NMR signals for each enantiomer as in Equation 3.1, where A_L and A_S represent the areas of the larger and smaller signals, respectively. The remaining percentage is considered as racemic. A sample of 94% ee contains 97% of one and 3% of the other enantiomer.

$$\frac{A_L - A_S}{A_L + A_S} \times 100 = \% \text{ enantiomeric excess} \tag{3.1}$$

There is one direct physical measurement that allows for a differentiation between enantiomers and can be used to determine ratios of enantiomers: rotation of polarized light. Light emerging from a polarizer may be understood as follows. Each of the multitude of parallel rays is a pair of dipoles perpendicular to the line of travel that rotate as they advance, one spiraling to the right and one spiraling to the left. These dipoles alternately reinforce and cancel each other such that their vector sum is an oscillating wave in one plane. Coming through the same polarizer, all the rays summarize in parallel planes, which means that the right and left dipole pairs all have the same phase relationship. When such a bundle of rays travels through a liquid aracemic material, the speed of one of the rotating dipoles is retarded more than the other. One enantiomer is interacting more with one spiral, while the other enantiomer will interact more with the other (mirror image) spiral. This is analogous to the selectivity of diastereomeric interactions discussed earlier. This results in a change in their phase relationship and their vector sum is in a new plane rotated from the original. Further penetration through the substance causes further rotation of the plane.

Polarimetry allows for measurement of the amount of this rotation, α, which is a characteristic of the substance. A *specific rotation*, $[\alpha]_D^T$, is defined for a sample of density 1 g/mL with a 10-cm pathlength, where T represents temperature and D is indicative of the wavelength of the light used for the measurement, most typically the sodium D-line at 589 nm. The sign of the rotation is + for clockwise as viewed from the end where the light emerges and – for counterclockwise. Of course, enantiomers give equal but opposite rotations. Measurements made with different concentrations and/or pathlengths are proportioned to the specific value. Samples that rotate the plane of polarized light are called *optically active*. Racemic material is not optically active. If a sample is 75% of one enantiomer and 25% of the other, the rotation will be one-half of the maximum and that sample will be labeled 50% optically pure or as having 50% enantiomeric excess. Other techniques useful in the study of chiral molecules include optical rotatory dispersion and circular dichroism (CD) [16].

As mentioned before, the melting point of one enantiomer of a chiral substance will be identical to that of the other enantiomer. However, if both enantiomers are together in a sample, the melting point is likely to be different from that of a pure enantiomer. Consider rows of people shaking hands using all right hands (or all left hands). A certain fit will exist. Consider instead random shaking, including right with left. This produces a different fit. By analogy, interactions between (+) and (–) enantiomers will be different from (+) and (+), and we can expect a different melting point for the mixture. If solutions or melts of various ratios of enantiomers are cooled to produce a solid, we find one of the following three possible behaviors [17]:

1. The enantiomers may crystallize separately, giving a mixture of (+) and (–) crystals called a *conglomerate*. The melting point of a

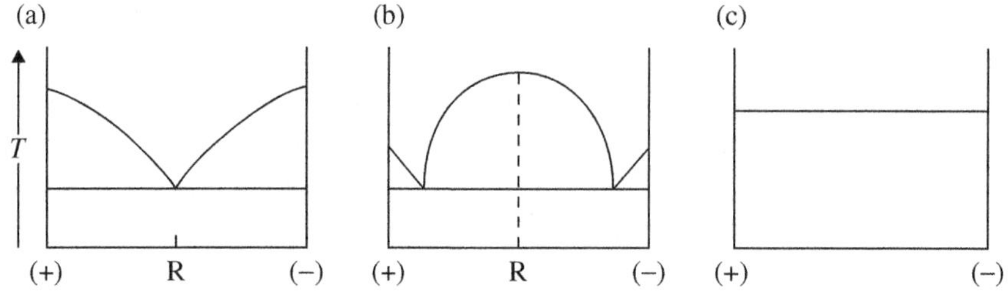

FIGURE 3.6 Melting-point behavior of various enantiomeric pairs. Reproduced from Eliel et al. [1]/with permission of John Wiley & Sons.

conglomerate is then a mixed melting point and is lower than that of a pure enantiomer. Various ratios of (+) and (−) give the melting points graphed in Figure 3.6a.

2. In other cases, a stronger attraction exists between the (+) and (−) pairs and a *racemic compound* is formed. The crystals will contain both enantiomers in equal amounts. The melting point of the racemate is depressed by adding a small amount of either enantiomer (Fig. 3.6b). The racemate may melt either lower or higher than a pure enantiomer.
3. If the difference between enantiomers is small, the substance may crystallize as an ideal solid solution, known as a *pseudoracemate*, where the ratio of (+) to (−) has no effect on the melting point, and the graph is flat (Fig. 3.6c). A few give nonideal solutions with maxima or minima.

The most common type of behavior observed is exhibited in Figure 3.6b and is shown by the broad-spectrum antibiotic ofloxacin [18]. The pure (−) enantiomer (levofloxacin) and the pure (+) enantiomer melt at 225–227°C, while the racemic material melts at 250–257°C.

3.4 RESOLUTION OF ENANTIOMERS

The diastereomeric compounds or complexes discussed in Section 3.3 are used to separate racemic materials into enantiopure components, a process called *resolution* [19]. Chromatographic separation of racemates on various chiral substrates has been demonstrated [20, 21]. One enantiomer interacting with the chiral group on the silica support by hydrogen bonding, π-donor complexation, and/or dipole orientation

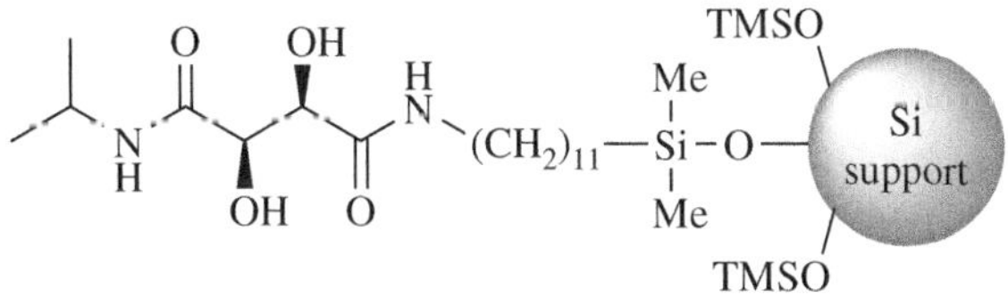

FIGURE 3.7 A chromatographic resolving agent.

is diastereomeric with the complex from the other enantiomer and will differ in dissociation equilibrium constant, thus eluting at different rates. An example of a chiral stationary phase is shown in Figure 3.7 [22]. In this chiral stationary phase, forming trimethylsilyl (TMS) derivatives with the unreacted silyl groups on the silica support helps block the nonselective interactions with the material to be separated, improving enantioselectivity. Columns with chiral stationary phases are increasingly commercially available.

Hydrolytic enzymes are often able to catalyze the esterification of one enantiomer of a racemic alcohol, thus providing one enantiomer of the alcohol and the ester of the other enantiomer, readily separable by extraction or other means. *Candida rugosa* lipase, immobilized on DEAE-Sephadex-A-25, has been used in such a resolution of (±)-menthol (Eq. 3.2) [23].

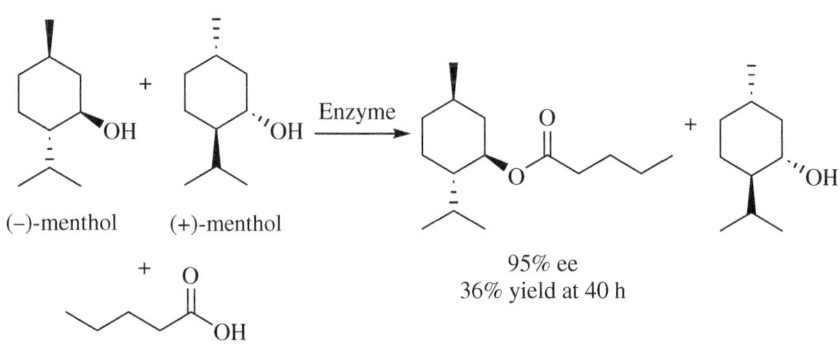

(3.2)

The resolutions described above are based on the formation of diastereomeric complexes with a column stationary phase or an enzyme. The more common alternative is to bond the enantiomers covalently to a chiral resolving agent to make stable diastereomeric molecules, separate those diastereomers by chromatography or recrystallization, and then disassemble each purified diastereomer to obtain the resolved enantiomers.

Racemic *N,N′*-dimethyl-1,2-diphenylethylenediamine was treated with l-(+)-tartaric acid to afford a pair of diastereomeric salts which were separated overnight by selective precipitation from ethanol. The (*R,R*)-diamine was separated by filtration, followed by extraction with a base to remove the tartaric acid. Removal of the solvent produced the crude diamine with 90% yield and greater than 99% ee. The (*S,S*)-diamine was isolated by the concentration of the mother liquor and the same extraction procedure to produce the crude mixture in quantitative yield and greater than 99% ee. Recrystallization from pentane gave narrower melting ranges for the two isolated solids [24].

Both enantiomers of mandelic acid are commercially available, and these are suitable resolving agents for a variety of functional groups. They are often used to isolate chiral alcohols. A sequential use of (*R*)- and (*S*)-mandelic acid allowed the resolution of racemic amino alcohols (Eq. 3.3) [25]. Subsequent extraction gave 90% recovery of the amino alcohols with 99% ee or better. Mandelic acid enantiomers were recovered in 93% yield.

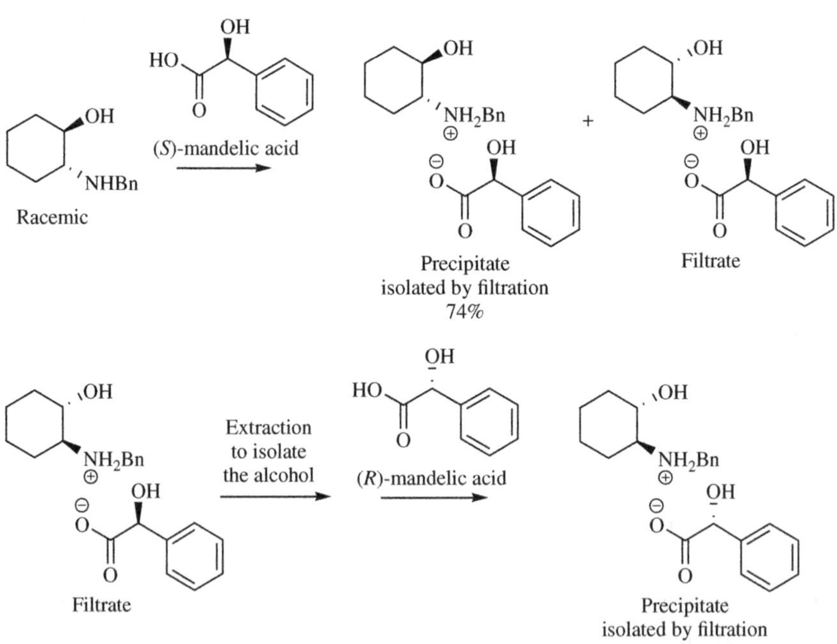

(3.3)

The diastereomeric differentiations in the preceding cases involve physical interactions in chromatography or crystallization. Another possibility is to use a chemical reaction that is fast with one stereoisomer and slow with the other (*kinetic resolution*). A strain of baker's yeast engineered to express cyclohexanone monooxygenase was used to catalyze a Baeyer–Villiger reaction for a variety of 2-alkylcyclohexanones, an example of which is shown in Equation 3.4 [26]. The ketone and lactone are readily separated by silica gel chromatography, and the reaction results in improved enantiopurity for both the ketone and the lactone. Since the initial discovery of these reactions, a wide variety of enzymatic catalysts have been developed and mechanistic studies reported [27].

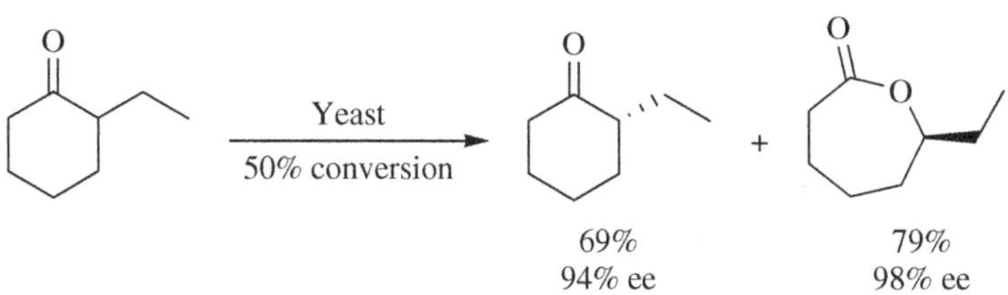

(3.4)

Some racemates that give conglomerates may be resolved with no external diastereomeric influences [28]. Direct crystallization of individual enantiomers from saturated solutions of racemates may be localized by seeding with pure enantiomers, especially if large crystals will grow as, for example, with hydrobenzoin. Good separations of conglomerate mixtures have been accomplished with the use of a magnetic surface as the crystallization substrate, taking advantage of differential spin polarization in chiral molecules in the presence of charge polarization [29].

3.5 ENANTIOSELECTIVE SYNTHESIS

In Section 3.4, pure enantiomers were obtained by the separation of racemic material. Another alternative is to begin with an achiral compound and generate a single enantiomer of a chiral compound from it [30]. This requires a chiral influence from another component in the chemical reaction. If a resolution procedure or an asymmetric synthesis gives a material of perhaps 95% enantiomeric excess, simple recrystallization [31] or sublimation [32] will often give enantiopure material. Resolution and asymmetric synthesis depend on the energy difference between a pair of diastereomeric complexes or between crystals of diastereomers. These differences are very small compared to reaction enthalpies and are, therefore, not as generalizable. You should not expect routine application of these techniques to new cases but rather much trial-and-error development. On the other hand, partially resolved materials can serve very well in studies of reaction stereochemistry or in correlations of configuration.

3.5.1 Chiral Templates

A chiral template may be temporarily attached to an achiral molecule and a new stereogenic center made under that influence, and then the original template is removed [33]. Such a strategy was employed in a key step in the synthesis of (+)-lactacystin, as shown in Equation 3.5. Both enantiomers of the starting tricycloiminolactone are readily available and are useful in a variety of synthetic reactions [34].

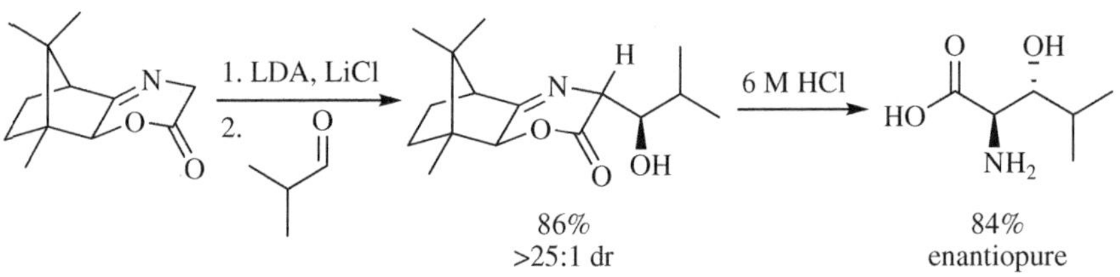

(3.5)

Templates have also been used in amide enolate hydroxylations [35], conjugate additions [36], Diels–Alder reactions [37], Simmons–Smith reactions [38], Claisen rearrangements [39], and many others.

3.5.2 Chiral Catalysts

The chiral influence in these reactions may be a catalyst, in which case a small amount of enantiopure material can lead to large amounts of aracemic product. Methyl acetoacetate was hydrogenated using a soluble enantiopure ruthenium complex as a catalyst [40]. This gave the (*R*)-hydroxyester in 97% enantiomeric excess (Eq. 3.6). The catalyst is sensitive to air and must be prepared under an inert atmosphere. The other enantiomer of the catalyst is also available; therefore, the enantiomeric

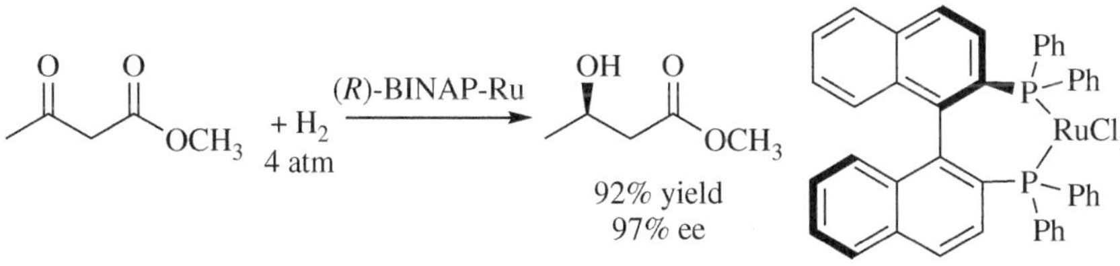

(*R*)-BINAP-Ru

(3.6)

alcohol is as easily prepared. This is a highly efficient multiplication of aracemic material since the substrate-to-catalyst mole ratio is greater than 1000. The (*S*)-enantiomer of ethyl 3-hydroxybutanoate may also be prepared from the ketone by reduction with baker's yeast and sucrose with 59% yield and 85% ee [41].

3.5.3 Chiral Reagents

In these reactions, the reagent itself may be chiral. Enantiopure forms of 1,1′-bi-2-naphthol are readily available [42] and can be complexed with lithium aluminum hydride (LAH) to form the selective reducing agent BINAL-H, which reduces a variety of ketones, including the precursor to Pirkle's alcohol, **3.13**, a chiral solvating agent mentioned in Section 3.3 (Eq. 3.7) [43].

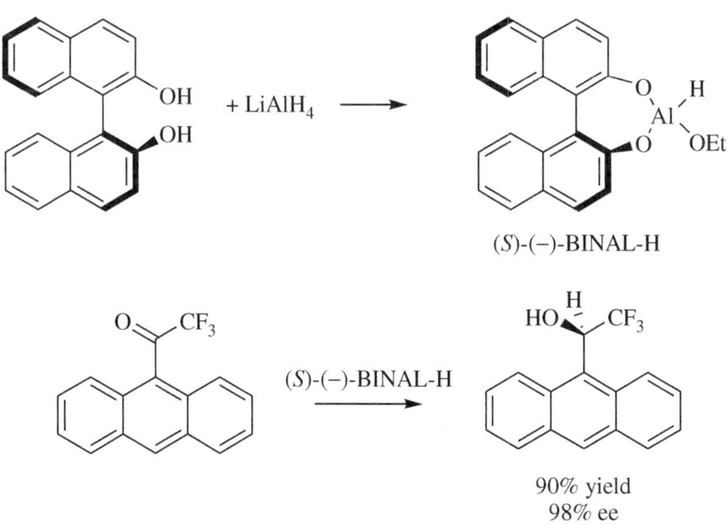

(3.7)

3.5.4 Substrate-Directed Enantioselective Reactions

Over the years, several models have evolved to explain enantioselective reactions of nucleophiles in carbonyl compounds. Cram's rule [44] states that the asymmetric carbon next to the carbonyl group will direct the incoming nucleophile based on steric effects. The Felkin–Anh model [45] further describes the transition state in a way that takes into account the steric crowding between the substituent on the carbonyl group (R) and the largest group on the asymmetric carbon (R_L). This model can also incorporate assumptions for polar groups on the asymmetric center. The presence of chelating agents and chelating substituents can further control the stereochemistry of these reactions [46]. It is also possible that the nucleophile can transfer from the chelating group, giving further stereocontrol. A comparison of these three transition states is shown in Figure 3.8.

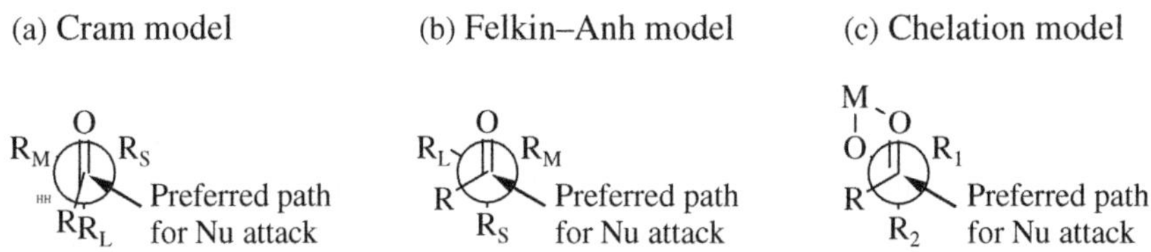

FIGURE 3.8 Comparison of enantioselective transition states in carbonyl additions.

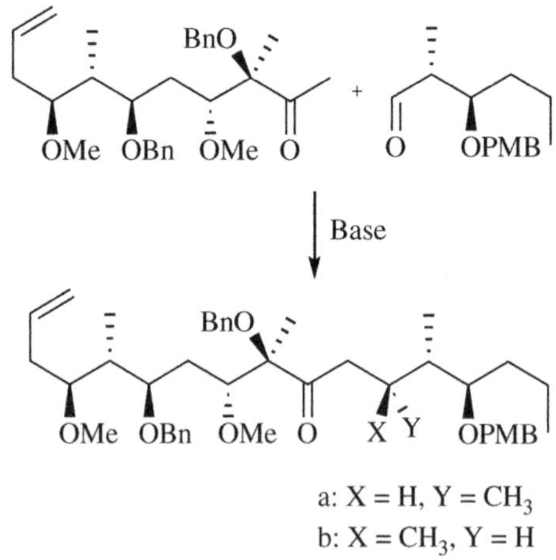

(3.8)

Base	a:b	Base	a:b
LDA	1:10	NaHMDS	2:1
LiHMDS	1:15	KHMDS	4:1

To choose one example from among many, Felkin-moderated selectivity is found in an Aldol condensation step in the total synthesis of des-thiomethyllooekeyolide, a stereocomplex macrocyclic natural product, shown in Equation 3.8 [47].

3.6 REACTIONS AT A STEREOGENIC ATOM

3.6.1 Racemization

In molecules that contain only one tetrahedral stereogenic atom, certain conditions will lead to a loss of optical activity, eventually giving racemic material. These conditions lead to the formation of an intermediate structure where a plane of symmetry passes through the former stereogenic atom. For instance, the removal of a proton from the stereogenic atom to give a carbanion allows a planar or rapidly inverting pyramidal structure to exist at that atom. The return of a proton will then be equally probable on either face, leading to either enantiomer. After sufficient time, equal amounts of enantiomers will be present and the result is called *racemization.*

Other circumstances where a plane of symmetry may occur at an intermediate stage include nucleophilic substitution, neighboring-group participation, rearrangements, and carbocation formation. 1-Bromoethylbenzene shows an optical half-life in solution in hexamethylphosphoric triamide–pentane of only 8 h at 27°C [48]. In this ionizing solvent, a relatively stable, flat intermediate carbocation may be responsible, or perhaps a small amount of bromide ion impurity may give nucleophilic displacement of bromide via a symmetric transition state.

3.6.2 Epimerization

If a proton is removed from a stereogenic atom in a molecule that contains a second, nonreacting, tetrahedral stereogenic atom, no symmetry plane is possible and the returning proton may favor one side more than the other. This may give finally unequal amounts of diastereomers, but, of course, no enantiomers. Since the second stereogenic atom is preserved, optical activity will change, but not to zero. Such a process is called *epimerization.* An example is shown in Equation 3.9, in which epigallocatechin gallate (EGCG), an antioxidant found in tea, is converted to gallocatechin gallate (GCG). The epimerization was found to be time- and temperature-dependent with a 37% conversion observed after 30 minutes at 90°C [49].

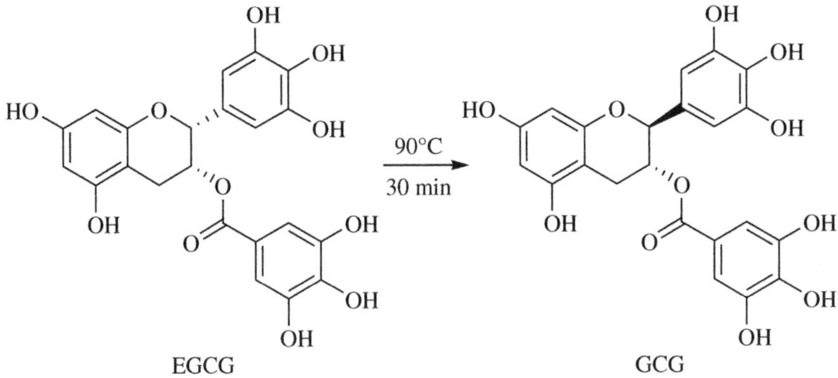

(3.9)

3.6.3 Inversion

Inversion is the replacement of a leaving group on a stereogenic atom by a new group, not in the same position but approaching from the opposite side of the stereogenic atom, causing the remaining three groups to spread through a planar condition and resume tetrahedral angles on the opposite side. One of the first examples in which inversion was known to occur in a particular step is shown in Equation 3.10 [50]. The overall three-step process gave the alcohol of inverted configuration as found by rotation

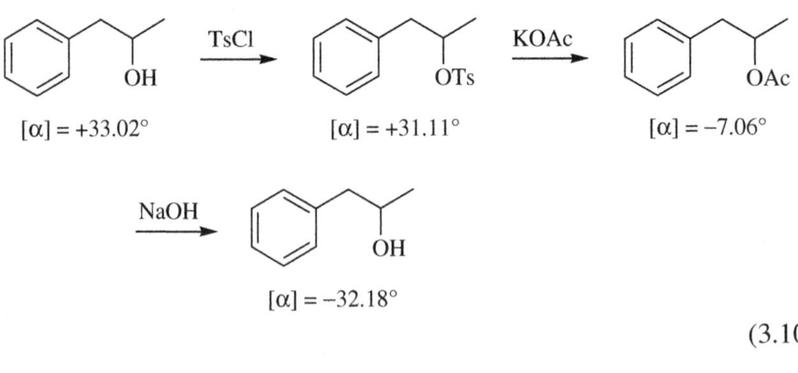

(3.10)

measurements. The first and third steps did not involve bonds to the stereogenic atom and could not give inversion. Thus, inversion must have occurred in the second step. Note that this conclusion was made without the knowledge of the absolute configurations (Section 3.7). Nucleophilic substitution reactions that follow clean second-order kinetics (S_N2 reactions) generally give complete inversion of configuration, owing to the simultaneous bond formation and breakage at the stereogenic center.

3.6.4 Retention

Retention of configuration occurs when an incoming group replaces a leaving group on a stereogenic atom directly (front side) without inversion. Retention is also found when a two-step substitution occurs; that is, a temporary group arrives with inversion and is, in turn, replaced by a final group with a second inversion. The one-step front-side substitution occurs when the incoming group is attached to the leaving group and thus held to the front side in a three-membered ring transition state, as is often the case in 1,2 rearrangements. The Beckmann [51], Hofmann [52], Curtius [53], Schmidt [54], Wolff [55], Lossen, and Baeyer–Villiger [56] rearrangements generally give retention (Section 4.4.4). The Beckmann rearrangement shown in Equation 3.11 [57] is promoted by polyphosphate ester (PPE) with no loss of stereochemistry.

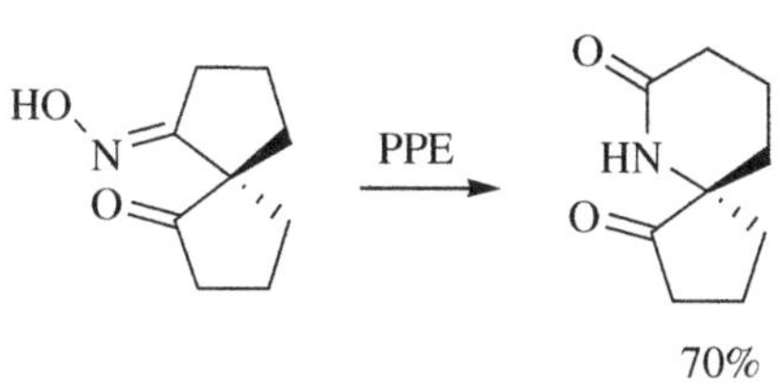

(3.11)

The retention of configuration was demonstrated in the Baeyer–Villiger reaction, as shown in Equation 3.12 [58]. Perbenzoic acid initiates migration of the norbornyl group in a concerted fashion, allowing the *exo* stereochemistry of the starting material to be maintained.

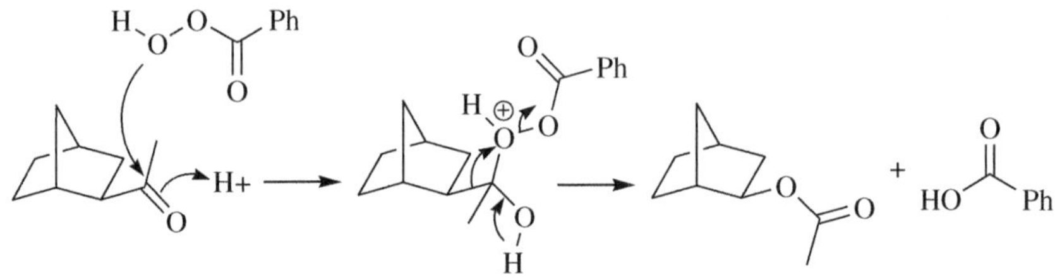

(−)-exo-acetylnorbornane

(−)-exo-norbornyl acetate
81% yield
94.2-100% retention

(3.12)

3.6.5 Transfer

A new tetrahedral stereogenic atom may be formed stereospecifically, while an original stereogenic atom flattens. The Claisen rearrangement (Section 8.7) of aracemic allylic alcohols shows such a transfer (Eq. 3.13) [59]. The (*R*)-*cis* allylic alcohol and the ortho ester react to give a ketene

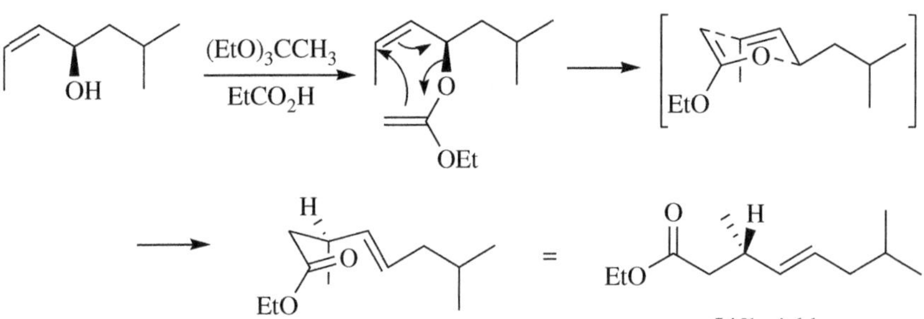

54% yield
97.8% chiral transfer

(3.13)

acetal intermediate that undergoes stereospecific rearrangement via the chair–six-membered ring transition state in which the isobutyl group is equatorial, forming the (*S*)-*trans* ester. As in retention, the leaving group and the incoming group are bound together and thus disconnect from and connect to the same face of the allylic system. By close examination of the corresponding transition state, you can see that the (*S*)-*trans* allylic alcohol gives the same stereoisomeric product.

3.7 RELATIVE AND ABSOLUTE CONFIGURATIONS

If models or drawings of an enantiomeric pair are made, we can label each with an *R* or an *S* for each tetrahedral stereogenic atom. If two actual samples of the enantiomeric materials are on hand, we can make a measurement of rotation direction and label each as (+) or (−). Now, which go together? Does the (+) sample have the *R* or *S* structure? Conventional X-ray structural analysis affords interatomic distances and angles but does not provide any indication of which mirror image is present in a crystal of a chiral substance. Until the work of Bijvoet in 1951 [60], there was no way of determining this. He showed that by choosing X-rays of a wavelength that excites an element in a crystal, a phase lag effect is produced that indicates which enantiomer is present. Since then, many such determinations have been done. For a particular substance, correlating the sign of rotation with the configurational designation of structure gives the *absolute configuration.* Relative configuration is what it sounds like: the configuration of the molecule is assigned *relative* to other molecules to which it is related, likely by stereospecific reactions.

For example, the absolute configuration of (+)-tartaric acid was determined to be (2*R*,3*R*) by Bijvoet [61]. Once this relationship is known, the absolute configuration of 2,3-butanediol produced by the sequence in Equation 3.14 is also known. Because none of the reactions in the sequence involve the chiral carbons, the original stereochemistry is preserved [62].

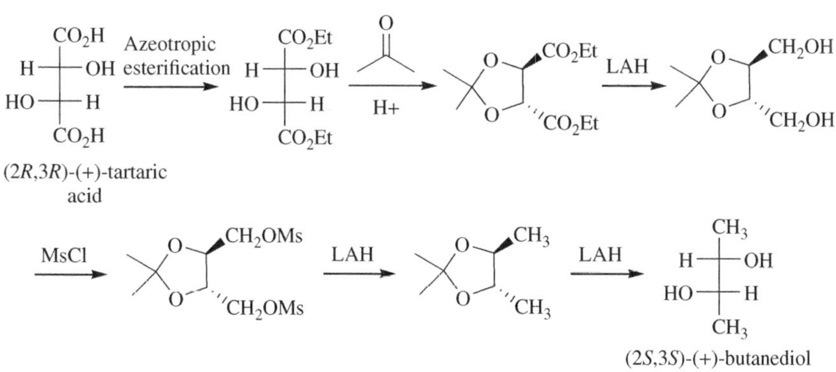

(3.14)

Similarly, in Equation 3.10, if the absolute configuration is known for one of the compounds, it is known for all of them.

Ordinary X-ray crystallography of diastereomeric substances wherein the absolute configuration at one stereogenic atom is known allows for assignment of absolute configuration at the other stereogenic atoms present in the crystal structure. For example, the absolute configuration of 2,8-dimethyl-6*H*,12*H*-5,11-methanodibenzo[*b*,*f*]diazocine, Tröger's base, is assigned (5*S*,11*S*)-(+) based on X-ray analysis of the salt, **3.16** [63], in which the configuration of the acid component is already known [64].

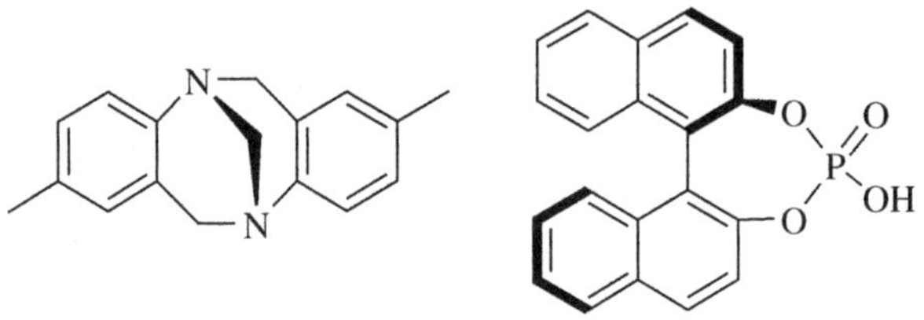

3.16

Assignment of absolute configuration requires some precautions. Bijvoet X-ray crystal structures are unambiguous but can be difficult to obtain in the absence of a heavy atom such as a metal. Direct conversion of compounds to others of known absolute configuration as in the preceding cases allows for assignment of absolute configuration, but care must be taken in drawing analogies from the literature because the sign of the specific rotation can change with wavelength, temperature, solvent, and concentration, and impurities can cause errors [65]. It is interesting to note that the use of configurational information such as proof of inversion in S_N2 reactions was made with relative configurations before absolute ones were available. In fact, the absolute configurations are not useful in themselves, except as another means of obtaining more relative configurations [66].

Keep in mind that two compounds with the "same" configuration may have different configurational designations and/or they may have opposite signs of rotation as in Equation 3.15.

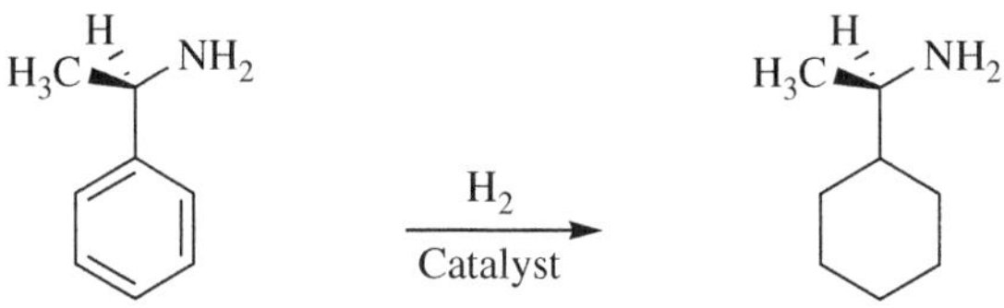

(3.15)

(*R*)-(+)-1-Phenylethylamine (*R*)-(−)-1-Cyclohexylethylamine

The preceding reactions demonstrating configurational correlation did not involve bonding changes at the stereogenic carbon and are quite reliable. Many other correlations involve reactions at the stereogenic atom but with known stereochemistry, such as S_N2 with inversion. These are also reasonably reliable. Other correlations have been made by observing a constancy of direction of rotation in a family of compounds such as *n*-alkyl secondary alcohols where the *S* isomer is generally (+). Still other correlations and statements of absolute configuration have been made on the basis of the order of elution in chromatography [67] and on generalizations on stereospecificity in asymmetric syntheses.

An extensive, referenced, illustrated list of stereochemical correlations and absolute configurations [68] and a list of 6000 selected absolute configurations [69] are available.

3.8 TOPISM

Up to this point, we have considered whole molecules differing as stereoisomers. We now turn to an atom or a group (A/G) within a molecule and examine the three-dimensional shape of the environment of that A/G within the molecule [70]. An A/G that resides in a chiral environment in a molecule is called *chirotopic*. All atoms in a chiral molecule are chirotopic, but some A/G in achiral molecules can be chirotopic also. For example, bromochloromethane is not chiral and has no stereoisomers, but the environment of each of the hydrogen atoms is chiral (Fig. 3.9); therefore, those hydrogen atoms are chirotopic.

In most molecules, the tetrahedral stereogenic atoms are chirotopic as well. This is not always the case as exemplified by carbon 3 in the

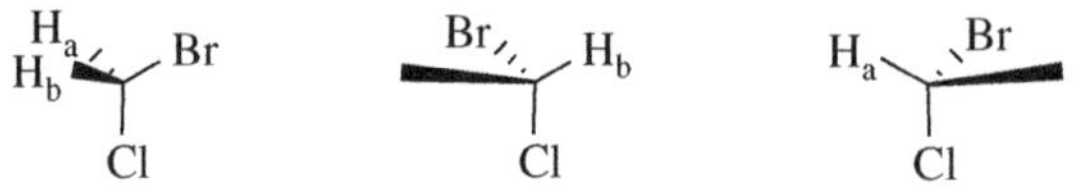

FIGURE 3.9 Environments in bromochloromethane.

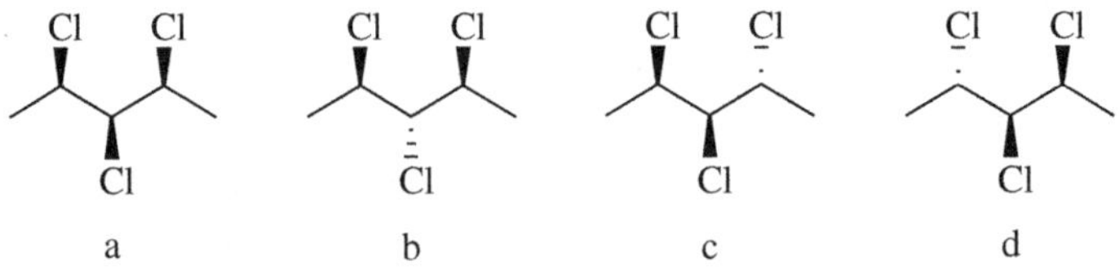

FIGURE 3.10 The stereoisomers of 2,3,4-trichloropentane.

stereoisomers of 2,3,4-trichloropentane (Fig. 3.10). Isomers *a* and *b* are both meso, and carbon 3 is stereogenic but not chirotopic. Isomers *c* and *d* are racemic, and carbon 3 is chirotopic but not stereogenic.

It is useful to compare two chirotopic A/Gs within the same molecule that have the same bonding connectivity. We will find that they are either *enantiotopic* (have enantiomeric environments), *diastereotopic* (have environments that are more different than a mirror image), or *homotopic* (have identical environments). Consider again the two hydrogens in bromochloromethane. These hydrogen atoms reside in mirror-image environments and are thus enantiotopic to each other. There are no enantiotopic A/Gs in molecules that are chiral, but they exist in meso compounds and molecules as simple as butane.

Under ordinary circumstances, enantiotopic A/Gs exhibit identical character, but in association with chiral materials, their environments become more different than mere mirror images. This differentiation is demonstrated graphically for the general case in Figure 3.11. The molecule containing enantiotopic groups can attach to a chiral substrate with all complementary groups and sites matched using one of the enantiotopic groups; however, attempting to use the other enantiotopic group fails to give a match. They are thus differentiable. Observable differentiations of this sort occur in enzyme-catalyzed reactions. For example, *cis*-3,5-diacetoxycyclopentene can be hydrolyzed

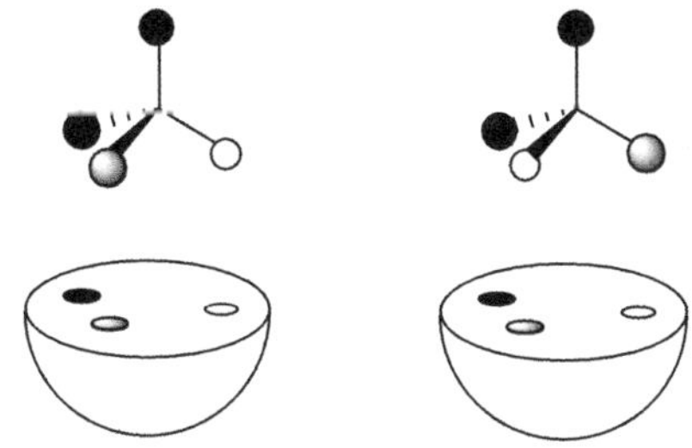

FIGURE 3.11 Two orientations for complexing enantiotopic groups on a chiral surface.

with electric eel acetyl cholinesterase (EEAC) to give enantiopure (1*R*,4*S*)-(+)-4-hydroxy-2-cyclopentenyl acetate (Eq. 3.16) [71]. Recrystallization improves the ee from 96% to 99% or better.

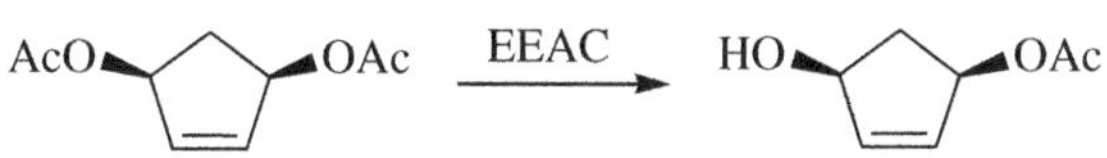

86% yield
>99% ee after recrystallization
[α] +73.8° after recrystallization

(3.16)

There is a simple thought test to determine whether certain groups in a molecule are enantiotopic. Simply imagine replacing one of the two groups with an atom X. Then do likewise with the other one instead (Fig. 3.12). If this produces a pair of enantiomers, the groups are enantiotopic. These distinguishable enantiotopic groups may each be labeled. If the X group has a higher priority (Cahn–Ingold–Prelog) than the group it replaced but

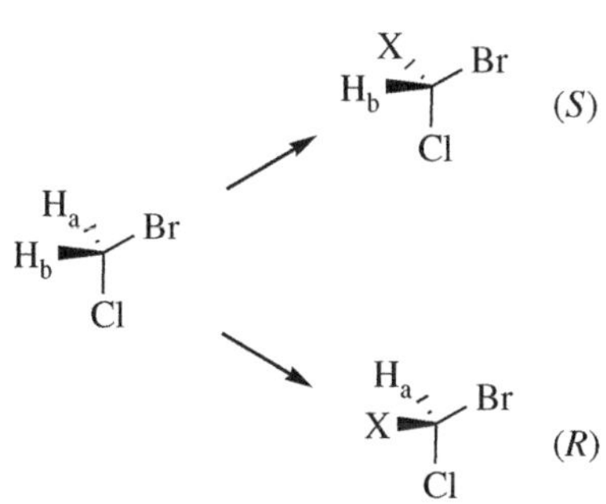

FIGURE 3.12 Identifying enantiotopic pairs.

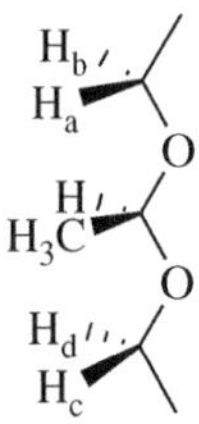

FIGURE 3.13 Stereotopic relationships in acetaldehyde diethyl acetal.

not higher than the next-higher original group, and the resulting stereogenic atom is of *R* configuration, the original enantiotopic group replaced is designated *pro-R*. If the other enantiotopic group had been replaced, the stereogenic atom would have been *S*; therefore, the other enantiotopic group is designated *pro-S* [72]. Thus, H_a in Figure 3.9 is *pro-S* and H_b is *pro-R*.

Pairs of A/G that reside in diastereomeric environments are called *diastereotopic*. These, too, can be identified by the thought test. If replacing each of a pair of A/G with X produces a pair of diastereomers, the original pair of A/G is diastereotopic. Since diastereotopic A/Gs are not related by molecular symmetry, they may react at different rates and they will not be equivalent in NMR spectra. The H_a and H_b in acetaldehyde diethylacetal (Fig. 3.13) have this relationship. The replacement of H_a with X generates two new stereogenic atoms. The replacement of H_b gives a structure diastereomeric with the first one; thus H_a and H_b are diastereotopic. The environments within the molecule of H_a and H_b are diastereomeric, and H_a and H_b give separate signals in the ^1H NMR spectrum (Section 10.5.2.6). In Figure 3.13, the two ethyl groups are enantiotopic. You should build molecular models to assure yourself of these relationships and also those listed in Figure 3.13. In contrast, the diethyl acetal of formaldehyde contains no diastereotopic A/G. Diastereotopic A/Gs are found in most chiral molecules, meso molecules, and molecules that include an atom that bears two identical A/G and two other different A/G, such as bromochloromethane.

Pairs of A/Gs that have identical environments and can be interchanged by rotation of the molecule or rotations within the molecule are absolutely indistinguishable and are called *homotopic*. Substitution of X for either of them gives the same identical molecule. For example, the hydrogens of the

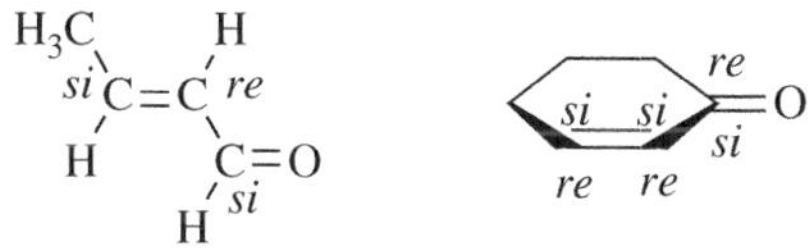

FIGURE 3.14 *trans*-2-Butenal shown with the 1*si*, 2*re*, and 3*si* faces toward you, and 2-cyclohexen-1-one with the 1*re*, 2*si*, and 3*si* faces upward.

free rotating methyl group in 1-chloro-1-bromoethane are homotopic and the chlorine atoms in (*R*,*R*)-2,3-dichlorobutane are homotopic.

The two faces of a flat molecular site may be enantiotopic. If an sp^2 carbon has three different groups bonded to it, the faces are enantiotopic as exemplified in Figure 3.14. Looking directly at one face, if the three groups give a clockwise decrease in priority, the face toward you is designated *re*. If it is turned over, the face now toward you will give a counterclockwise decrease and be designated *si* [72].

As with enantiotopic groups, enantiotopic faces are differentiated in reactions by chiral reagents or catalysts. The chiral hydride-reducing agent in Equation 3.7 selectively adds a hydride to the *si* face of the carbonyl carbon.

RESOURCES

1. Basic Terminology of Stereochemistry. https://iupac.qmul.ac.uk/stereo/. A web version of the original document published by IUPAC [73]. The original document is available for download on the web page as well.
2. ChemSpider. https://www.chemspider.com/. This chemical database provides three-dimensional structures for molecules.

PROBLEMS

Questions marked with an asterisk are answered in Appendix C.

3.1* Identify the mechanistic steps in Equation 3.10 that explain the observed stereochemistry.

3.2 How would you convert (*S*)-7-methyl-6-nonen-3-ol to (*S*)-7-methyl-6-nonen-3-ol acetate? How would you convert (*S*)-7-methyl-6-nonen-3-ol to (*R*)-7-methyl-6-nonen-3-ol acetate? [74]

3.3 Suppose that you needed one pure enantiomer of *exo*-bicyclo[2.2.1]heptan-2-ol. How would you prepare it from bicyclo[2.2.1]hept-2-ene (norbornene)? [75]

3.4* How would you carry out the following conversion to give mostly one enantiomer from the meso diester? [76]

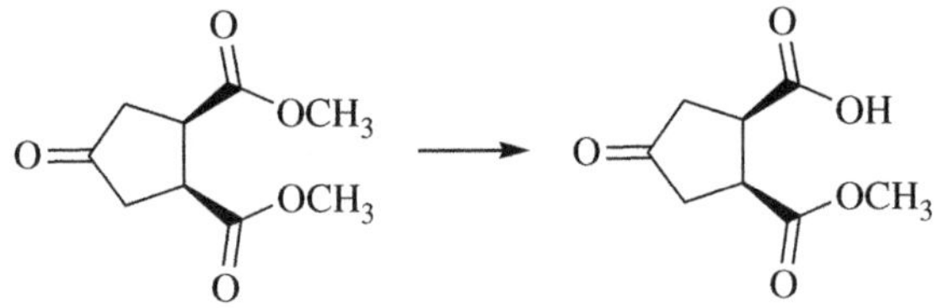

3.5* The separate enantiomers of 2-hydroxyheptanoic acid were needed for the determination of the absolute configuration of verticilide, a potential insecticide [77]. Suggest a possible synthetic route to enantiopure 2-hydroxyheptanoic acid, starting from (*R*)- or (*S*)-malic acid [78].

3.6 Verticilide, below, was hydrolyzed to give four moles of 2-hydroxyheptanoic acid and four moles of *N*-methyl-L-alanine. Chiral high-performance liquid chromatography (HPLC) was used to determine the absolute configuration of the acid to be (*R*)-(+), using the enantiopure materials from Problem 3.5. Draw (*R*)-(+)-2-hydroxyheptanoic acid and explain how the configuration was determined.

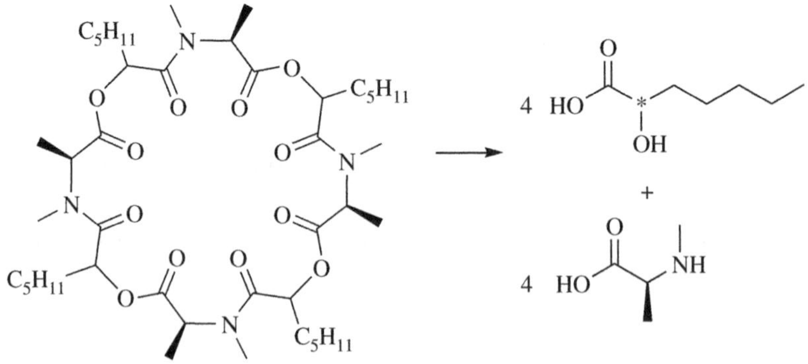

3.7 (+)-3-Tetradecanol is known to have the *S* configuration. Treatment of (+)-1,2-epoxytridecane with methyllithium gives (+) 3 tetradecanol. What is the absolute configuration of (+)-1,2-epoxytridecane? Draw a three-dimensional representation of it. How would you determine the absolute configuration of (−)-2-tridecanol? [79]

3.8 The specific rotation of the (+)-3-tetradecanol prepared in Problem 3.7 was $[\alpha]_D^{25} + 6.7^\circ$. The previously reported rotation for this compound was $[\alpha]_D^{25} + 5.1^\circ$. What is the maximum possible enantiomeric excess in the 5.1° rotating sample? What is the maximum percent of (+)-enantiomer in the 5.1° rotating sample?

3.9 The epoxidation of *trans*-2-buten-1-ol with *tert*-butyl hydroperoxide catalyzed by titanium tetraisopropoxide in the presence of (+)-diisopropyl L-tartrate, (+)-DIPT, gave aracemic *trans*-2,3-epoxybutanol, $[\alpha]_D^{20} - 50.0^\circ$. Treatment of this epoxide with $(CH_3)_2CNCuLi_2$ gave some 2-methyl-1,3-butanediol, which was converted to a sulfonate ester selectively at the primary alcohol, and thence to the iodide (1-iodo-2-methyl-3-butanol) with sodium iodide in acetone. The iodo compound was reduced with $NaBH_4$ to 3-methyl-2-butanol of rotation $[\alpha]_D^{20} + 5.00^\circ$. Pure (*S*)-3-methyl-2-butanol has $[\alpha]_D^{20} + 5.34^\circ$. Draw and name the major stereoisomer of the epoxide and give the percent of the enantiomer present with it. What would you expect if the d-tartrate were used instead? [80]

3.10* Dynamic thermodynamic resolution has been used to synthesize (*R*)-(+)-*β*-conhydrine by the sequence shown [81]. Quenching the lithiated species with propanal leads to a 70:30 mixture of diastereomers with 96:4 enantiomeric ratio. What is the origin of the enantioselectivity in this reaction?

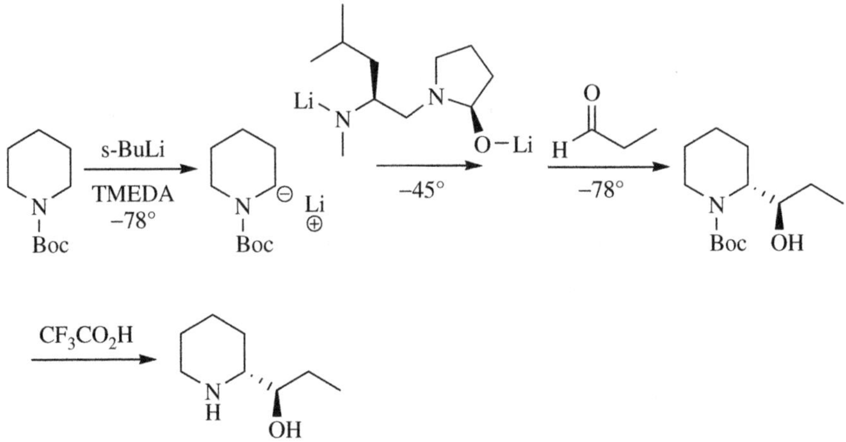

3.11* Draw a three-dimensional representation of 4-methylheptane. Identify a pair of enantiotopic atoms and a pair of diastereotopic atoms.

3.12* What is the relationship between the two methoxy groups in a molecule of 3-bromobutanal dimethyl acetal?

3.13 What is the relationship between the two benzylic hydrogens in *meso*-2,5-diphenylhexane? What is the relationship between the two benzylic hydrogens in (*R*,*R*)-2,5-diphenylhexane?

3.14* The following steps were used to convert the (*R*)-alkynol to the (*R*)-γ,δ-unsaturated ester. How would you convert the (*R*)-alkynol into the (*S*)-γ,δ-unsaturated ester? Red-Al, sodium *bis*(2-methoxyethoxy) aluminum hydride, $NaAlH_2(OCH_2CH_2OCH_3)_2$, is used to reduce alkynols to *trans* allylic alcohols [82].

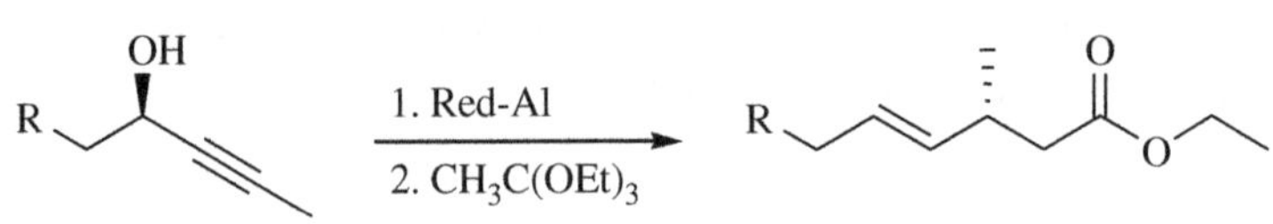

3.15 Draw three-dimensional representations of all the possible stereoisomeric esters formed between (*R*)-*O*-methyl mandelic acid and all the possible stereoisomers of the alcohol below [83]. How many stereoisomers from this combination are there?

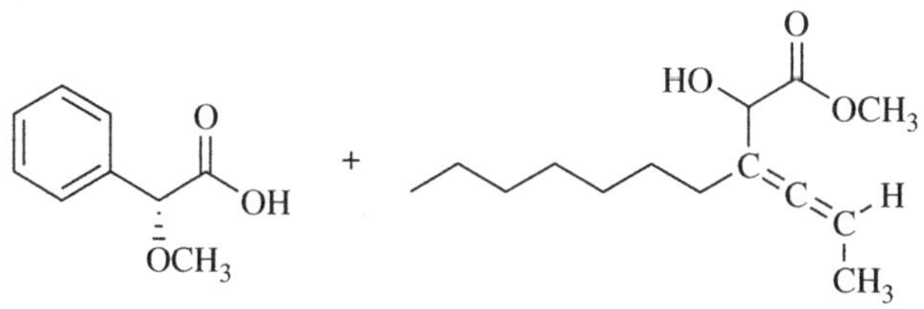

3.16 The photochemical chlorination of (*S*)-(+)-2-bromobutane with *t*-butyl hypochlorite at −78°C gave the following, among other products:

	Yield (%)	Enantiomeric purity
2-Bromo-2-chlorobutane	53	Racemic
erythro-2-Bromo-3-chlorobutane	20	Enantiopure
threo-2-Bromo-3-chlorobutane	6	Racemic
3-Bromo-1-chlorobutane	4	Enantiopure
1-Bromo-2-chlorobutane	3	Enantiopure

Draw three-dimensional representations of each of these products and explain in terms of intermediates why three of these are enantiopure and two are not [84].

3.17 The two stereoisomers of 3-methyl-2,4-dibromopentane shown below were cyclized by treatment with zinc in 1-propanol−water. The products were analyzed for the ratio of *cis*- to *trans*-trimethylcyclopropanes; the results are as follows:

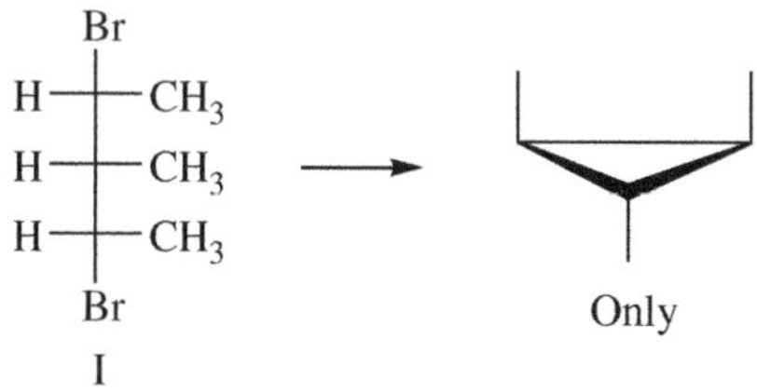

I

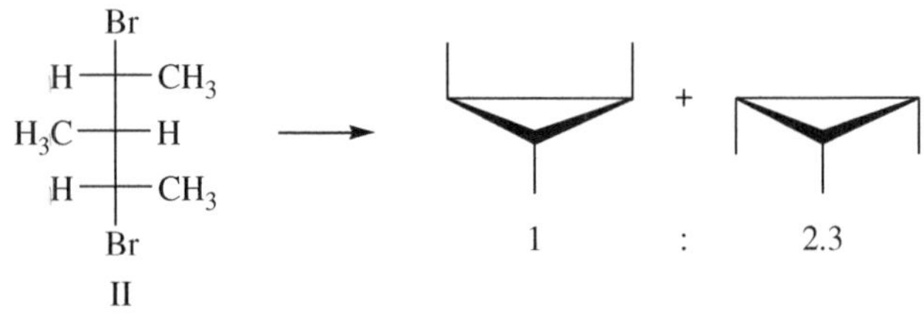

The overall stereochemical possibilities are retention at both sites, retention at one and inversion at the other site, and inversion at both sites. Considering all the results, what is the overall stereochemistry for the process or processes that give(s) the major product from the first isomer? What is the overall stereochemistry for the formation of the minor product from the first isomer? Explain how you reached your conclusions with drawings. Molecular models may be helpful [85].

3.18 In terms of *re* and *si*, what face of the double bond received the hydrogen atoms in Equation 3.6?

3.19* Chiral imines can be used in Michael additions, giving a high degree of stereoselectivity. In the reaction below, racemic 2-methylcyclohexanone gives the octalone product with 44% overall yield and 96% ee [86, 87]. The reaction proceeds through the enamine. Sketch the intermediate, including the likely stereochemistry, that explains the selectivity of this reaction.

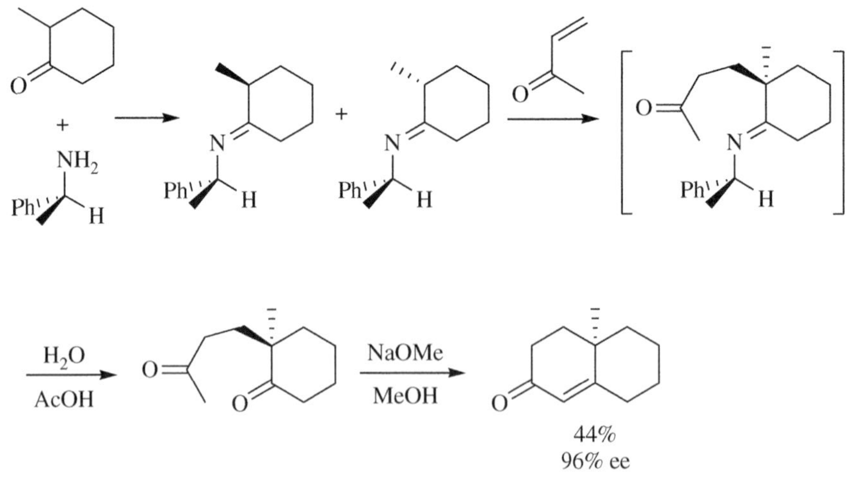

3.20* A carboxylic acid of unknown absolute configuration was converted to the amide of (*S*)-1-phenylethylamine. Conventional X-ray crystallography gave a structure for the amide that might be either of the following mirror images. Which of the two is the correct amide? What is the absolute configuration of the acid at each stereogenic atom? [88]

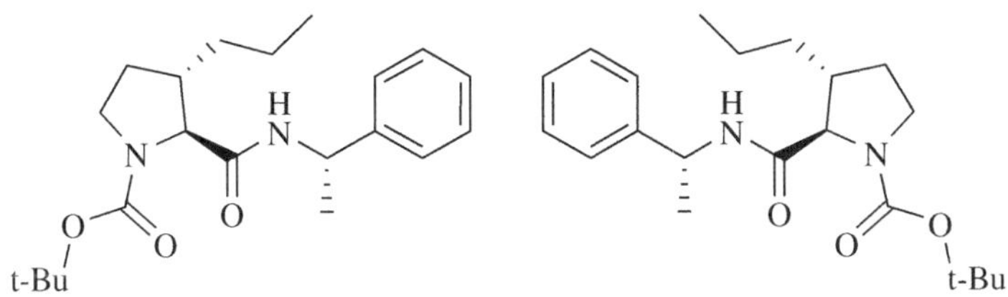

3.21 The reaction to form bedaquiline results in the formation of two chiral centers and four possible stereoisomers. Draw the stereoisomers. Label each chiral center as *R* or *S*. Label the pairs of compounds as enantiomers or diastereomers [89].

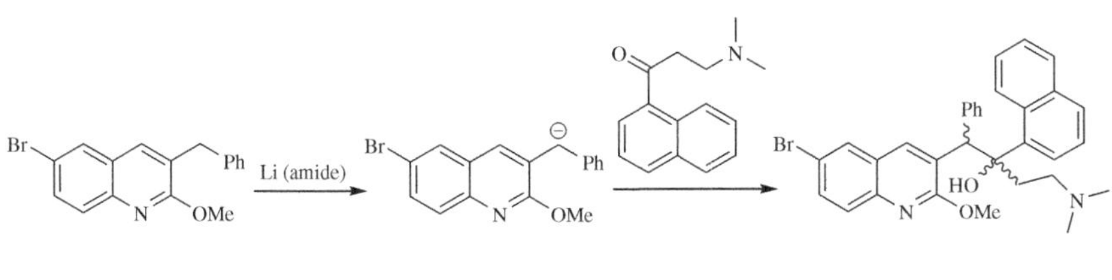

3.22* Bedaquiline is an important treatment option for multi-drug-resistant tuberculosis. The *R*,*S* isomer is shown to be the most effective. The original procedure, shown in Problem 3.21, used lithium diisopropylamide (LDA) in the deprotonation step. Improvements in the procedure replace LDA with a chiral amide, resulting in better diastereoselectivity. For example, when lithium (*R*)-2-methoxymethylpyrrolidide is used in place of LDA, a dr of 13.6:1 and an er of 3.6:1 with 56% ee are obtained. What is the source of this improvement [90, 91]?

REFERENCES

1. Eliel, E. L.; Wilen, S. H.; Mander, L. N. *Stereochemistry of Organic Compounds*; Wiley: New York, 1994. A helpful glossary is included at the end of this text.
2. Rhodes, G. Stereoviewing. https://spdbv.unil.ch/TheMolecularLevel/0Help/StereoView.html (accessed May 22, 2024).
3. Black, K. A. *J. Chem. Educ.* **1990**, *67*, 141–142.
4. Cahn, R. S.; Ingold, C. K.; Prelog, V. *Angew. Chem. Int. Ed.* **1966**, *5*, 385–415. Cahn, R. S. *J. Chem. Educ.* **1964**, *41*, 116–125. Prelog, V.; Helmchen, G. *Angew. Chem. Int. Ed.* **1982**, *21*, 567–583.
5. Maitland, P.; Mills, W. H. *Nature* **1935**, *135*, 994.
6. Li, Y.; Yagi, A.; Itami, K. *Chem. Sci.* **2019**, *10*, 5470–5475.
7. Hassan, Z.; Spuling, E.; Knoll, D. M.; Brase, S. *Angew. Chem. Int. Ed.* **2020**, *59*, 2156–2170.
8. Cope, A. C.; Banholzer, K.; Jones, F. N.; Keller, H. *J. Am. Chem. Soc.* **1966**, *88*, 4700–4703.
9. Wenzel, T. J. *Tetrahedron Asymmetry.* **2017**, *28*, 1212–1219.
10. Su, B.-N.; Park, E. J.; Mbwambo, Z. H.; Santarsiero, B. D.; Mesecar, A. D.; Fong, H. H. S.; Pezzuto, J. M.; Kinghorn, A. D. *J. Nat. Prod.* **2002**, *65*, 1278–1282.
11. Chi, Y.; Peelen, T. J.; Gellman, S. H. *Org. Lett.* **2005**, *7*, 3469–3472.
12. Pirkle, W. H.; Hoover, D. J. *Top. Stereochem.* **1982**, *13*, 263–331.
13. Goering, H. L.; Eikenberry, J. N.; Koermer, G. S.; Lattimer, C. J. *J. Am. Chem. Soc.* **1974**, *96*, 1493–1501.
14. Inamoto, A.; Ogasawara, K.; Omata, K.; Kabuto, K.; Sasaki, Y. *Org. Lett.* **2000**, *2*, 3543–3545.
15. Sullivan, G. R. *Top. Stereochem.* **1978**, *10*, 287–329.
16. Rodger, A.; Marshall, D. *Biochemistry* **2021**, *43*(2), 58–64 https://doi.org/10.1042/bio_2020_105.
17. Srisanga, S.; ter Horst, J. H. *Cryst. Growth Des.* **2010**, *10*, 1808–1812.
18. *Merck Index*, 12th ed.; Budavari, S., Ed.; Merck & Co., Inc.: Whitehouse Station, NJ, **1996**, p. 6875.
19. Jacques, J.; Collet, A.; Wilen, S. H. *Enantiomers, Racemates and Resolutions*; John Wiley & Sons: New York, **1981**.
20. Ward, T. J.; Ward, K. D. *Anal. Chem.* **2012**, *84*, 626–635.
21. Pirkle, W. H.; Pochapsky, T. C. *Chem. Rev.* **1989**, *89*, 347–362.

22. Dobashi, Y.; Hara, S. *J. Org. Chem.* **1987**, *52*, 2490–2496.

23. Wang, D.-L.; Nag, A.; Lee, G.-C.; Shaw, J.-F. *J. Agric. Food Chem.* **2002**, *50*, 262–265.

24. Alexakis, A.; Aujard, I.; Kanger, T.; Mangeney, P. *Org. Synth.* **1999**, *76*, 23–36.

25. Schiffers, I.; Bolm, C. *Org. Synth.* **2008**, *85*, 106–117.

26. Stewart, J. D.; Reed, K. W.; Zhu, J.; Chen, G.; Kayser, M. M. *J. Org. Chem.* **1996**, *61*, 7652–7653.

27. Leisch, H.; Morley, K.; Lau, P. C. K. *Chem. Rev.* **2011**, *111*, 4165–4222. Harwood, L. A.; Wong, L. L.; Robertson, J. *Angew. Chem. Int. Ed.* **2021**, *60*, 4434–4447.

28. Collet, A.; Brienne, M. J.; Jacques, J. *Chem. Rev.* **1980**, *80*, 215–230.

29. Bhowmick, D.; Sang, Y.; Santra, K.; Halbauer, M.; Capua, E.; Paltiel, Y.; Naaman, R.; Tassinari, F. *Crys. Growth Des.* **2021**, *21*, 2925–2931.

30. Corey, E. J.; Kürti, L. *Enantioselective Chemical Synthesis: Methods, Logic and Practice*; Direct Book Publishing; Dallas, TX: LLC, **2010**.

31. Denis, J.-N.; Correa, A.; Greene, A. E. *J. Org. Chem.* **1990**, *55*, 1957–1959.

32. Soloshonok, V. A.; Ueki, H.; Yasumoto, M.; Mekala, S.; Hirschi, J. S.; Singleton, D. A. *J. Am. Chem. Soc.* **2007**, *129*, 12112–12113.

33. Li, Q.; Yang, S.-B.; Zhang, Z.; Li, L.; Xu, P.-F. *J. Org. Chem.* **2009**, *74*, 1627–1631.

34. Luo, Y.-C.; Zhang, H.-H.; Wang, Y.; Xu, P.-F. *Acc. Chem. Res.* **2010**, *43*, 1317–1330.

35. Lubin, H.; Tessier, A.; Chaume, G.; Pytkowicz, J.; Brigaud, T. *Org. Lett.* **2010**, *12*, 1496–1499.

36. Meyers, A. I.; Shipman, M. *J. Org. Chem.* **1991**, *56*, 7098–7102.

37. Sanyal, A.; Snyder, J. K. *Org. Lett.* **2000**, *2*, 2527–2530.

38. Mash, E. A.; Torok, D. S. *J. Org. Chem.* **1989**, *54*, 250–253.

39. Kallmerten, J.; Gould, T. J. *J. Org. Chem.* **1986**, *51*, 1152–1155.

40. Kitamura, M.; Tokunaga, M.; Ohkuma, T.; Noyori, R. *Org. Synth.* **1992**, *71*, 1–13.

41. Seebach, D.; Sutter, M. A.; Weber, R. H.; Züger, M. F. *Org. Synth.* **1985**, *63*, 1–9.

42. Kazlauskas, R. J. *Org. Synth.* **1992**, *70*, 60–67.

43. Chong, J. M.; Mar, E. K. *J. Org. Chem.* **1991**, *56*, 893–896.

44. Cram, D. J.; Elhafez, F. A. A. *J. Am. Chem. Soc.* **1952**, *74*, 5825–5835.

45. Cherest, M.; Felkin, H.; Prudent, N. *Tetrahedron Lett.* **1968**, *9*, 2199–2204. Anh, N. T.; Eisenstein, O.; Lefour, J. M.; Tran Huu Dau, M. E. *J. Am. Chem. Soc.* **1973**, *95*, 6146–6147. Mengel, A.; Reiser, O. *Chem. Rev.* **1999**, *99*, 1191–1223.
46. Reetz, M. T. *Acc. Chem. Res.* **1993**, *26*, 462–468.
47. Liu, J.; Chem, Y.; Luesch, H.; Ye, T. *Org. Lett.* **2022**, *24*, 7260–7264.
48. Hutchins, R. O.; Masilamani, D.; Maryanoff, C. A. *J. Org. Chem.* **1976**, *41*, 1071–1073.
49. Suzuki, M.; Sano, M.; Yoshida, R.; Degawa, M.; Miyase, T.; Maeda-Yamamoto, M. *J. Agric. Food Chem.* **2003**, *51*, 510–514.
50. Phillips, H. *J. Chem. Soc.* **1923**, *123*, 44–59.
51. Gawley, R. E. *Org. React.* **1988**, *35*, 1–420.
52. Wallis, E. S.; Lane, J. F. *Org. React.* **1946**, *3*, 267–306.
53. Smith, P. A. S. *Org. React.* **1946**, *3*, 337–449.
54. Wolff, H. *Org. React.* **1946**, *3*, 307–336.
55. Kirmse, W. *Eur. J. Org. Chem.* **2002**, 2193–2256.
56. Krow, G. R. *Org. React.* **1993**, *43*, 251–798.
57. Hilmey, D. G.; Paquette, L. A. *Org. Lett.* **2005**, *7*, 2067–2069.
58. Berson, J. A.; Suzuki, S. *J. Am. Chem. Soc.* **1959**, *81*, 4088–4094.
59. Chan, K.-K.; Cohen, N.; De Noble, J. P.; Specian, A. C. Jr.; Saucy, G. *J. Org. Chem.* **1976**, *41*, 3497–3505.
60. Trommel, J.; Bijvoet, J. M. *Acta Cryst.* **1954**, *7*, 703–709. Bijvoet, J. M. *Endeavour* **1955**, *14*, 71–77.
61. Bijvoet, J. M.; Peerdeman, A. F.; van Bommel, A. J. *Nature* **1951**, *168*, 271–272.
62. Rapoport, H.; Plattner, J. J. *J. Am. Chem. Soc.* **1971**, *93*, 1758–1761.
63. Wilen, S. H.; Qi, J. Z.; Willard, P. G. *J. Org. Chem.* **1991**, *56*, 485–487.
64. Jacques, J.; Fouquey, C.; Viterbo, R. *Tetrahedron Lett.* **1971**, *12*, 4617–4620.
65. Archelas, A.; Furstoss, R. *J. Org. Chem.* **1999**, *64*, 6112–6114.
66. Fiaud, J. C.; Horeau, A.; Kagan, H. B. *Determination of Configurations by Chemical Methods*; Georg Thieme: Stuttgart, 1977.
67. Doolittle, R. E.; Heath, R. R. *J. Org. Chem.* **1984**, *49*, 5041–5050.
68. Klyne, W.; Buckingham, J. *Atlas of Stereochemistry*, 2nd ed., Volume 1–2; Oxford University Press: New York, 1978. (Volume 1 is available online at www.archive.org.)
69. Jacques, J.; Gros, C.; Bourcier, S. *Absolute Configuration of 6000 Selected Compounds with One Asymmetric Carbon Atom*; Georg Thieme: Stuttgart, 1977.

70. Mislow, K.; Raban, M. *Top. Stereochem.* **1967**, *1*, 1–38. Mislow, K.; Siegel, J. *J. Am. Chem. Soc.* **1984**, *106*, 3319–3328.
71. Deardorff, D. R.; Windham, C. Q.; Craney, C. L. *Org. Synth.* **1996**, *73*, 25–35.
72. Hanson, K. R. *J. Am. Chem. Soc.* **1966**, *88*, 2731–2742.
73. Moss, G. P. *Pure Appl. Chem.* **1996**, *68*, 2193–2222.
74. Johnston, B. D.; Oehlschlager, A. C. *J. Org. Chem.* **1986**, *51*, 760–763.
75. Whitesell, J. K.; Reynolds, D. *J. Org. Chem.* **1983**, *48*, 3548–3551.
76. Gais, H.-J.; Bülow, G.; Zatorski, A.; Jentsch, M.; Maidonis, P.; Hemmerle, H. *J. Org. Chem.* **1989**, *54*, 5115–5122.
77. Monma, S.; Sunazuka, T.; Nagai, K.; Arai, T.; Shiomi, K.; Matsui, R.; Ōmura, S. *Org. Lett.* **2006**, *8*, 5601–5604.
78. Dutton, F. E.; Lee, B. F.; Johnson, S. S.; Coscarelli, E. M.; Lee, P. H. *J. Med. Chem.* **2003**, *46*, 2057–2073.
79. Coke, J. L.; Richon, A. B. *J. Org. Chem.* **1976**, *41*, 3516–3517.
80. White, J. D.; Theramongkol, P.; Kuroda, C.; Engebrecht, J. R. *J. Org. Chem.* **1988**, *53*, 5909–5921.
81. Beng, T. K.; Gawley, R. E. *J. Am. Chem. Soc.* **2010**, *132*, 12216–12217.
82. Chan, K.-K.; Specian, A. C. Jr.; Saucy, G. *J. Org. Chem.* **1978**, *43*, 3435–3440.
83. Marshall, J. A.; Wang, X. *J. Org. Chem.* **1990**, *55*, 2995–2996.
84. Skell, P. S.; Pavlis, R. R.; Lewis, D. C.; Shea, K. J. *J. Am. Chem. Soc.* **1973**, *95*, 6735–6745.
85. Applequist, D. E.; Pfohl, W. F. *J. Org. Chem.* **1978**, *43*, 867–871.
86. Revial, G.; Pfau, M. *Org. Synth.* **1992**, *70*, 35–46.
87. Sevin, A.; Tortajada, J.; Pfau, M. *J. Org. Chem.* **1986**, *51*, 2671–2675.
88. Chung, J. Y. L.; Wasicak, J. T.; Arnold, W. A.; May, C. S.; Nadzan, W. A.; Holladay, M. W. *J. Org. Chem.* **1990**, *55*, 270–275.
89. Lubanyana, H.; Arvidsson, P. I.; Govender, T.; Hruger, H. G.; Naicker, T. *ACS Omega.* **2020**, *5*, 3607–3611.
90. Mear, S. J.; Lucas, T.; Ahlqvist, G. P.; Robey, J. M. S.; Dietz, J.-P.; Khairnar, P. V.; Maity, S.; Williams, C. L.; Snead, D. R.; Nelson, R. C.; Opatz, T.; Jamison, T. F. *Chem. Eur. J.* **2022**, *28*, e202201311. doi: 10.1002/chem.202201311.
91. Robey, J. M.; Maity, S.; Aleshire, S. L.; Ghosh, A.; Yadaw, A. K.; Roy, S.; Mear, S. J.; Jamison, T. F.; Sirasani, G.; Senanayake, C. H.; Stringham, R. W.; Gupton, B. F.; Donsbach, K. O.; Nelson, R. C.; Shanahan, C. S. *Org. Process Res. Dev.* **2023**, *27*, 2146–2159.

4

MECHANISMS AND PREDICTIONS

When planning a new reaction in organic chemistry, we look at the accumulated information on similar reactions in order to predict the best conditions for it. The more we know about the details of the reaction process at the molecular level, the better will be our predictions. A particular reaction may be described as an ordered sequence of bond breaking and making and a series of structures that exist along the way from the starting material to the product. The description includes the concurrent changes in potential energy. Structures at energetic maxima are called transition states and structures at minima are called intermediates. The complete description is called the mechanism of the reaction.

4.1 REACTION COORDINATE DIAGRAMS AND MECHANISMS

The energy–structure relationship is sometimes illustrated with a plot of potential energy versus progress along the pathway of the lowest

Understanding Organic Chemistry: Bridging Fundamental and Advanced Courses, Fourth Edition. Ann M. Fabirkiewicz and John C. Stowell.

Companion website: www.wiley.com/go/fabirkiewiczorganic4e

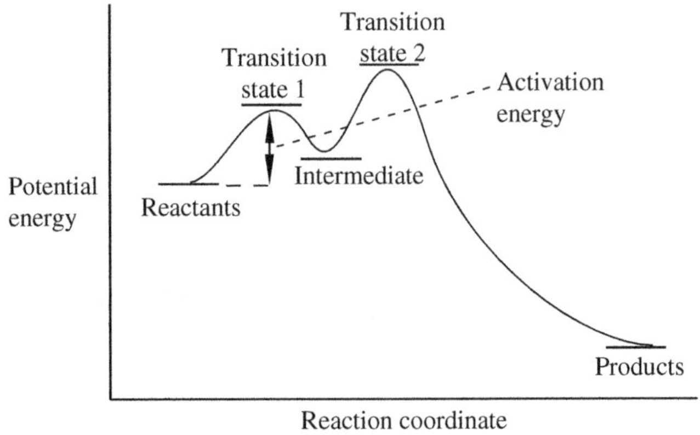

FIGURE 4.1 Reaction coordinate diagram.

maximum potential energy. This is exemplified in a general way in Figure 4.1. The plot shows a two-step reaction leading from reactants through a first transition state to intermediates. The intermediates pass through a second transition state to products. The net overall descent for this reaction corresponds to an exothermal process.

A reaction that requires a higher rise to a transition state (activation energy) will be slower than the one requiring a lesser rise (if the probability factors are similar) because a smaller fraction of collisions will have sufficient potential energy to get to the transition state.

The energy values for such plots are derived from measurements of overall exo- or endothermicity and from measurements of the effect of varying the temperature on the rate of the reaction (Section 4.3.6).

Many techniques have been developed for determining mechanisms including complete product (and sometimes intermediate) identification, isotope labeling, stereochemistry, and kinetics [1–3], as are covered in Section 4.3. Two or more alternative mechanisms are proposed, and their differences are probed with these techniques. When an observation is made that is incompatible with one of the proposed mechanisms, that mechanism is eliminated from consideration, eventually leaving one. Reasonable mechanisms that have withstood various experimental tests gain some acceptance and are then very useful in predicting the possible range of applications of the reaction and for suggesting changes in reaction conditions that will improve its yield and efficiency.

Since molecules are individually too small and too fast for direct observation, our pictorial mechanisms are not the final word. Published mechanisms vary from unsupported conjecture to highly tested near certainty, and you should maintain a healthy skepticism in order to improve on the information available.

Even where the goal, a well-defined singular mechanism, is not yet attained for a given reaction, observations can be used to make predictions. For example, using the Hammett equation (Section 4.3.7), we can predict from measured rates of some cases of a reaction how fast a new case with a different substitution will occur.

4.2 THE HAMMOND POSTULATE

Let us begin our discussion with the transition state. In 1955, Hammond proposed that the transition state will most closely resemble the species (product or reactant) closest to it in energy [4]. This means that for an endothermic reaction, the energy of the transition state is close to the energy of the product. In competing endothermic reactions, the energy difference between the two transition states is as different from the reactants as the products are, and a high degree of selectivity for the more stable product is expected (Fig. 4.2a). In an exothermic reaction, the transition state is similar in structure to the reactant, and the energy difference between them is small. Thus, in competing exothermic reactions, little selectivity is expected in the products (Fig. 4.2b).

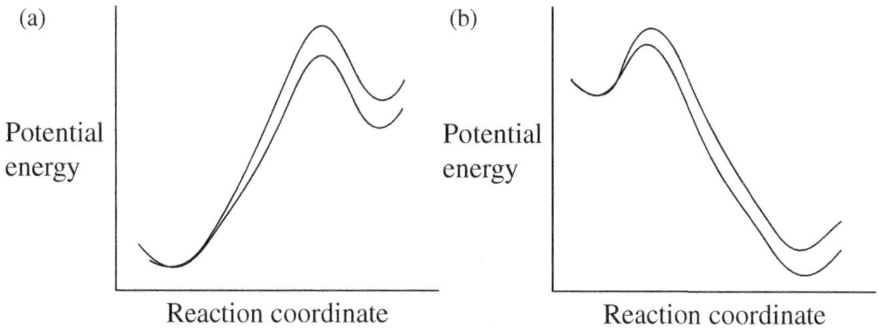

FIGURE 4.2 Reaction coordinate diagrams for (a) competing endothermic reactions and (b) competing exothermic reactions.

Comparisons can be made for less extreme cases as well. Comparing two endothermic reactions, we can predict that the more endothermic one will be the more selective. For example, hydrogen atom abstraction from propane by bromine atoms is more endothermic and more selective than by chlorine atoms [5].

4.3 METHODS FOR DETERMINING MECHANISMS

4.3.1 Identification of Products and Intermediates

The first step in determining a mechanism is a thorough identification of the products of the reaction under investigation. If a mechanism proposal includes a temporary intermediate compound that may have some stability, attempts should be made to isolate some of it from the reaction by premature interruption. If it is not stable enough for this, it may be detectable spectroscopically while the reaction is in progress. It may also be possible to divert the intermediate by adding a new reactant to the mixture. Finally, if the intermediate is a stable compound and can be prepared by another means, it can be used as a substitute starting material to determine whether it gives the same products overall, at least as rapidly.

4.3.2 Isotope Tracing

If competing mechanistic proposals differ in what atom in the starting material becomes a particular atom of the product, a determination may be made using isotope labeling. A starting material may be synthesized with an uncommon isotope in a specific site in the molecule. A reaction may then be carried out using this labeled material, and the product examined for the presence or absence of the uncommon isotope, or to determine the site in the product wherein the isotope resides. Isotopes available for tracing the origin of atoms in the product include ^{2}H, ^{3}H, ^{13}C, ^{15}N, and ^{18}O. The presence of radioactive isotopes in the products or degradation derivatives of the products is determined by decay counting. Mass spectra will show the presence and sometimes the location of isotopes if the product is compared with the unlabeled compound. Nuclear magnetic resonance (NMR) spectroscopy is especially useful for isotope tracing.

If a deuterium is present in a product, the signal for that site will be missing or smaller in the ^{1}H NMR spectrum of the product. The deuterium atoms themselves may be detected at their resonance frequency. The low natural abundance of ^{13}C gives weak ^{13}C spectra, but if a compound is enriched in that isotope at a particular site, the signal for that site will be obviously stronger in the spectrum.

Carbon-13 NMR was used to determine which group migrated in the reaction, as shown in Equation 4.1 [6]. Treatment of the ortho-substituted biphenyl derivative with $AlCl_3$ gave the corresponding meta compound. It is not obvious whether the phenyl group migrated or the aminomethyl group migrated. This was determined by synthesizing a sample of the ortho compound with a greater-than-natural amount of ^{13}C at the * site shown in Equation 4.1. The ^{13}C NMR spectrum (without proton decoupling) of the

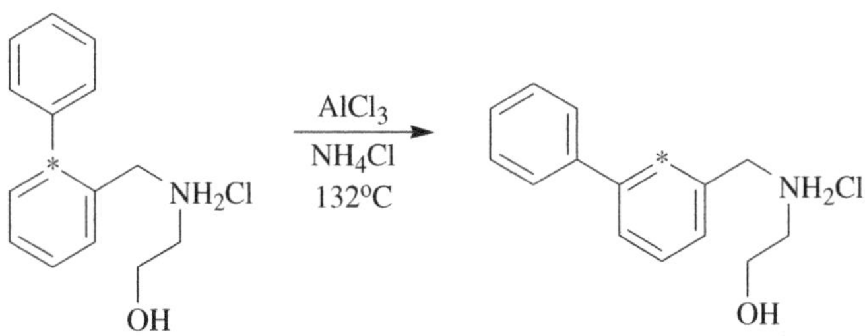

(4.1)

labeled starting material showed the expected weak signals plus a strong singlet at 142.36 ppm for the labeled carbon site. The product of the reaction showed weak signals plus a strong doublet at 129.07 ppm, $J_{CH} = 158.2$ Hz. The starting singlet indicated that there was no hydrogen directly attached to the labeled site, as expected, but the product doublet indicated that the labeled carbon now had a hydrogen attached. One may conclude with confidence that the phenyl group migrated. (^{13}C NMR spectroscopy is presented in Section 10.6.)

4.3.3 Stereochemical Determination

If a reaction is carried out on a particular stereoisomer of the starting material and the products are stereoisomerically identified (Chapter 3), a choice among mechanisms can often be made. In substitution and rearrangement reactions, an inversion of configuration will indicate a back-side attack, retention will indicate a front-side attack, and racemization will indicate the formation of a flattened (achiral) intermediate such as a carbocation. If two steps have occurred between the starting material and the product, the interpretations will differ. Retention could be the result of two inversions, and racemization could be the result of some inverting attack by the former leaving group.

It is possible to determine whether additions to alkenes are syn, anti, or mixed. For example, trifluoromethyl hypochlorite adds to alkenes and the mechanisms in Equations 4.2, 4.3, and 4.4 can be proposed.

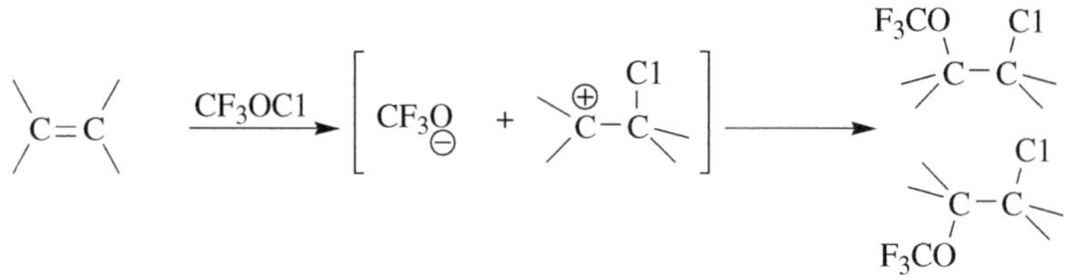

(4.2)

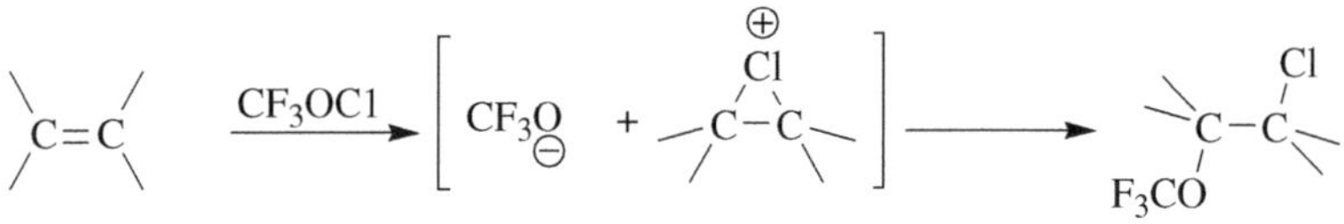

(4.3)

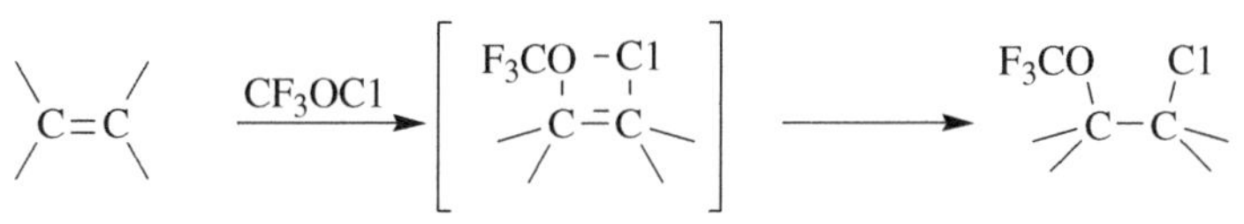

(4.4)

When this addition was carried out with pure *cis*-2-butene, only the erythro product was obtained, and with pure *trans*-2-butene, only the threo product was obtained [7]. If Equation 4.2 were the mechanism, both erythro and threo diastereomeric products would be expected; if Equation 4.3 were the mechanism, the anti-addition product would be obtained, with *cis*- giving the threo product and *trans*- giving the erythro product. Equation 4.4 shows the syn addition product. Only Equation 4.4 is in accord with the stereochemical results, that is, a concerted (or nearly so) syn addition.

In elimination reactions, a similar comparison of the stereochemistry of starting materials and products can indicate syn, anti, or mixed processes. One method for changing the stereochemistry of a double bond includes halogenation to the dihalide followed by heating with sodium iodide to affect first an S_N2 reaction and then anti-elimination, illustrated for one iodination product in Equation 4.5 [8]. Kinetic data support this sequence of reaction steps, leading to the trans alkene as the only product.

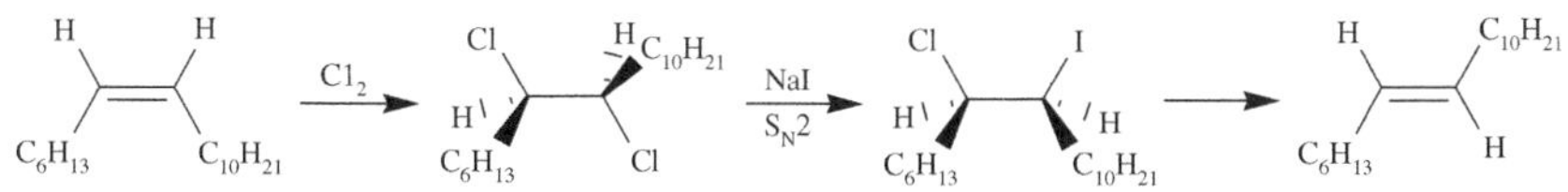

(4.5)

4.3.4 Concentration Dependence of Kinetics

The measurement of rates of reactions under various conditions gives several different kinds of mechanistic information [9]. The concentration dependence of rates can give information on the number of steps in a mechanism and the species involved in reaction collisions. In these experiments, other variables such as temperature and choice of solvent are held constant.

4.3.4.1 Rate Equations For reactions in homogeneous solutions, the concentrations of components are measured as time progresses; the usual units are moles of solute per liter of solution per second. At any point in time during the progress of a reaction, a certain number of moles of product per liter of solution appear per unit time. We can also measure the disappearance of the reactant in moles per liter per time. This is called the rate of the reaction and it generally decreases as the reactant is depleted. This is shown graphically in Figure 4.3 for the simple reaction in Equation 4.6. The rate at any one moment is the slope of the graph of [B]. At the same time, [A] decreases; thus, the slope of the curve for [A] is equal but opposite in sign to that for [B], so two alternative rate expressions may be written (Eq. 4.7).

$$A \rightarrow B \tag{4.6}$$

$$\text{rate} = \frac{d[\text{B}]}{dt} \quad \text{or} \quad \text{rate} = \frac{-d[\text{A}]}{dt} \tag{4.7}$$

The rate of a reaction depends on collision frequency and will thus vary with [A]. The rate constant, k, reflects that proportionality (Eq. 4.8).

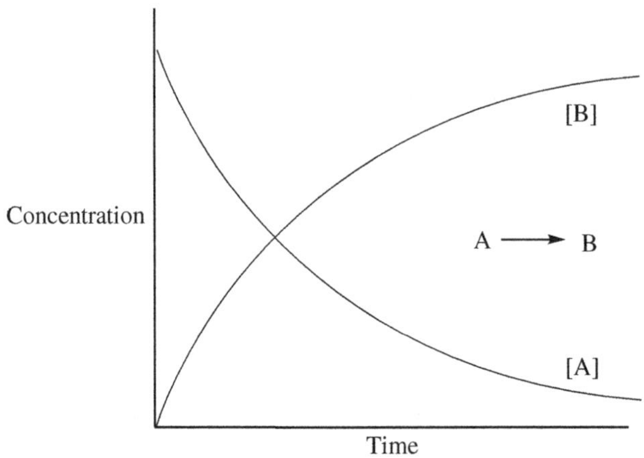

FIGURE 4.3 Change in concentration with time.

The rate constant is also affected by changes in temperature and the

$$\text{rate} = k[\text{A}] = \frac{-d[\text{A}]}{dt} \tag{4.8}$$

structure of A, which are discussed in Sections 4.3.6 and 4.3.7. Determining the value of k can give insights into the reaction mechanism, so we can use mathematical methods to convert the differential equations that we have thus far into equations that result in a straight-line plot, thus providing both readily determined values for k and a validity check for the data obtained.

Experimentally, a solution of known concentration of starting material A is prepared, and as the reaction proceeds, the concentration of A and/or B is measured repeatedly, with the choice determined primarily by convenience.

The simple reaction in Equation 4.6 is termed first order or unimolecular because the rate depends only on the concentration of one reaction component. This simple first-order reaction involves only collisions of the reactant with unreactive solvent molecules that provide the kinetic energy necessary for bond reorganization. Integrating and rearranging Equation 4.8 gives Equation 4.9, which takes the form of the equation of

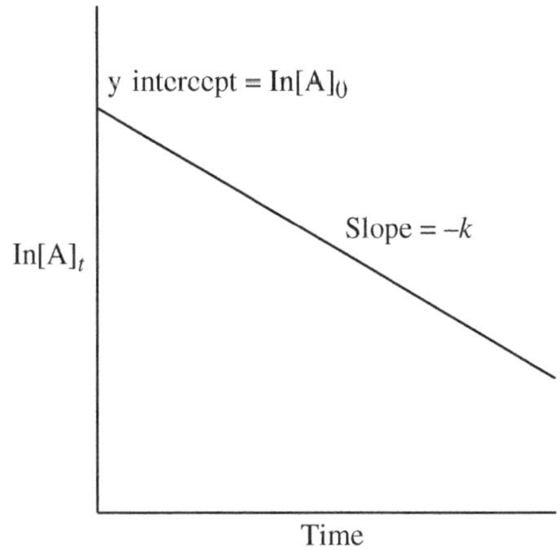

FIGURE 4.4 Linear plot of kinetic data for a first-order reaction.

a straight line. Plotting $\ln[A]_t$ versus time gives a straight line, with $-k$ as the slope and $\ln[A]_0$ as the y-intercept (Fig. 4.4).

$$\int_{[A]_0}^{[A]_t} \frac{d[A]}{[A]} = -k \int_0^t dt$$

$$\ln [A]_t - \ln [A]_0 = -kt \quad (4.9)$$

The value of k from this linear plot is based on the whole set of points graphically and least-squares-optimized. Also note that the actual value of [A] does not need to be measured, only a value that is proportional to it. This sort of number is easily obtained spectroscopically as integration values in a series of NMR spectra or absorbances from a spectrometer. And, if the data do not give a straight line in such a plot, the reaction is not first order.

Dimerization reactions (Eq. 4.10) are bimolecular and give a different kinetic result. Here, a collision between two molecules of A is necessary; therefore, as [A] decreases, the rate drops faster than is the case in Equation 4.6. The rate expression in this case is given in Equation 4.11. The integration and rearrangement of Equation 4.11 gives Equation 4.12.

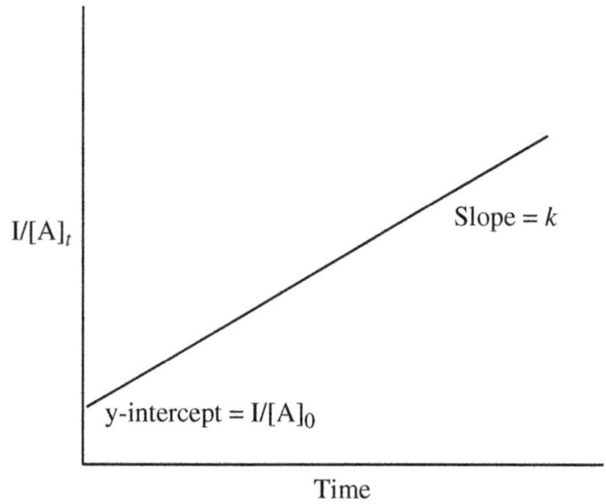

FIGURE 4.5 Linear plot of kinetic data for a dimerization reaction.

A plot of $1/[A]_t$ versus time gives k as the slope and $1/[A]_0$ as the y-intercept (Fig. 4.5).

$$A + A \rightarrow B \tag{4.10}$$

$$\text{rate} = \frac{-d[A]}{dt} = k[A]^2 \tag{4.11}$$

$$\int_{[A]_0}^{[A]_t} \frac{d[A]}{[A]^2} = -k \int_0^t dt$$

$$\frac{1}{[A]_t} - \frac{1}{[A]_0} = kt \tag{4.12}$$

A simple bimolecular reaction with two different starting materials A and B (Eq. 4.13) gives a rate equation (Eq. 4.14) showing that it is first order in A and first order in B or second order overall.

$$A + B \rightarrow C \tag{4.13}$$

$$\text{rate} = \frac{-d[A]}{dt} = k[A][B] \tag{4.14}$$

There are three ways to analyze data for this kind of reaction to confirm the order and determine the value of k:

1. One may begin with equal concentrations of A and B and use Equation 4.11.

2. One may use a large excess of B so that as $[A]_t$ diminishes, the change in [B] is negligible, that is, $[B]_0 = [B]_t$. The integration and rearrangement of Equation 4.14 assuming [B] as a constant gives Equation 4.15. Plotting $\ln[A]_t$ versus t, as in Equation 4.9, gives a straight line of slope k[B].

$$\ln [A]_t - \ln [A]_0 = -k[B]t \tag{4.15}$$

   Dividing the slope by the known [B] gives the second-order rate constant. This use of a large excess of B is called pseudo-first-order conditions. Some authors tabulate k[B] values labeled as k_{obs} or k_{pseudo}. It is necessary to repeat the reaction with one or more additional concentrations of B to be certain that the reaction is first-order in [B] (not zero or second-order in [B], which would still give pseudo-first-order results).
3. It is possible to integrate the rate equation with both [A] and [B] as variables, with several ways to rearrange the resulting equation. One such solution is shown in Equation 4.16, which when plotted gives a straight line as shown in Figure 4.6.

$$\frac{1}{[A]_0 - [B]_0} = \ln \frac{[B]_0[A]_t}{[A]_0[B]_t} = kt \tag{4.16}$$

The order in a reactant can easily be determined by solving the general rate equation (Eq. 4.17) for n. Because any two rates are related to each other by k, measuring the rate for two different concentrations of A will allow for the calculation of n using Equation 4.18.

$$\text{rate} = k[A]^n \tag{4.17}$$

$$\log k = \log \text{rate}_1 - n \log [A]_1 = \log \text{rate}_2 - n \log [A]_2$$

$$\log \frac{\text{rate}_1}{\text{rate}_2} = n \log \frac{[A]_1}{[A]_2} \tag{4.18}$$

Similarly, pseudo-first-order conditions can be used regardless of the order in the reactant used in excess. The k_{pseudo} values obtained at two different concentrations of B can be used to determine the order in B

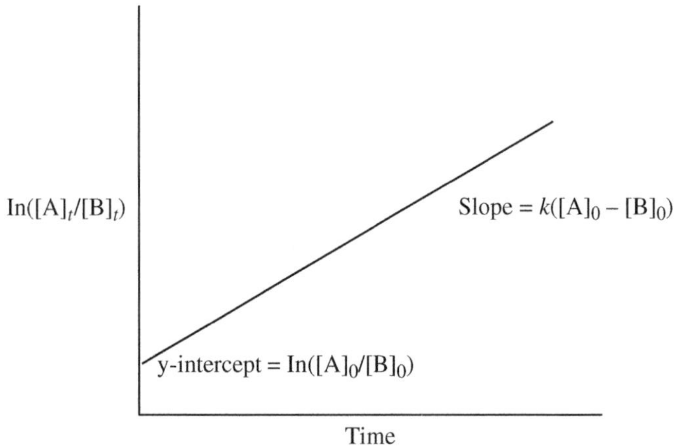

FIGURE 4.6 Linear plot of kinetic data for a bimolecular reaction.

using Equation 4.19, which is derived from $k_{\text{pseudo}} = k[\text{B}]^n$ in analogy to Equation 4.18.

$$\log \frac{k_{\text{pseudo1}}}{k_{\text{pseudo2}}} = n \log \frac{[\text{B}]_1}{[\text{B}]_2} \quad (4.19)$$

Most reactions are uni- or bimolecular. Termolecular reactions are rare as they would require simultaneous collisions of three species. Some reactions give kinetic orders that do not match the stoichiometry of the process. A reactant may not appear in the kinetic expression (zero order) or may appear with an order higher or lower than the number of equivalents consumed in the process. Some reactions may total third-order or higher. Such kinetic characteristics indicate multistep mechanisms. On the other hand, some multistep mechanisms may exhibit simple first- or second-order kinetics and require other evidence to delineate the mechanism.

4.3.4.2 Intermediates Consider the two-step process shown in Equation 4.20, where an intermediate compound I is formed in the first step and consumed in the second.

$$\text{A} + \text{B} \xrightarrow{k_1} \text{I}$$

$$\text{I} + \text{C} \xrightarrow{k_2} \text{D} \quad (4.20)$$

If the first step is relatively slow and the second step fast, then I will be consumed as rapidly as it is formed. The concentration of I will remain very low and practically constant. Assuming that constancy (known as the steady-state approximation [10]), we can equate the rate of formation of I and the rate of consumption of I, which also equals the rate of formation of D (Eq. 4.21). Thus, in terms of measurable concentrations, the reaction is second order just as for Equation 4.14, and reactant C does not affect the rate.

$$\frac{d[D]}{dt} = k_1[A][B] = k_2[I][C] \tag{4.21}$$

In other reactions, the first step may be fast and reversible, followed by a slow step (Eq. 4.22). The first step provides a low concentration of I that is related by an equilibrium constant to [A] and [B] (Eq. 4.23). The rate of formation of C is proportional to [I] which from Equation 4.23 is proportional to [A][B] (Eq. 4.24). Once again, we would observe simple second-order kinetics indistinguishable from Equations 4.13 and 4.20. Detection of I and determination of the equilibrium constant would distinguish these.

$$\begin{gathered} A + B \underset{k_{-1}}{\overset{k_1}{\rightleftharpoons}} I \\ I \xrightarrow{k_2} C \end{gathered} \tag{4.22}$$

$$K = \frac{[I]}{[A][B]} \tag{4.23}$$

$$\frac{d[C]}{dt} = k_2[I] = k_2K[A][B] \tag{4.24}$$

4.3.4.3 Examples As an example, consider the reaction of 1,1-diphenyl-2,2,2-trichloroethanol with sodium hydroxide [11]. This reaction was studied to shed light on the oxidative cleavage of a metabolite of dichlorodiphenyltrichloroethane (DDT), a closely related compound. The proposed mechanism is shown in Equation 4.25.

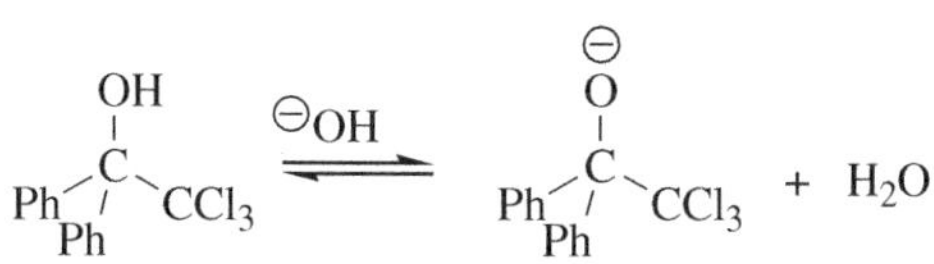

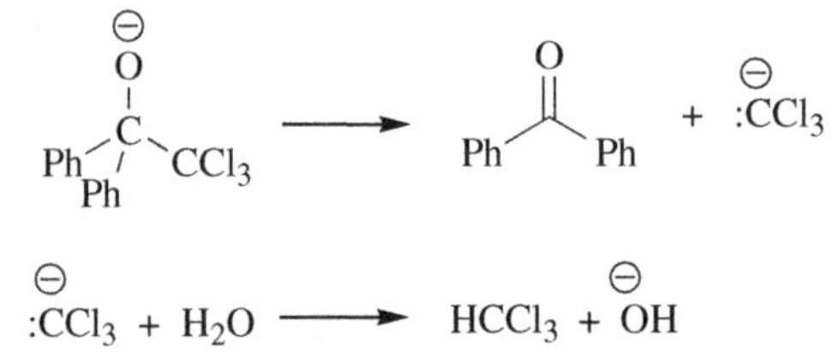

(4.25)

In this reaction, the anionic intermediates were present in low concentrations and not detected directly. The hydroxide concentration in this reaction does not change because as it is used in step 1, it is regenerated in step 3. Therefore, pseudo-first-order conditions exist at any concentration of hydroxide. Measurements indicate that the reaction is first order in hydroxide and pseudo first order in the starting chloroalcohol.

The reaction was run in the sample cell of a spectrometer, where the temperature was controlled to ±0.1°C. The formation of the benzophenone product was followed by tracking absorbance at 258 nm, and plotting this change provided the rates for a range of hydroxide concentrations.

A plot of log k_{exp} versus pH was linear with a slope of 1, indicating a first-order reaction in $[OH^-]$, over a pH range of 9.5–12. Above pH 12, the plot leveled off. Because the chloroalcohol has a pK_a of 12, it is largely converted to the alkoxide above pH 12 and additional hydroxide would do little to change this, reflected in the flattening of the curve above pH 12. This supports the mechanism given in Equation 4.25, with the second step as rate determining.

The mechanism in Equation 4.25 is called "ElcB" because it is an elimination (of $CHCl_3$) from the conjugate base of the chloroalcohol. The 1 signifies that the reaction is first order in the conjugate base (step 2).

Another sort of preliminary equilibrium is illustrated in Equation 4.26. Let us first consider the case where A dissociates rapidly and reversibly to give a low concentration of intermediate I and by-product B. The second step, the consumption of I by C, is the rate-determining step and gives the rate expression (Eq. 4.27). The unmeasurable [I] is replaced using the

$$\mathrm{A} \underset{k_{-1}}{\overset{k_1}{\rightleftharpoons}} \mathrm{I} + \mathrm{B}$$

$$\mathrm{I} + \mathrm{C} \xrightarrow{k_2} \mathrm{D} \tag{4.26}$$

$$\frac{d[\mathrm{D}]}{dt} = k_2[\mathrm{I}][\mathrm{C}] \tag{4.27}$$

equilibrium expression for the first step (Eq. 4.28). This reaction will follow second-order kinetics initially, but because [B] is in the denominator of the rate expression, the reaction will slow down more as B accumulates than it would if simple bimolecular kinetics (Eq. 4.14) were operating.

$$\begin{aligned} K &= \frac{[\mathrm{I}][\mathrm{B}]}{[\mathrm{A}]} \\ \frac{d[\mathrm{D}]}{dt} &= \frac{k_2K[\mathrm{A}][\mathrm{C}]}{[\mathrm{B}]} \end{aligned} \tag{4.28}$$

In fact, the preliminary equilibrium step may be detected simply by adding extra B to the initial reaction solution and noting the slower rate.

Now let us consider the case in Equation 4.26 where the first step is slow and the second is fast, as in the familiar S_N1 reactions. In this case, [I] will remain low and practically constant, and we can use the steady-state approximation which assumes that the rate of formation of I is equal to the rate of its consumption (Eq. 4.29). Substituting Equation 4.27 into Equation 4.29 gives the rate expression for the reaction (Eq. 4.30). Early in the reaction, the concentrations of I and B are very small, and the reaction follows first-order kinetics. A two-step mechanism is suggested because [C] does not appear in the rate expression, so C must react later in the fast step.

$$\text{Production of I} = k_1[\mathrm{A}]$$

$$\text{Consumption of I} = k_2[\mathrm{I}][\mathrm{B}] - k_{-1}[\mathrm{I}][\mathrm{C}]$$

$$k_1[\mathrm{A}] = k_{-1}[\mathrm{I}][\mathrm{B}] + k_2[\mathrm{I}][\mathrm{C}] \tag{4.29}$$

$$\frac{d[\mathrm{D}]}{dt} = k_1[\mathrm{A}] - k_{-1}[\mathrm{I}][\mathrm{B}] \tag{4.30}$$

The kinetic treatments developed here cover a variety of well-known reactions. More complicated reactions and kinetics are covered in the references [3, 9, 12].

4.3.5 Isotope Effects in Kinetics

The ground-state vibrational energy of a bond is lower for a bond to a heavier isotope than it is for a lighter isotope. Therefore, the activation energy required to break the bond with the heavier isotope will be greater than that required for the lighter isotope. The higher activation energy process is slower. The largest differences are found with hydrogen, where the mass ratios of the isotopes are greatest. For deuterium, rate constant ratios k_H/k_D range up to about 10. The smaller mass ratio for ^{12}C to ^{13}C gives k_{12C}/k_{13C} up to about 1.1. These ratios are called primary kinetic isotope effects [13]. Heavier isotopes located near a reaction site but not involved in bond breaking give a smaller secondary kinetic isotope effect.

If the rate-determining step of a reaction mechanism involves breaking a C—H bond, the corresponding C—D compound will react substantially slower. The maximum effect occurs if the transition state happens at the midpoint of hydrogen transfer, and if the outgoing bonding partner, the incoming bonding partner, and the hydrogen are colinear. If the C—H bond is broken in a step subsequent to the rate-determining one, k_H/k_D will be 1.

A distinction was made between the two proposed mechanisms for the fragmentation of an oxaziridine, using a combination of the primary kinetic isotope effect and stereochemistry [14]. Amines can behave as bases and attack at hydrogen, or they can be nucleophiles and attack at carbon or nitrogen; E2 and S_N2 reactions are examples. In Equation 4.31, a tertiary amine (brucine) is shown in the role of base in a step resembling an E2 reaction. In Equation 4.32, the amine is shown attacking the nitrogen as in an S_N2 reaction.

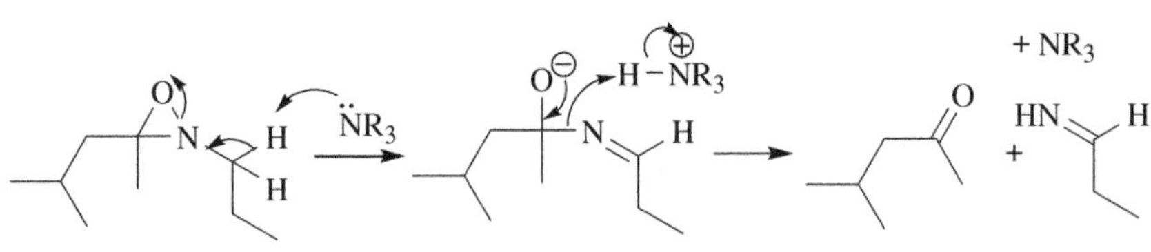

(4.31)

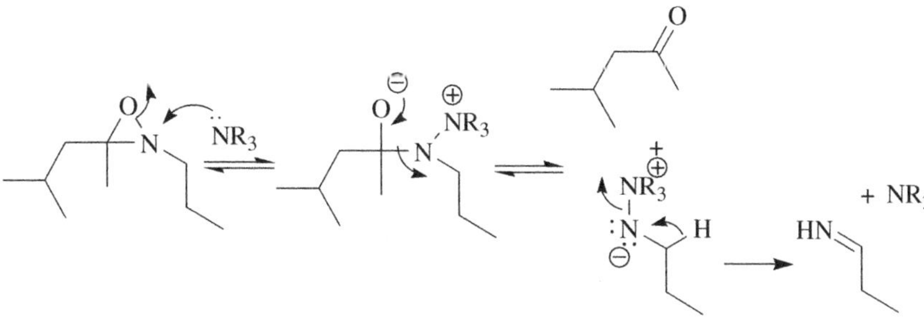

(4.32)

In order to disprove one of these alternatives, the dideutero analog **4.1** was prepared and the rates were measured for the fragmentation of both the hydrogen and deutero compounds with various concentrations of brucine

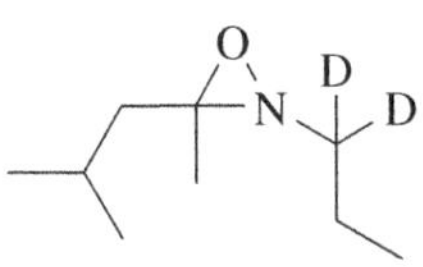

4.1

in refluxing acetonitrile. The ratio of the pseudo-first-order rate constants k_H/k_D was found to be 4.25, thus a bond to H or D is broken in or before the rate-determining step. This is in accord with Equation 4.31, with the first step rate-determining, and in accord with Equation 4.32 if the last step were rate-determining. If the last step were the slow one, either intermediates would accumulate (they do not in this case) or the early steps must be fast and reversible, giving only a low concentration of intermediate. The fact that ring opening was not reversible was proved stereochemically. Inversion of configuration on nitrogen does not readily occur when the nitrogen is part of a three-membered ring; therefore, the *Z* and *E* isomers of the oxaziridine could be prepared and isolated. One isomer was treated with brucine until 43% of it was converted to products, and the remaining oxaziridine was examined by ^{1}H NMR. This showed the complete absence of the other diastereomer, which would have been present if some of the acyclic intermediates had reclosed. Thus, Equation 4.32 is not in accord with the results, and Equation 4.31 remains as the most reasonable mechanism.

4.3.6 Temperature Effects on Kinetics

In any reaction step involving some bond breaking, there will be a high potential energy point called a transition state, as described in Section 4.1. The higher the temperature, the larger the proportion of molecules with sufficient kinetic energy to successfully reach the transition-state condition. Determining how steeply the rate of a reaction increases with increasing temperature gives two kinds of information about the transition state: (1) the enthalpy of activation, $\Delta H^{\ddagger}$, which is the difference in enthalpy between the starting material and the activated complex at the transition state (related to the difference in structure), and (2) the entropy, $\Delta S^{\ddagger}$, which is related to the change in the extent of organization of the atoms and solvent molecules as they proceed to the transition state. Reactions with a larger $\Delta H^{\ddagger}$ are accelerated to a greater extent by a temperature increase (if $\Delta S^{\ddagger}$ is similar).

In Figure 4.1, the activation energy, E_a, is shown as the energy needed by a molecule to get to the transition state. The Arrhenius equation [15, 16] relates the rate constant to the temperature, as shown in Equation 4.33. In this expression, A is a preexponential factor related to the frequency of successful collisions (which is related to $\Delta S^{\ddagger}$), E_a is the activation energy (related to $\Delta H^{\ddagger}$), k is the rate constant, R is the gas constant, and T is the Kelvin temperature. Simplifying and rearranging Equation 4.33 gives a linear form, Equation 4.34. A plot of ln k versus $1/T$ allows the calculation of E_a and A (Fig. 4.7).

$$k = Ae^{\frac{-E_a}{RT}} \tag{4.33}$$

$$\ln k = \frac{-E_a}{R}\left(\frac{1}{T}\right) + \ln A \tag{4.34}$$

It is also possible to calculate E_a if rate constants at two temperatures are known using Equation 4.35, which is derived from Equation 4.34.

$$ln\left(\frac{k_1}{k_2}\right) = \frac{E_a}{R}\left(\frac{1}{T_2} - \frac{1}{T_1}\right) \tag{4.35}$$

Calculation of thermodynamic values for $\Delta H^{\ddagger}$ and $\Delta S^{\ddagger}$ can provide useful information regarding the pathway of a reaction. Enthalpy in this case is

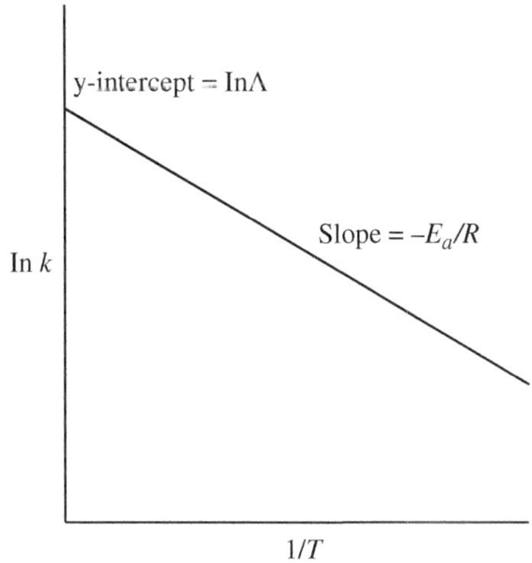

FIGURE 4.7 An Arrhenius plot.

the energy difference between the starting material and the transition state. This value provides some clue as to the degree of change in the structure required for the pathway to proceed as proposed. It can be obtained from Equation 4.36.

$$\Delta H^{\ddagger} = E_a - RT \tag{4.36}$$

The entropy of activation, while often referred to as a measure of disorder in a reaction process, maybe more accurately understood as a dispersal of molecular energy in a process. It can be calculated from Equation 4.37, in which R is the gas constant, T is the Kelvin temperature, N_0 is Avogadro's number, and h is Planck's constant. For unimolecular reactions,

$$A = \left(\frac{RT}{N_0 h}\right) e^{\left(\frac{\Delta S^{\ddagger}}{R}\right)} \tag{4.37}$$

$\Delta S^{\ddagger}$ indicates whether the transition state has either more or less freedom of motion than the starting molecule. For example, the large positive $\Delta S^{\ddagger}$ for the thermal decomposition of the azo compound in Equation 4.38 [17] is in accord with a simple fragmentation mechanism where the transition state is a more loosely bound, more freely moving structure than the starting molecule.

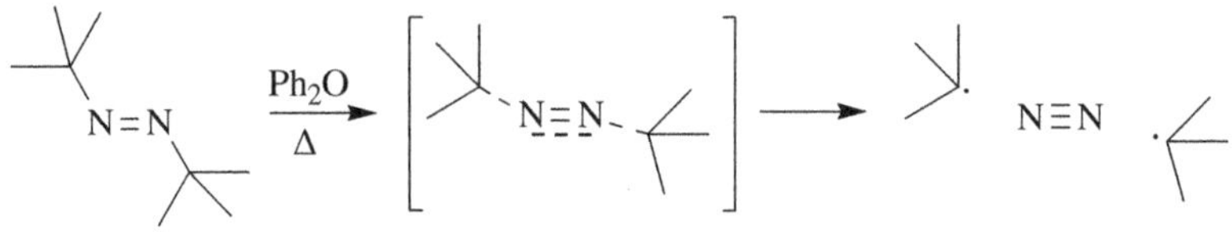

$\Delta S^{\ddagger}$ = 16.2 cal/mol-K

(4.38)

In contrast with this, the large negative $\Delta S^{\ddagger}$ for the reaction in Equation 4.39 [18] requires that some freedom of motion in the starting molecule, such as rotation of the ring substituent bonds, is restricted in the transition state.

(4.39)

$\Delta S^{\ddagger}$ = −11.7 cal/mol-K

The $\Delta S^{\ddagger}$ values calculated from second-order rate constants are not independent of the concentration units [19], although they are generally negative. They can be used to compare similar reactions [20].

Equations 4.36 and 4.37 are derived from transition state theory which is well described in physical chemistry texts and other resources [21].

4.3.7 Substituent Effects on Kinetics

The order of bond breaking or bond forming in most reactions leads to the temporary development of a positive or negative electrostatic charge at a particular site in the transition-state structure. Determining whether the charge is positive or negative and whether it is relatively large or small allows postulation of the sequence of bond changes. How can we determine this?

Certain substituents are known to withdraw electron density from a reaction site, making it easier to develop a negative charge at that site. If a negative charge is partially formed on progressing from the starting material to the transition state of a reaction, the substance with the electron-withdrawing substituent will react at a faster rate than one with a hydrogen atom. Other substituents are known to donate electron density to a reaction site (electron-donating groups), and these would intensify a developing negative charge, thus slowing the reaction. In other reactions, a positive charge is developed at the reaction site. Here the substituents would have the opposite effect, that is, the electron-donating substituents will give faster reactions for sites developing a positive charge.

For those reactions where the substituents are on a benzene ring and far enough from the reaction site to be sterically out of the way, the substituent effects have been correlated quantitatively. The Hammett equation (Eq. 4.40) states that the log of the ratio of the rate constant for a reaction with a substituent present (k) to the rate constant with no substituent (k_0) is equal to the product of a factor indicating the sensitivity of that reaction to substituent effects (ρ) and the characteristic electronic influence of the substituent (σ) [22].

$$\log_{10}\frac{k}{k_0} = \rho\sigma \tag{4.40}$$

Table 4.1 lists the σ values for a selection of common substituents in meta and para positions [23]. The ρ value for a reaction is determined by first measuring rate constants for a series of examples of the reaction with various substituents present. These are used to plot $\log_{10} k/k_0$ versus σ. The slope of this plot is ρ. With the value of ρ and the table of σ values, one may then predict k values of other examples with different substituents.

Where did the σ values come from? Hammett chose to define them from a particular reaction, the ionization of benzoic acid. In this case, and many others, the equilibrium constants rather than rate constants were measured. For equilibria, the Hammett equation is written in terms of equilibrium constants (Eq. 4.41). Since the ionization of a series of substituted benzoic

$$\log_{10}\frac{K}{K_0} = \rho\sigma \tag{4.41}$$

acids are used to define σ values, the ρ for this reaction was defined as 1.000. To find the σ_m for the chloro substituent, the K_a values for *m*-chlorobenzoic acid, 1.48×10^{-4}, and benzoic acid, 6.27×10^{-5}, are entered in Equation 4.41 and it is solved for σ, which equals 0.37, the table value.

TABLE 4.1 Selected Hammett σ Values[a]

Substituent	σ_{meta}	σ_{para}	σ^+	σ^-
NMe_2	−0.16	−0.83	−1.70	−0.12
O^-	−0.47	(−0.81)[b]	−2.30	−0.82
NH_2	−0.16	−0.66	−1.30	−0.15
CO_2^-	−0.10	0.00	−0.02	0.31
CH_3	−0.07	−0.17	−0.31	−0.17
C_2H_5	−0.07	−0.15	−0.30	−0.19
H	0	0	0	0
Ph	0.06	−0.01	−0.18	0.02
OH	0.12	−0.37	−0.92	−0.37
OCH_3	0.12	−0.27	−0.78	−0.26
SCH_3	0.15	0.00	−0.60	−0.06
$NHCOCH_3$	0.21	0.00	−0.60	−0.46
OPh	0.25	−0.03	−0.50	−0.10
SH	0.25	0.15	−0.03	—
$CONH_2$	0.28	0.36	—	0.61
F	0.34	0.06	−0.07	−0.03
I	0.35	0.18	0.14	0.27
CHO	0.35	0.42	0.73	1.03
Cl	0.37	0.23	0.11	0.19
CO_2H	0.37	0.45	0.42	0.77
CO_2Et	0.37	0.45	0.48	0.75
$COCH_3$	0.38	0.50	—	0.84
Br	0.39	0.23	0.15	0.25
$OCOCH_3$	0.39	0.31	−0.19	—
CF_3	0.43	0.54	0.61	0.65
CN	0.56	0.66	0.66	1.00
NH_3^+	0.86	0.60	—	−0.56
NO_2	0.71	0.78	0.79	1.27
NMe_3^+	0.88	0.82	0.41	0.77

[a] Adapted from Ref. [23].
[b] Value uncertain.

TABLE 4.2 Selected Hammett ρ Values[a]

Reaction	ρ
$ArCO_2CH_2CH_3 + OH^- \rightarrow ArCO_2^-$	2.193
$ArCH{=}CHCO_2CH_2CH_3 + OH^- \rightarrow ArCH{=}CHCO_2^-$	1.329
$ArCO_2H \rightarrow ArCO_2^- + H^+$	1.000
$ArCO_2CH_2CH_3 + H^+ \rightarrow ArCO_2H$	0.144
$ArCO_2H + CH_3OH + H^+ \rightarrow ArCO_2CH_3$	−0.229
$ArCH_2Cl + H_2O \rightarrow ArCH_2OH$	−2.178

[a]Adapted from Ref. [24].

In the ionization of benzoic acid, a negative charge develops on the substituent-carrying molecule. Any other reaction in which a negative charge develops would be affected in the same direction by the substituents and it would be found to have a positive value for ρ. In reactions where a positive charge develops in the transition state or the product, the ρ values will be negative. In reactions where the sensitivity to substituents is low, the ρ can approach zero, and where it is high, the ρ value can be as high as ± 5 or more. This sensitivity is related to the distance separating the charge and the substituent, the transmitting ability of the intervening atoms, and also to the size of the charge, usually a fraction of a unit charge in transition states. The representative ρ values given in Table 4.2 demonstrate this point.

Estimates of the magnitude of the charge in a transition state have been made by taking a ratio of ρ values for a rate and an equilibrium in very similar reactions [25].

In some reactions, electron-donating substituents in the para position can have a greater effect than their σ values would predict. These are cases where a direct resonance contribution can be made as in benzylic carbocation formation (Eq. 4.42).

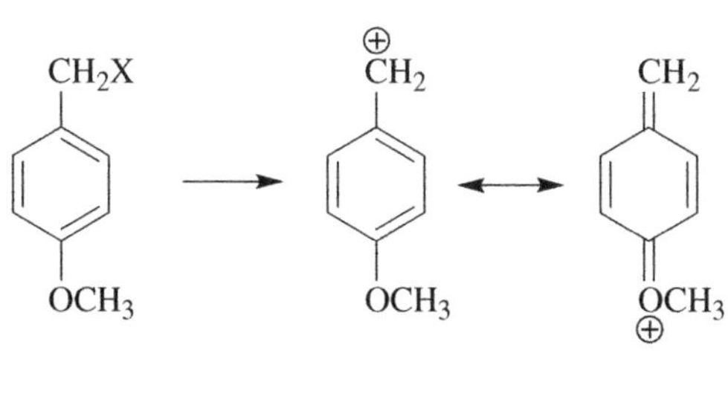

(4.42)

For such reactions, another set of substituent values, σ^+, was developed [26]. The rates of solvolysis of meta- and para-substituted cumyl chlorides in 90% aqueous (aq) acetone (Eq. 4.43) were measured. The ρ for the reaction was found to be −4.54 by plotting only the meta isomers against σ. This was then used to obtain σ^+ values according to Equation 4.44 with the para isomers. Representative values are found in Table 4.1. Note that the σ^+ values are similar to the σ_{para} values for the electron-withdrawing substituents, while the σ^+ are larger than the σ_{para} for the electron-donating substituents, as would be expected for resonance contributions to the stability of carbocations. Similarly, σ^- values provide better correlation for reactions in which the intermediates are electron-rich, and electron-withdrawing substituents can participate directly by resonance. The σ^- values were determined by analyzing the ionization of phenols and anilines [24] and are also included in Table 4.1.

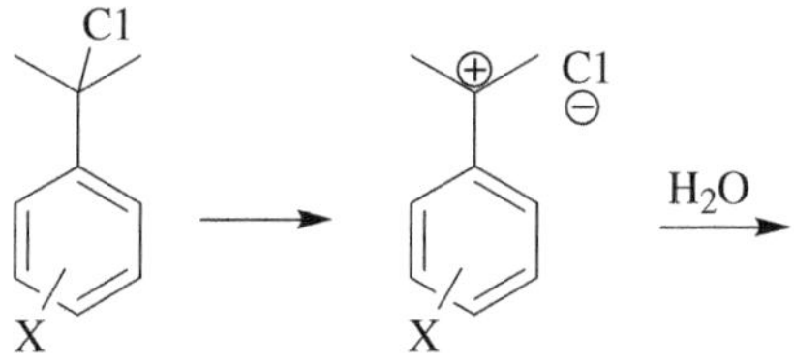

(4.43)

$$\log_{10}\frac{k}{k_0} = -4.54\sigma^+ \tag{4.44}$$

The familiar bromination of substituted benzenes with Br_2 in acetic acid correlates with σ^+, and ρ is −12, indicating substantial plus-charge development very close to the substituents as shown for anisole in Equation 4.45 [26]. The use of σ^+ constants is justified by the resonance participation of the substituent.

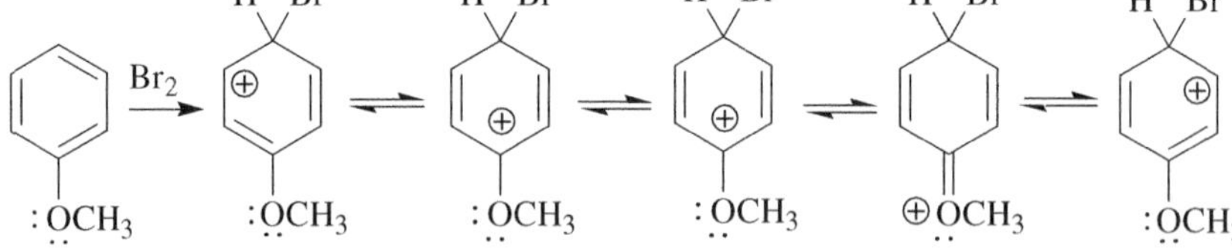

(4.45)

A study of the reactions of aryldichloromethide carbanions [27] is presented here to illustrate the use of the Hammett equation in exploring mechanistic investigations. For the reaction in Equation 4.46, the disappearance of the initially formed carbene is followed by ultraviolet

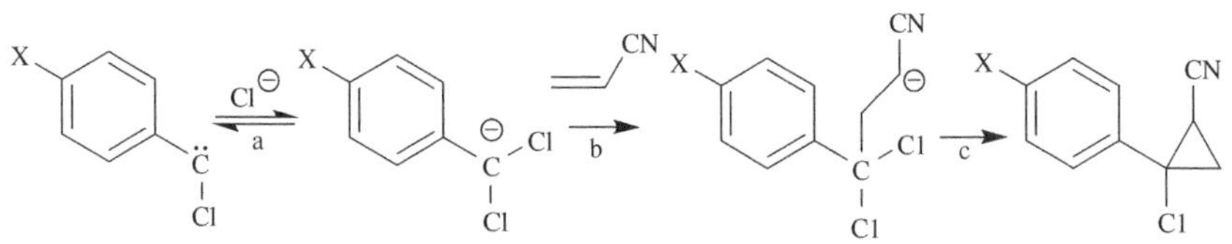

(4.46)

(UV) spectroscopy and the rate of the reaction is determined as a function of chloride concentration. The disappearance of the resulting anion is also monitored spectroscopically as a function of acrylonitrile concentration. The Hammett plot for the carbene reaction with chloride (reaction a) has a small positive ρ value of +0.86, indicating that substituents on the aromatic ring are not as significant to the progress of the reaction and that the transition state occurs early in the reaction pathway. The Hammett plot for the anion reaction (reaction b), in contrast, has a ρ value of −2.65. This larger negative value indicates that substituents on the aromatic ring are significant in the pathway and that the transition state occurs later in the process.

The Hammett equation is a linear free-energy relationship. The log K and log k values are proportional to ΔG and $\Delta G^{\ddagger}$ values, respectively (Eq. 4.47), for equilibria and rates. The change in ΔG or $\Delta G^{\ddagger}$ with changing substituents for many reactions is linearly related to the σ scale.

$$\Delta G^0 = -RT \ln K \tag{4.47}$$

Therefore, the change in ΔG or $\Delta G^{\ddagger}$ with changes in substituents in one reaction is linearly related to the change in ΔG or $\Delta G^{\ddagger}$ in another reaction for the same changes in substituents where $\Delta\Delta G_1$ is the change in free energy for reaction 1 for a change in substituents and $\Delta\Delta G_2$ is the change in free energy for reaction 2 for the same change in substituents (Eq. 4.48).

$$\frac{\Delta\Delta G_1}{\rho_1} = \frac{\Delta\Delta G_2}{\rho_2} \tag{4.48}$$

4.4 WORKED EXAMPLES

After decades of mechanistic investigations and the discerning of generalities, it is now common practice to propose reasonable mechanisms for new reactions, considering their relationships to ones that have been investigated mechanistically. At the simplest level, these proposals consist of a succession of reaction intermediates without great attention to transition states. A sampling of such proposals is gathered in this section without supporting mechanistic data to illustrate possible reaction pathways. For simplicity, only one resonance form of each ion or radical is given in most cases.

Some working principles to keep in mind in writing such proposals include the following:

1. Sites of unlike charge attract one another and often become bonded together.
2. Elements of different electronegativity that are bonded together carry partial charges.
3. In most ionic and concerted reactions, electrons remain paired throughout the process.
4. An odd, unpaired electron exists on radical species temporarily.
5. Complete electron octets are maintained on all C, N, O, or F atoms except at carbocation, radical, carbene, and nitrene sites where fewer than eight electrons reside temporarily.
6. In most steps, reactivity should decrease on progressing to the product. For example, highly basic compounds give products of lower basicity, and unstable 1° carbocations rearrange to 2° and to 3° where possible.
7. Overall stability should increase. For example, conjugation may increase, especially aromatization, single bonds may form at the expense of the double bonds (as in the Diels–Alder reaction), and double bonds to oxygen may form at the expense of double bonds to carbon (as in the Claisen rearrangement).

4.4.1 Reactions in Basic Solutions

In basic solutions, it is common for the base to remove a proton from carbon to give a nucleophilic species that attacks an electrophile to give new bonding. If the initial base is a carbanion generated by a redox process, it, too, attacks electrophiles. In other circumstances, a leaving group departs from the carbanion to generate an electrophile such as a carbene or benzyne, which then leads to other products. There are endless variations of this, as may be appreciated by viewing many examples. A few are given here.

Example 4.1

Propose a mechanism for the base-catalyzed reaction below.

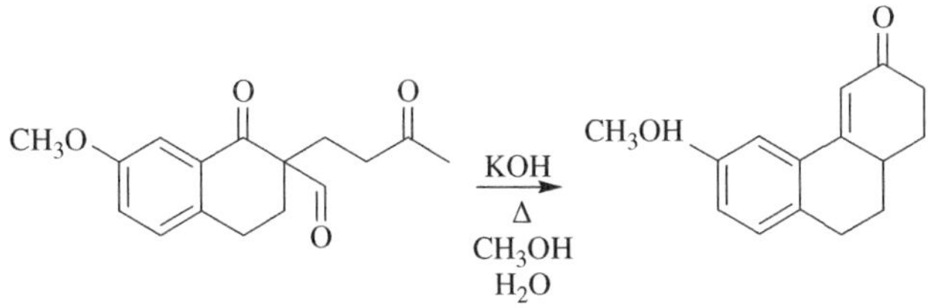

Mechanism:

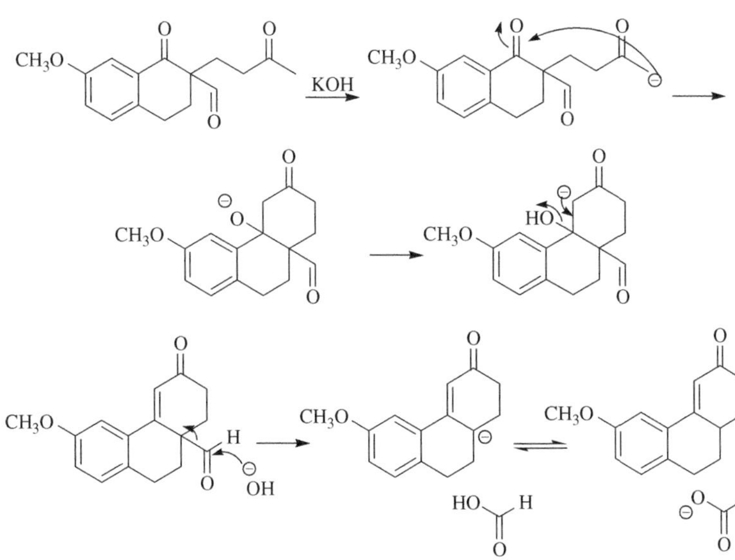

Treatment of a ketone with the relatively weak base OH^- generates small equilibrium amounts of enolate anions at sites alpha to the carbonyl group. While there are several possibilities, only one leads to the observed product. The nonbonding electron pair on the carbanion site is attracted to the partially positive carbonyl carbon and closes a six-membered ring. The resulting alcohol eliminates readily even in basic solution, in this case because conjugation is connected from the aromatic ring to the carbonyl group. The hydroxide attacks the formyl group directly with the loss of formic acid to give an enolate ion that is delocalized to the α carbon and carbonyl oxygen. The final weak base present is formate [28].

Example 4.2

Propose a mechanism for the base-catalyzed reaction below.

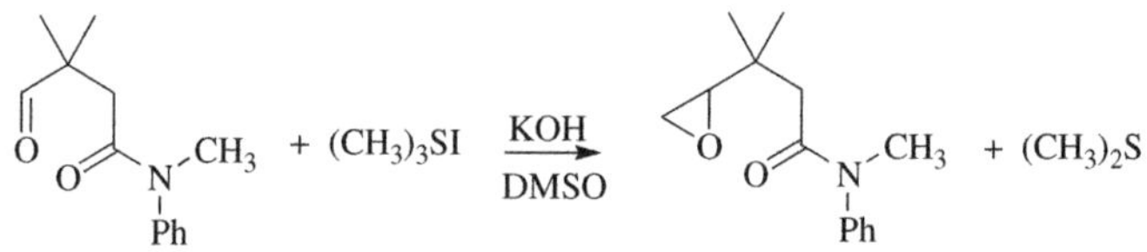

Mechanism:

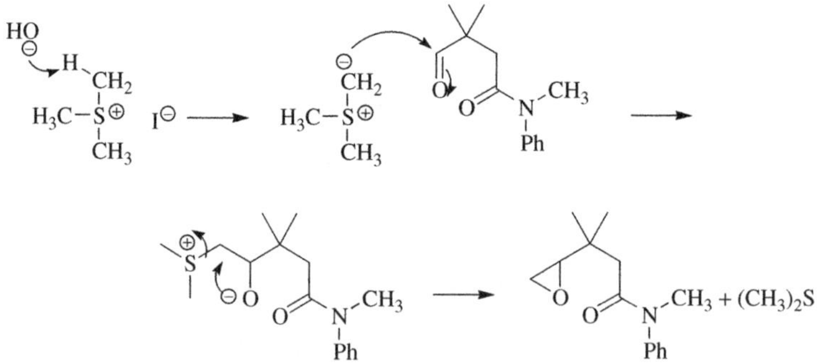

To begin, identify the origin of the new carbon in the product. The source of that carbon must be trimethylsilyl iodide. The hydroxide removes a proton from the trimethylsulfonium ion to give a 1,2-dipolar species called an ylide, which has nucleophilic character in the CH_2 group [29]. The ylide adds to the aldehyde to give an alkoxide that displaces the dimethyl sulfide leaving group.

Example 4.3

Propose a mechanism for the carbene reaction below.

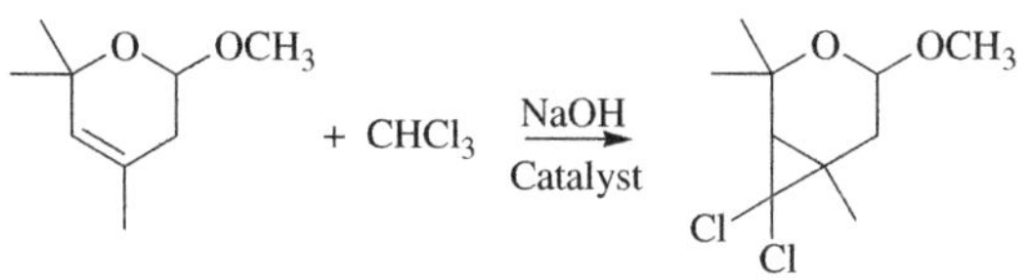

Mechanism:

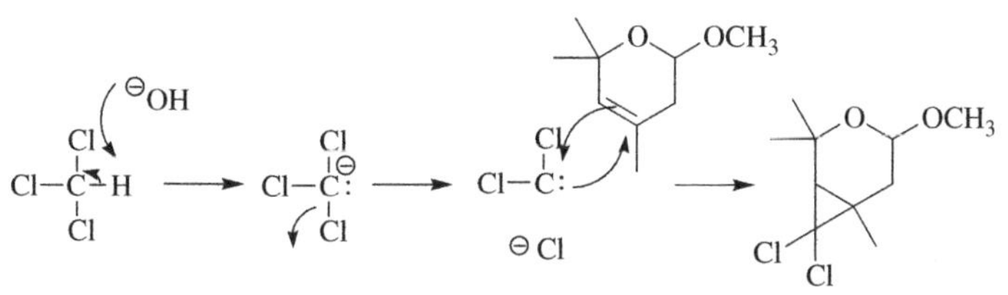

The first step again is to identify the origin of the new carbon in the product, which must be chloroform. The hydroxide removes a proton from chloroform to give the trichlorocarbanion which loses a chloride ion to give the neutral dichlorocarbene [30]. This is electrophilic and forms two σ bonds to the alkene site.

Some other electrophiles that convert alkenes to cyclopropanes are not free carbenes but have metals coordinated with their electrophilic site. These are called carbenoids and include the Simmons–Smith reaction and the copper-, rhodium-, or palladium-catalyzed decomposition of diazoketones and esters (Section 7.3). That the metal atom is present in the electrophile is shown by the variation of the stereoselectivity of the reaction with changes in the other ligands on metal.

Example 4.4

Propose a mechanism for the base-catalyzed reaction below.

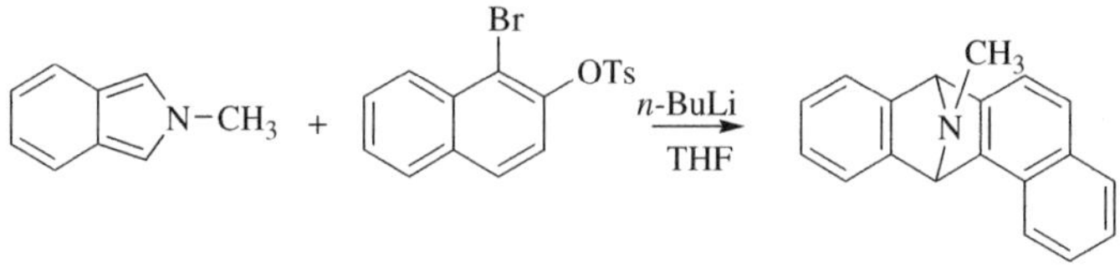

Mechanism:

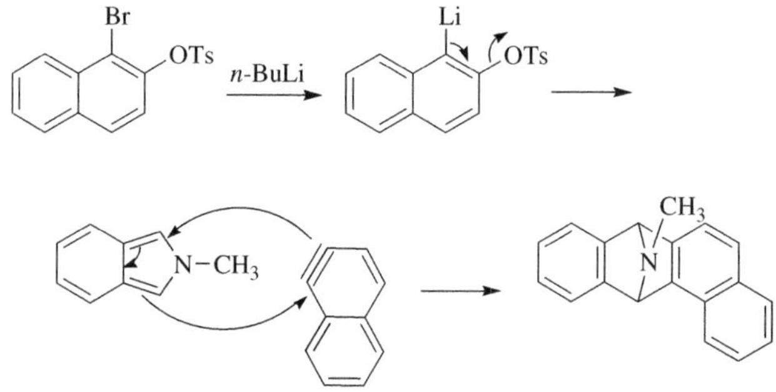

Compare the starting material to the product to determine the origin of the new bonds. The *n*-butyllithium exchanges with the bromo compound to give *n*-butyl bromide and an aryllithium. Lithium–halogen exchange reactions were reported early in the twentieth century, and investigations continue [31]. Elimination of lithium tosylate gives the naphthalyne, which combines with the isoindole in a Diels–Alder reaction [32] (Section 5.3.1).

4.4.2 Reactions in Acidic Solutions

In acidic solutions, generally, a proton or other Lewis acid attaches to an electron-rich site, giving a carbocation that may lead to a series of other carbocations, often ending finally with the loss of the Lewis acid or a proton. The carbocations are frequently stabilized by resonance, with much of the charge residing on an oxygen atom.

Example 4.5

Propose a mechanism for the acid-catalyzed reaction below.

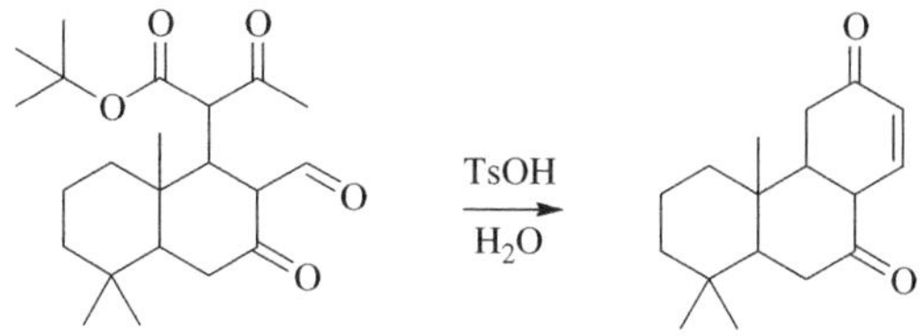

Mechanism:

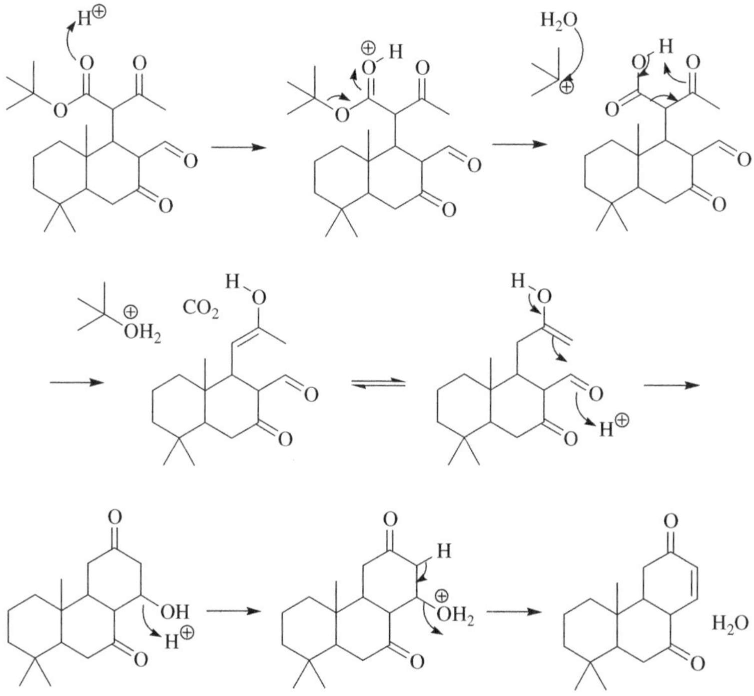

It is possible to write different steps within this proposed mechanism, and it is likely that other equilibria exist, particularly the protonation of the other carbonyl groups [33]. These equilibria do not affect the formation of the final product however.

Example 4.6

Propose a mechanism for the acid-catalyzed reaction below.

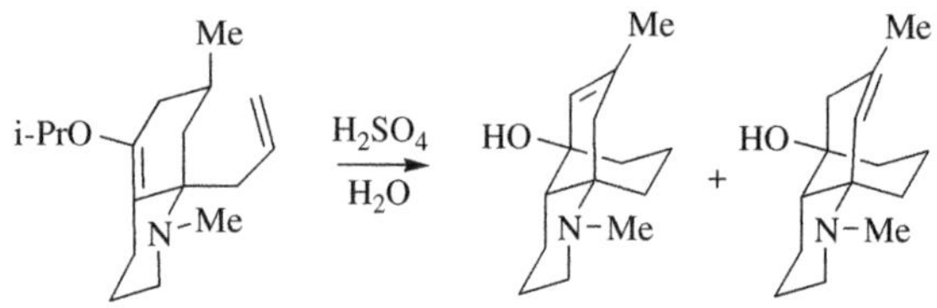

Mechanism:

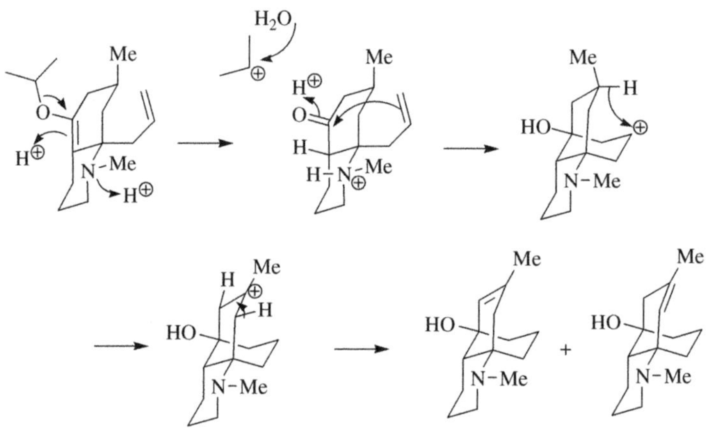

The Prins reaction leads to a secondary carbocation which undergoes a 1,5-hydride shift to form the more stable tertiary carbocation. Loss of a hydrogen on either side of the carbocation (only one of these steps is shown) leads to the two possible alkene products [34].

Example 4.7

Propose a mechanism for the acid-catalyzed reaction below.

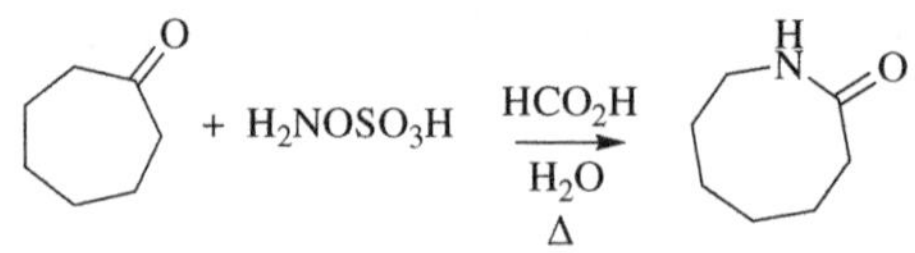

Mechanism:

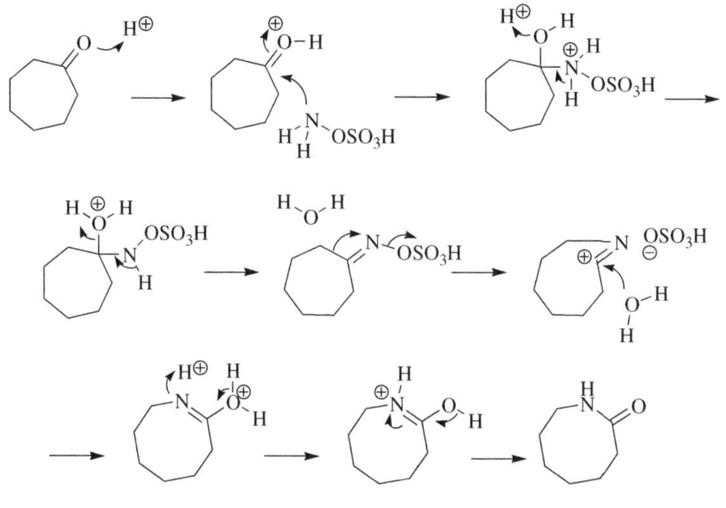

This is a variation of the classic Beckmann rearrangement of oximes [35, 36].

Example 4.8

Propose a mechanism for the acid-catalyzed reaction below.

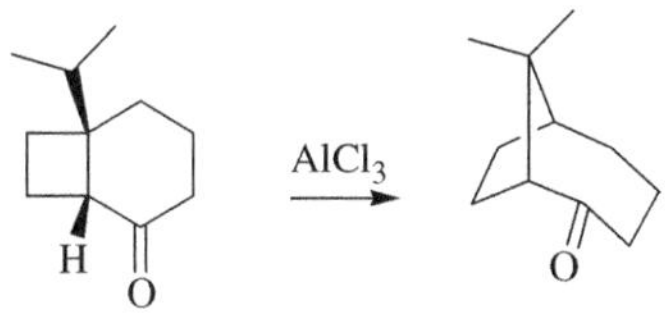

Mechanism:

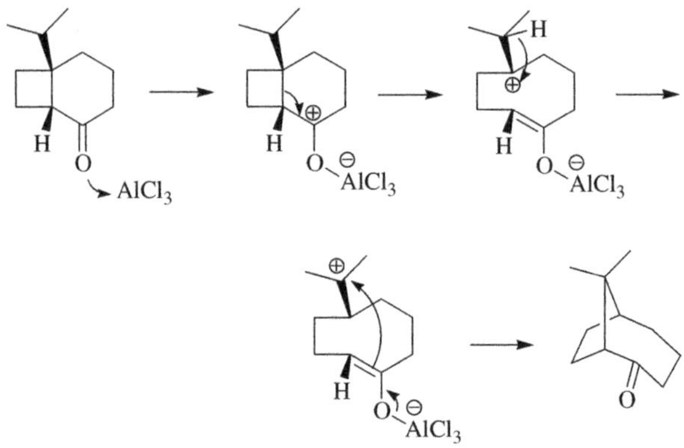

Aluminum trichloride polarizes the carbonyl group, leading to ring-opening of the four-membered ring. A 1,2-hydride shift followed by cyclization leads to the product [37].

4.4.3 Free Radical Reactions

Certain classes of compounds will decompose thermally or photochemically in a homolytic way, that is, a bonding pair of electrons will unpair. Common initiators for such reactions include aliphatic azo compounds like azobisisobutyronitrile (AIBN) and peroxides such as benzoyl peroxide (BPO). These can be used in catalytic amounts to initiate radical chain reactions of other molecules.

Example 4.9

Propose a mechanism for the free radical reaction below.

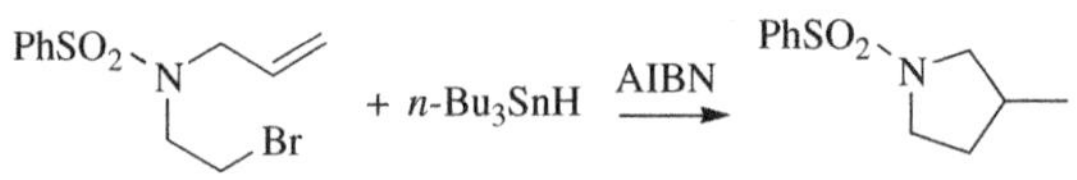

Mechanism:

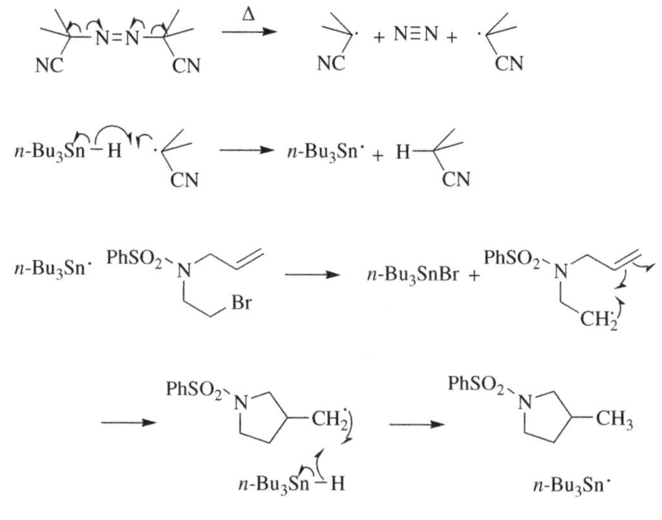

AIBN is used to initiate the radical process and the ring closure occurs in a propagation step [38].

Example 4.10

Propose a mechanism for the free radical reaction below.

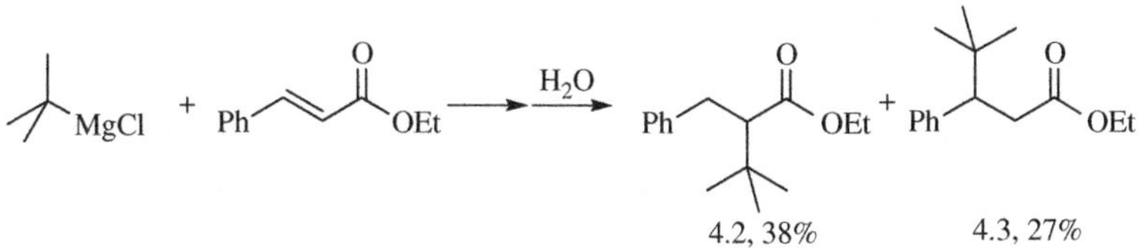

4.2, 38% 4.3, 27%

Mechanism to form product 4.2:

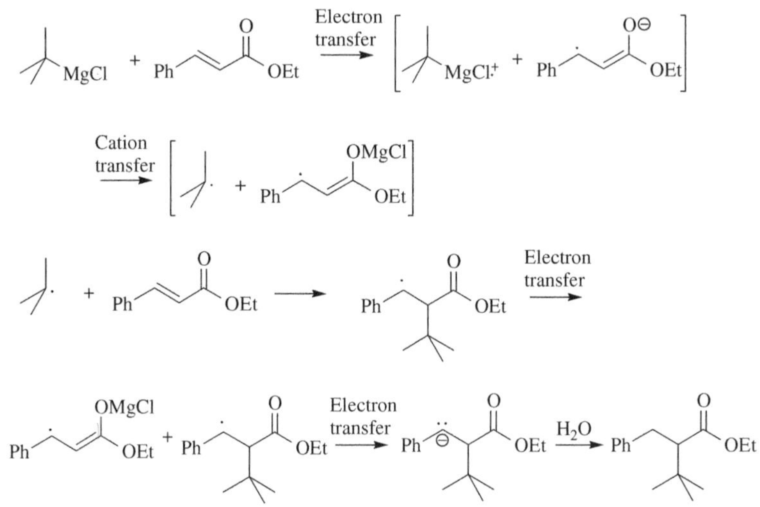

Product **4.3**, but not **4.2**, would be expected from a carbanion mechanism via the more stable enolate intermediate [39]. However, product **4.2** might arise from the more stable benzylic free radical. This is a single-electron-transfer mechanism where the easily reduced cinnamate ester gains a single electron from the Grignard reagent to become a radical anion, leaving the Grignard reagent as a radical cation. Transfer of the magnesium cation leaves a *tert*-butyl radical that attacks another ethyl cinnamate molecule alpha to the ester to give a benzylic radical. This radical then takes back the electron originally transferred to the first cinnamic acid, giving the benzylic carbanion, which gives product **4.2** on protonation. Product **4.3** may arise from competing conventional carbanion conjugate addition or from the combination of the cinnamate radical anion with the *tert*-butylmagnesium radical cation. The addition of α-methylstyrene to the original reaction mixture suppresses the formation of **4.2** by capturing the *tert*-butyl radicals, providing evidence that it is a radical process.

4.4.4 Molecular Rearrangements

4.4.4.1 **Rearrangement to Electron-Deficient Nitrogen** The Lossen and Hofmann rearrangements involve an α elimination from an amide nitrogen to leave the nitrogen neutral but short of an octet of electrons. The loss of N_2 in the Curtius and Schmidt rearrangements gives a similar circumstance on nitrogen.

Example 4.11

Propose a mechanism for the molecular rearrangement below.

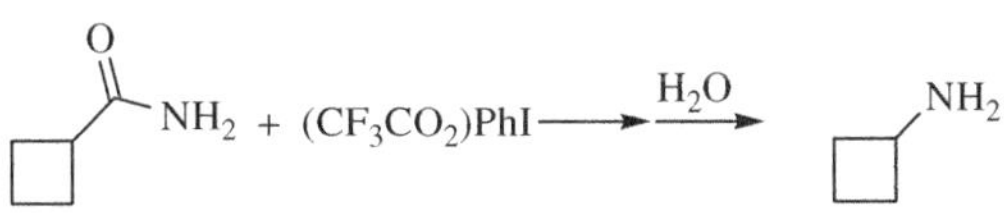

Mechanism:

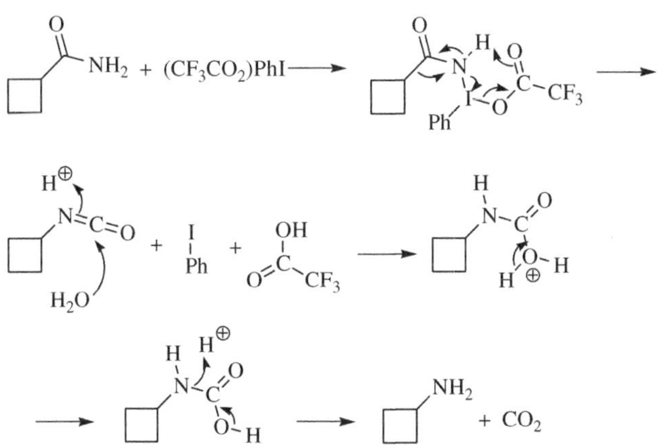

This reaction constitutes an acid-catalyzed Hofmann rearrangement [40, 41]. The trifluoroacetic acid formed in the reaction hydrolyzes the intermediate isocyanate and protonates the amine as it forms, preventing further reaction.

4.4.4.2 **Rearrangement to Electron-Deficient Oxygen** The Baeyer–Villiger oxidation, like the Hock rearrangement, is an example of a molecular rearrangement to oxygen. Examples are shown in Equations 6.8 and 6.9. The stereoselectivity of this reaction has been improved by the use of metal catalysts. Enzymatic catalysis has also improved the selectivity of the reaction as well, as shown in Example 4.12.

Example 4.12

Propose a mechanism for the molecular rearrangement below.

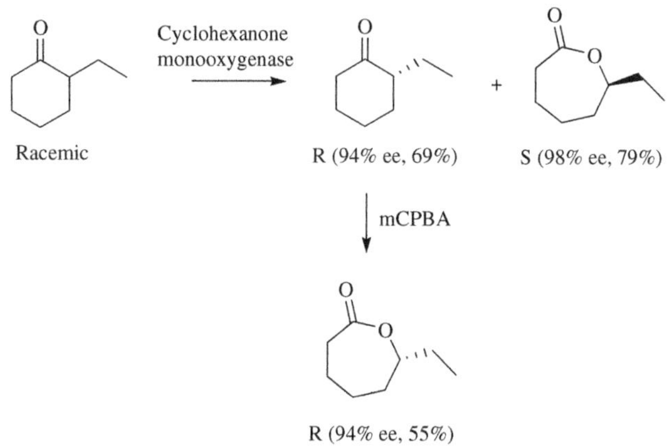

Mechanism:

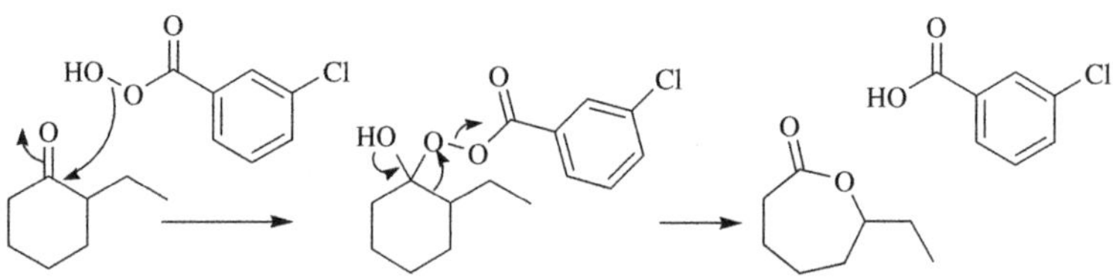

The mechanism shown uses *m*-chloroperoxybenzoic acid (mCPBA) as the oxidant. The enzymatically catalyzed reaction uses molecular oxygen. The reaction is considered a kinetic resolution, discussed further in Section 3.4. The enzyme reacts faster with the *S* enantiomer, allowing the formation of the *S* lactone and separation of the unreacted *R* ketone, which can be converted to the *R* lactone using traditional Baeyer–Villiger conditions [42].

4.4.4.3 **Rearrangement of Electron-Deficient Carbon** Carbocation rearrangements are well known. One named example is the pinacol rearrangement, the mechanism and stereochemistry of which have been well studied.

Example 4.13

Propose a mechanism for the molecular rearrangement below.

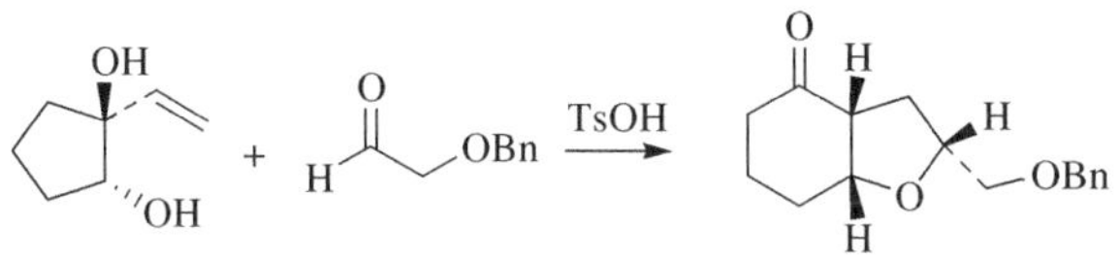

Mechanism:

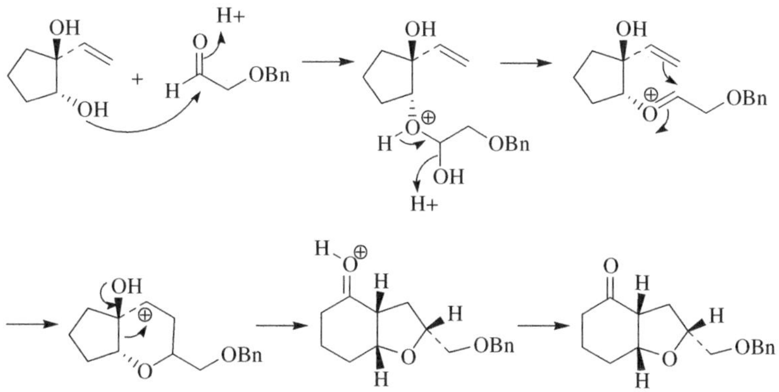

The cascade reaction shown couples the Prins cyclization with a pinacol rearrangement to allow the formation of several bonds in one reaction with predictable stereochemistry [43].

RESOURCE

1. The NIST Kinetics Database lists rate parameters for more than 38,000 reactions. https://kinetics.nist.gov/kinetics/welcome.jsp (J. A. Manion, R. E. Huie, R. D. Levin, D. R. Burgess Jr., V. L. Orkin, W. Tsang, W. S. McGivern, J. W. Hudgens, V. D. Knyazev, D. B. Atkinson, E. Chai,

A. M. Tereza, C.-Y. Lin, T. C. Allison, W. G. Mallard, F. Westley, J. T. Herron, R. F. Hampson, and D. H. Frizzell, NIST Chemical Kinetics Database, NIST Standard Reference Database 17, Version 7.0 (Web Version), Release 1.4.3, Data version 2008.12, National Institute of Standards and Technology, Gaithersburg, Maryland, 20899-8320. Web address: https://kinetics.nist.gov/kinetics/.)

PROBLEMS

Questions marked with an asterisk are answered in Appendix C.

4.1* In determining the mechanism for the isomerization of γ-hydroxy-α,β-alkynoates, the following observations were made [44]:

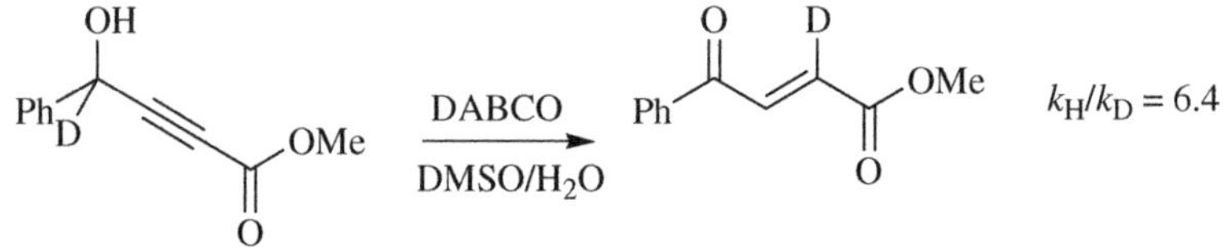

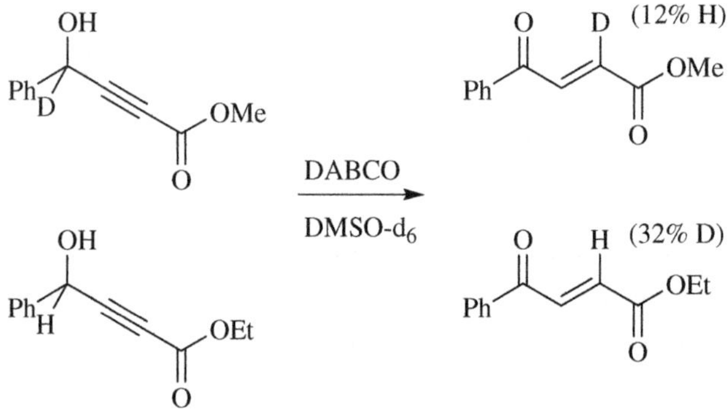

Write a mechanism consistent with these findings.

4.2* The mechanism of the ene reaction was studied with a variety of alkenes using the reaction below. The reaction was found to be insensitive to the effect of the solvent and first order in both reagents. In all cases, the E product was highly favored. Write a mechanism and intermediate consistent with this data [45].

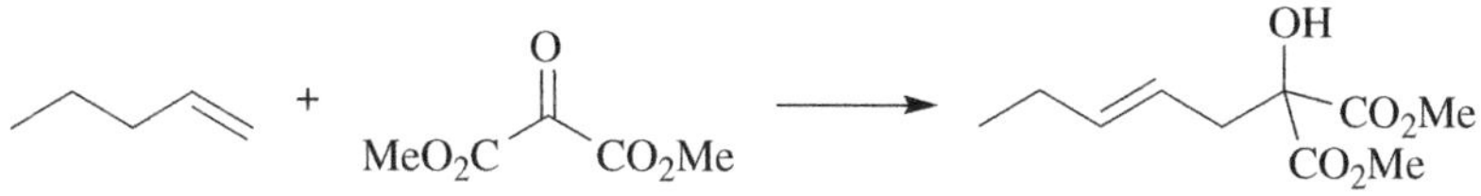

4.3 For each reaction in Table 4.2, consider the likely mechanism and explain both the magnitude and sign of the ρ value.

4.4 The rates of reaction of substituted benzoyl chlorides with excess methanol at 0°C were measured. The pseudo-first-order rate constant for *p*-methylbenzoyl chloride is 0.0178 mol/L min and for *p*-nitrobenzoyl chloride is 0.413 mol/L min. Calculate the rate constant for *m*-bromobenzoyl chloride, assuming these two examples give a good correlation [46]. (Of course, in a proper study, a larger set of examples would be used to obtain a correlation.)

4.5 Nornicotine analogs have been explored as green catalysts for aqueous Aldol reactions, such as the one shown below. To clarify the proposed mechanism, a Hammett study was conducted and the data below obtained. Make a Hammett plot of the data and calculate the ρ value. What does the value of ρ tell you about the mechanism? Propose a mechanism that is consistent with this information [47].

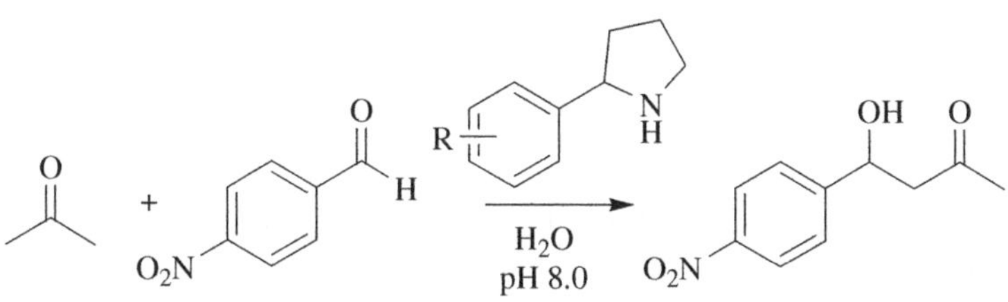

R	k_{obs} (10^{-3}/min)	R	k_{obs} (10^{-3}/min)
4-MeO	3.1	4-Br	9.1
H	4.6	3-Br	12.5
4-F	4.8	4-CF_3	18
3-MeO	6.3	3-NO_2	23.8
4-Cl	8.4		

4.6* The rate of the following reaction at three concentrations of NaOH in aqueous acetonitrile at 0°C was determined from absorbance measurements at 294 nm. The initial concentration of the starting material was 1.13×10^{-4} M. The results for three runs are tabulated [48].

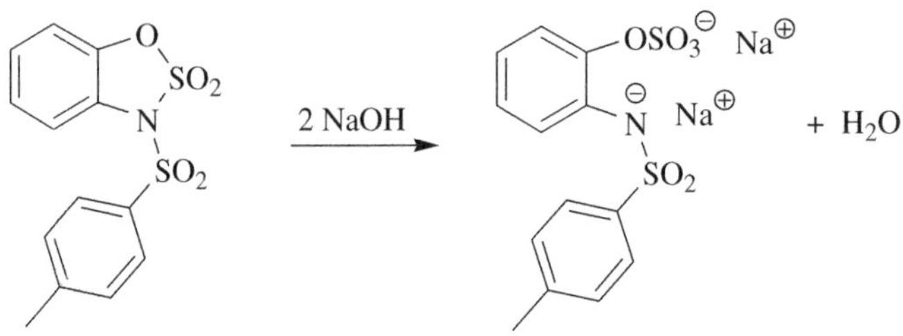

[OH−] (mol/L)	k_{obs} (sec^{-1})
0.050	8.67×10^{-2}
0.060	10.8×10^{-2}
0.10	17.6×10^{-2}

(a) The rate constants provided are pseudo first order. How do you know?

(b) What is the overall order of the reaction?

(c) Write a rate expression for the reaction.

(d) Calculate the rate constant.

4.7* The presence of bulking agents such as polyethylene glycol 400 (PEG400) in pharmaceuticals is a potential source of hydrogen peroxide, which can lead to loss of potency in medications through oxidative reactions. Hydrogen peroxide is also part of the cellular defense system. Its reaction with pyruvate is fast and efficient, and it is a possible means of protecting these pharmaceuticals from oxidative damage. The proposed mechanism below is supported by isotope studies. Direct observation of the intermediate is possible by ^{13}C NMR, and the rates of the two steps were studied at the low temperatures necessary to observe the intermediate [49]. An Eyring plot of $\ln(kh/k_BT)$ versus $1/T$ gives $\Delta H^{\ddagger}/R$ as the slope and $\Delta S^{\ddagger}/R$

as the y-intercept. Use an Eyring plot to calculate $\Delta H^{\ddagger}$ and $\Delta S^{\ddagger}$ for the two steps of the reaction. Do these values support the proposed mechanism?

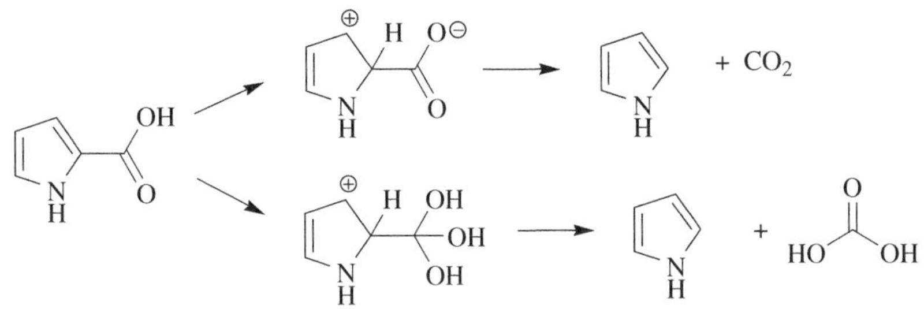

T(°K)	$k_1(10^{-3}/s)$	$k_2(10^{-4}/s)$
259	13.4	49.9
254		28.7
245	4.33	7.66
242	3.30	5.81
240	3.13	3.77
239		3.92
238	2.43	2.58

4.8* The following reaction occurs in CCl_4 at 25°C [50]:

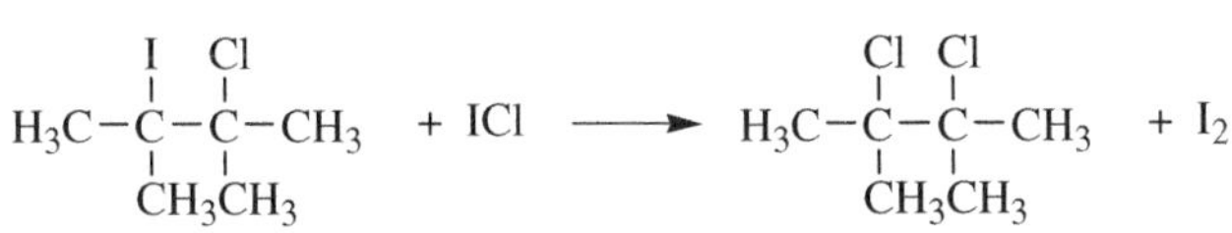

Kinetic measurements were made following the concentration of ICl, using a large excess of $C_6H_{12}ICl$. The data fit a linear plot as shown below. Write a rate expression for the reaction and calculate the rate constant.

	Initial concentrations (M)	
	Run 1	Run 2
$[C_6H_{12}ICl]_0$	0.0349	0.0174
$[ICl]_0$	0.00441	0.00160

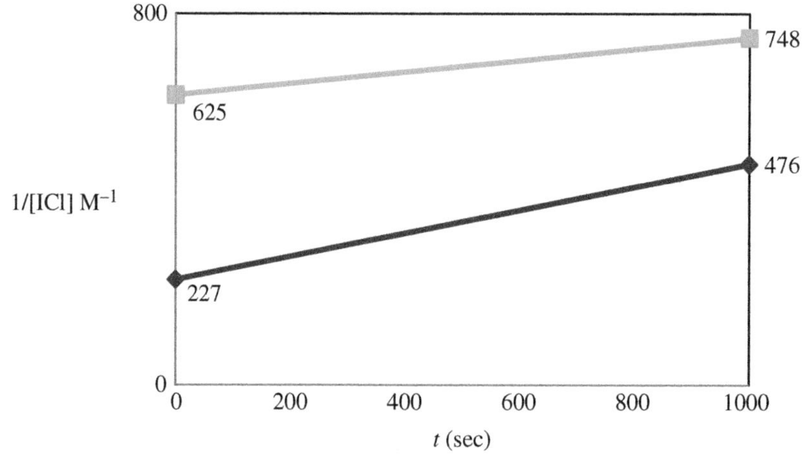

4.9 Using the kinetic data tabulated below, determine the order and write the rate expression for the illustrated reaction, rounding to the nearest whole order(s). Explain how you did this. Calculate the rate constant including the units [51].

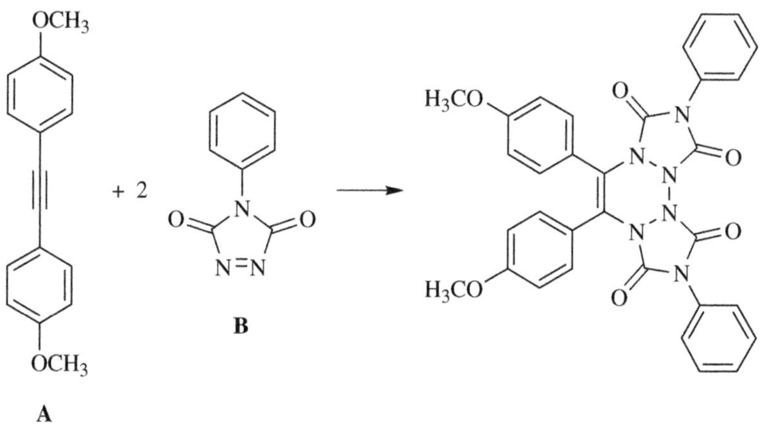

Run	Initial molarity of A	Initial molarity of B	Initial rate (mol/L/s)
1	3.85×10^{-4}	9.95×10^{-3}	7.78×10^{-8}
2	1.96×10^{-4}	10.3×10^{-3}	4.18×10^{-8}

Run	Initial molarity of A	Initial molarity of B	Initial rate (mol/L/s)
3	0.99×10^{-4}	10.2×10^{-3}	2.04×10^{-8}
4	$1.96 \times 10_{-4}$	4.87×10^{-3}	1.89×10^{-8}
5	1.96×10^{-4}	6.74×10^{-3}	2.73×10^{-8}

Which of the following reaction schemes fit the preceding results? Explain.

I $A + B \rightarrow I$ slow
$I + B \rightarrow C$ fast

II $B + B \rightarrow I$ slow
$I + A \rightarrow C$ fast

III $A + B \rightarrow I$ fast
$I + B \rightarrow C$ slow

IV $A + B + B \rightarrow C$

4.10* The rate of the following reaction was measured with varying concentrations of $PdCl_4^{-2}$, H^+, and Cl^-, and the results are given in the table [52]

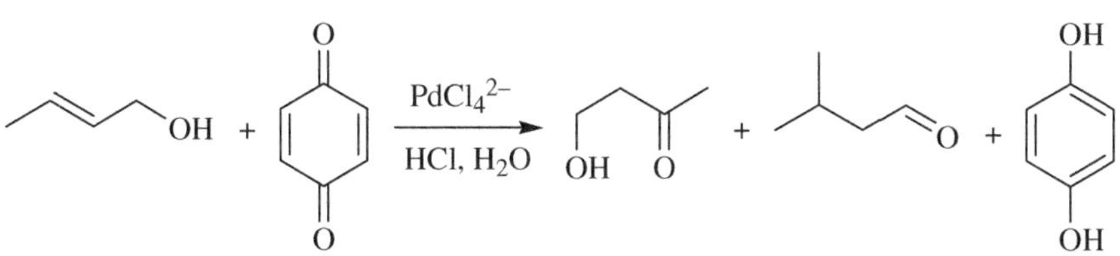

Run[a]	$[PdCl_4^{-2}]$	$[H^+]^b$	$[Cl^-]^c$	k_{obs}, s^{-1}
1	0.005	0.2	0.6	0.52×10^{-5}
2	0.0125	0.2	0.6	1.7×10^{-5}
3	0.025	0.2	0.6	2.7×10^{-5}
4	0.050	0.2	0.6	5.1×10^{-5}
5	0.025	0.2	0.4	5.1×10^{-5}
6	0.025	0.2	0.9	1.5×10^{-5}
7	0.025	0.2	1.2	0.77×10^{-5}
8	0.025	0.4	0.6	1.6×10^{-5}
9	0.025	0.6	0.6	1.2×10^{-5}

Run[a]	$[PdCl_4^{-2}]$	$[H^+]^b$	$[Cl^-]^c$	k_{obs}, s^{-1}
10	0.025	0.8	0.6	0.84×10^{-5}

[a]All runs are in an aqueous solution at 25°C. $LiClO_4$ was added to bring the ionic strength to 2.0. Initial 2-buten-1-ol and quinone concentrations are 0.005 molar.
[b]Added as $HClO_4$.
[c]Added as LiCl.

The tabulated k_{obs} is from the pseudo-first-order expression

$$-\frac{d[\text{butenol}]}{dt} = k_{obs}[\text{butenol}]$$

The benzoquinone simply oxidizes the Pd(0) back to Pd(II) to maintain a constant concentration of Pd(II). Note the effect of changing concentrations on the observed rate and write a rate expression for the oxidation of 2-buten-1-ol in terms of all involved concentrations (except benzoquinone, which does not affect the rate). Round off all orders to the nearest whole number. Select a run from the table and calculate an actual rate constant for the reaction and give the units for it.

4.11 Treatment of 1-chloro-2-methylcyclohexene with 5 eq of methyllithium in tetramethylethylenediamine-tetrahydrofuran (TMEDA–THF) followed by aqueous workup gave a 47% yield of 1-methylbicyclo[4.1.0]heptane [53]. Two mechanisms were proposed for this process as outlined below. What experiment(s) would you use to determine which mechanism is incorrect? Tell what you would do, what results you might expect, and how you would draw your conclusions.

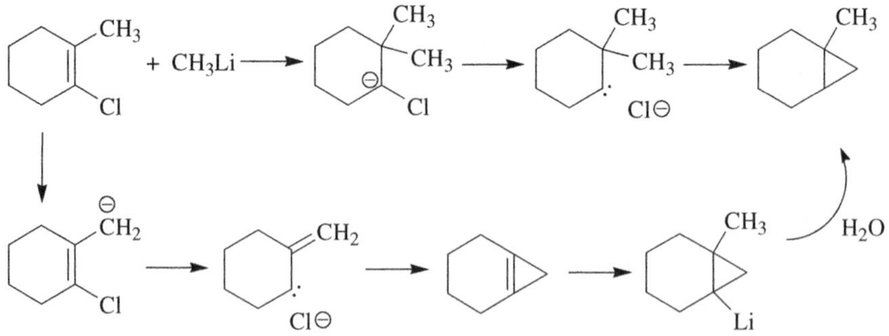

4.12* The reaction of alkenes with nitroso compounds gives hydroxylamine products [54]. The process was thought to occur by either a one-step ene reaction (Eq. 4.49, below) or a two-step reaction via a transient aziridine oxide intermediate (Eq. 4.50). The following deuterium isotope effects were measured for the reaction of 2,3-dimethyl-2-butene and pentafluoronitrosobenzene. Explain the similarities and differences among these results in terms of one of these mechanisms. Use the results to exclude one mechanism. The first three cases are intramolecular competitions where k_H is the rate of formation of the $C{=}CH_2$ product and k_D is the rate of formation of the $C{=}CD_2$ alternative product. The last case is an intermolecular competition. The results were determined by ^1H NMR analysis of the products.

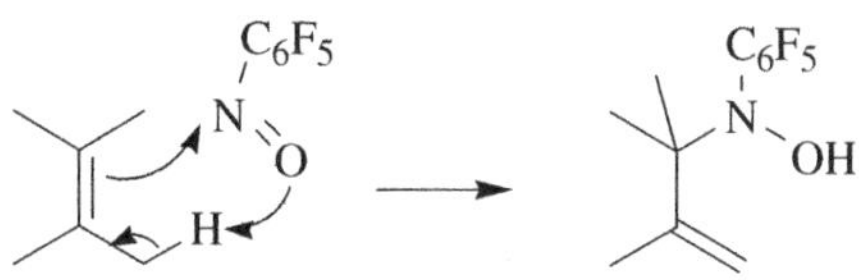

(4.49)

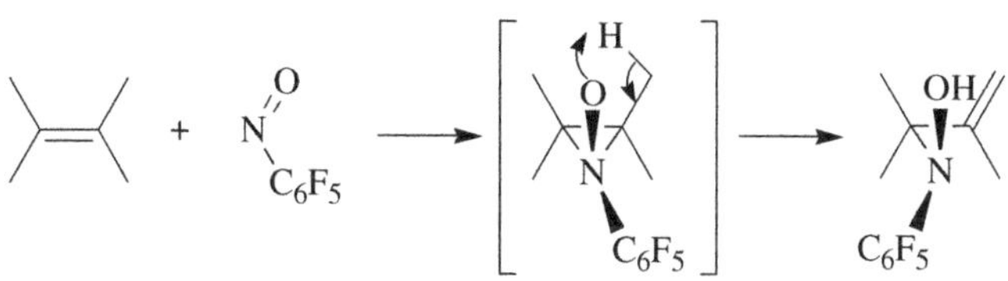

(4.50)

Alkene	k_H/k_D	Alkene			k_H/k_D
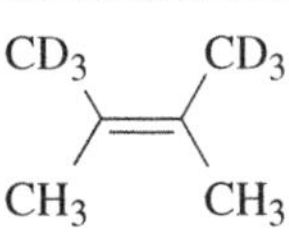	1.2 ± 0.2	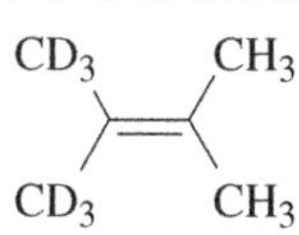			4.5 ± 0.2
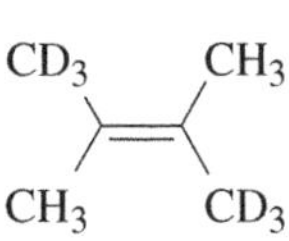	3.0 ± 0.2	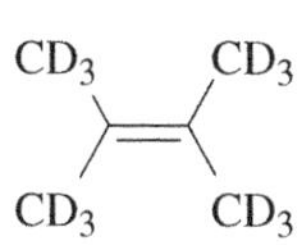	and	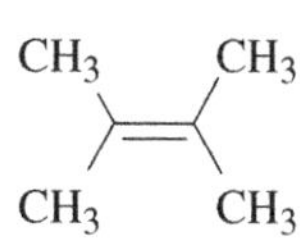	1.03 ± 0.05

4.13 Explain in terms of mechanism why one of the following sulfonates is more reactive [55]:

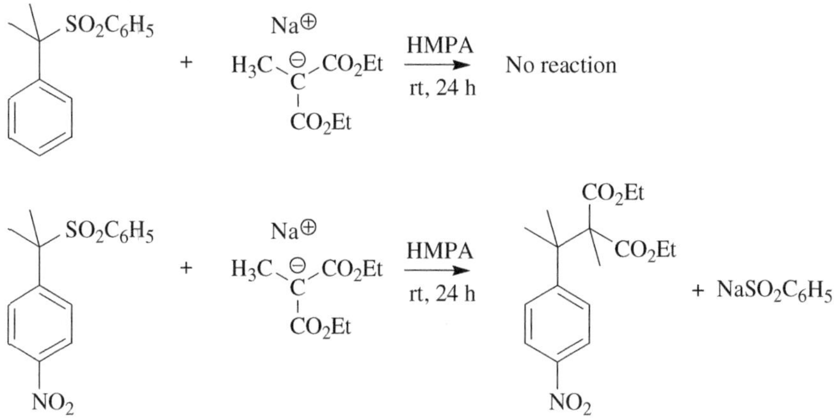

In other closely related reactions, small amounts of sulfur or *p*-dinitrobenzene prevent the reaction from occurring.

4.14 A SmI_2 reduction of nitriles was reported [56] which produces primary amines in high yield under much milder conditions than the metal hydride reagents frequently used in this reaction. Several studies were undertaken to elucidate the mechanism. The kinetic isotope effect, k_H/k_D, was determined to be 1.33. Data for the Hammett study are reported in the table below. Plot this data and determine whether the Hammett relationship is valid. Calculate ρ for this reaction and discuss what it and the kinetic isotope effect suggest about the mechanism of this reaction.

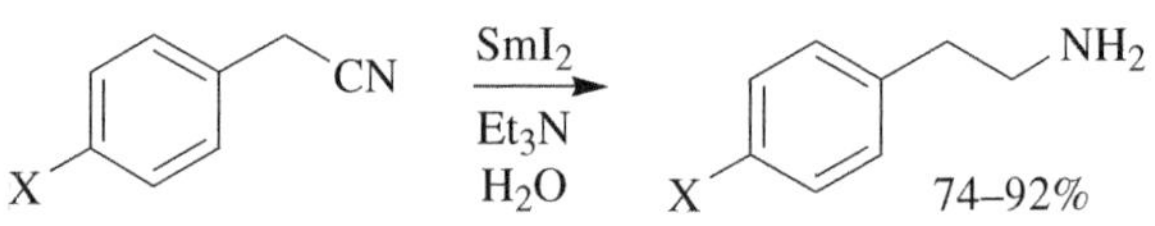

X–	k_X/k_H	X–	k_X/k_H
CF_3–	1.84	F–	1.24
Cl–	1.46	H–	1.00
Br–	1.43	MeO–	0.94

4.15* In the synthesis of (+)-hexachlorosulfolipid, a toxin linked to food poisoning via the Adriatic mussel, the sequence below was used [57]. The first and last steps are protecting and deprotecting steps, respectively. How might the stereochemistry of the dichloro compound be confirmed chemically?

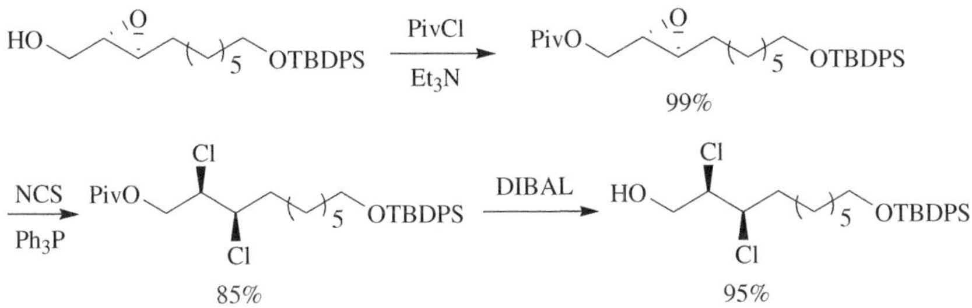

4.16* Write the mechanism for the formation of Compound **4.3**.

Draw likely intermediates in the mechanism of each of the reactions in the problems below.

4.17*

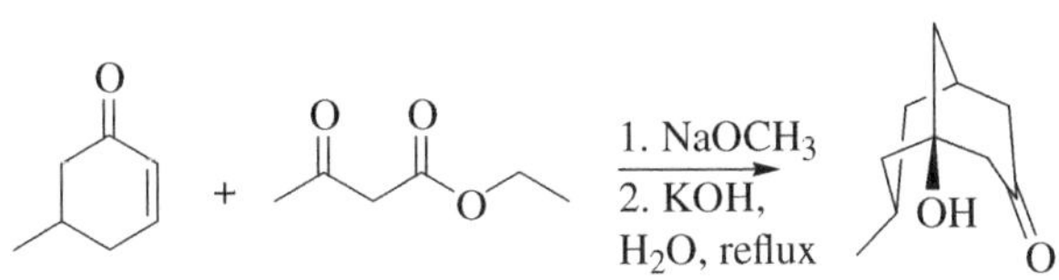

Ref. [58]

4.18

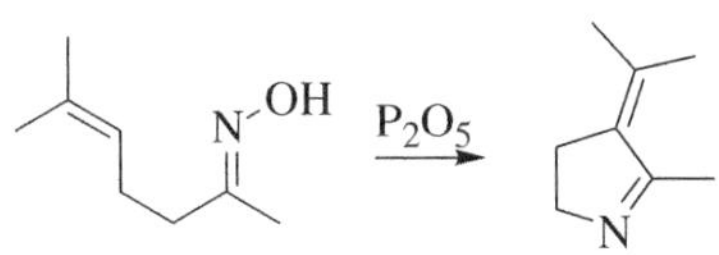

Ref. [59]

4.19*

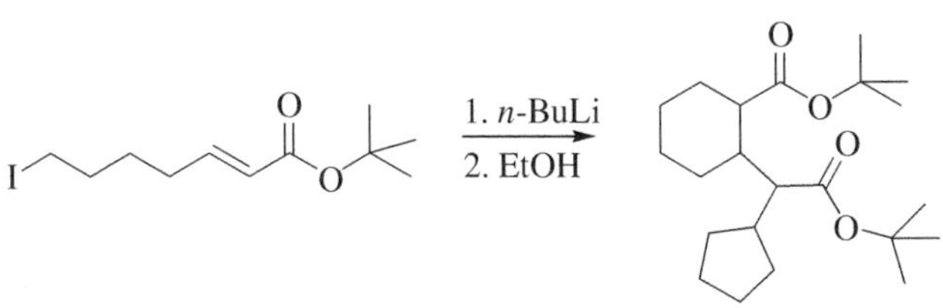

Ref. [60]

4.20*

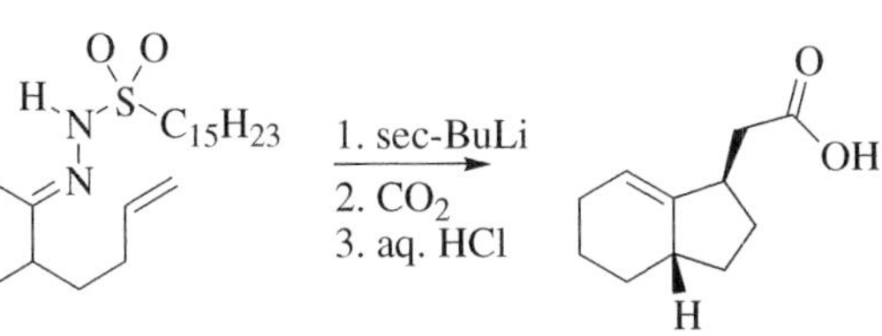

Ref. [61]

4.21

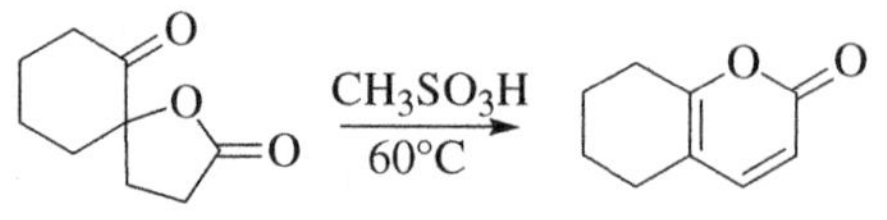

Ref. [62]

4.22

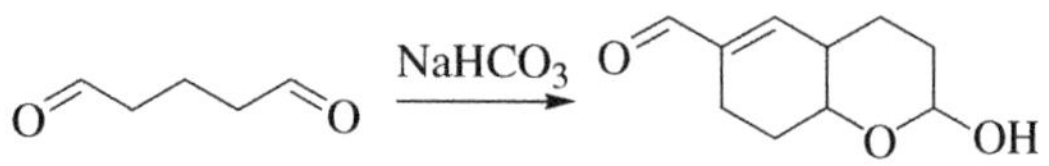

Ref. [63]

4.23*

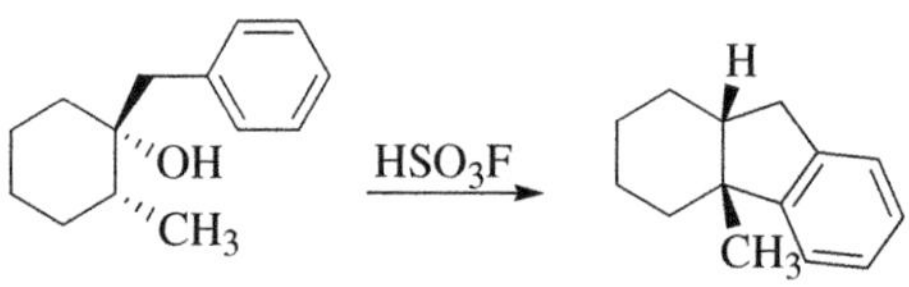

Ref. [64]

4.24

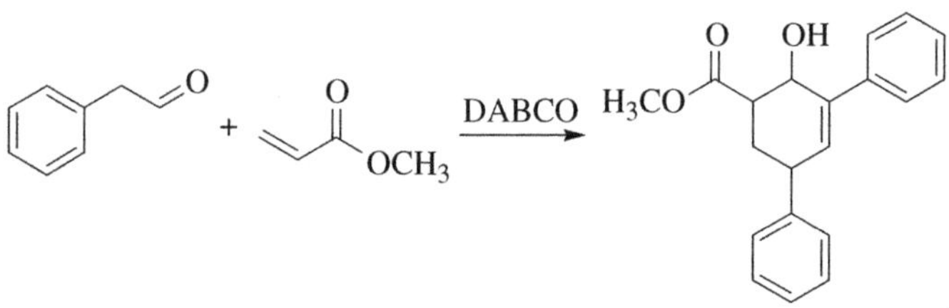

Ref. [65]

4.25

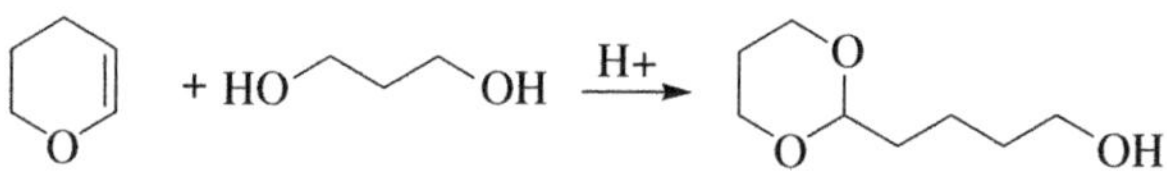

Ref. [66]

4.26*

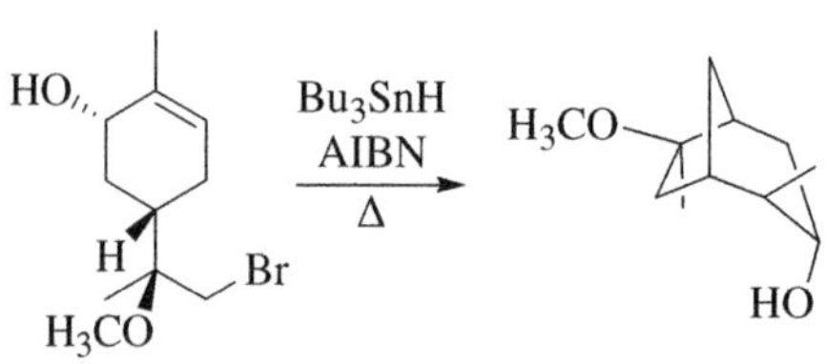

Ref. [67]

4.27*

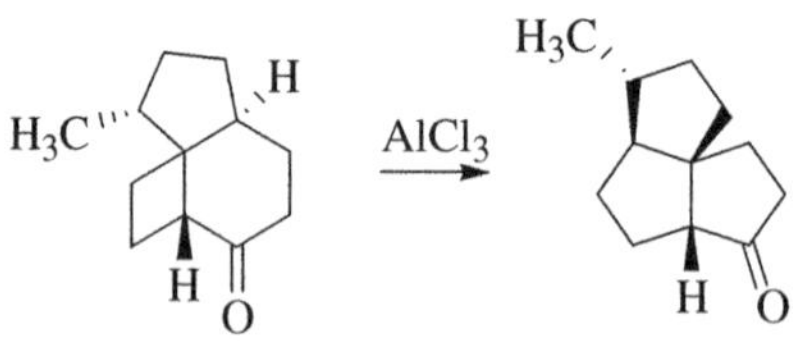

Ref. [37]

4.28

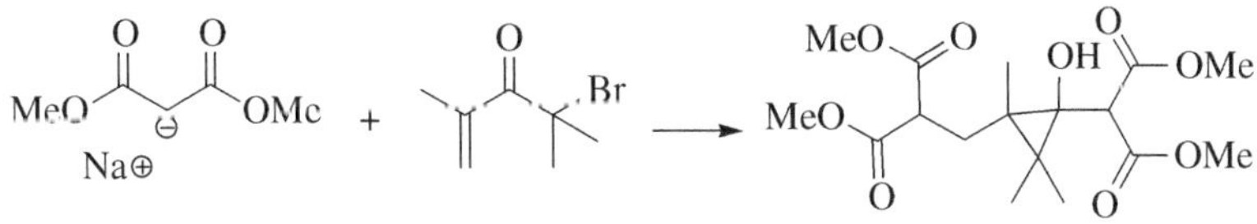

Ref. [68]

4.29

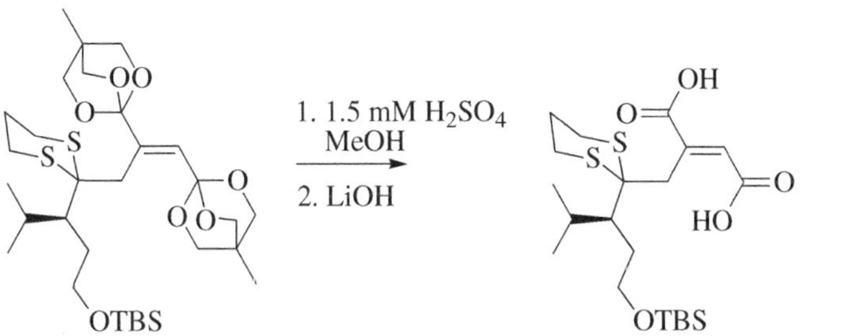

Show the mechanism for deprotection of one of the acid groups

Ref. [69]

4.30* Write the mechanism for step 2 of Problem 3.14.

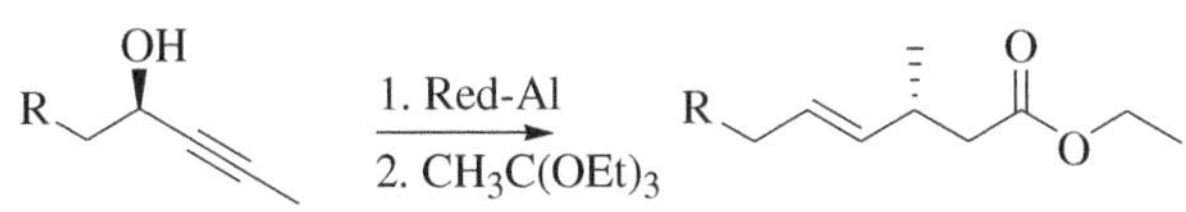

4.31* Hydroxamic acids have found use as histone deacetylase inhibitors, an enzyme that is upregulated in several types of cancer. Several hydroxamic acid-derived compounds have been approved for the treatment of T-cell lymphoma and multiple myeloma. However, hydroxamic acids are known to be mutagenic, which argues against their wider medical use. The mechanism of this toxicity is believed to result from a Lossen rearrangement. Write the mechanism for the Lossen rearrangement of Belinostat, used in the treatment of T-cell lymphoma [70].

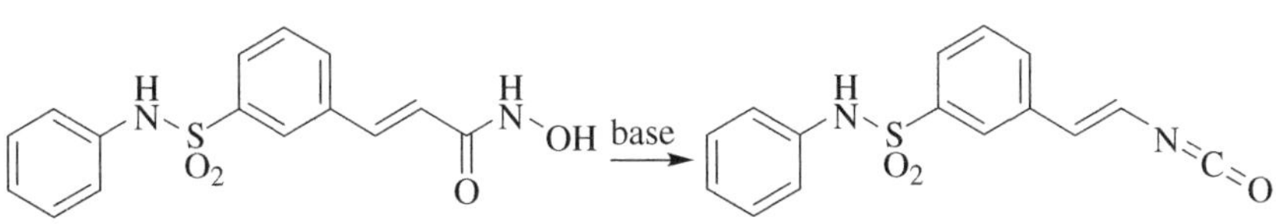

REFERENCES

1. Carpenter, B. K. *Determination of Organic Reaction Mechanisms*; Wiley-Interscience: New York, 1984.
2. Sykes, P. *A Guidebook to Mechanism in Organic Chemistry*, 6th ed.; Longman Scientific and Technical copublished in the United States with Wiley: New York, 1986.
3. Capellos, C.; Bielski, B. H. J. *Kinetic Systems: Mathematical Description of Chemical Kinetics in Solution*; Wiley-Interscience: New York, 1972.
4. Hammond, G. S. *J. Am. Chem. Soc.* **1955**, *77*, 334–338.
5. Scala, A. A. *J. Chem. Educ.* **2004**, *81*, 1661–1664.
6. Mendelson, W.; Pridgen, L.; Holmes, M.; Shilcrat, S. *J. Org. Chem.* **1989**, *54*, 2490–2492.
7. Johri, K. K.; DesMarteau, D. D. *J. Org. Chem.* **1983**, *48*, 242–250.
8. Sonnet, P. E.; Oliver, J. E. *J. Org. Chem.* **1976**, *41*, 3284–3286.
9. Moore, J. W.; Pearson, R. G. *Kinetics and Mechanism*, 3rd ed.; Wiley-Interscience: New York, 1981.
10. Raines, R. T.; Hansen, D. E. *J. Chem. Educ.* **1988**, *65*, 757–759.
11. Nome, F.; Erbs, W.; Correia, V. R. *J. Org. Chem.* **1981**, *46*, 3802–3804.
12. Keszei, E. *Reaction Kinetics: An Introduction*; Springer: Switzerland, 2021.
13. Melander, L.; Saunders, W. H. Jr. *Reaction Rates of Isotopic Molecules*; Wiley-Interscience: New York, 1980.
14. Rastetter, W. H.; Wagner, W. R.; Findeis, M. A. *J. Org. Chem.* **1982**, *47*, 419–422.
15. Laidler, K. J. *J. Chem. Educ.* **1984**, *61*, 494–498.
16. Kuntzleman, T. S.; Swanson, M. S.; Sayers, D. K. *J. Chem. Educ.* **2007**, *84*, 1776–1778.
17. Martin, J. C.; Timberlake, J. W. *J. Am. Chem. Soc.* **1970**, *92*, 978–983.
18. Hammond, G. S.; DeBoer, C. *J. Am. Chem. Soc.* **1964**, *86*, 899–902.
19. Robinson, P. J. *J. Chem. Ed.* **1978**, *55*, 509–510.
20. Sanjeev, R.; Padmavathi, D. A.; Jagannadham, V. *World J. Chem. Educ.* **2018**, *6*, 78–81.
21. LibreTexts Chemistry. Physical and Theoretical Chemistry. Supplemental Modules. Modeling Reaction Kinetics. https://chem.libretexts.org/Bookshelves/Physical_and_Theoretical_Chemistry_Textbook_Maps/Supplemental_Modules_(Physical_and_Theoretical_Chemistry)/Kinetics (accessed June 18, 2025).
22. Hammett, L. P. *J. Am. Chem. Soc.* **1937**, *59*, 96–103.

23. Hansch, C.; Leo, A.; Taft, R. W. *Chem. Rev.* **1991**, *91*, 165–195.
24. Jaffe, H. H. *Chem. Rev.* **1953**, *53*, 191–261.
25. Poh, B.-L. *Can. J. Chem.* **1979**, *57*, 255–257.
26. Brown, H. C.; Okamoto, Y. *J. Am. Chem. Soc.* **1958**, *80*, 4979–4987.
27. Moss, R. A.; Tian, J. *Org. Lett.* **2006**, *8*, 1245–1247.
28. Turner, R. B.; Nettleton, D. E. Jr.; Ferebee, R. *J. Am. Chem. Soc.* **1956**, *78*, 5923–5927.
29. Majewski, M.; Snieckus, V. *J. Org. Chem.* **1984**, *49*, 2682–2687.
30. Slessor, K.; Oehlschlager, A. C.; Johnston, B. D.; Pierce, H. D. Jr.; Grewal, S. K.; Wickremesinghe, L. K. G. *J. Org. Chem.* **1980**, *45*, 2290–2297.
31. Seyferth, D. *Organometallics* **2006**, *25*, 1–24.
32. Gribble, G. W.; LeHoullier, C. S.; Sibi, M. P.; Allen, R. W. *J. Org. Chem.* **1985**, *50*, 1611–1616.
33. Meyer, W. L.; Manning, R. A.; Schroeder, P. G.; Shew, D. C. *J. Org. Chem.* **1977**, *42*, 2754–2761.
34. Lin, H.-Y.; Causey, R.; Garcia, G. E.; Snider, B. B. *J. Org. Chem.* **2012**, *77*, 7143–7156.
35. Olah, G. A.; Fung, A. P. *Org. Synth.* **1985**, *63*, 188–189.
36. Gawley, R. A. *Org. React.* **1988**, *35*, 1–420.
37. Kakiuchi, K.; Fukunaga, K.; Matsuo, F.; Ohnishi, Y.; Tobe, Y. *J. Org. Chem.* **1991**, *56*, 6742–6744.
38. Padwa, A.; Nimmesgern, H.; Wong, G. S. K. *J. Org. Chem.* **1985**, *50*, 5620–5627.
39. Holm, T.; Crossland, I.; Madsen, J. O. *Acta Chem. Scand. B* **1978**, *32*, 754–758.
40. Almond, M. R.; Stimmel, J. B.; Thompson, E. A.; Loudon, G. M. *Org. Synth.* **1988**, *66*, 132–141.
41. Boutin, R. H.; Loudon, G. M. *J. Org. Chem.* **1984**, *49*, 4277–4284.
42. Stewart, J. D.; Reed, K. W.; Zhu, J.; Chen, G.; Kayser, M. M. *J. Org. Chem.* **1996**, *61*, 7652–7653.
43. Overman, L. E.; Pennington, L. D. *J. Org. Chem.* **2003**, *68*, 7143–7157.
44. Sonye, J. P.; Koide, K. *Org. Lett.* **2006**, *8*, 199–202.
45. Achmatowicz, O. Jr.; Szymoniak, J. *J. Org. Chem.* **1980**, *45*, 1228–1232 Achmatowicz, O. Jr.; Szymoniak, J. *J. Org. Chem.* **1980**, *45*, 4774–4776.
46. Norris, J. F.; Young, H. H. Jr. *J. Am. Chem. Soc.* **1935**, *57*, 1420–1424.
47. Rogers, C. J.; Dickerson, T. J.; Brogan, A. P.; Janda, K. D. *J. Org. Chem.* **2005**, *70*, 3705–3708.

48. Andersen, K. K.; Bray, D. D.; Chumpradit, S.; Clark, M. E.; Habgood, G. J.; Hubbard, C. D.; Young, K. M. *J. Org. Chem.* **1991**, *56*, 6508–6516.

49. Asmus, C.; Mozziconacci, O.; Schoneich, C. *J. Phys. Chem. A* **2015**, *119*, 966–977.

50. Schmid, G. H.; Gordon, J. W. *J. Org. Chem.* **1983**, *48*, 4010–4013.

51. Cheng, C.-C.; Greene, F. D.; Blount, J. F. *J. Org. Chem.* **1984**, *49*, 2917–1922.

52. Zaw, K.; Henry, P. M. *J. Org. Chem.* **1990**, *55*, 1842–1847.

53. Gassman, P. G.; Valcho, J. J.; Proehl, G. S.; Cooper, C. F. *J. Am. Chem. Soc.* **1980**, *102*, 6519–6526.

54. Seymour, C. A.; Greene, F. D. *J. Org. Chem.* **1982**, *47*, 5226–5227.

55. Kornblum, N.; Ackermann, P.; Manthey, J. W.; Musser, M. T.; Pinnick, H. W.; Singaram, S.; Wade, P. A. *J. Org. Chem.* **1988**, *53*, 1475–1481.

56. Szostak, M.; Sautier, B.; Spain, M.; Procter, D. *J. Org. Lett.* **2014**, *16*, 1092–1095.

57. Yoshimitsu, T.; Fujumoto, N.; Nakatani, R.; Kojima, N.; Tanaka, T. *J. Org. Chem.* **2010**, *75*, 5425–5437.

58. Kraus, G. A.; Hon, Y.-S. *J. Org. Chem.* **1985**, *50*, 4605–4608.

59. Gawley, R. E.; Termine, E. J. *J. Org. Chem.* **1984**, *49*, 1946–1951.

60. Cooke, M. P. Jr. *J. Org. Chem.* **1984**, *49*, 1144–1146.

61. Chamberlin, A. R.; Bloom, S. H.; Cervini, L. A.; Fortsch, C. H. *J. Am. Chem. Soc.* **1988**, *110*, 4788–4796.

62. Mandal, A. K.; Jawalkar, D. G. *J. Org. Chem.* **1989**, *54*, 2364–2369.

63. Tashima, T.; Imai, M.; Kuroda, Y.; Yagi, S.; Nakagawa, T. *J. Org. Chem.* **1991**, *56*, 694–697.

64. Barrow, C. J.; Bright, S. T.; Coxon, J. M.; Steel, P. J. *J. Org. Chem.* **1989**, *54*, 2542–2549.

65. Poly, W.; Schomburg, D.; Hoffmann, H. M. R. *J. Org. Chem.* **1988**, *53*, 3701–3710.

66. Kozikowski, A. P.; Stein, P. D. *J. Org. Chem.* **1984**, *49*, 2301–2309.

67. Srikrishna, A.; Hemamslini, P. *J. Org. Chem.* **1990**, *55*, 4883–4887.

68. Barbee, T. R.; Hedeel, G.; Heeg, M. J.; Albrizati, K. F. *J. Org. Chem.* **1991**, *56*, 6773–6781.

69. Ichige, T.; Okano, Y.; Kanoh, N.; Nakata, M. *J. Org. Chem.* **2009**, *74*, 230–243.

70. Citarella, A.; Moi, D.; Pinzi, L.; Bonanni, D.; Rastelli, G. *ACS Omega.* **2021**, *6*, 21843–21849.

5

ELECTRON DELOCALIZATION, AROMATIC CHARACTER, AND PERICYCLIC REACTIONS

By the mid-1960s, ionic and radical reactions were well studied and their mechanisms largely understood. There were a number of reactions that did not fit into either of these categories. One example is the Diels–Alder reaction, identified in the 1930s. This reaction is notable for its stereocontrol, implying a highly ordered transition state and a concerted mechanism. Another example, electrocyclic ring closure, was identified by R. B. Woodward in his work on the synthesis of vitamin B_{12}. These reactions and others like them were called "no-mechanism" reactions.

In a matter of a few years and many publications, patterns were identified that allowed these reactions to be described and largely understood. In this chapter, you will see reactions that occur when the reactant(s) contain(s) $4n + 2$ π electrons, and different reactions that occur when there are $4n$ π electrons. Recognition of these patterns has allowed accurate predictions with regard to stereochemistry and reaction conditions. These patterns are rationalized in terms of molecular orbitals.

Understanding Organic Chemistry: Bridging Fundamental and Advanced Courses, Fourth Edition. Ann M. Fabirkiewicz and John C. Stowell.

Companion website: www.wiley.com/go/fabirkiewiczorganic4e

5.1 MOLECULAR ORBITALS

Matter has a dual character, showing the properties of particles and of waves. Diffraction of electromagnetic radiation gives patterns of interference and reinforcement caused by overlapping waves that are either out of phase or in phase, establishing the wave nature of radiation. Similar patterns of diffraction have been produced by directing a beam of electrons through a crystal of nickel, showing the wave nature of electrons. Electrons confined by their electrostatic charge to the very small volume of an atom or molecule are best described as standing waves, rather than as particles. Each standing wave is approximated mathematically by a wavefunction known as the Schrödinger equation, which gives the shape and energy of the wave. The higher the energy state of the electron, the more nodes in the wave. The square of the wavefunction at any location around the nucleus is directly proportional to the electron density at that location.

For a p orbital, the wavefunction values are high at two points on opposite sides of the nucleus and diminish with distance from these points, to zero at a plane through the nucleus or at infinite distance from the atom. A plot of the wavefunction for the $2p_x$ orbital is shown in Figure 5.1a. The sign of the function (phase of the wave) is opposite on each side of that zero (nodal) plane through the nucleus. The square of this wavefunction is the electron density function, and indicates the relative electron density, or the probability of finding the electron, along that axis. A contour plot, as shown in Figure 5.1b, gives the traditional dumbbell shape associated with the p_x orbital. When representing such orbitals in discussions of bonding, a simplified form, Figure 5.1c, is common [1].

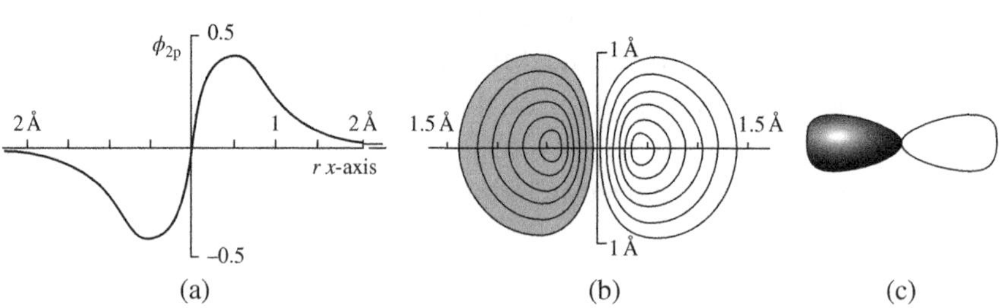

FIGURE 5.1 The $2p_x$ orbital. (a) Wavefunction for a $2p_x$ orbital. (b) Contour plot for the wave function. (c) Common representation. Reproduced with permission of Fleming [1]/John Wiley & Sons.

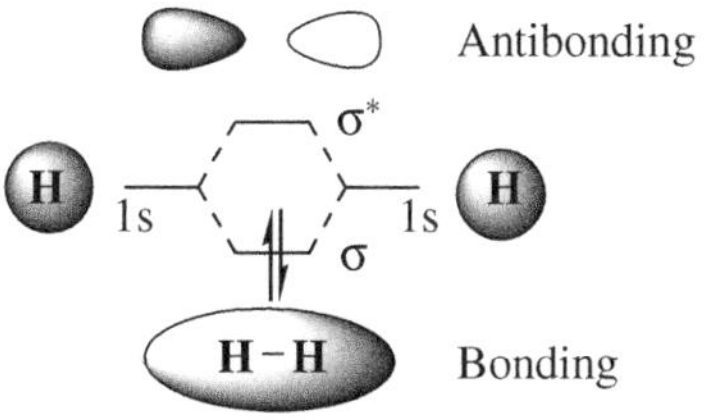

FIGURE 5.2 The bonding and antibonding orbitals for H_2.

Molecular orbitals are formed when two atomic orbitals overlap. As an example, consider the overlap that occurs between the s orbitals when a molecule of hydrogen forms. Overlap occurs in two ways: with reinforcement in the space between the nuclei (σ orbital) or with interference between the nuclei (σ^* orbital), illustrated in Figure 5.2. You can see that the electron density between nuclei in the σ^* orbital would be very small and the nuclei would repel each other, thus, this molecular orbital is termed antibonding. However, the electron density is largely concentrated between the nuclei in the σ orbital, and the nuclei are attracted to it and, therefore, together. This is the bonding molecular orbital.

In ethylene, besides the σ and σ^* orbitals, there are π molecular orbitals described as laterally overlapping p orbitals. Here again, there are two ways of overlapping: with reinforcement (π) and with interference (π^*) in the space above and below the nodal plane. A conventional cross-sectional representation of these is shown in Figure 5.3, where opposite phases are indicated by differences in shading. The bonding pair of electrons resides in the lower-energy π orbital, where the attractive forces effectively bond the carbon atoms together.

The change in energy when one electron in an isolated carbon p orbital spreads to occupy an ethylene π orbital is called β. The π-bonding

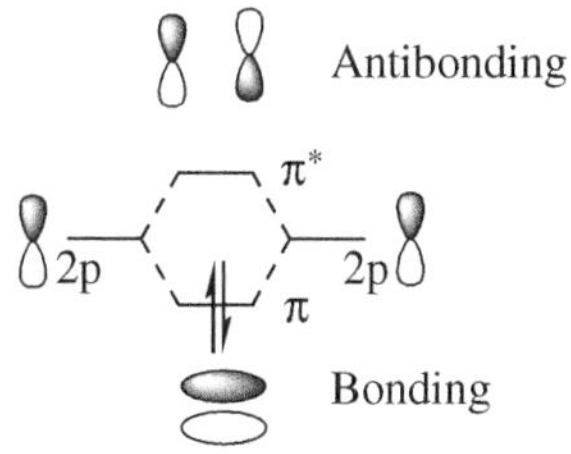

FIGURE 5.3 The bonding and antibonding π orbitals for ethylene.

energy, E_π, of ethylene with two electrons is 2β according to the Hückel approximation, where the interelectron repulsion is neglected. This energy gap is approximately −250 kJ/mol for ethylene and has been confirmed by ultraviolet (UV)-spectroscopy [2].

Conjugated systems consist of three or more p orbitals, where the bonding energy is more than β for each electron in the lowest bonding molecular orbital. As more p orbitals are mixed in a linear array, a matching number of π molecular orbitals are formed and the lowest energy orbital is lower, approaching a minimum of 2β per electron (Fig. 5.4). Each molecular orbital, Ψ, is composed of portions of the atomic p orbitals, ϕ, and the extent to which each atomic orbital contributes to each molecular orbital is given as a coefficient, c. This is called a linear combination of atomic orbitals and can be expressed mathematically by Equation 5.1, in which two atomic orbitals are considered. The square of the coefficient is

$$\Psi = c_1\phi_1 + c_2\phi_2 \qquad (5.1)$$

the electron density at that atom if one electron occupies the molecular orbital with that coefficient.

In butadiene, the sum of the squares of all the coefficients on carbon 1 (or any other carbon in Figure 5.4) is one atomic orbital. Likewise, the sum of the squares of the coefficients of the four atoms contributing to one molecular orbital is also one. In each set of molecular orbitals shown in Figure 5.4, the lowest (ψ_1) has no nodes (beyond the one in the plane of the nuclei), the next-higher (ψ_2) has one node, ψ_3 has two, and so forth. The nodes occur symmetrically about the center of the system. Electrons in the lowest orbital will bond all the atoms together. Where a node occurs between a pair of carbons, antibonding occurs for that pair if the orbital is occupied. Where a node occurs at an atom, the coefficient is zero, and there is a nonbonding relationship with the flanking carbons. For simplicity, all the contributions of atomic orbitals are drawn the same size and in a distorted, small, narrow shape in the figure. The net bonding molecular orbitals are shown below the zero-energy line and antibonding above on a scale of energy in units of β. The π-bonding energy per electron for each molecular orbital is also given in units of β.

In Figure 5.5, the butadiene molecular orbitals are represented with the size of the atomic orbital contributions proportional to the coefficients.

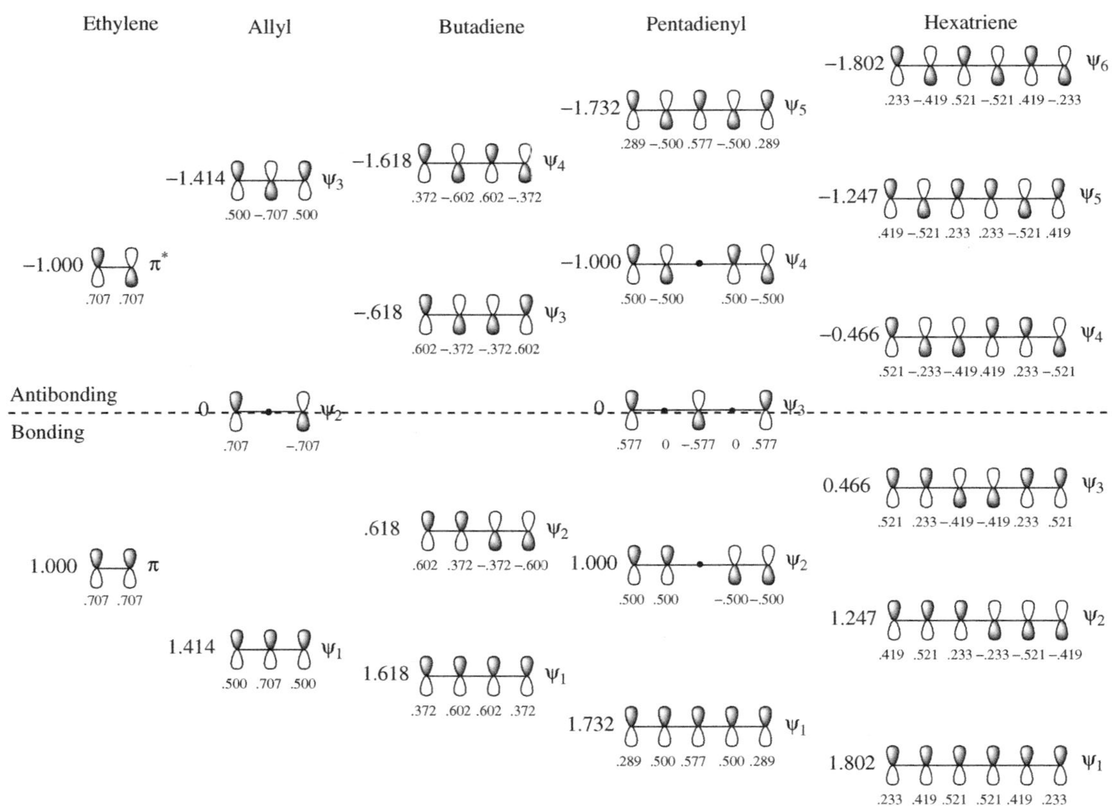

FIGURE 5.4 Molecular orbitals of linear C-2–C-6 π systems. The π-bonding energy per electron in each molecular orbital is given to the left in units of β. The orbitals are arranged on an energy scale between the values of + and −2β. The values are from the Hückel approximation.

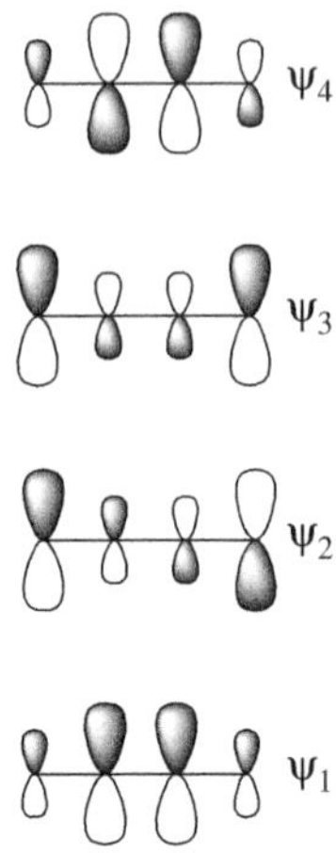

FIGURE 5.5 Butadiene π molecular orbitals.

The shapes approximate a series of waves. Neutral butadiene contains four electrons, which populate ψ_1 and ψ_2. Those in ψ_1 are bonding throughout and particularly so between C-2 and C-3, where the coefficients are large. Those in ψ_2 bond C-1 to C-2 and C-3 to C-4, but give antibonding between C-2 and C-3. This antibonding is relatively weak owing to the small coefficients at C-2 and C-3. The net π bonding in butadiene is greater than in a pair of isolated ethylenes, which is in accord with the observation of greater thermodynamic stability in conjugated systems compared to nonconjugated. This is observed experimentally in the lower heat of hydrogenation of conjugated dienes compared to nonconjugated analogs and in the selective formation of conjugated dienes in elimination reactions.

Using the π-bonding energies per electron given in Figure 5.4, we find that two ethylenes with two electrons each have less bonding energy (4β) than does butadiene with four electrons in ψ_1 and ψ_2 ($2 \times 1.618\beta + 2 \times 0.618\beta = 4.472\beta$). The additional 0.472β, the delocalization energy, is the value of conjugation. The orbitals above the zero β line are net antibonding, and if they were occupied, the molecule would weaken.

The electrons in the highest occupied molecular orbital (HOMO) are the most easily moved to new bonding relationships and can be considered the molecular valence electrons [3]. Just as we can write electron configurations for atoms, successively filling progressively higher orbitals with electron pairs, so we can use the molecular orbitals established here to

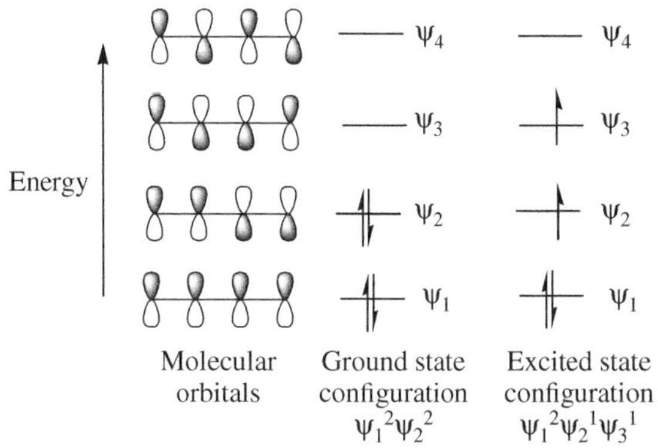

FIGURE 5.6 The electron configuration of butadiene.

provide electron configurations for molecules. The electron configuration for butadiene is shown in Figure 5.6 for both the ground state and the excited state of the molecule.

The systems containing an odd number of conjugated carbons are reactive intermediate carbocations, radicals, and carbanions. The allyl carbocation has two electrons in $\psi_{1,}$ where the coefficient is high on the central carbon; therefore, the positive charge (electron deficiency) will be on the first and third carbons. The allyl radical contains three π electrons evenly spread to all three carbons (no charges). The allyl anion contains four π electrons with the negative charge distributed evenly to the first and third carbons, as the structure of ψ_2 implies. The electron configurations are shown in Figure 5.7.

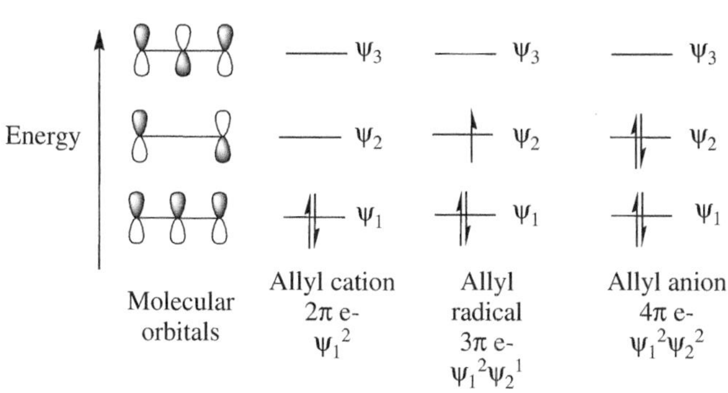

FIGURE 5.7 Electron configurations for the allyl system.

The charge at each carbon can be calculated by taking the squares of the coefficients at that carbon and multiplying each square by the number of electrons in that molecular orbital. Summing these for the orbitals that are occupied gives the total π electron density on that atom. If it is 1.00, the atom is neutral; if less than 1.00, the shortage is the plus charge; and if more than 1.00, the excess is the negative charge. This is shown for all carbons in the pentadienyl cation and anion in Table 5.1. The result is that the plus charge in the cation is evenly divided by 3, with 1/3 on each of the alternating carbons 1, 3, and 5. The anion is divided in the same way. This is the first alternation rationalized by molecular orbitals.

In Section 5.3, reactions of polyenes are examined with particular attention to the phase relationships in the HOMO and lowest unoccupied molecular orbitals (LUMO). Anticipating this, notice that for the neutral polyenes the HOMO is a row of alternating bonding and antibonding relationships whatever the length (shaded 2 up, 2 down, 2 up, etc.), and that the C_{4n} have the opposite phase at the first and last carbons while the C_{4n+2} have the same phase at the first and last carbons.

5.2 AROMATIC CHARACTER

The lowest molecular orbital in the linear polyenes approaches a π-bonding energy of 2β as the chain becomes longer and the end carbons become a smaller fraction of the conjugated system. If the system is cyclic and conjugated all the way around, there are no ends, and the lowest energy becomes 2β for all ring sizes. As with acyclic cases, there are as many molecular orbitals as there were contributing atomic orbitals. Unlike acyclic compounds, however, there are degenerate (equal-energy) pairs of molecular orbitals above the lowest level. These orbitals are diagrammed on a π-bonding energy scale in Figure 5.8 for rings with up to eight members.

Each vertical step in these orbitals involves an additional nodal plane, which cuts the ring at two sites. In the ground state, the electrons populate the lower energy orbitals following Hund's rule. For example, there are four π electrons in cyclobutadiene. The lowest energy pair gives bonding among all the carbons (Fig. 5.8, ψ_1, cyclobutadiene). However, the next two electrons are in the nonbonding orbitals ψ_2 and ψ_3, contributing nothing to the stability of the molecule. In the tricyclic dimer of cyclobutadiene,

TABLE 5.1 Calculation of the Charge on Each Atom of the Pentadienyl Cation and Anion

	C-1	C-2	C-3	C-4	C-5	
ψ_3	0.577	0	−0.577	0	0.577	Coefficients from Figure 5.4
ψ_3^2	0.333	0	0.333	0	0.333	Probability
$2\psi^2$	0.666	0	0.666	0	0.666	
ψ_2	0.500	0.500	0	−0.500	−0.500	Coefficients from Figure 5.4
ψ_2^2	0.250	0.250	0	0.250	0.250	Probability
$2\psi_2^2$	0.500	0.500	0	0.500	0.500	
ψ_1	0.289	0.500	0.577	0.500	0.289	Coefficients from Figure 5.4
ψ_1^2	0.0835	0.250	0.333	0.250	0.0835	Probability
$2\psi_1^2$	0.167	0.500	0.667	0.500	0.167	
$2\psi_1^2 + 2\psi_2^2$	0.667	1.000	0.667	1.000	0.667	Electron density for cation
$1.00 - 2\psi_1^2 + 2\psi_2^2$	+0.333	0	+0.333	0	+0.333	Charges on cation
$2\psi_1^2 + 2\psi_2^2 + 2\psi_3^2$	1.333	1.000	1.333	1.000	1.333	Electron density for anion
$1.000 - (2\psi_1^2 + 2\psi_2^2 + 2\psi_3^2)$	−0.333	0	−0.333	0	−0.333	Charges on anion

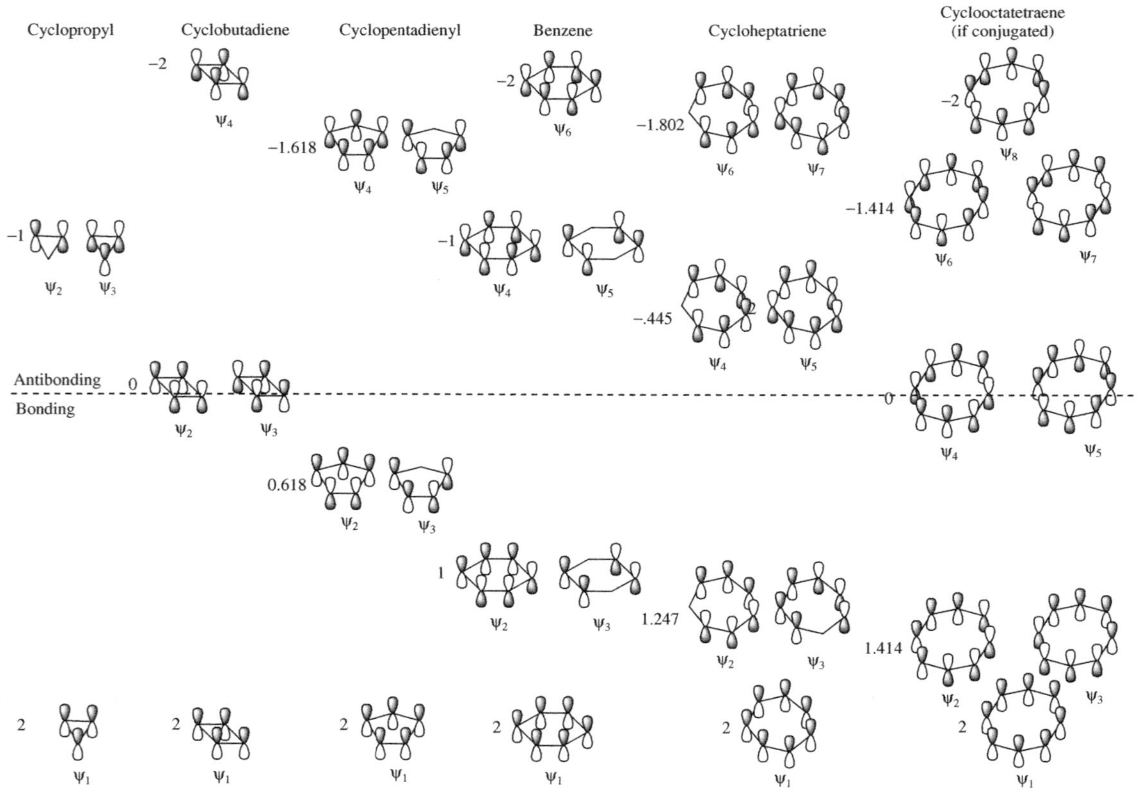

FIGURE 5.8 Cyclic π molecular orbitals for three- to eight-membered rings. These are arranged on a vertical scale of π-bonding energy per electron from −2β to +2β. The energy per electron is given to the left of each orbital representation. Neutral cyclooctatetraene is not planar and does not have the π molecular orbital arrangement shown.

all electrons are in bonding orbitals; thus, cyclobutadiene is an unstable transient molecule that rapidly dimerizes (Eq. 5.2) [4]. In benzene, all the π electrons are in orbitals of net bonding, giving a stable system.

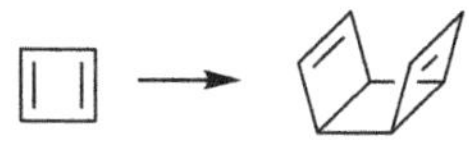
(5.2)

In cyclooctatetraene (if it were conjugated), two electrons would occupy nonbonding orbitals. This again would be unstable, and it is relieved by not conjugating. The molecule is tub-shaped, and the π electrons are all in bonding "ethylene" orbitals, avoiding the angle strain present in a planar octagonal ring. These circumstances alternate through the cyclic vinylogous series, continuing to larger rings. Those with $4n + 2$ π electrons have the highest occupied energy level filled with four electrons (closed-shell occupation) and, if these are all bonding orbitals, they have greater stability than found in the acyclic cases. The greater stability in the $4n + 2$ series is called aromatic character and is manifest in numerous physical and chemical properties.

Among the odd-membered rings, aromatic ions are readily prepared. Cyclopentadiene is deprotonated by alkoxide bases while cycloheptatriene is not, even with stronger bases. On the other hand, bromocycloheptatriene is ionic while 5-bromocyclopentadiene is not. Tripropylcyclopropenyl perchlorate exists largely as the carbocation in aqueous acetonitrile at pH ≤ 7 [5]. Electron configurations for the cyclopentadienyl anion, benzene, and the cycloheptatrienyl cation are shown in Figure 5.9. For simplicity, the molecular orbitals are represented by horizontal lines.

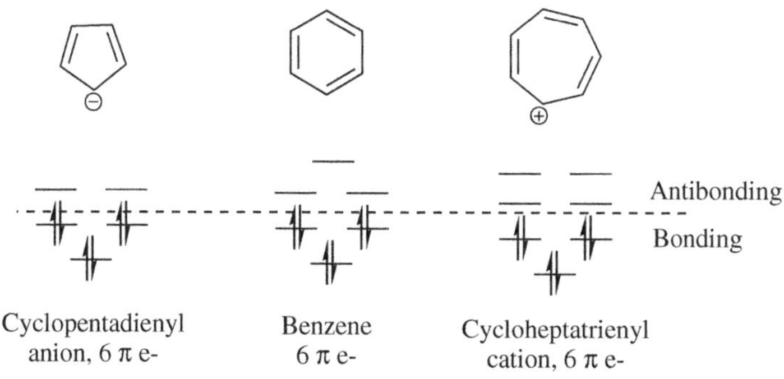

FIGURE 5.9 Some aromatic π systems.

Molecules with $4n$ π electrons in a ring of p orbitals have a partially filled energy level, and the four- and five-membered rings have less stability than the acyclic analogs. They are called antiaromatic [6]. The stability differences can be seen by calculating and comparing the π bonding energy for cyclic and acyclic compounds with the same number of π electrons using the orbital energy values in Figures 5.4 and 5.8. The $(CH)_{16}$ and $(CH)_{24}$ rings have been prepared and show the opposite ^1H NMR chemical-shift effects from those found for the $4n + 2$ compounds [7].

At the lowest energy level in carbocyclic compounds, there is a single orbital, and above this, the degenerate pairs begin. Thus, with $4n + 2$ electrons, the 2 fills the lowest level and the $4n$ fills succeeding levels. Other numbers give incomplete levels. Those monocyclic aromatic systems where n is greater than 1 are not nearly as chemically resistant as benzene, although some of them undergo electrophilic substitution, and some show large nuclear magnetic resonance (NMR) shifts. For example, the outer protons on the $(CH)_{18}$ ring appear at 8.8 ppm, and the inner protons are at −1.8 ppm. Fused-ring polycyclic compounds with $4n + 2$ π electrons show aromatic character, but there is less stabilization per ring than in benzene [8].

Some heterocyclic compounds show aromatic character despite their lack of degenerate orbitals. This may involve one π electron from the heteroatom, as in pyridine, or two, as in pyrrole and furan, making a set of six π electrons in a ring [9]. Many fused-ring heterocyclic compounds are aromatic also. A ten π electron monocyclic example, 1,4-dioxocin, does not show aromatic character [10].

The orbitals described thus far are called Hückel molecular orbitals. As seen in Figure 5.8, ψ_1 of benzene has no nodal planes perpendicular to the plane of the ring, while ψ_2 and ψ_3 each have one nodal plane perpendicular to the ring. Such nodal planes intersect the ring in two places called nodal zones. Now imagine opening a benzene ring, giving it a half twist, and rejoining it (Fig. 5.10). The lowest π molecular orbital would

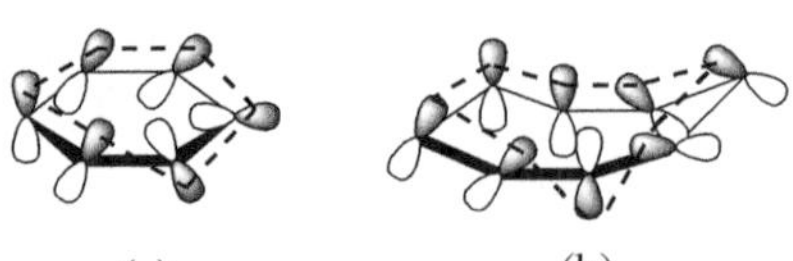

(a) (b)

FIGURE 5.10 (a) Möbius benzene and (b) Möbius cyclooctatetraene.

(11)

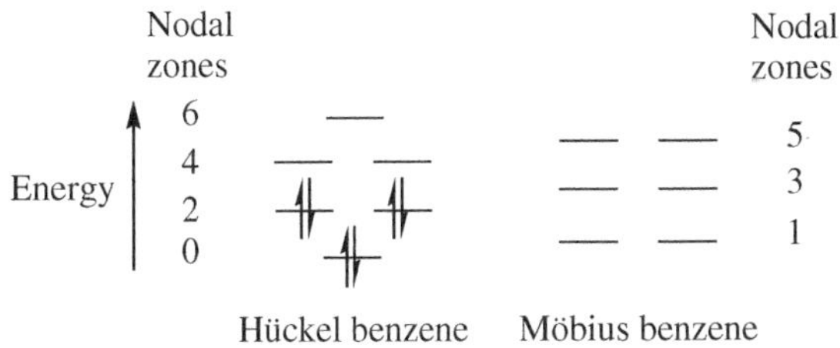

FIGURE 5.11 Benzene π molecular orbitals.

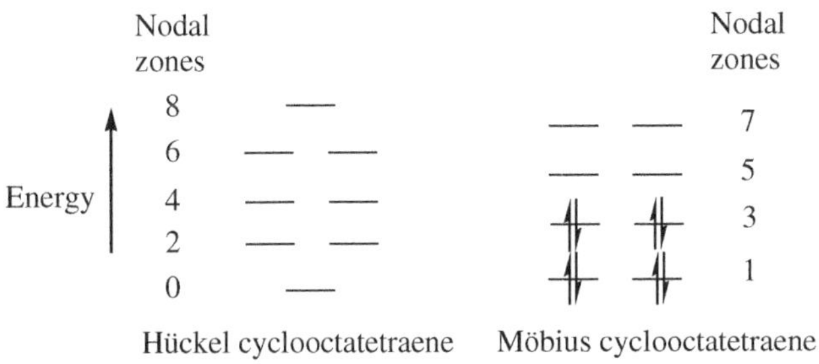

FIGURE 5.12 Cyclooctatetraene π molecular orbitals.

now have one nodal zone in it. Of the next higher pair of molecular orbitals, one would gain a nodal zone and one would lose one, and likewise for the remaining pairs. The entire set would now be degenerate pairs (Fig. 5.11), and a closed shell would require $4n$ electrons. Six electrons in Möbius benzene would give four bonding and two nonbonding and an incomplete shell. With eight electrons, Möbius cyclooctatetraene can now be neutral and aromatic (Fig. 5.12).

Actually, a twisted benzene or cyclooctatetraene ring would have very poor overlap from both lobes of each p orbital, but research into "Möbius-aromatic" molecules continues [11]. This is an alternative approach, rationalized as having closed-shell occupation of degenerate bonding molecular orbitals.

5.3 PERICYCLIC REACTIONS

Numerous organic reactions proceed through a cyclic transition state in a single step involving no intermediates. These are called pericyclic reactions [12–14]. Certain bonds break while others form, in concert.

These reactions are divided into two groups, those with $4n + 2$ electrons and those with $4n$ electrons. Taking stereochemical choices into account, examples from one group have been observed under thermal conditions, while examples from the other group are found only under photochemical conditions. In the thermally observable reactions, the phase relationships of the combining orbitals are such that the bonding electrons maintain bonding character from starting materials, through the transition state, and in the products. These concerted processes are therefore favored over nonconcerted ones, where bonds must first be broken to give radicals or other intermediates, which subsequently give new bonds. A smaller activation energy is necessary when the exothermal formation of the new bonds is underway during the endothermal breaking of the original bonds. These concerted reactions are stereospecific, while stepwise reactions involving preliminary breakage of bonds often lose stereochemical integrity.

Selection rules were put forth by Woodward and Hoffmann based on the conservation of orbital symmetry [15]. Molecular orbitals in the starting material are correlated with orbitals in the product according to their symmetry elements. If all the ground-state occupied molecular orbitals in the starting material correlate with ground-state occupied molecular orbitals in the product, the reactions will be allowed thermally. If, instead, a ground-state occupied orbital in the starting material correlates with an antibonding orbital in the product, photochemical excitation will be necessary. In photochemically allowed reactions, the lowest excited state of the starting material correlates with the lowest excited state of the product. It is the HOMO that is pivotal in these choices.

Fukui examined the interaction of the HOMO and LUMO alone (the frontier orbitals) and rationalized the same rules [1, 16]. Basically, each reaction is viewed as the coalescing (or dissociation) of two sets of molecular orbitals intra- or intermolecularly. The HOMO of one reactant is matched with the LUMO of the other, and if the overlap at both sites of projected new bond formation between them is in-phase (a bonding overlap), the reaction is allowed.

Dewar [17] and Zimmerman [18] proposed selection rules based on the cyclic transition state. A Hückel transition state is one in which there are zero or an even number of phase transitions perpendicular to the plane of the reaction, and is the allowed condition for thermal cases in which there are $4n + 2$ electrons. These reactions occur in a suprafacial manner. A Möbius transition state, in which there is one or an odd number of phase transitions perpendicular to the plane of the reaction, predicts thermally allowed reactions when there are $4n$ electrons involved. These reactions have an antarafacial component. The advantage of this transition state analysis is that all types of concerted reactions are covered by basically one selection rule. If a continuous loop of overlap through all the atoms involved in the transition state can be drawn with no nodes, the reaction will be allowed thermally if it involves $4n + 2$ electrons. If the continuous loop requires one nodal zone, the reaction will be thermally allowed if it involves $4n$ electrons. Photochemical reactions are the opposite. The rules governing these reactions are summarized in Table 5.2.

All these analyses view the reactions as reversible, and allowedness applies in both directions. Other factors, such as ring strain, steric hindrance, and bond energies, determine the energetically favorable direction.

5.3.1 Cycloaddition Reactions

A cycloaddition reaction [19] is the joining together of two independent π-bonding systems to form a ring with two new σ bonds. The reverse is called a retrocycloaddition reaction, and the selection rules apply in both directions of a given reaction.

If butadiene and an appropriately substituted ethylene approach and begin to overlap as in Equation 5.3, there is a favorable phase relationship using the HOMO of the diene and the LUMO of ethylene (the frontier molecular orbitals) for a face-to-face joining. This is, of course, the familiar Diels–Alder reaction, and it is thermally allowed. With respect to both components of the reaction, the reaction occurs from the same "face," termed suprafacial addition. Because there are four π electrons in butadiene and two π electrons in ethylene, the Diels–Alder reaction is named as a $[_{\pi}4_{s} + _{\pi}2_{s}]$ reaction. The stereochemical consequences of this approach are further illustrated in Section 8.6.

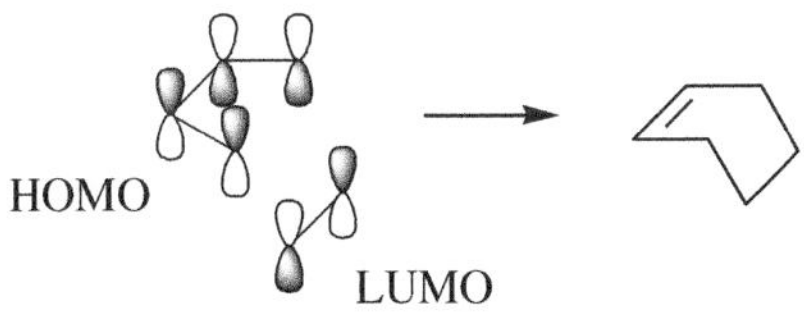

(5.3)

The energy gap of 1.618β (Fig. 5.4) between the HOMO of butadiene and the LUMO of ethylene (or the LUMO of butadiene and the HOMO of ethylene) is too great for favorable interaction, and no reaction occurs. Substituents can narrow this gap. Electron-withdrawing substituents lower the energy level of LUMOs and electron-donating substituents raise the energy level of HOMOs. Thus, the lower-energy LUMO in maleic anhydride is close enough in energy to the HOMO of butadiene for favorable reaction [20]. An electron-rich alkene such as ethyl vinyl ether has a higher-energy HOMO than ethylene and will react with electron-poor dienes as in Equation 7.40. Here the closer frontier orbitals are the LUMO of the diene and the HOMO of the dienophile. These are called reverse-electron-demand reactions. Either way, the phase relationships predict a thermally allowed reaction.

In unsymmetric cases, the regiochemistry may be predicted from the frontier orbital coefficients at the joining atoms [21]. An electron-donating substituent on carbon 1 of a diene or alkene will enlarge the coefficient of the HOMO at the remote end. An electron acceptor group on carbon 1 will enlarge the coefficient of the LUMO at the remote end. In the reaction, the greater orbital overlap interaction occurs when the atoms with larger coefficients overlap with each other and those with the smaller coefficients overlap with each other, as opposed to overlaps of smaller with larger. Thus, in Diels–Alder reactions where an electron donor is on carbon 1 of the diene and an electron acceptor is on carbon 1 of the alkene, the donor and acceptor will selectively appear on adjacent cyclohexene ring atoms (Eq. 5.4).

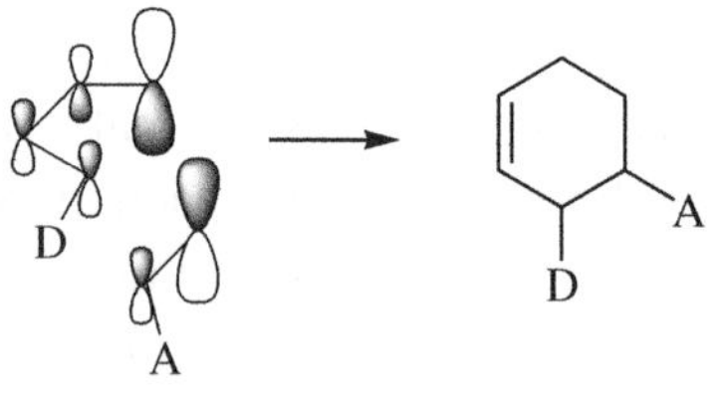

(5.4)

The same is true when the donor and acceptor are on alkene and diene, respectively. A donor on carbon 2 of a diene enlarges the HOMO coefficient at carbon 1, and an acceptor on carbon 2 enlarges the LUMO coefficient at carbon 1. This favors a 1,4-disubstituted cyclohexene product. The increase in regioselectivity brought about with Lewis acid catalysis is a result of coordination of the catalyst with a receptor substituent, making it a stronger receptor.

Photons generally excite an electron from the HOMO to the next-higher molecular orbital. This higher orbital was the LUMO, but it becomes the highest singly occupied molecular orbital (HSOMO). In butadiene, for example, the ground-state HOMO was ψ_2, but after photoexcitation, the HSOMO is ψ_3 (Fig. 5.6). The phase relationship of the HSOMO will be the opposite of the former HOMO since there is one more node; therefore, the stereochemistry will be opposite that of the ground-state thermal reactions. In the case of a photochemical 4 + 2 cycloaddition, the relevant orbitals include the HSOMO of the diene and the LUMO of the dienophile (or the HSOMO of the dienophile and the LUMO of the diene), which are shown in Equation 5.5. Orbital overlap cannot occur for these orbitals in an s + s fashion, so the reaction is symmetry forbidden. Orbital overlap is theoretically possible considering a suprafacial approach of the dienophile and an antarafacial approach of the diene, but such a reaction is geometrically impossible in a developing six-membered ring. In fact, photochemical Diels–Alder reactions are rarely observed, and nonconcerted mechanisms predominate in those cases.

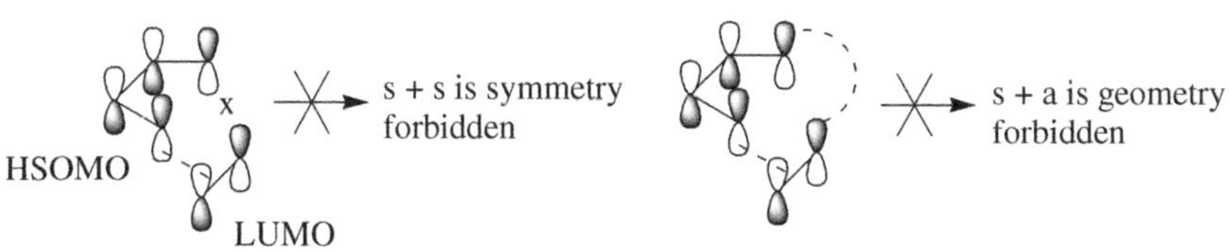

(5.5)

Turning now to cycloadditions in which there are four π electrons, we can see that if two ethylenes are brought together suprafacially, the HOMO—LUMO phase relationship is unfavorable for bond formation (Eq. 5.6), thus the $[_{\pi}2_s + {}_{\pi}2_s]$ reaction is symmetry forbidden in the thermal case.

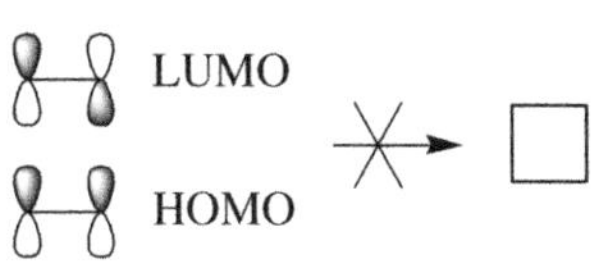

(5.6)

Under photochemical conditions, however, this is a valuable route to cyclobutanes (Section 8.4). Photochemical excitation of one of the ethylene components allows access to the HSOMO, and the phase relationship of the HSOMO with the LUMO predicts bonding as in Equation 5.7. The stereochemical implications of the predicted concerted reaction are supported by experimental evidence and shown in Equation 5.8 [22].

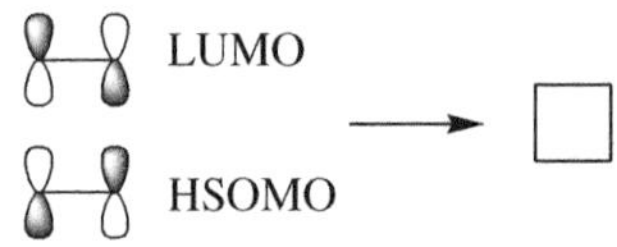

(5.7)

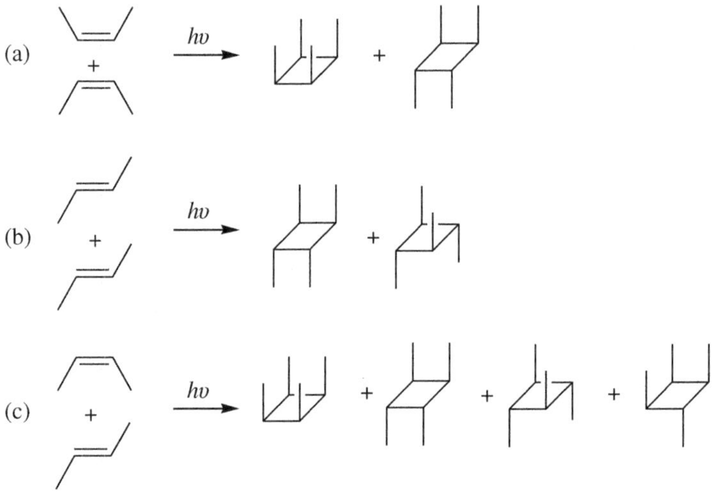

(5.8)

A triene and an ethylene combining face to face at their ends is again not thermally favorable (Eq. 5.9), as is a diene combining with a diene (Eq. 5.10), since the phase relations are unfavorable to bonding. These components may instead combine in a Diels–Alder reaction to give a vinylcyclohexene. Note that in both cases, there are eight ($4n$) π electrons.

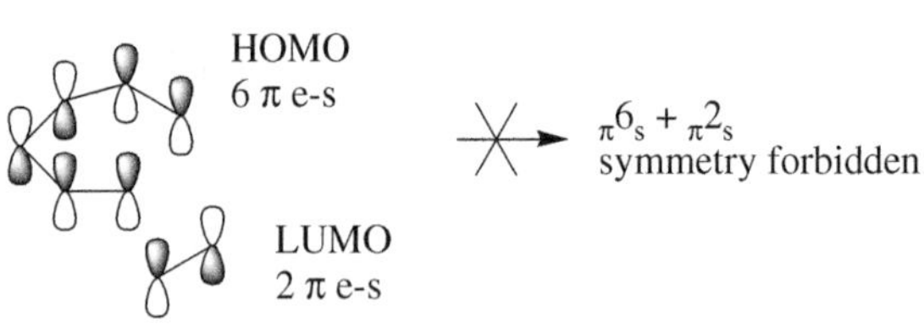

(5.9)

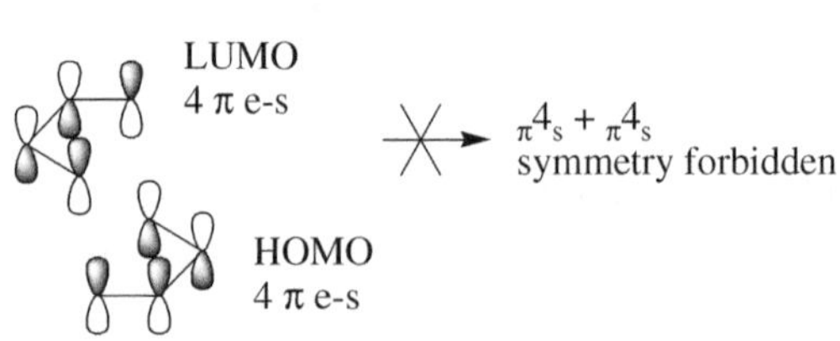

(5.10)

Going one step further, the combination of a diene and a triene (Eq. 5.11), or a tetraene and an ethylene, is favorable face to face. Note that in these reactions, there are a total of 10 π electrons ($4n + 2$). A specific example is shown in Equation 5.12 [23].

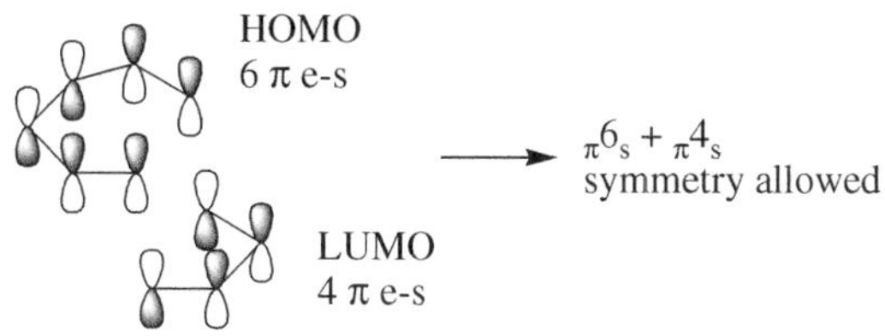

(5.11)

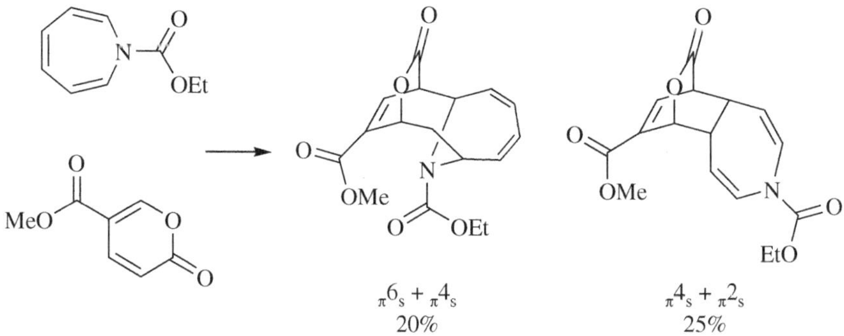

(5.12)

It was stated earlier that antarafacial addition with respect to either component in a cycloaddition reaction can be geometrically difficult or impossible to accomplish. If one of the reactants has a large enough ring system, however, the reactions can occur. For example, heptafulvalene combines with tetracyanoethylene this way (Eq. 5.13) (cited in [15])).

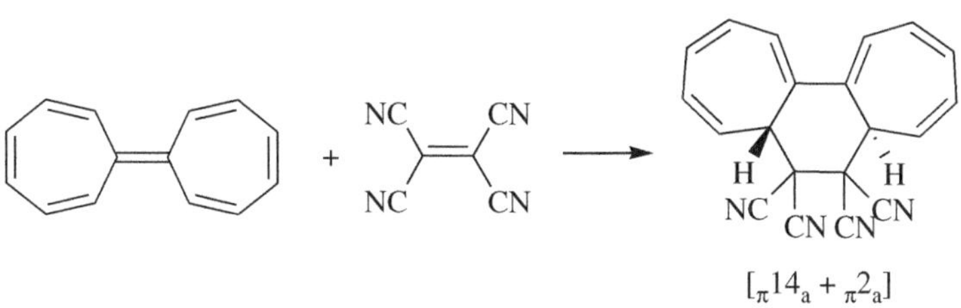

(5.13)

Heptafulvalene has 14 π electrons, and the face-to-face reaction with an ethylene at the carbons shown is thermally forbidden, but by attaching at the top face at one end and the bottom at the other, the opposite phase interaction is reached and the reaction is thermally allowed. On the tetracyanoethylene side, it is a suprafacial addition, but the reaction occurs antarafacially on the heptafulvalene side. The process of Equation 5.13 is abbreviated $[_{\pi}14_{a} + {}_{\pi}2_{s}]$. Antarafacial cycloadditions occur in polyenes that are highly strained or twisted to allow such access.

5.3.2 Electrocyclic Reactions

An electrocyclic reaction [24] is the closure of a conjugated polyene to give a cyclic compound with one less π bond, or the reverse. For the first example, consider the thermal cyclization reaction of *E,Z,E*-2,4,6-octatriene to *cis*-5,6-dimethylcyclohexadiene (Eq. 5.14). A σ bond forms and new π bonds develop concurrently.

(5.14)

There are six π electrons in this system and the reaction occurs via the HOMO of the triene (Fig. 5.4, ψ_3, hexatriene) as shown in Equation 5.15. In order for the new σ bond to form, the orbital at C-2 must rotate 90° counterclockwise, while the orbital at C-7 must rotate clockwise. This is

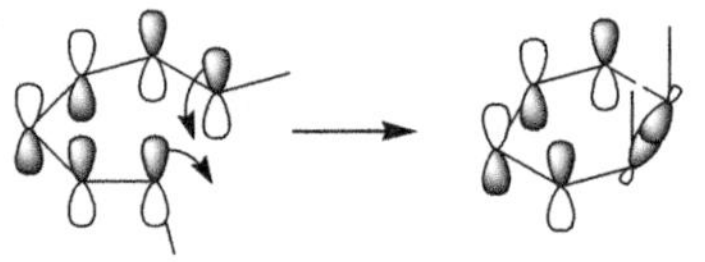

(5.15)

called disrotatory motion and results in the stereochemistry observed in the cyclization of the isomers of 2,4,6-octatriene [25]. The *E,Z,E* isomer gave only the *cis* ring compound (Eq. 5.14). The *Z,Z,E* and *Z,Z,Z* isomers were in thermal equilibrium, but rate measurements indicate that the *Z,Z,E* isomer likely gave the *trans* ring compound (Eq. 5.16).

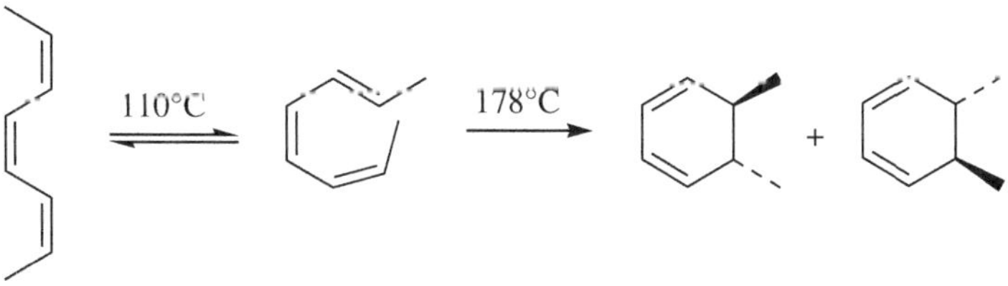

(5.16)

The choice of phase indication in the molecular orbital is arbitrary, and it is equally valid to choose to rotate together the unshaded orbitals in Equation 5.15, that is, 90° clockwise rotation at C-2 paired with 90° counterclockwise rotation at C-7. Both choices are equally valid for consideration of allowedness and both indicate disrotatory motion, but each may lead to different stereoisomeric products. Two allowed motions should be considered in all pericyclic reactions. Which of the two disrotatory motions occurs will be determined in most cases by steric or electronic effects among the substituents, or both may occur. Conrotation, in which both orbitals rotate in the same direction, is forbidden for this reaction because it leads to antibonding overlap.

Electrocyclic reactions can be labeled with the suprafacial and antarafacial terminology used for cycloadditions. The disrotatory opening of a cyclohexadiene has the front lobes on both ends (or back lobes on both ends) of the σ bond attaching to the same face of the π system, and this is a ${}_{\sigma}2_{s}$ process (Eq. 5.17).

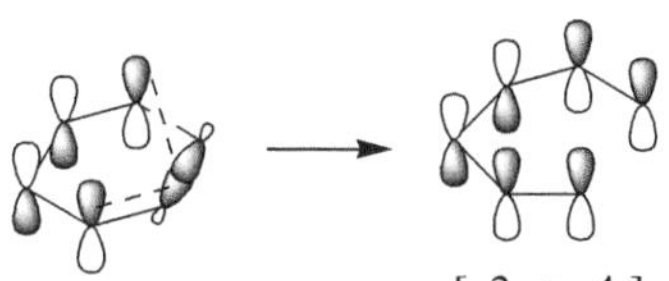

$[{}_{\sigma}2_{s} + {}_{\pi}4_{s}]$

(5.17)

In electrocyclic reactions, there are two ways to abbreviate each process because the π system could be considered antarafacially instead. Note that Equation 5.17 could be abbreviated $[{}_{\sigma}2_{a} + {}_{\pi}4_{a}]$. Both s designations are reversed, but the physical meaning remains the same. In considering these reactions, it is also important to note that whenever a σ bond is broken in an antara fashion in a concerted reaction wherein both carbons are

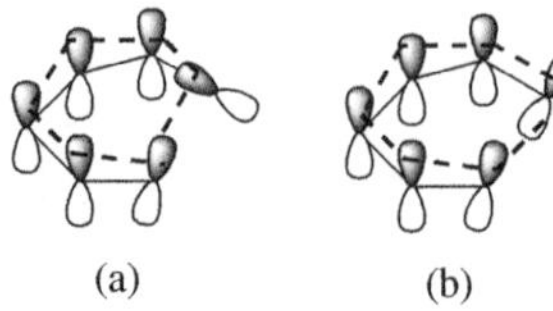

FIGURE 5.13 Transition states for closure of hexatriene: (a) disrotatory, allowed and (b) conrotatory, forbidden.

sp^3-hybridized at the finish, one end will be stereochemically inverted and the other retained. If the σ bond is broken in a supra fashion, both ends will be retained, or both ends will be inverted.

The electrocyclic ring closure of hexatriene (Eq. 5.13) can also be considered in terms of its transition state, shown in Figure 5.13a, where a complete loop of overlap is followed with a dashed line from atom to atom through all six. This, as drawn with a minimum of nodal zones, resembles the lowest molecular orbital of aromatic benzene since there are six electrons and no nodes. This transition state is favored by aromatic stabilization and is reached only by disrotation. Conrotation would have given a transition state with one nodal zone (Fig. 5.13b) that would be part of a Möbius orbital set, but six electrons do not give a closed shell in this set, and the transition state does not have aromatic stabilization and is not allowed.

Under photochemical conditions, we find the opposite results. In hexatriene, the ground-state HOMO was ψ_3 (Fig. 5.4), but after photoexcitation, the HSOMO is ψ_4. The photochemical reaction can be illustrated with the example in Equation 5.18 [26]. In order for the new σ bond to

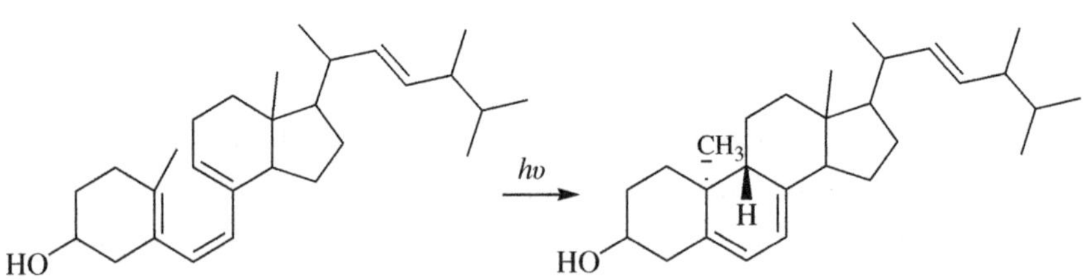

(5.18)

form, conrotatory motion of the termini is required (Eq. 5.19), and this is supported in the experimentally observed stereochemistry.

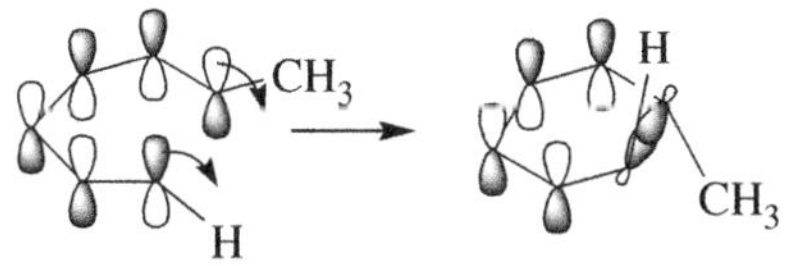

(5.19)

The ring opening of cyclobutenes can be analyzed in a similar fashion. This system contains four π electrons and occurs to the HOMO of the diene (Eq. 5.20).

(5.20)

In contrast to the thermal cyclohexadiene case, conrotatory motion is necessary here for in-phase overlap. As in the previous case, we can invert the phase indication of one of the molecular orbitals and see conrotatory motion in the other direction. Ring closure can be seen to require conrotatory motion in the HOMO to result in bonding in the product (Eq. 5.21).

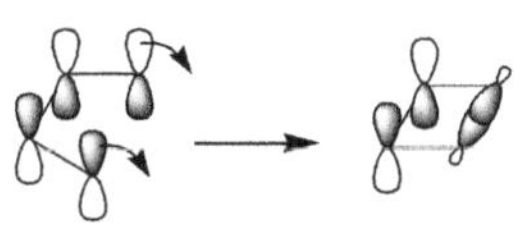

(5.21)

The predictions above have been borne out by experimental observations. Both directions of conrotation in the opening of *cis*-3,4-dimethylcyclobutene lead to *Z,E*-2,4-hexadiene (Eq. 5.22) [27]. The *trans*-3,4-dimethylcyclobutene might lead to the *E,E* and *Z,Z* dienes, but only the *E,E* was found.

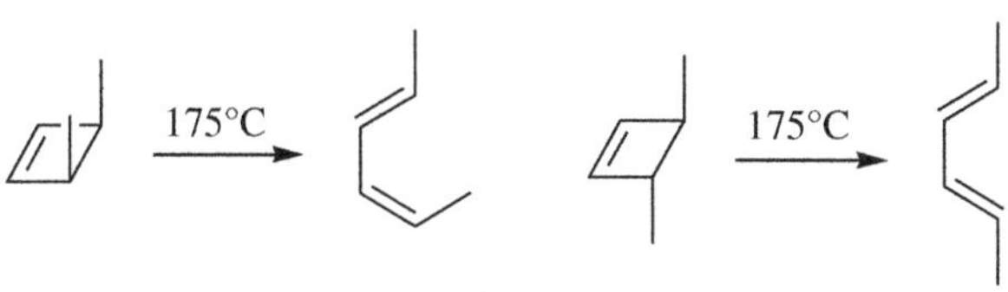

(5.22)

There are always two directions of conrotatory opening or closing to consider. In the opening of cyclobutenes, electron-donating substituents are strongly inclined toward outward rotation, while electron-withdrawing groups rotate inward [28]. Computational and experimental evidence indicate that the effect is electronic rather than steric and due to stabilization of the transition state by the orbitals of the substituent. The opening of 3-formylcyclobutene at 25–70°C, for example, gives exclusively *Z*-pentadienal, the product of inward rotation of the formyl group (Eq. 5.23) [29].

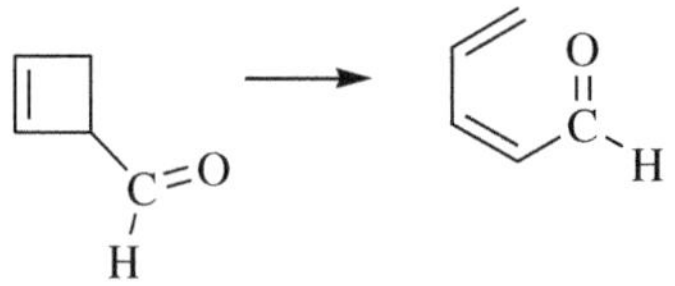

(5.23)

Equation 5.24 shows a reaction in which the outward preference of the donor methoxy group even overcomes the steric strain of a tert-butyl group rotating inward [30].

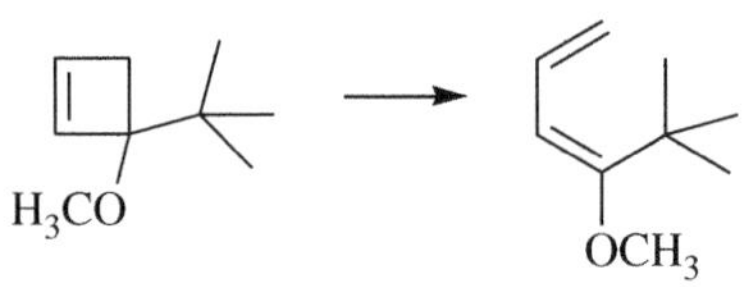

(5.24)

The thermal conrotatory opening of a cyclobutene has one face of a π bond attaching to a front lobe from a σ bond at one end and a back lobe at the other end, so the conrotatory motion is a $_{\sigma}2_{a}$ process. Equation 5.25 indicates the new bonding with dashed lines.

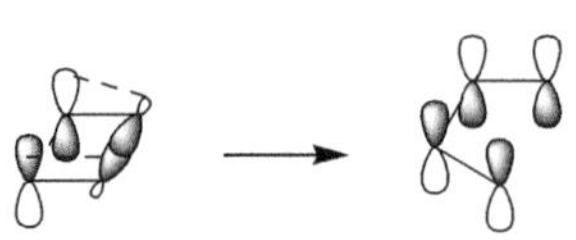

$[_{\sigma}2_{s} + _{\pi}2_{s}]$

(5.25)

The transition state for the closure (or the reverse) of a butadiene is shown in Figure 5.14. The disrotatory closure allows a loop with no nodes (a Hückel orbital), but four electrons cannot give a closed-shell occupation, and the transition state is forbidden. However, the conrotatory closure can be drawn with one nodal zone and therefore belongs to the Möbius group, where the four electrons will give a closed shell with aromatic stabilization; that is, this transition state is stabilized and allowed.

Photochemical conditions yield stereochemistry opposite to that of the thermal reactions. In butadiene, the ground-state HOMO is ψ_2 (Fig. 5.4),

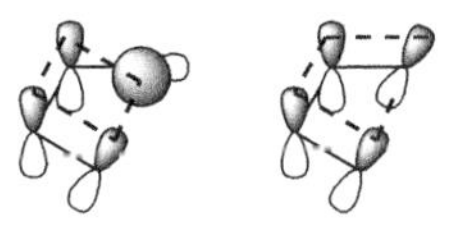

(a) (b)

FIGURE 5.14 Transition states for closure of a butadiene: (a) disrotatory, forbidden and (b) conrotatory, allowed.

but after photoexcitation, the HSOMO is ψ_3. These orbitals can be used to explain the disrotatory ring closure observed in 2-pyrone (Eqs. 5.26 and 5.27) [31].

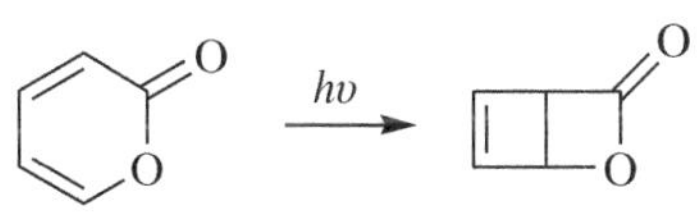

(5.26)

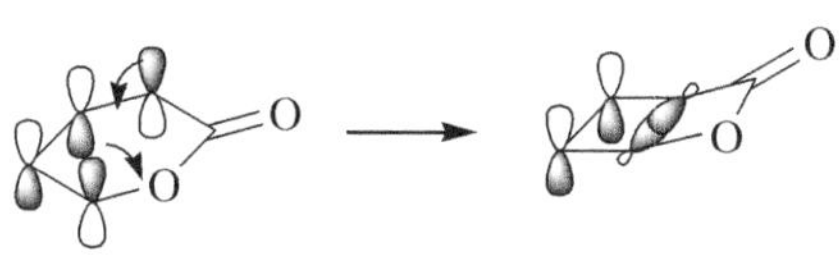

(5.27)

Structural constraints can prevent reactions that would be allowed by these rules. For example, the very strained bicyclo[2.2.0]hex-2-ene (Dewar benzene) opens slowly at 130°C to cyclohexadiene [32], while the less strained bicyclo[4.2.0]octa-2,4-diene is in rapid equilibrium with cyclooctatriene at 80–100°C [33]. The allowed conrotatory movement in the first case would lead to cyclohexadiene with a *trans* double bond, which is impossible. However, the latter opening is disrotatory, giving all-*cis* double bonds in the cyclooctatriene. The higher-temperature reaction, disallowed by these rules, is a higher activation energy process, perhaps involving radicals, but not a concerted mechanism.

Higher vinylogs continue the alternation. The closure (or opening) of $4n + 2$ systems occurs in a disrotatory way, and for $4n$ systems, conrotation is the thermally allowed process. The rules reverse for the photochemical reactions.

5.3.3 Sigmatropic Reactions

In a sigmatropic reaction [34], an allylic σ bond cleaves, and a new one forms further along the π chain as exemplified in Equations 5.28 and 5.29.

In Equation 5.28, the Cope rearrangement, a carbon three positions along

(5.28)

from the detaching one attaches to a site three positions along in the other chain. This is called a [3,3] sigmatropic shift. In Equation 5.29, a hydrogen detaches from the first carbon on the chain and reattaches to the fifth position. This is a [1,5] sigmatropic shift.

(5.29)

Consider first the [1,5] sigmatropic shift. The 1s orbital of the migrating hydrogen atom necessarily has only one phase and the reaction occurs from the HOMO of the pentadienyl system (Fig. 5.4, ψ_3). Examining the phase relationship of the HOMO with the migrating hydrogen, we find an in-phase overlap (Eq. 5.30) when the hydrogen reattaches to the same face of the π system, that is, a suprafacial shift. This is a thermally

(5.30)

allowed reaction. In contrast, a [1,3] suprafacial shift of a hydrogen is not allowed, as indicated by the antibonding relationship shown in Equation 5.31, considering the migrating hydrogen and the HOMO of the allyl system.

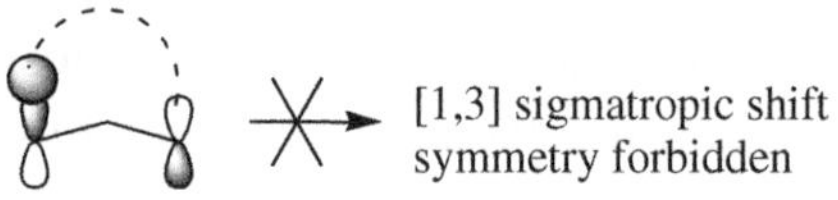

(5.31)

If the [1,3] migrating group is a carbon instead of a hydrogen, there is a back lobe of opposite phase, and the σ bond can migrate antarafacially to give in-phase overlap, but the migrating carbon will undergo inversion of configuration (Eq. 5.32). An example of this is the thermal rearrangement of the *exo*-bicyclo[2.1.1] compound in Equation 5.33.

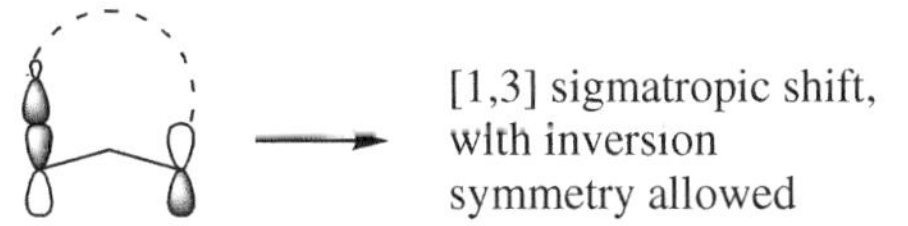

(5.32)

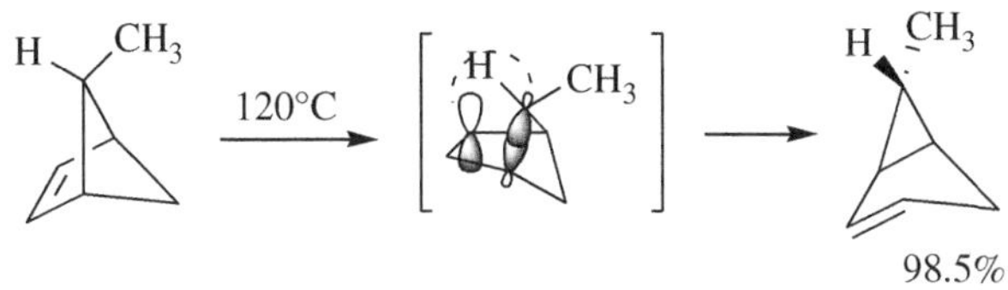

(5.33)

The endo isomer (Eq. 5.34) required a higher temperature and was not as clean in accord with the selection rules, but it gave mostly allowed products, including some migration of the CH_2 bridge. A proposed explanation involves consideration of the potential energy surfaces for these reactions [35].

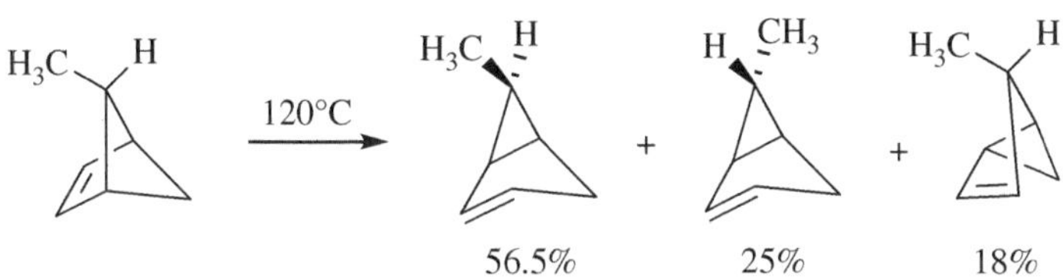

(5.34)

A concerted [1,5] shift of carbon is allowed suprafacially with retention of configuration, as shown in Equation 5.35 [36]. A concerted [1,7] shift of hydrogen is thermally allowed if it can reach the opposite face of the

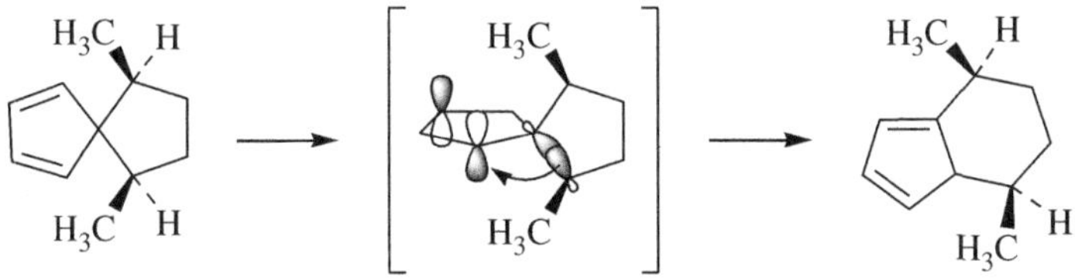

(5.35)

π system antarafacially, which is possible via a helical conformation (Fig. 5.15). An example is shown in Equation 5.36 [25].

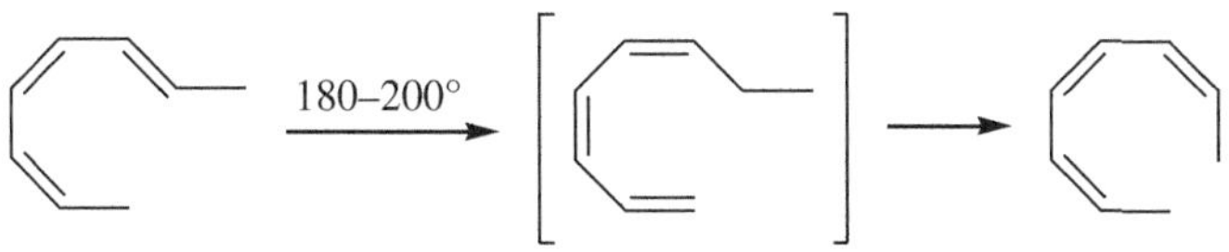

(5.36)

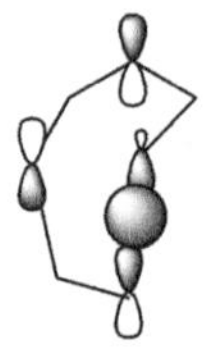

FIGURE 5.15 An antarafacial [1,7] sigmatropic hydrogen shift.

The [3,3] shift of Equation 5.28 can be considered as the sum of its parts: a σ bond and two π bonds, as shown in Equation 5.37. Evidence for the chair-like structure of the transition state was provided in early experiments to determine the mechanism [37], which is supported by mathematical methods as well [38]. The reaction is thermally allowed and an example of a $4n + 2$ system.

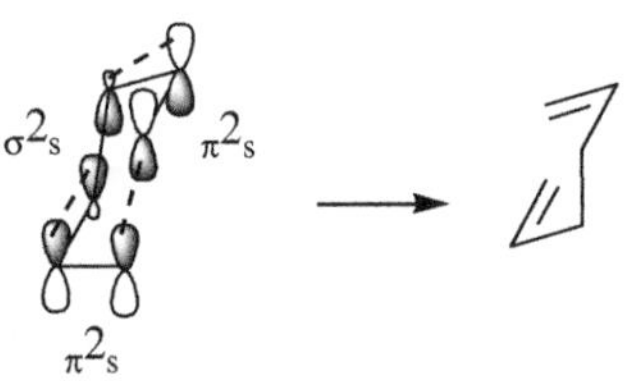

(5.37)

Photochemical sigmatropic reactions have been observed, as exemplified by the [1,3] shift in Equation 5.38 [39]. In these reactions, several competing processes are observed, and thus, operant mechanisms are complicated. When photochemical sigmatropic shifts occur, evidence supports the prediction that they occur suprafacially with retention of configuration [40].

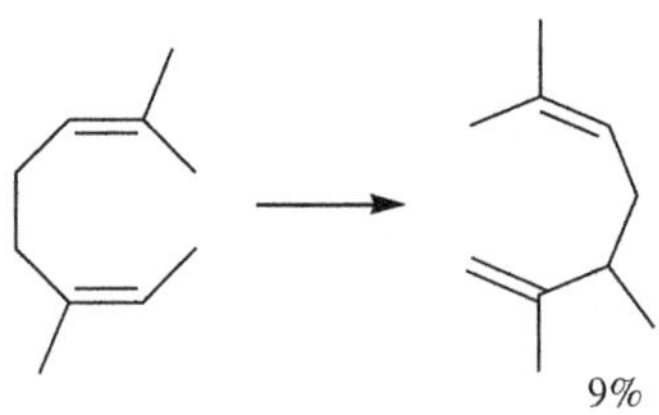

(5.38)

5.3.4 Ene Reactions

The ene reaction [41] can be a pericyclic reaction, conferring the stereo- and regioselective advantages of these concerted reactions. The ene reaction combines the formation of a new C–C bond with a [1,5] sigmatropic shift (Eq. 5.39). It occurs from the HOMO of the ene and the LUMO of the enophile (Eq. 5.40) and is considered a ${}_{\sigma}2_s + {}_{\pi}2_s + {}_{\pi}2_s$ reaction. Often, high temperatures are required for these reactions, which may open the possibility of free radical processes. Like the related Diels–Alder reaction (Section 5.3.1), the ene reaction is facilitated by electron-withdrawing groups on the enophile. Lewis acid catalysts are also effective in supporting the concerted reaction. As the polarity of the components increases, however, the mechanism may become stepwise.

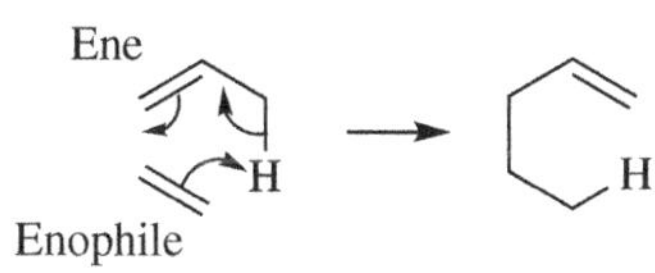

(5.39)

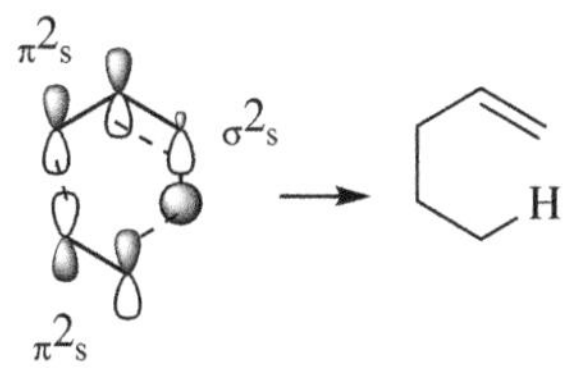

(5.40)

In its simplest form, the olefin-ene reaction is energetically difficult and often requires high enough temperatures that may trigger free radical reactions. Equation 5.41 shows an ene reaction that requires extreme heat, even with the advantage of an intramolecular process.

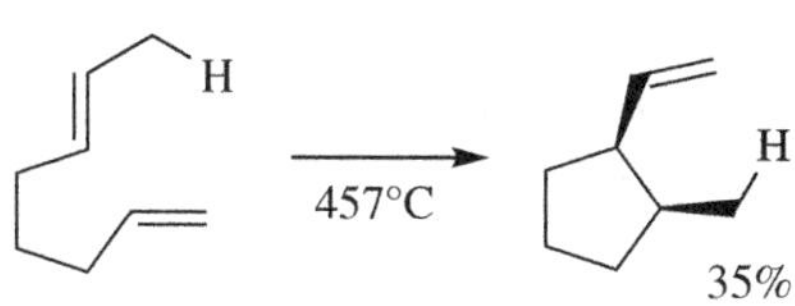

(5.41)

It is possible to improve the reaction conditions using subcritical ethanol as the solvent [42]. In Equation 5.42, subcritical ethanol, instead of the more typical trichlorobenzene, was used to increase the polarity, which assists the pericyclic process while suppressing the radical reaction. Note that this example has electron-withdrawing groups on the eneophile, which further supports the reaction.

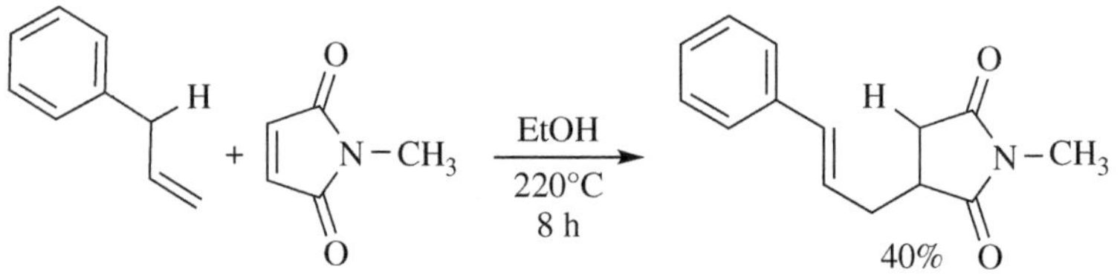

(5.42)

In the carbonyl-ene reaction, the eneophile is replaced by a carbonyl group. These transformations happen even more readily, and when supported by a Lewis acid catalyst, can occur with significant stereocontrol under mild conditions. The pericyclic reaction occurs suprafacially in all components, resulting in cis addition of the transferred hydrogen. The Lewis acid catalyst coordinates the ene and enophile in a six-membered ring transition state, which also supports the cis transfer of hydrogen [43].

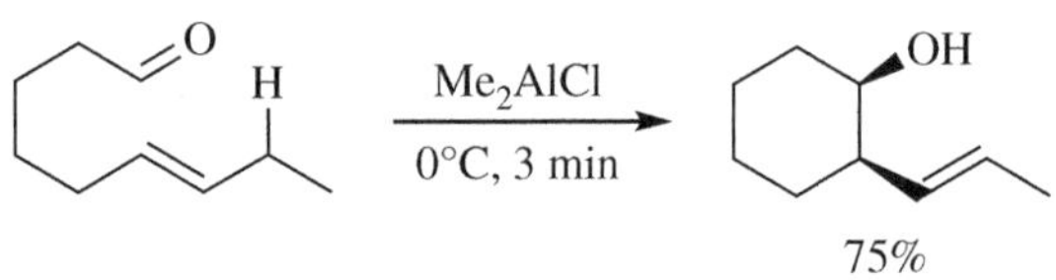

(5.43)

5.3.5 Selection Rules for Pericyclic Reactions

The pericyclic reactions covered in this chapter are governed by rules based on the number of electrons in the transition state of the intermediate. Aromatic systems, with $4n + 2$ electrons, are governed by one set of rules, and antiaromatic systems, with $4n$ electrons, operate under opposite conditions. Photochemical conditions will use the HSOMO and thus invert the allowed reactions from thermal conditions. This is summarized in the table below.

TABLE 5.2 Selection Rules for Allowed Pericyclic Reactions

No. of π Electrons	Allowed Thermal Process	Allowed Photochemical Process
$4n$ (antiaromatic)	Conrotatory/antarafacial[a]	Disrotatory/suprafacial
$4n + 2$ (aromatic)	Disrotatory/suprafacial	Conrotatory/antarafacial[a]

[a]Antarafacial reactions can be geometrically forbidden.

There are more rules covering rearrangements of carbocations and carbanions beyond the scope of this book [15].

It must be kept in mind that although these theories have very broad predictive powers, there are many limitations. Factors such as structural constraints can prevent a reaction that is allowed on an orbital basis. Which direction an allowed reaction will go is not predicted by this theory. Often, more than one allowed reaction is possible from given starting materials, and again, these theories seldom make a distinction. Furthermore, nonconcerted mechanisms may operate to give products that are forbidden by these theories, but as has been demonstrated with the examples in this chapter, molecular orbitals can be used to rationalize the pericyclic reactions that were once termed "no-mechanism" reactions.

RESOURCES

1. This website is maintained by Dr. Mark Winter at the University of Sheffield and allows visualization of the molecular orbitals discussed in the introductory material: https://winter.group.shef.ac.uk/orbitron/
2. This programmed text allows the reader to work through orbital symmetry relationships using a series of answered questions: Bellamy, A. J. An Introduction to Conservation of Orbital Symmetry: A Programmed Text; Longman Group Ltd.: London, 1974.
3. WebMO (free) and WebMO Premium are apps that allow the user to build molecules, optimize structures, and calculate molecular orbitals, among other functionalities.

PROBLEMS

Questions marked with an asterisk are answered in Appendix C.

5.1* The series of reactions below demonstrates a number of pericyclic reactions. For each, indicate the type of reaction, the number of electrons involved, and the allowed stereochemistry [44].

(a)

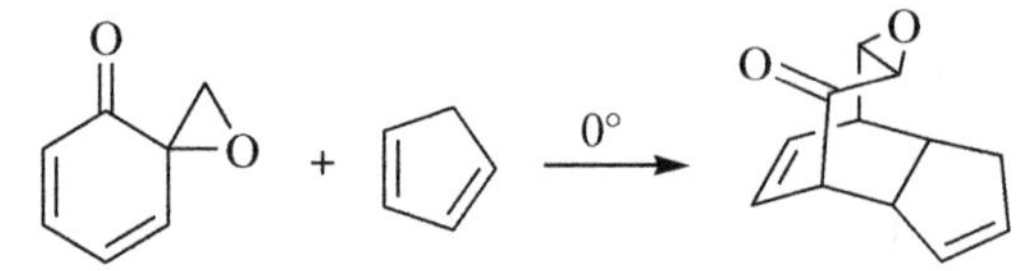

(b)

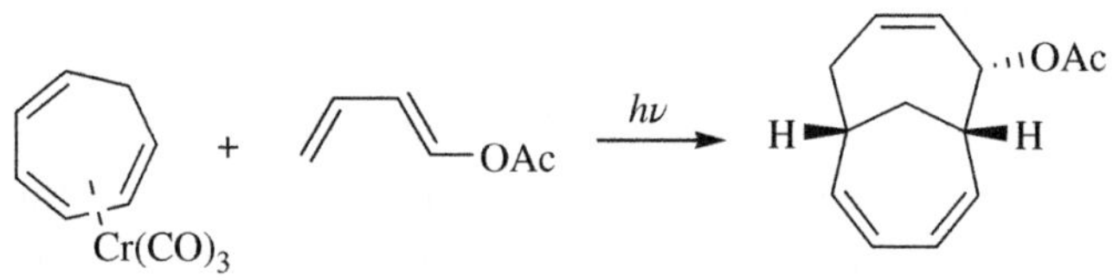

(c)

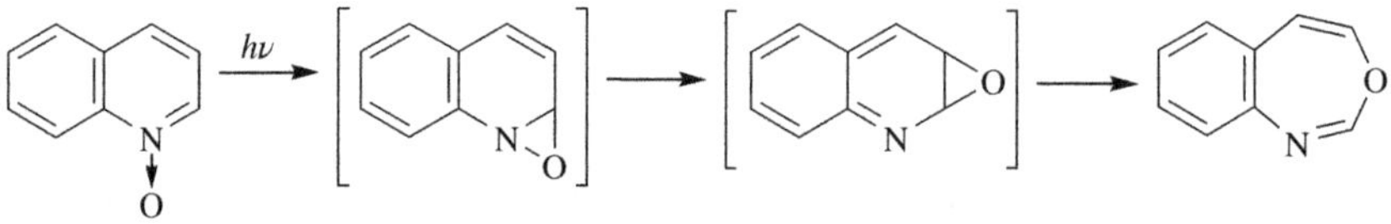

(d)

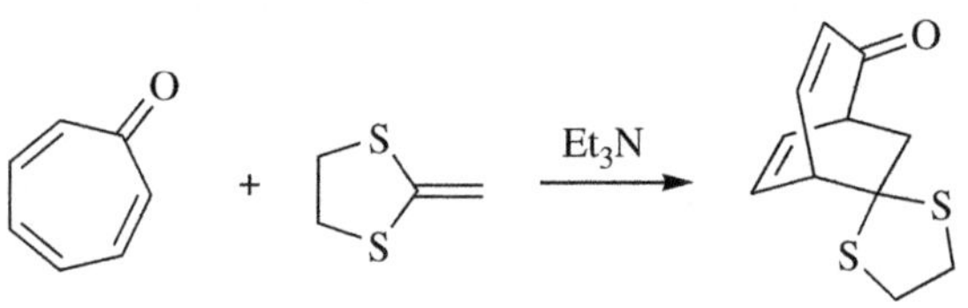

(e)

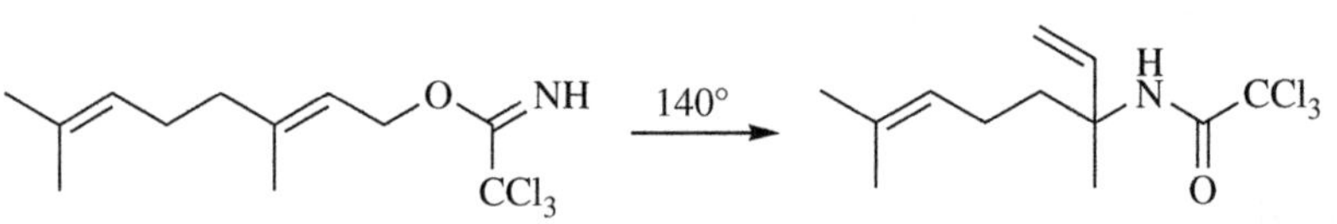

(f)

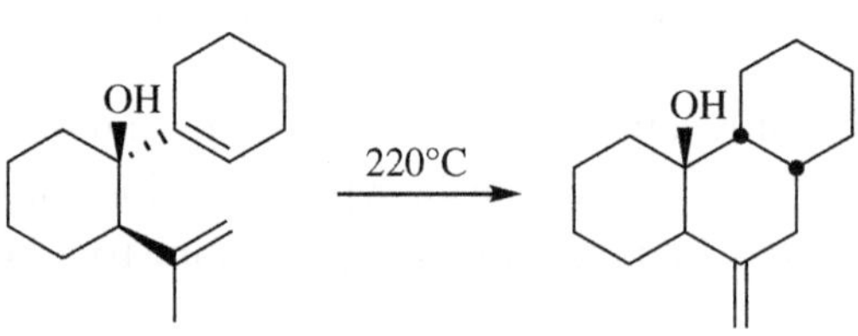

5.2* Table 5.1 shows the calculations for each atom in the cyclopentadienyl cation and anion. Outline the calculations at C-1 for the cyclopentadienyl cation and anion and show the resonance structures for each.

5.3 Bullvalene is a fluxional molecule that undergoes a series of [3,3] sigmatropic shifts, which make all 10 carbons in the molecule equivalent. Use arrows to show at least one of the [3,3] sigmatropic shifts. Can you explain the observation that at 120° the proton NMR spectrum of this molecule shows only one peak at 4.2 ppm? [45, 46]

5.4* Three of the following structures were products from heating 2-chlorotropone with cyclopentadiene at 105°C. Which are the likely ones? Explain [47].

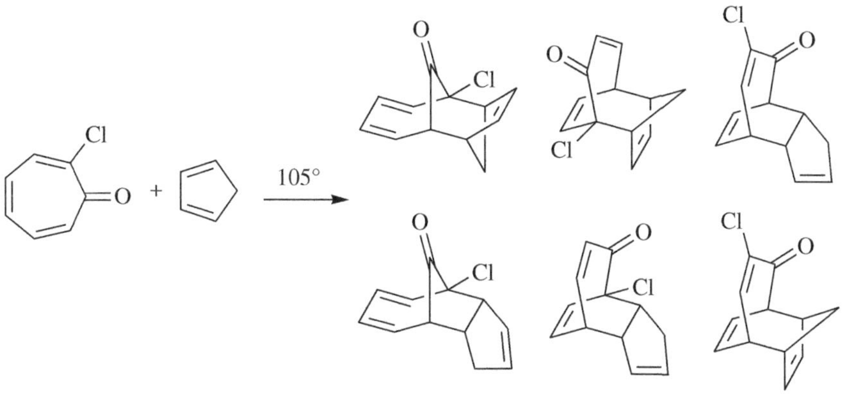

5.5 Which isomer of 9,10-dihydronaphthalene is produced in each reaction? Explain [48].

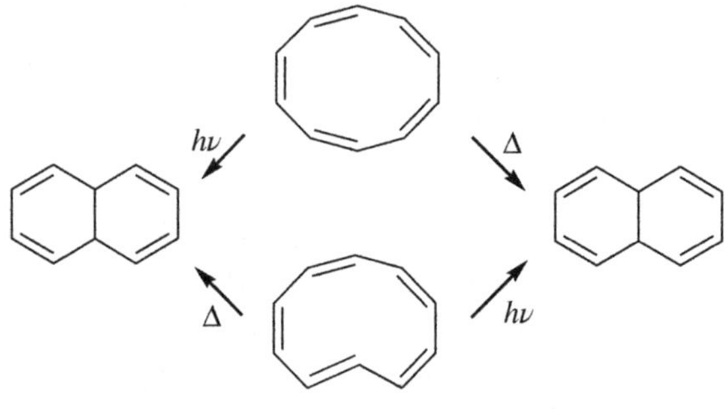

5.6 Heating the following acid–ester gave two acyclic products. Draw the likely structures [49].

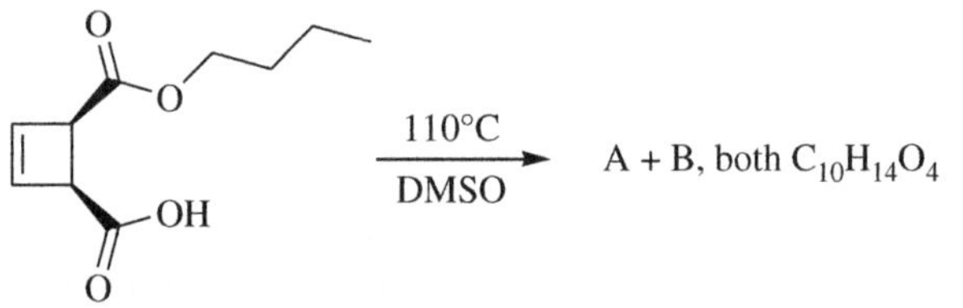

5.7* The following reaction occurred on standing in the dark at 25°C [50]. Indicate the stereochemistry at the newly joined carbons; that is, are the hydrogens *cis* or *trans* to each other? Explain how you made your prediction.

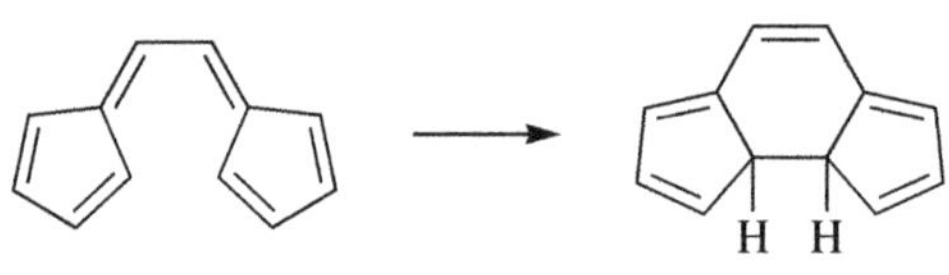

5.8 Although tropones give [4 + 2] adducts with maleic anhydride, the sulfur analog, cycloheptatrienethione, gives [8 + 2] cycloaddition [51]. Draw the structure of that product, including stereochemical representation.

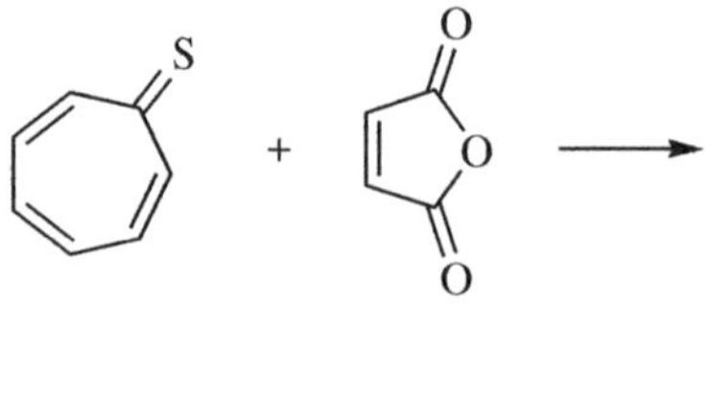

5.9* Explain in terms of one or more intermediates how the following reaction occurred [52]:

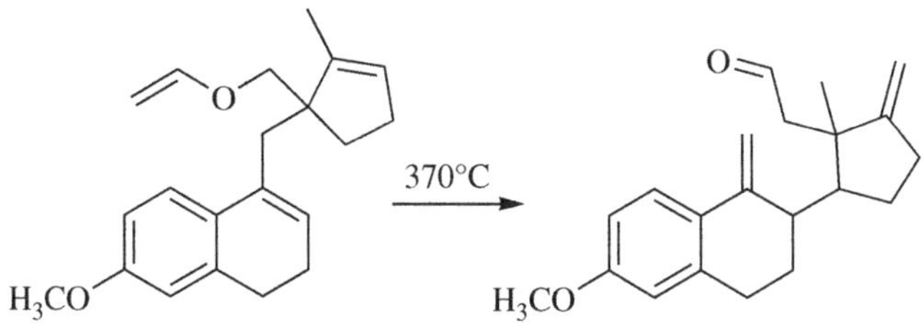

5.10 Draw the intermediate(s), and predict the complete stereochemistry of the product of the following reaction [53]:

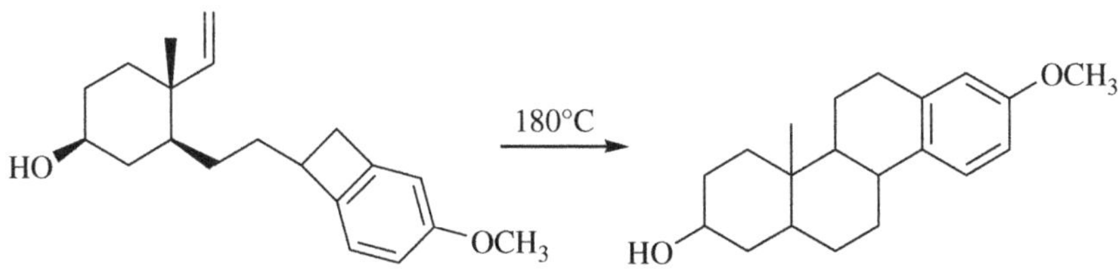

5.11 The cyclobutenone below undergoes thermal rearrangement to form the products shown via two possible ketene intermediates. Show the curved arrows and label the pericyclic processes involved in the formation of the intermediate and both products. Which product should predominate? Explain [54].

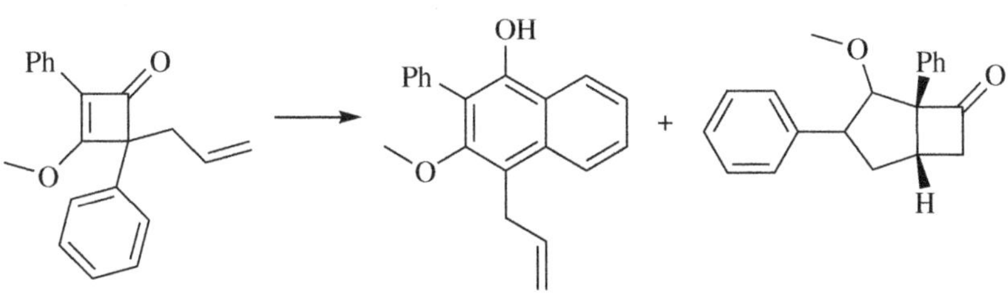

5.12 Is the following reaction photochemical or thermal? Explain how you drew your conclusion [55].

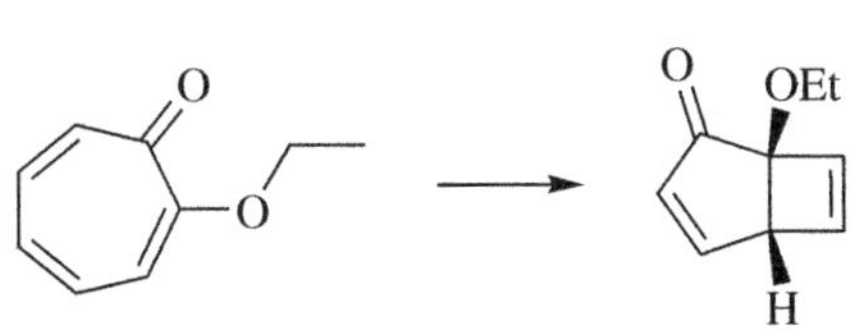

5.13* Heating the following compound gave an intramolecular [6 + 4] cycloaddition in 81% yield. Draw the structure of the product, including the stereochemistry [56].

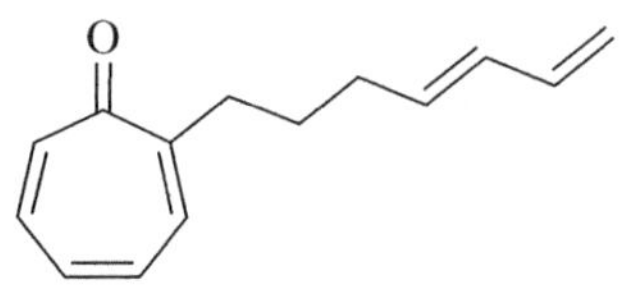

5.14* Give the expected stereochemistry (*Z* or *E*) about the double bonds in the product of the reaction sequence shown. Draw the transition state, including the orbitals involved, for the concerted step [57].

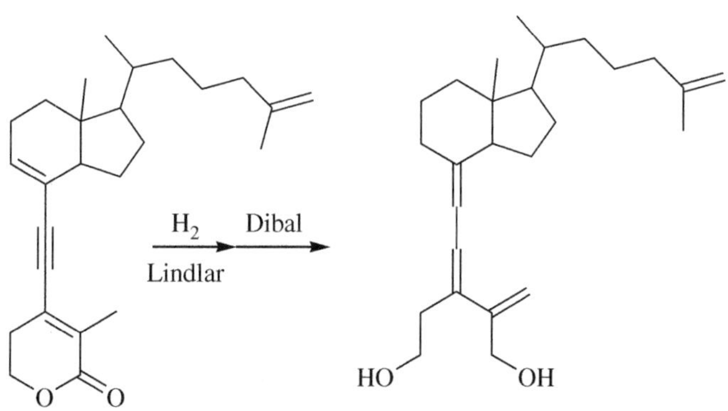

5.15 Propose a series of three pericyclic reactions that explain the following result [58].

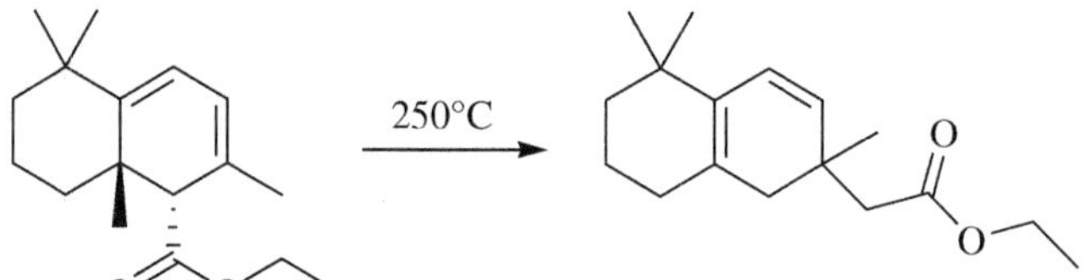

5.16 A variation on the Fischer indole synthesis allows the formation of the 4-azaindole as shown below. The intermediate hydrazone undergoes a [3,3] sigmatropic shift to form the product indole [59]. Write a mechanism for the reaction and identify the [3,3] shift.

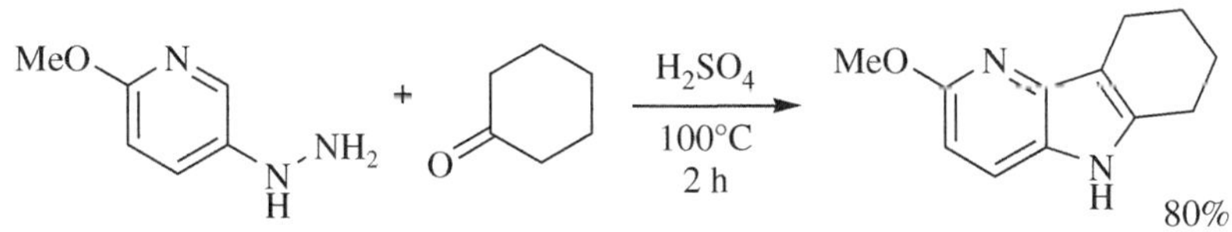

5.17* Heating the following compound gave a product isomeric with the starting material but showing five vinyl hydrogens in the NMR spectrum. What is the structure of the product? [60]

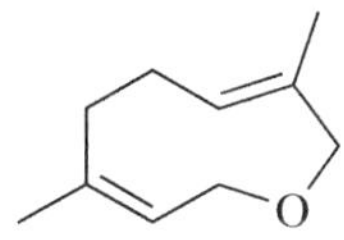

5.18 Naphthopyran dyes show an interesting photochromic effect. Exposure to UV light converts these colorless compounds reversibly to colored compounds, as shown below. These compounds can be used in self-darkening eyeglasses. Identify the pericyclic reaction and describe the allowed stereochemistry [61].

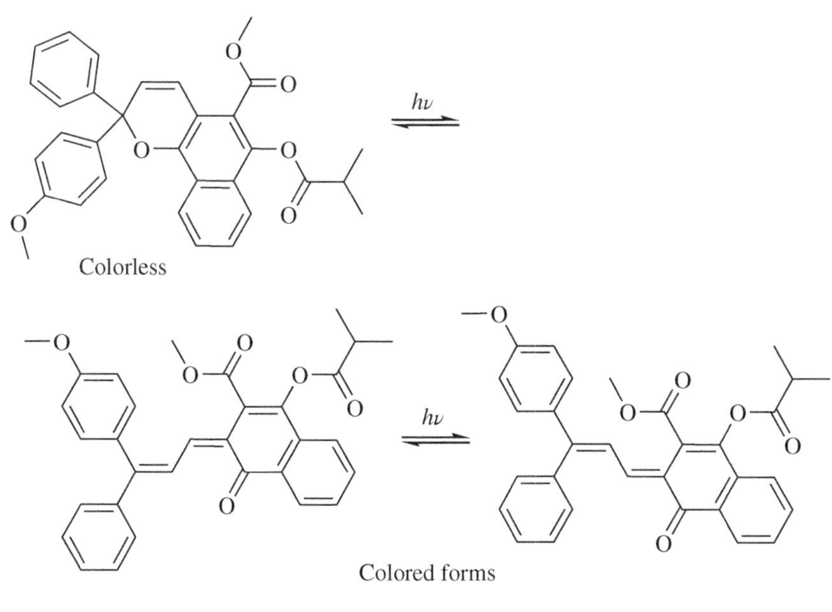

5.19* An enzymatic Diels–Alder reaction has been reported in the biosynthesis of solanapyrones [62]. When this reaction was carried out in the lab, the endo product predominated in all cases, more so as solvent polarity increased. Propose an explanation based on the orbitals involved.

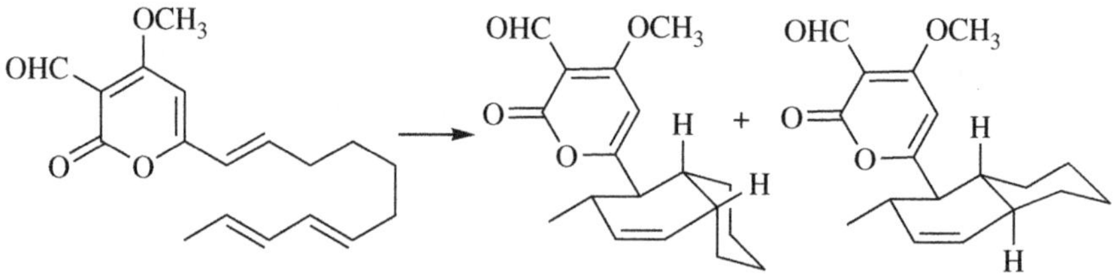

5.20 An aryne-ene reaction has been reported, and evidence supports it as a pericyclic reaction [63]. Write a mechanism for the reaction below. When the allylic methyl groups are substituted with deuterium, greater than 95% deuterium transfer is observed. How does this support the conclusion that the reaction is concerted?

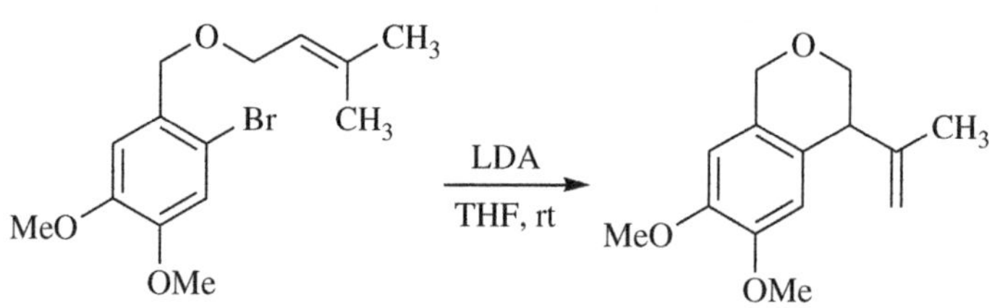

5.21 A Knoevenagel condensation was the key step in forming the intermediate for the reaction below. A subsequent pericyclic reaction formed the product [64]. Write the structure of the intermediate and identify the pericyclic reaction.

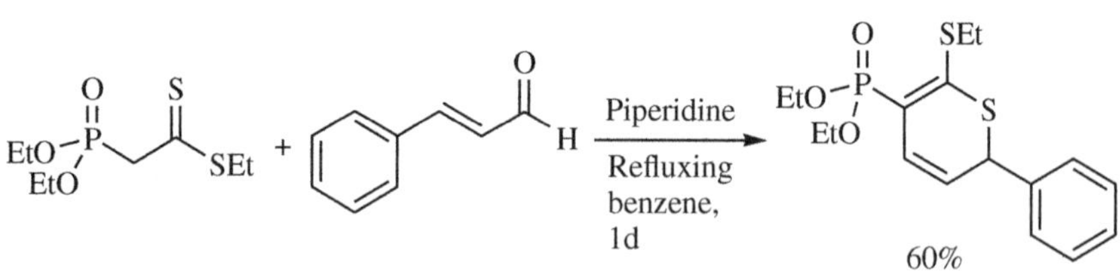

5.22* Draw the HOMO and LUMO for the reactions in Equation 5.8 to demonstrate how each product arises.

5.23* Isochorismate pyruvate lyase catalyses the reaction below. Mechanistic studies of the reaction show quantitative transfer of D to pyruvate, and give $k_{H,\,cat} = 1.01\ sec^{-1}$, and $k_{D,cat} = 0.43\ sec^{-1}$. Write a mechanism for the reaction that fits these observations [65].

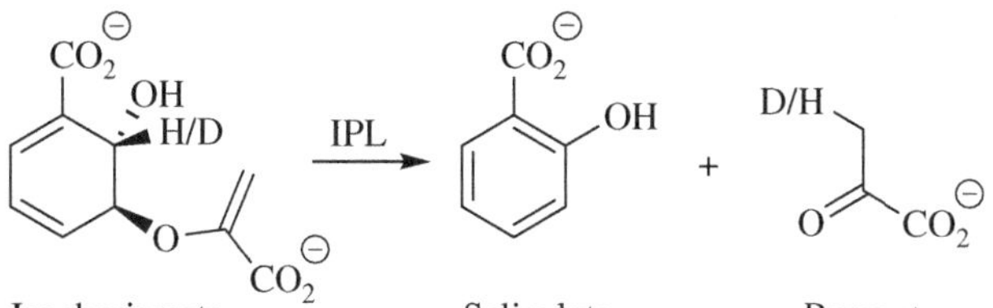

5.24* Propose an explanation for the experimental observations in the investigation of an ene reaction between two differently substituted double bonds and diethyl oxomalonate [66].

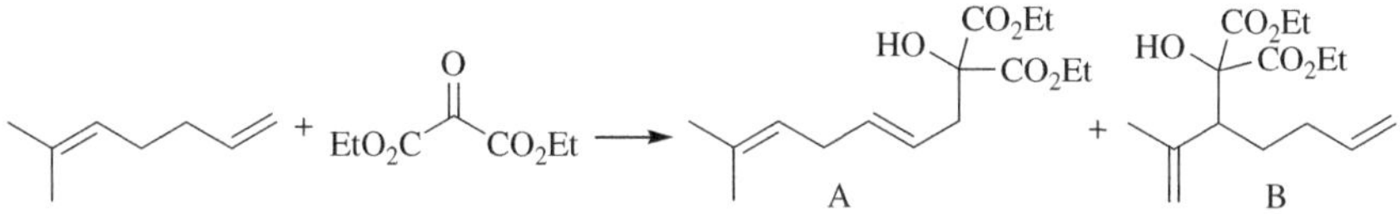

Catalyst	Reaction conditions	A:B	Yield
None	180°C, 48 h	11:1	40%
$SnCl_4$	0°C, 5 min	1:30	75%

REFERENCES

1. Fleming, I. *Molecular Orbitals and Organic Chemical Reactions*; John Wiley & Sons: Chichester, UK, 2010.
2. Streitweiser, A. Jr. *Molecular Orbital Theory for Organic Chemists*; John Wiley & Sons: New York, 1961.
3. Fukui, K. *Angew. Chem. Int. Ed.* **1982**, *21*, 801–809 Available online at https://www.nobelprize.org/uploads/2018/06/fukui-lecture.pdf (accessed January 22, 2025).
4. Li, Y.; Houk, K. N. *J. Am. Chem. Soc.* **1996**, *118*, 880–885.
5. Breslow, R.; Höver, H.; Chang, H. W. *J. Am. Chem. Soc.* **1962**, *84*, 3168–3174.

6. Breslow, R. *Acc. Chem. Res.* **1973**, *6*, 393–398.
7. Sondheimer, F. *Acc. Chem. Res.* **1972**, *5*, 81–91.
8. Slayden, S. W.; Liebman, J. F. *Chem. Rev.* **2001**, *101*, 1541–1566.
9. Katritzky, A. R.; Jug, K.; Oniciu, D. C. *Chem. Rev.* **2001**, *101*, 1421–1450.
10. Vogel, E.; Altenbach, H.-J.; Cremer, D. *Angew. Chem. Int. Ed.* **1972**, *11*, 935–937.
11. Guo, Q. H.; Stoddart, J. F. *Chem* **2022**, *8*, 2076–2078.
12. Gilchrist, T. L.; Storr, R. C. *Organic Reactions and Orbital Symmetry*; Cambridge University Press: London, 1979.
13. Fleming, I. *Pericyclic Reactions*; Oxford University Press: Oxford, 1999.
14. Houk, K. N.; Gonzalez, J.; Li, Y. *Acc. Chem. Res.* **1995**, *28*, 81–90.
15. Woodward, R. B.; Hoffmann, R. *Angew. Chem. Int. Ed.* **1969**, *8*, 781–853. Woodward, R. B.; Hoffmann, R. *The Conservation of Orbital Symmetry*; Verlag Chemie GmbH, Academic Press, Inc.: Weinheim, Germany, 1970.
16. Fukui, K. *Acc. Chem. Res.* **1971**, *4*, 57–64.
17. Dewar, M. J. S. *The Molecular Orbital Theory of Organic Chemistry*; McGraw-Hill Book Company: New York, 1969.
18. Zimmerman, H. *Acc. Chem. Res.* **1971**, *4*, 272–280.
19. Hoffmann, R.; Woodward, R. B. *J. Am. Chem. Soc.* **1965**, *87*, 2046–2048.
20. Herndon, W. C. *Chem. Rev.* **1972**, *72*, 157–179.
21. Houk, K. N. *Acc. Chem. Res.* **1975**, *8*, 361–369.
22. Yamazaki, H.; Cvetanovic, R. J. *J. Am. Chem. Soc.* **1969**, *91*, 520–522.
23. Saito, K.; Ida, S.; Mukai, T. *Bull. Chem. Soc. Jpn.* **1984**, *57*, 3483–3487.
24. Woodward, R. B.; Hoffmann, R. *J. Am. Chem. Soc.* **1965**, *87*, 395–397.
25. Marvell, E. N.; Caple, G.; Schatz, B.; Pippin, W. *Tetrahedron* **1973**, *29*, 3781–3789.
26. Havigna, E.; Schlatmann, J. L. M. A. *Tetrahedron* **1961**, *16*, 146–152.
27. Winter, R. E. K. *Tetrahedron Lett.* **1965**, *6*, 1207–1212.
28. Rondan, N. G.; Houk, K. N. *J. Am. Chem. Soc.* **1985**, *107*, 2099–2111.
29. Rudolf, K.; Spellmeyer, D. C.; Houk, K. N. *J. Org. Chem.* **1987**, *52*, 3708–3710.
30. Houk, K. N.; Spellmeyer, D. C.; Jefford, C. W.; Rimbault, C. G.; Wang, Y.; Miller, R. D. *J. Org. Chem.* **1988**, *53*, 2125–2127.
31. Corey, E. J.; Streith, J. *J. Am. Chem. Soc.* **1964**, *86*, 950–951.
32. Van Tamelen, E. E.; Pappas, S. P.; Kirk, K. L. *J. Am. Chem. Soc.* **1971**, *93*, 6092–6101. Johnson, R. P.; Daoust, K. J. *J. Am. Chem. Soc.* **1996**, *118*, 7381–7385.

33. Cope, A. C.; Haven, A. C. Jr.; Ramp, F. L.; Trumbull, E. R. *J. Am. Chem. Soc.* **1952**, *74*, 4867–4871.

34. Woodward, R. B.; Hoffmann, R. *J. Am. Chem. Soc.* **1965**, *87*, 2511–2513.

35. Leber, P. A.; Baldwin, J. E. *Acc. Chem. Res.* **2002**, *35*, 279–287. Carpenter, B. K. *Acc. Chem. Res.* **1992**, *25*, 520–528.

36. Spangler, C. W. *Chem. Rev.* **1976**, *76*, 187–217.

37. Doering, W. v. E.; Roth, W. R. *Tetrahedron* **1962**, *18*, 67–74.

38. Houk, K. N.; Gustafson, S. M.; Black, K. A. *J. Am. Chem. Soc.* **1992**, *114*, 8565–8572.

39. Manning, T. D. R.; Kropp, P. J. *J. Am. Chem. Soc.* **1981**, *103*, 889–897.

40. Bernardi, F.; Olivucci, M.; Robb, M. A.; Tonachini, G. *J. Am. Chem. Soc.* **1992**, *114*, 5805–5812 and references cited therein.

41. Mikami, K.; Shimizu, M. *Chem. Rev.* **1992**, *92*, 1021–1050.

42. Laitinen, A.; Takebayashi, Y.; Kylänlahti, I.; Yli-Kauhaluoma, J.; Sugetea, T.; Otake, K. *Green Chem.* **2004**, *6*, 49–52.

43. Snider, B. B.; Karras, M.; Price, R. T.; Rodini, D. *J. Org. Chem.* **1982**, *47*, 4538–4545.

44. a) Singh, V.; Porinchu, M.; Vedantham, P.; Sahu, P. K. *Org. Synth.* **2005**, *81*, 171–177. b) Rigby, J. H.; Fales, K. R. *Org. Synth.* **2000**, *77*, 121–134. c) Albini, A.; Bettinetti, G. F.; Minoli, G. *Org. Synth.* **1983**, *61*, 98–102. d) Dahnke, K. R.; Paquette, L. A. *Org. Synth.* **1993**, *71*, 181–188. e) Clizbe, L. A.; Overman, L. E. *Org. Synth.* **1978**, *58*, 4–11. f) Warrington, J. M.; Yap, G. P. A.; Barriault, L. *Org. Lett.* **2000**, *2*, 663–665.

45. Ault, A. *J. Chem. Educ.* **2001**, *78*, 924–927.

46. Klärner, F. G.; Jones, M. Jr.; Magid, R. M. *Acc. Chem. Res.* **2009**, *42*, 169–181.

47. Ito, S.; Sakan, K.; Fujise, Y. *Tetrahedron Lett.* **1969**, *10*, 775–778.

48. Masamune, S.; Hojo, K.; Bigam, G.; Rabenstein, D. L. *J. Am. Chem. Soc.* **1971**, *93*, 4966–4968.

49. Trost, B. M.; McDougal, P. G. *J. Org. Chem.* **1984**, *49*, 458–468.

50. Sauter, H.; Gallenkamp, B.; Prinzbach, H. *Chem. Ber.* **1977**, *110*, 1382–1402.

51. Machiguchi, T.; Hoshino, M.; Ebine, S.; Kitahara, Y. *J. Chem. Soc. Chem. Commun.* **1973**, 196–196.

52. Ziegler, F. E.; Lim, H. *J. Org. Chem.* **1982**, *47*, 5229–5230.

53. Kametani, T.; Suzuki, K.; Nemoto, H. *J. Chem. Soc. Chem. Commun.* **1979**, 1127–1128.

54. Tiedemann, R.; Turnbull, P.; Moore, H. W. *J. Org. Chem.* **1999**, *64*, 4030–4041.

55. Kaftory, M.; Yagi, M.; Tanaka, K.; Toda, F. *J. Org. Chem.* **1988**, *53*, 4391–4393.
56. Rigby, J. H.; Moore, T. L.; Rege, S. *J. Org. Chem.* **1986**, *51*, 2398–2400.
57. Barrack, S. A.; Gibbs, R. A.; Okamura, W. H. *J. Org. Chem.* **1988**, *53*, 1790–1796.
58. Fráter, G. *Helv. Chim. Acta* **1974**, *57*, 2446–2454.
59. Jeanty, M.; Blu, J.; Suzenet, F.; Guillaumet, G. *Org. Lett.* **2009**, *11*, 5142–5145.
60. Marshall, J. A.; Lebreton, J. *J. Org. Chem.* **1988**, *53*, 4108–4112.
61. Ercole, F.; Davis, T. P.; Evans, R. A. *Macromolecules* **2009**, *42*, 1500–1511.
62. Oikawa, H.; Kobayashi, T.; Katayama, K.; Suzuki, Y.; Ichihara, A. *J. Org. Chem.* **1998**, *63*, 8748–8756.
63. Candito, D. A.; Panteleev, J.; Lautens, M. *J. Am. Chem. Soc.* **2011**, *133*, 14200–14203.
64. Riu, A.; Harrison-Marchand, A.; Maddaluno, J.; Gulea, M.; Albadri, H.; Masson, S. *Eur. J. Org. Chem.* **2007**, 4948–4952.
65. DeClue, M. S.; Baldridge, K. K.; Künzler, D. E.; Kast, P.; Hilvert, D. *J. Am. Chem. Soc.* **2005**, *127*, 15002–15003.
66. Salomon, M. F.; Pardo, S. N.; Salomon, R. G. *J. Org. Chem.* **1984**, *49*, 2446–2454.

6

FUNCTIONAL GROUP TRANSFORMATIONS

A chemist who undertakes the synthesis of an organic compound of some complexity must consider three aspects: (1) the synthesis of the functional groups in the final molecule plus those needed at intermediate stages, (2) the formation of carbon–carbon bonds to develop larger molecules, and (3) the strategy of selecting starting materials and intermediate goals. A chapter is devoted to each; the first is concerned with functional groups.

The current practical alternatives for preparing each functional group include many classical reactions with relatively known mechanisms, plus many modern ones with complex or often unknown mechanisms. The introductory texts favor conceptually simple methods applied to small monofunctional molecules. A broader selection of methods is chosen here and new methods are published regularly. Online sources can be the most readily accessed and up-to-date [1]. The functional group syntheses that involve carbon–carbon bond-forming reactions are presented in Chapter 7.

Understanding Organic Chemistry: Bridging Fundamental and Advanced Courses,
Fourth Edition. Ann M. Fabirkiewicz and John C. Stowell.

Companion website: www.wiley.com/go/fabirkiewiczorganic4e

6.1 CARBOXYLIC ACIDS AND RELATED DERIVATIVES

The high oxidation state of carbon in which there are three bonds to electronegative atoms is a characteristic of carboxylic acids and the related acid chlorides, anhydrides, esters, ortho esters, amides, and nitriles. The transformations may involve oxidation from hydrocarbons or other partially oxidized substrates or exchange among various electronegative atoms on the carbon.

6.1.1 Carboxylic Acids

Benzylic sites that contain at least one hydrogen may be oxidized to the carboxylic acid by using strong agents including nitric acid or dichromate or permanganate salts. In polyalkyl benzenes, selectivity with moderate yields may be obtained using aqueous nitric acid. The order of ease of oxidation of some alkyl groups is isopropyl > ethyl > methyl ≫ *tert*-butyl [2]. For example, *p*-cymene was converted to *p*-toluic acid as shown in Equation 6.1 [3].

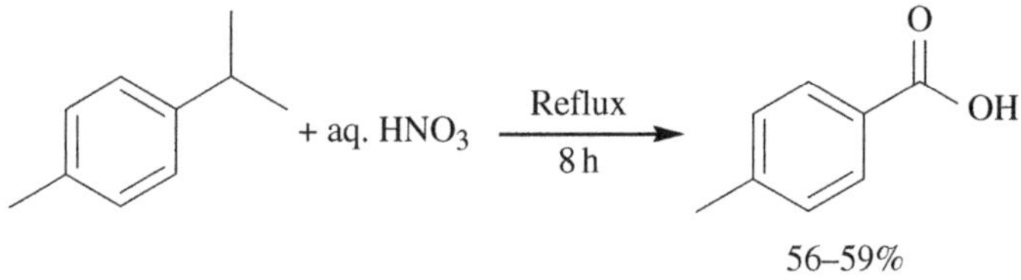

(6.1)

Carbons that are already partially oxidized such as alkenes, primary alcohols, aldehydes, and methyl ketones are more readily oxidized to the carboxylic acid. Appropriately substituted alkenes may be cleaved using ozone followed by treatment with hydrogen peroxide to give carboxylic acids. A variety of alkenes can be oxidatively cleaved to the corresponding acid using a catalytic amount of 3,4,5,6-tetramethyl-2-iodobenzoic acid and oxone. Oxone oxidizes the alkene to the dihydroxy-derivative, and oxidizes the tetramethyliodobenzoic acid to the tetramethyliodoxybenzoic acid, which cleaves the dihydroxy derivative. Oxone converts the resulting aldehyde to the acid (Eq. 6.2) [4].

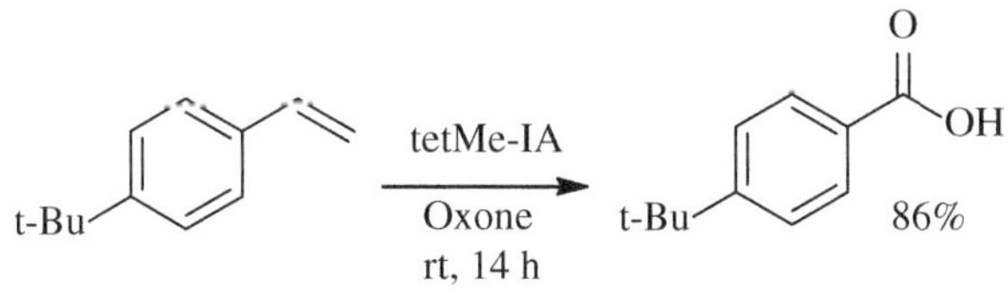

(6.2)

A phase-transfer catalyst (PTC) (Section 9.6.1) can be used in the environmentally benign synthesis of adipic acid from cyclohexene. Sodium tungstate and 30% hydrogen peroxide were used to accomplish the reaction in yields averaging 68% [5].

Primary alcohols are oxidized by the easily prepared pyridinium dichromate (PDC) [6] in dimethylformamide (DMF) [7] or by Jones reagent in acetone [8]. Secondary alcohols are also easily oxidized to ketones under these conditions. Primary alcohols and aldehydes are selectively oxidized using a hydrogen transfer reaction with 1-hydroxycyclohexyl phenyl ketone (1-HCPK) in good to excellent yields under mild conditions (Eq. 6.3). 1-HCPK can be recycled in 95% yield and reused. The reaction tolerates a wide variety of functionalities, including secondary alcohols and alkenes [9].

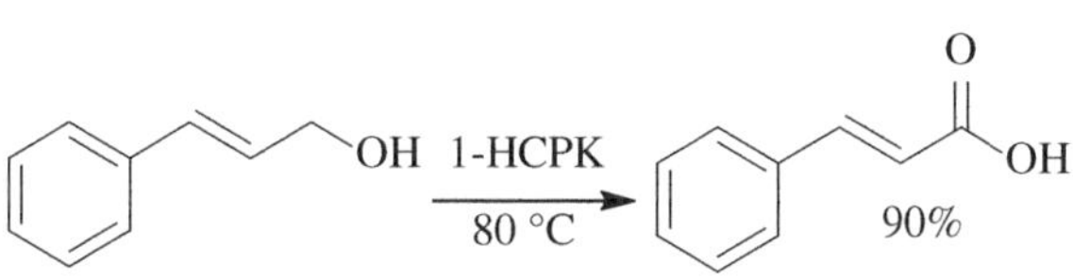

(6.3)

Aldehydes are more readily oxidized than alcohols and thus react with the reagents given above. Oxone has been used successfully to oxidize aromatic and aliphatic aldehydes to acids while tolerating a variety of functionalities (Eq. 6.4) [10]. The use of alcoholic solvents in this reaction results in the ester product. Very mild reagents such as freshly precipitated silver oxide [11] or sodium chlorite [12] serve well.

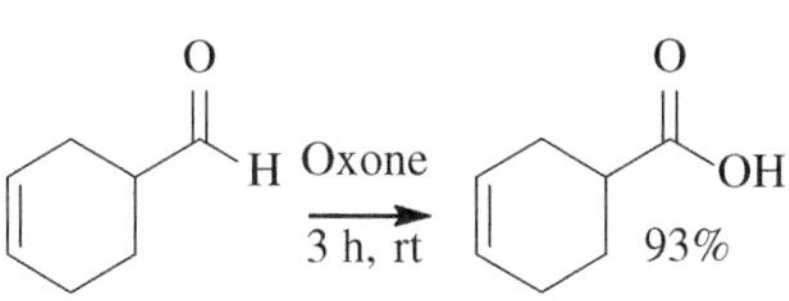

(6.4)

Within the same oxidation level, any of the acid derivatives may be hydrolyzed with aqueous acid or base, leading to the acid or its salt. Nitrile hydrolysis is particularly difficult, requiring prolonged heating in water–ethylene glycol (Eq. 6.5) [13].

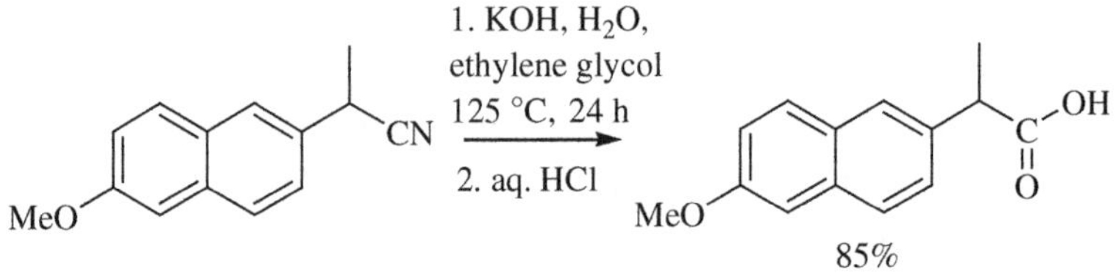

(6.5)

There are many syntheses of acids where a carbon–carbon bond is formed such as carbonation of Grignard reagents, malonic ester alkylation, and Reformatsky reactions. Some are covered in Chapter 7, and you can review others in your introductory text.

6.1.2 Carboxylic Esters

Carboxylic acids may be converted to esters directly by using a primary or secondary alcohol and a small amount of strong acid catalyst. This is a reversible equilibrium, where ester formation is favored by using excess alcohol or by removing the water produced. Azeotropic distillation of the water or consumption of the water by concurrent hydrolysis of an acetal is usually effective.

The highly reactive acid chlorides and anhydrides give esters irreversibly. Acetate esters of complex alcohols are routinely prepared by treating with acetic anhydride and pyridine.

Several methods are available that do not begin with alcohols. The sodium or potassium salts of carboxylic acids are sufficiently nucleophilic to displace primary iodides (Eq. 6.6) [14].

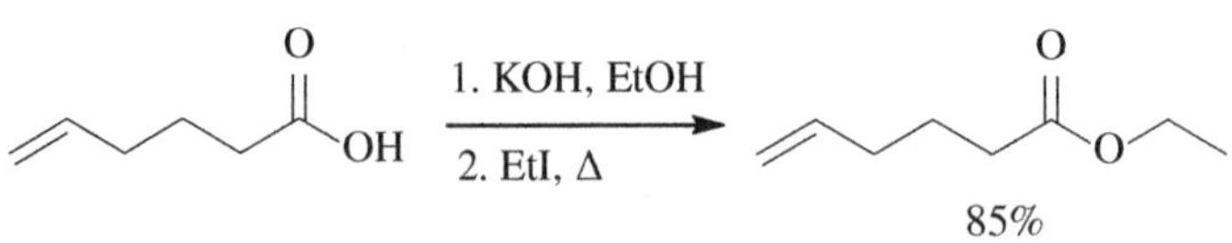

(6.6)

Diazomethane is commonly used to form methyl esters in high yield and purity under neutral conditions, but the reagent is volatile, toxic, and explosive [15]. Recently, imidazotetrazines have been reported as safer alternatives for this reaction (Eq. 6.7) [16]. Temozolomide (TMZ), a treatment for glioblastoma, can accomplish this reaction safely in good yields, which are higher when electron donor groups are present.

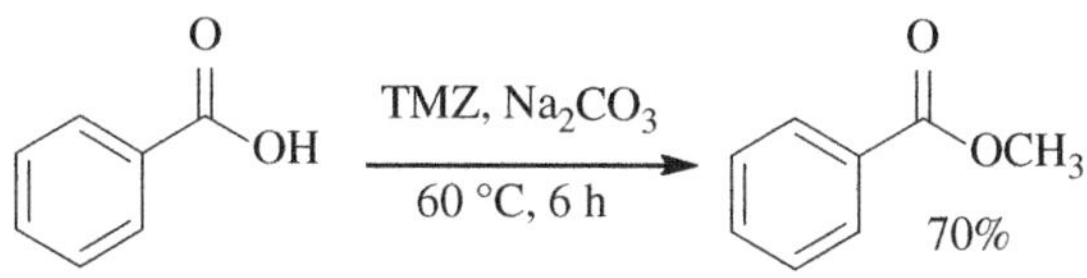

(6.7)

Ketones may be oxidized to esters by peracids or hydrogen peroxide, a process known as the Baeyer–Villiger oxidation. Unsymmetric ketones are oxidized selectively at the more substituted α carbon and that carbon migrates to oxygen with retention of configuration. Trifluoroperacetic acid generated *in situ* gave the double example in Equation 6.8 [17]. Cyclic ketones afford lactones (Eq. 6.9) [18].

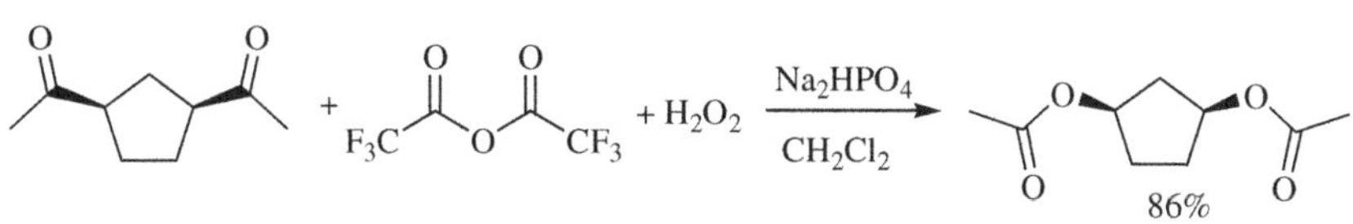

(6.8)

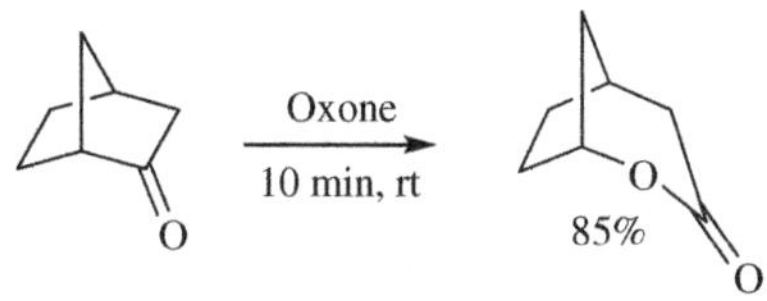

(6.9)

Enol esters, important synthetic intermediates, may be prepared from aryl alkynes using N-iodosuccinimide (NIS) followed by a coupling reaction. Orientation of the double bond is determined by the R- group on the alkyne, as shown in Equation 6.10 [19].

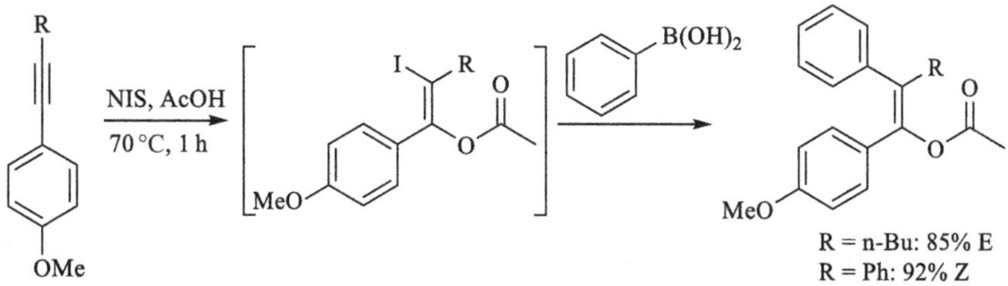

(6.10)

6.1.3 Carboxylic Amides

Heating a carboxylic acid with ammonia or urea gives a carboxamide. For example, heptanoic acid plus urea at 140–180°C gives heptanamide in 75% yield plus CO_2 and H_2O [20]. The highly reactive acid halides and anhydrides combine with ammonia or primary or secondary amines to give amides at ordinary temperatures. Aromatic and aliphatic nitriles can be readily converted to amides with trifluoroacetic acid (Eq. 6.11) [21]. This hydration may also be accomplished under neutral anhydrous conditions using acetaldoxime and a rhodium catalyst, which leaves other functionalities intact (Eq. 6.12) [22].

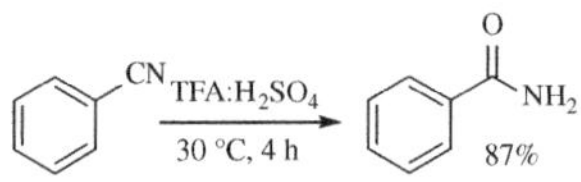

(6.11)

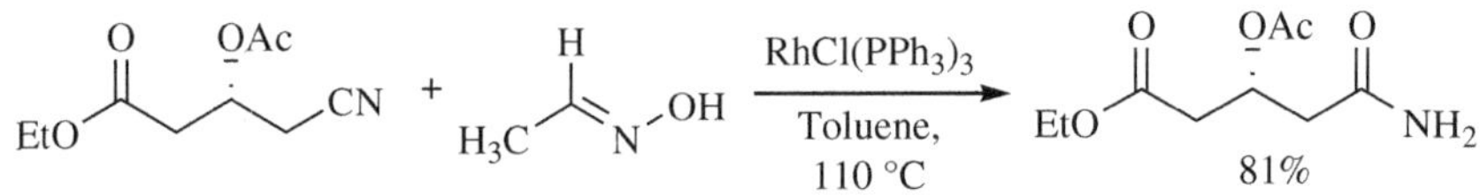

(6.12)

6.1.4 Carboxylic Acid Halides

Acid chlorides are commonly made from acids by exchange with an excess of thionyl chloride or oxalyl chloride. Brief heating gives the acid chloride plus gaseous by-products (Eq. 6.13) [23]. Phosphorus tri- and pentachlorides are used similarly. The acid bromides are made with phosphorus tribromide or oxalyl bromide [24]. Tetramethylammonium trifluoromethanethiolate, $(Me_4N)SCF_3$, conveniently forms acid fluorides in high yield at room temperature [25].

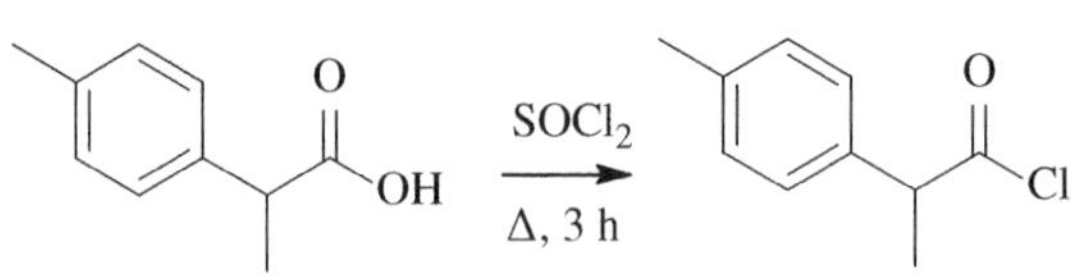

(6.13)

6.1.5 Carboxylic Anhydrides

Most anhydrides are prepared from carboxylic acids by exchange with the readily available acetic anhydride. Heating these and then distilling the acetic acid and excess acetic anhydride shifts the equilibrium toward the higher-boiling product (Eq. 6.14) [26]. Five- and six-membered cyclic anhydrides usually form simply by heating the dicarboxylic acid to about 120°C.

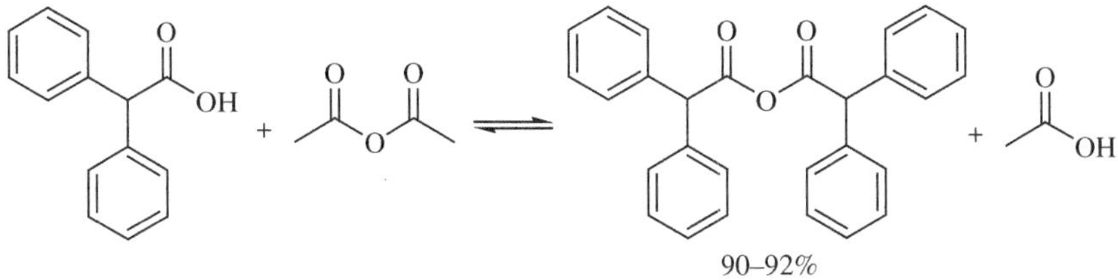

(6.14)

Unsymmetric anhydrides that react selectively on one side are useful. Although formic anhydride is unstable above −40°C, acetic formic anhydride can be prepared by stirring sodium formate with acetyl chloride in ether in 64% yield [27]. It is useful for the formylation of alcohols and amines. N-formylbenzotriazole is easily prepared and can selectively formylate alcohols, amines, and amides [28]. Ethyl chloroformate gives mixed anhydrides with various carboxylic acids which are then susceptible to nucleophilic substitution at the carboxylic carbonyl carbon.

6.1.6 Nitriles

Nitriles are prepared by a wide variety of methods [29]. Dehydration of amides has been accomplished using acidic reagents such as phosphorus pentoxide [30] or thionyl chloride [31]. Nitriles can be prepared from the corresponding amides and aldoximes under Swern oxidation conditions, as shown in Equation 6.15 [32]. Aromatic and aliphatic aldehydes are successfully converted to nitriles using hydroxylamine-O-sulfonic acid, as shown in Equation 6.16 [33].

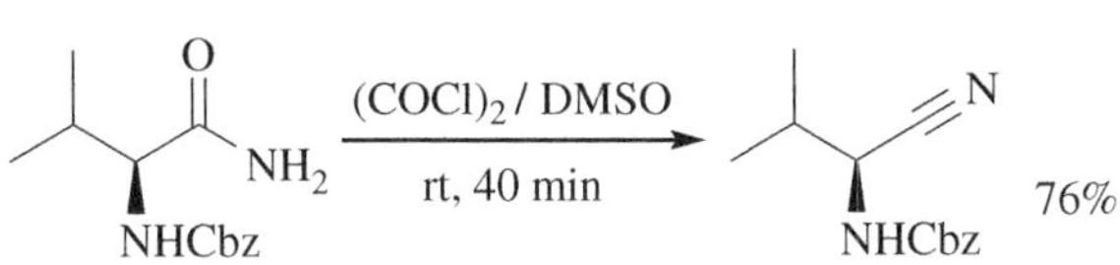

(6.15)

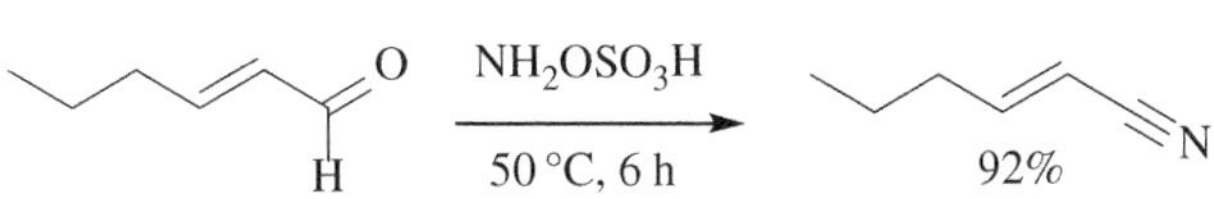

(6.16)

Nitriles are also commonly made by displacements with cyanide ion (Section 7.1.1, Eq. 7.14). A continuous flow process that avoids the use of cyanide salts has been developed for aromatic nitriles to avoid the hazards that occur on the scale-up of these reactions [34].

6.1.7 Ortho Esters

Ortho esters are acid derivatives in which the carboxyl carbon is sp^3-hybridized; however, most cannot be made from carboxylic acids. They are usually made by a two-stage alcoholysis of nitriles. Treatment of a nitrile with anhydrous hydrogen chloride in an alcohol gives the hydrochloride of an imidic ester. Treatment with an alcohol in a separate step (Eq. 6.17) leads to the ortho ester [35]. Other methods for the synthesis of ortho esters are available [36].

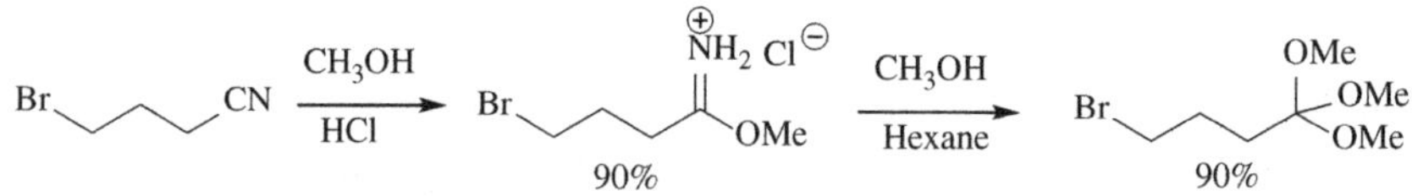

(6.17)

The alcohol parts of ortho esters may be exchanged under acidic conditions to give new ortho esters, especially where the incoming alcohol is a diol. This reaction is important in some Claisen rearrangements (Section 8.7). In contrast, the acid portion cannot be exchanged; that is, an acid cannot be converted to an ortho ester directly by transesterification. A few acids such as chloroacetic acid may be converted to bicyclic ortho esters by reaction with a triol with azeotropic removal of water. A general route to bicyclic ortho esters begins with commercially available 3-hydroxymethyl-3-oxetanemethanol, as shown in Equation 6.18 [37].

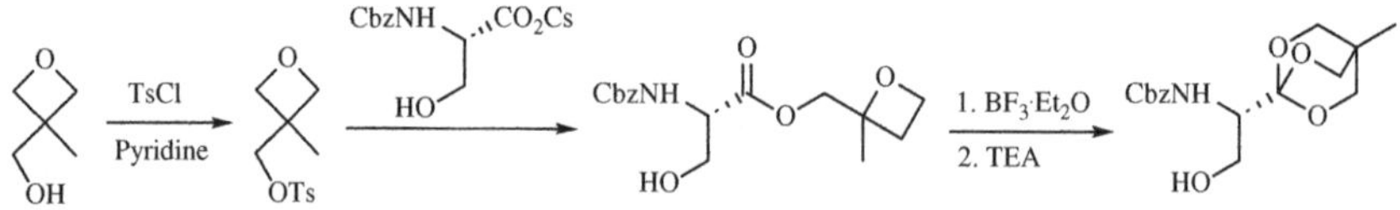

(6.18)

6.2 ALDEHYDES, KETONES, AND DERIVATIVES

The intermediate oxidation state of carbon in which there are two bonds to electronegative atoms is attained by the reduction of acid derivatives or oxidation of alcohols and hydrocarbons. Interconversions at the same oxidation level such as hydration of alkynes and hydrolysis of vinyl halides are also valuable.

6.2.1 Aldehydes

The reduction of acid derivatives to aldehydes requires control because aldehydes are more easily reduced than the acid derivatives. The hydrogenation of acid chlorides in the presence of quinoline-sulfur poisoned palladium catalyst, which is a modification of the Rosenmund reduction, shows this selectivity (Eq. 6.19) [38].

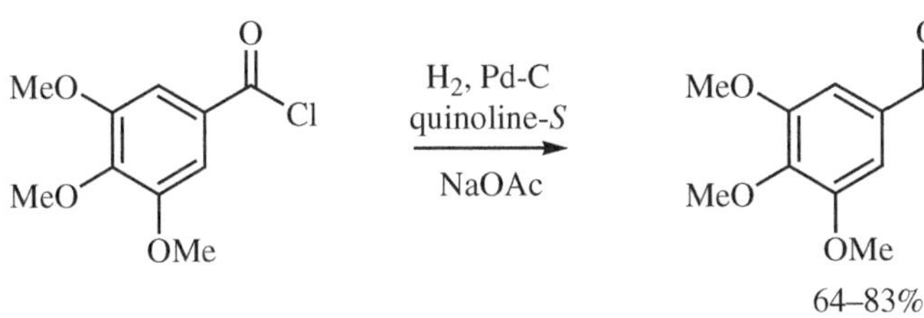

(6.19)

Thioesters are reduced selectively to aldehydes by the Fukuyama reduction [39], as shown in Equation 6.20 [40]. Even the alkene functionality can be preserved in this reaction by the use of Lindlar's catalyst and a sacrificial alkene, as shown in Equation 6.21 [41].

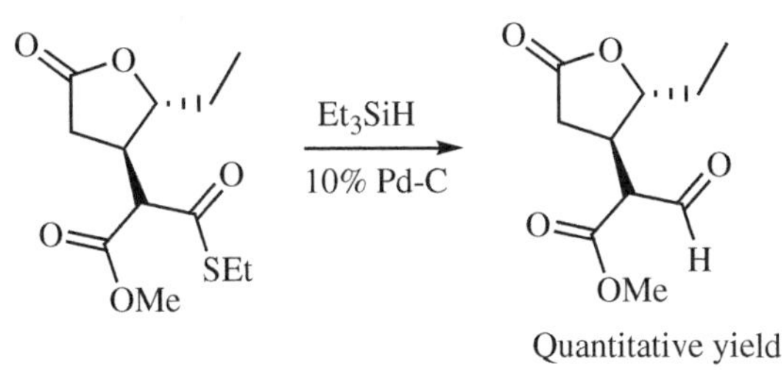

(6.20)

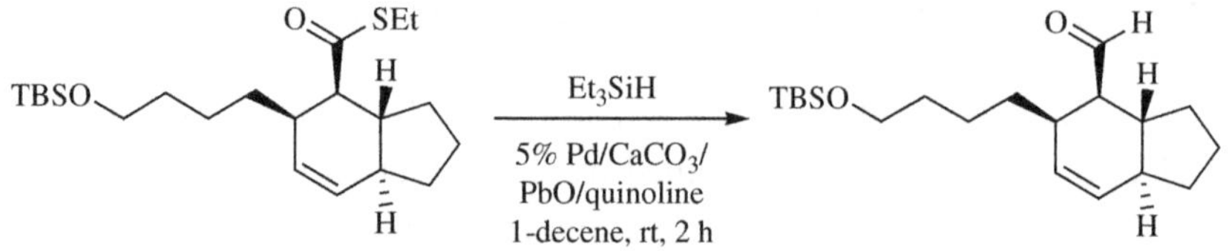

(6.21)

Alkoxyaluminum hydrides, such as diisobutylaluminum hydride (DIBAL), can be used to partially reduce nitriles (Eq. 6.22) [42] and esters (Eq. 6.23) [43] to the corresponding aldehydes. Overreduction to the amine or alcohol is a common problem that can be solved with the use of a flow reactor (Section 9.6.4) [44].

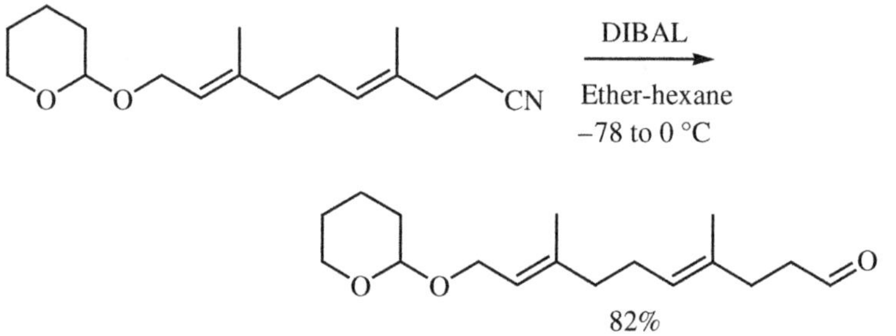

(6.22)

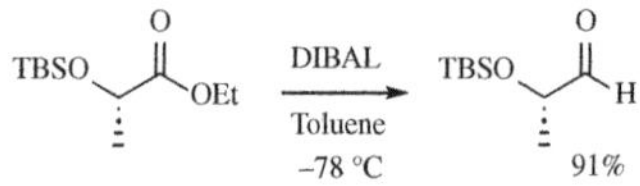

(6.23)

Oxidation of primary alcohols can give aldehydes, and again control is necessary because simple oxidizing agents such as chromic acid will carry on to the carboxylic acid stage. Many Cr^{6+} complexes are selective for this transformation. A solvent-free option using chromium trioxide on aluminum silicate confers advantages in time and workup as well as yield [45].

Another commonly used oxidizing agent is dimethylsulfoxide (DMSO) together with a dehydrating agent such as acetic anhydride or oxalyl chloride (Swern oxidation) [46], as shown in Equation 6.24 [47].

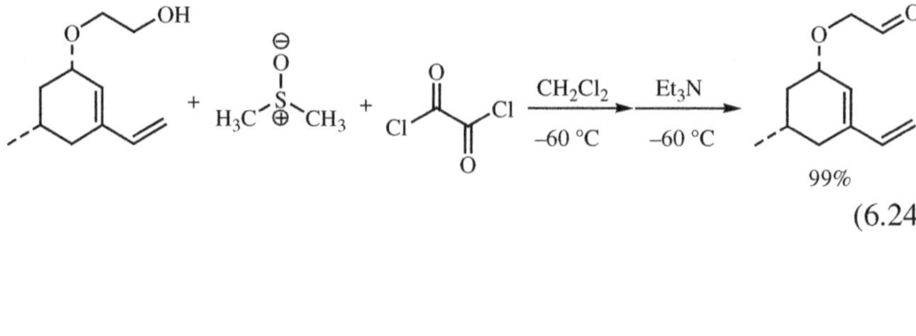

(6.24)

Dess–Martin periodinane (DMP), 1,1,1-triacetoxy-1,1-dihydro-1,2-benziodoxol-3(1H)-one, is commercially available and allows selectivity, mild conditions, and easy workup for the oxidations of primary and secondary alcohols (Eq. 6.25) [48].

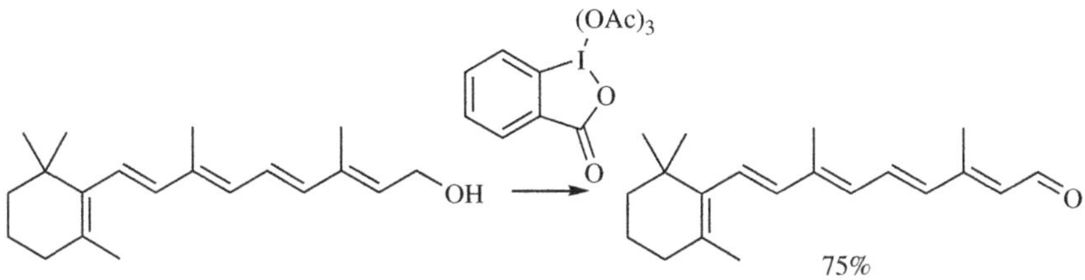

(6.25)

Aldehydes and ketones can be synthesized under mild conditions using (2,2,6,6-tetramethylpiperidin-1-yl)oxyl (TEMPO) in an ionic liquid. The solvent can be reused in subsequent reactions and the product is isolated by distillation when it is volatile enough, and by extraction otherwise (Eq. 6.26) [49].

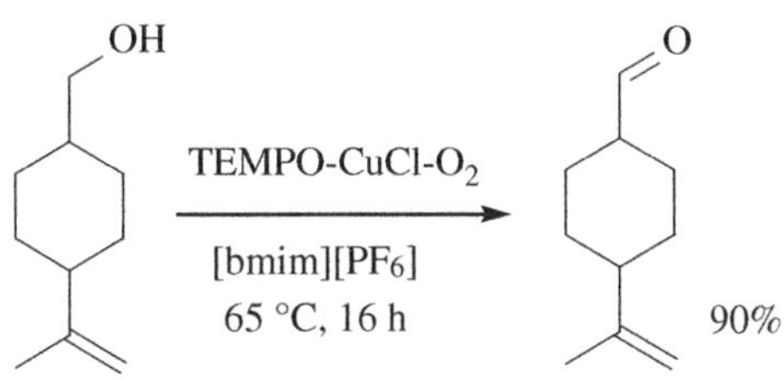

(6.26)

Oxidative cleavage of appropriate alkenes can give aldehydes. Where ozone is used, the intermediate ozonides have more oxidizing power that can oxidize the desired aldehydes to carboxylic acids during hydrolysis. To avoid this interference, dimethyl sulfide is added as a reducing agent [50]. The same overall result can be obtained using sodium periodate with a catalytic amount of osmium tetraoxide, which gives the glycol. Periodate cleaves the glycol to form the aldehyde and also regenerates the OsO_4. Pyridine can also be used to achieve these results in high yield with very short reaction times, as shown in Equation 6.27 [51].

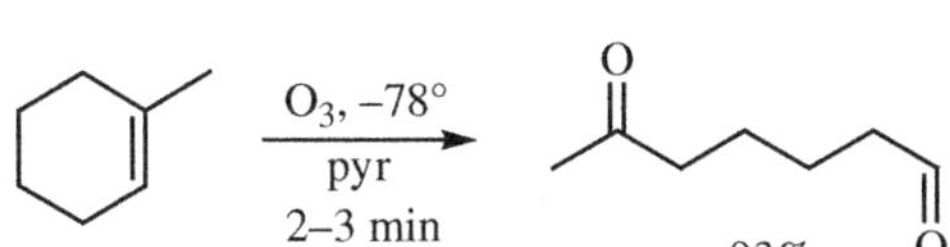

(6.27)

6.2.2 Ketones

Ketones are far less susceptible to overoxidation than aldehydes and are readily prepared by the oxidation of appropriately substituted alkenes and secondary alcohols. The conditions given in Sections 6.1.1 and 6.2.1 for the oxidation of alkenes to acids or aldehydes are applicable to ketones as well.

The oxidation of secondary alcohols is often accomplished by adding CrO_3 and H_2SO_4 in water (Jones reagent) [52] to a solution of the alcohol in acetone. If an excess is avoided, alkene sites are untouched. An inexpensive, high-yielding reagent is aqueous sodium hypochlorite in acetic acid [53]. This gives 2-octanone from the alcohol in 96% yield. Secondary alcohols usually undergo oxidation faster than primary ones, and the selectivity can be high as with the diol in Equation 6.28.

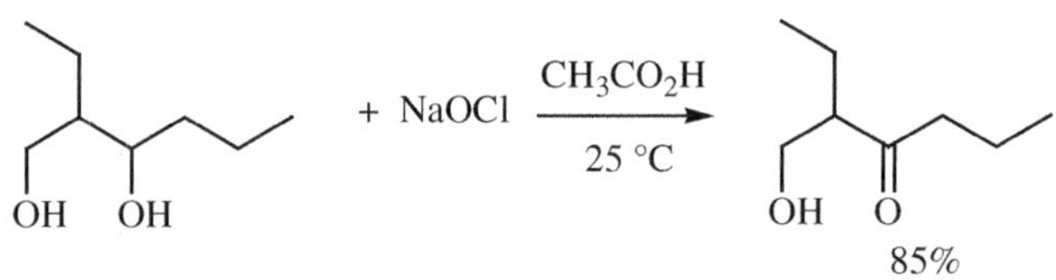

(6.28)

Vinyl halides can be converted to ketones using a cobalt catalyst at room temperature (Eq. 6.29) [54]. This process is valuable in synthesis because the vinyl halides are readily available by the alkylation of carbanions, including enolates, with 2,3-dihalopropene or 1,3-dihalo-2-butene (Eq. 8.8).

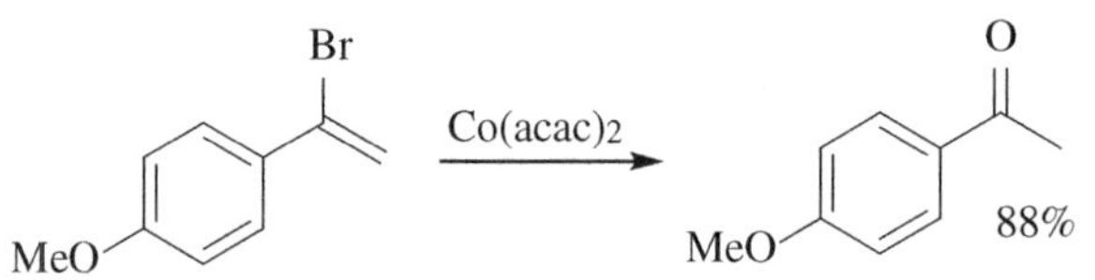

(6.29)

6.2.3 Imines and Enamines

Enamines and enolate anions from imines are useful in carbon–carbon bond formation (Eqs. 7.14, 7.15, and 8.13). Imines, the nitrogen analogs of ketones and aldehydes, are commonly prepared using primary amines and dehydrating conditions [55], as exemplified in Equation 6.30, where the water was removed as a benzene azeotrope [56].

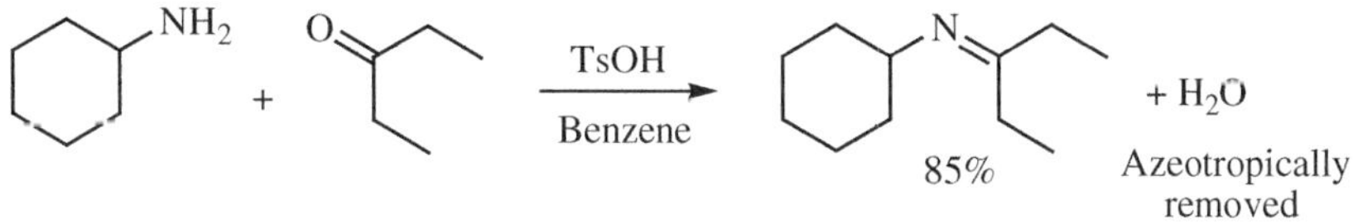

(6.30)

If a secondary amine, commonly pyrrolidine, piperidine, or morpholine, is used with a ketone or aldehyde, elimination of water cannot occur between the nitrogen and the former carbonyl carbon, so it occurs between the carbonyl carbon and an α carbon to form an enamine [57]. An example is shown in Equation 6.31 [58].

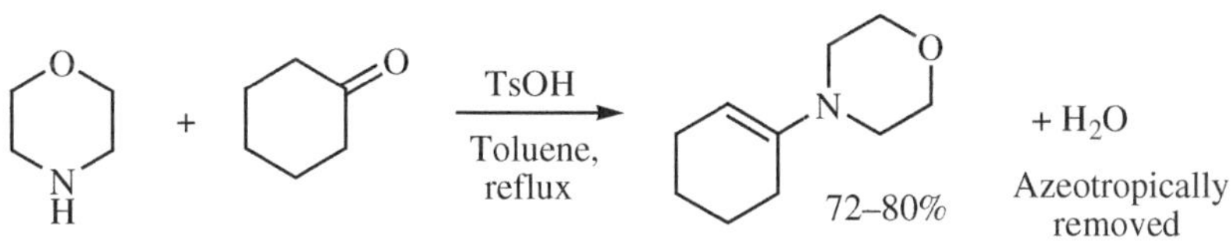

(6.31)

Sterically bulky enamines are difficult to form by this pathway but are easily formed starting from the nitrimine and a urea catalyst. This reaction can be performed without solvent at room temperature (Eq. 6.32) [59].

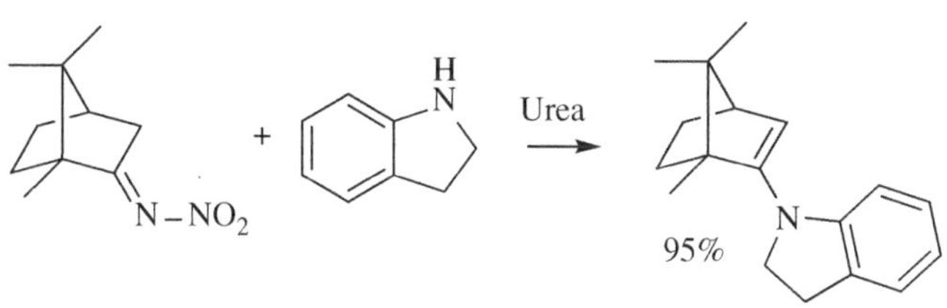

(6.32)

6.2.4 Acetals

Acetals are derivatives of aldehydes and ketones wherein the oxidation level remains the same, but the hybridization of the carbon changes to sp^3. This renders the former carbonyl carbon unattractive to nucleophiles and the acetal is therefore a temporary protecting group.

Aldehydes and ketones are converted to acetals and ketals, respectively, using an alcohol as the solvent with an acid catalyst. In Equation 6.33, trimethyl orthoformate is added and the water reacts with the ortho ester as it forms, eliminating the need to remove water [60]. This procedure results in high yields with a variety of acids, and the low concentration of acid allows success even in the presence of acid labile groups. If a 1,2- or 1,3-diol is used, the cyclic acetal forms readily, even exothermically, but can be difficult to remove.

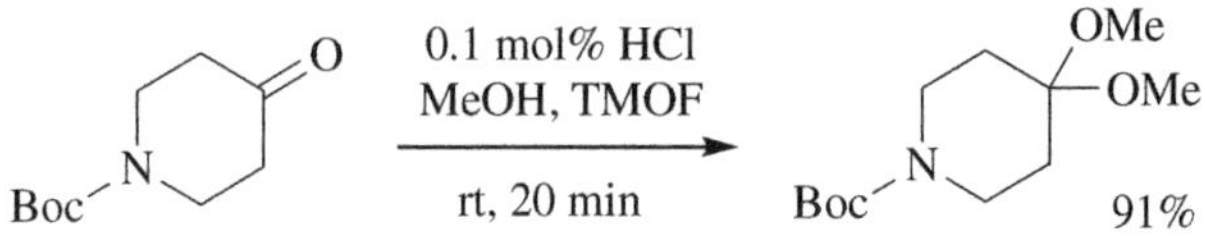

(6.33)

The greater steric hindrance in ketones makes acetal formation more difficult; thus, it is possible to selectively form the acetal on the aldehyde in the presence of the ketone using tetrabutylammonium tribromide as a catalyst (Eq. 6.34) [61]. In this example as well, the ortho ester is added to consume the water as it forms.

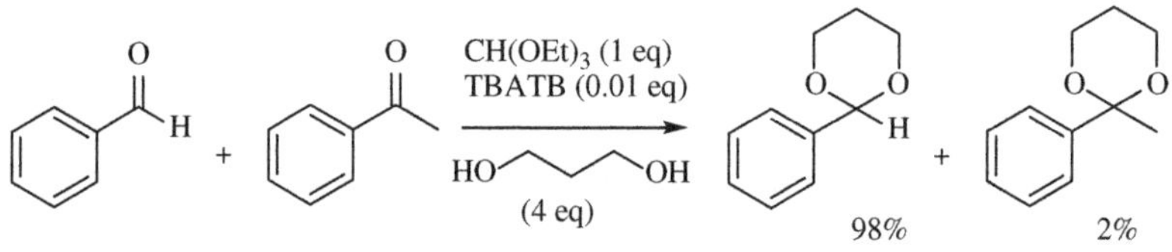

(6.34)

Transacetalization is also useful. Use of the commercially available chiral phosphonic acid catalyst, (S)-3,3′-Bis(2,4,6-triisopropylphenyl)-1, 1′-binaphthyl-2,2′-diyl hydrogenphosphate, ((S)-TRIP), allows enantioselective intramolecular transacetylation (Eq. 6.35) [62].

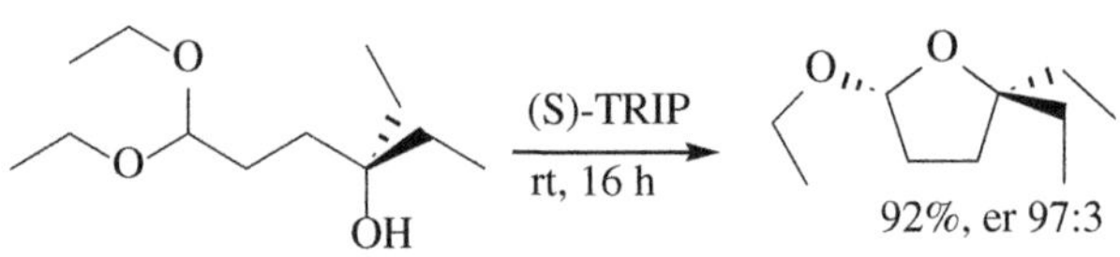

(6.35)

Acetals can also be used to protect alcohols. In these cases, an enol ether such as ethyl vinyl ether or dihydropyran is used in place of the aldehyde, as shown in Equation 6.36 [63]. Under basic conditions, an α-haloether will convert an alcohol to an acetal, as shown in Equation 6.37 [64].

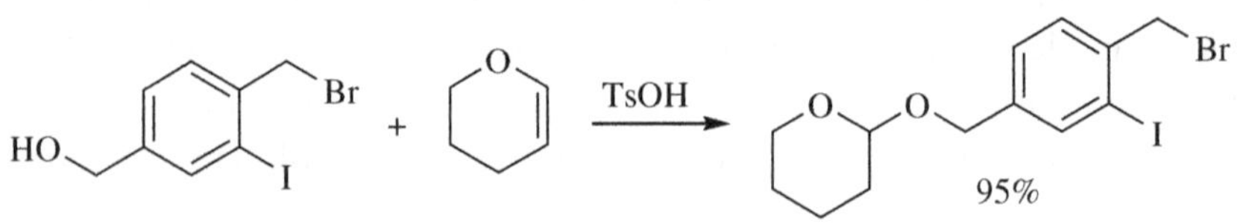

(6.36)

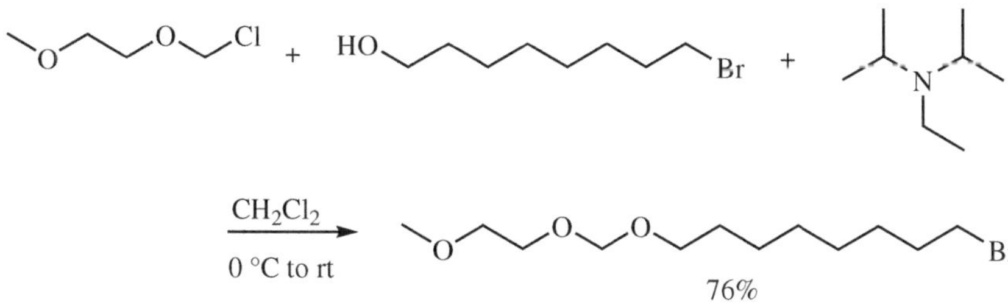

(6.37)

Where acetals are used as temporary protecting groups, they may be removed with aqueous acid to recover the ketone or aldehyde. Deprotection of alcohols is done in aqueous or alcoholic acid. The methoxyethoxyethers (Eq. 6.37) can be removed selectively by cerium (III) chloride in acetonitrile, conditions that leave other acetals intact [65]. This allows concurrent protection of two different alcohol functions in a molecule and the selective deprotection of one of them at an appropriate stage in a synthesis.

6.2.5 Vinyl Ethers

Vinyl ethers, also known as enol ethers, are generally prepared from acetals by acid-catalyzed elimination of an alcohol. The acetals can also be cleaved at −20°C to room temperature using trimethylsilyl trifluoromethanesulfonate and *N,N*-diisopropylethylamine, affording the vinyl ethers in 89–98% yields along with the alkyl trimethylsilyl ether [66].

An ester can be treated with a 1,1-dibromoalkane to give mostly the *Z*-vinyl ether in yields ranging from 52% to 96% depending on the substitution of the reactants, which also affects the *Z/E* ratio. An example is shown in Equation 6.38 [67].

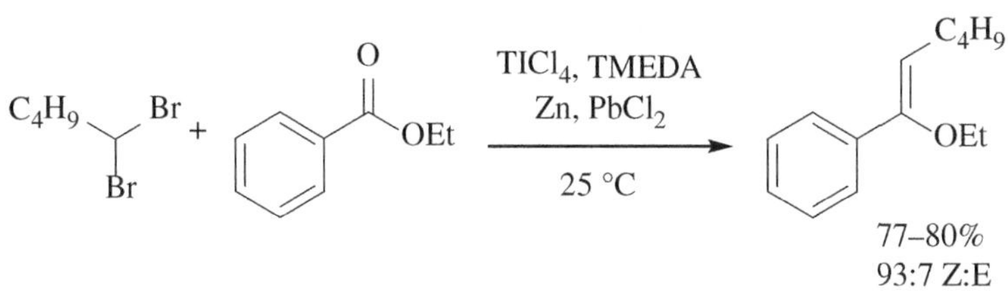

(6.38)

Vinyl ethers are readily formed by the iridium-catalyzed addition of alcohols to vinyl esters (Eq. 6.39) [68]. The catalyst, di-μ-chloro-*bis* (1,5-cyclooctadiene)diiridium(I), $[Ir(cod)Cl]_2$, is commercially available. A number of additional procedures for the synthesis of vinyl ethers are described in the literature [68].

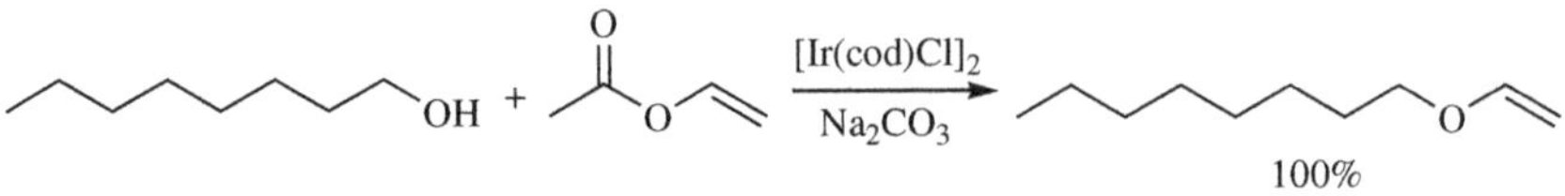

(6.39)

Silyl enol ethers are usually prepared by treating a ketone with trimethylsilyl chloride and triethylamine in refluxing DMF. In unsymmetrical ketones, this gives the more substituted double bond (Eq. 6.40) [69]. When the less substituted product is desired, it is made from the less

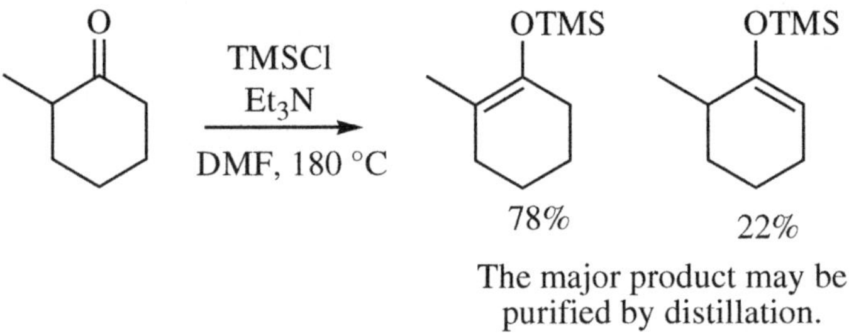

(6.40)

substituted enolate prepared under nonequilibrium conditions where the less hindered proton is removed with a bulky base (Eq. 6.41) [69].

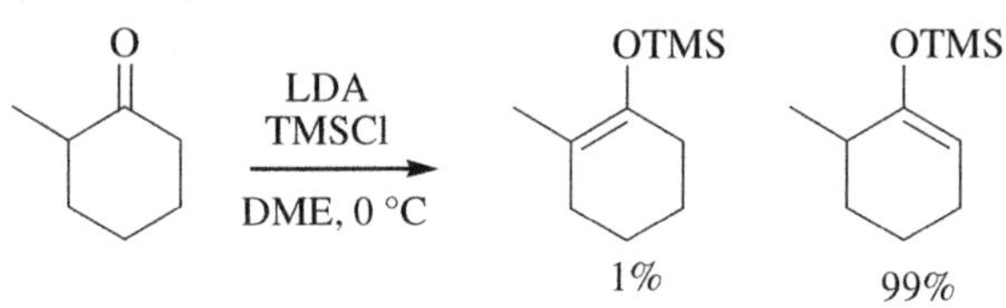

(6.41)

These enol derivatives are used for the formation of carbon–carbon bonds at the α position (Section 7.1.2). They also make the enol double bond available for oxidative cleavage or use in Diels–Alder reactions (Section 8.6).

6.3 ALCOHOLS

Reduction of acids, acid derivatives, aldehydes, and ketones gives alcohols. Esters and acids can be reduced with lithium aluminum hydride. Where selectivity is needed, carboxylic acids may be reduced with sodium borohydride and iodine even in the presence of esters (Eq. 6.42) [70]. Alkenes are unaffected. Diborane will also selectively reduce carboxylic

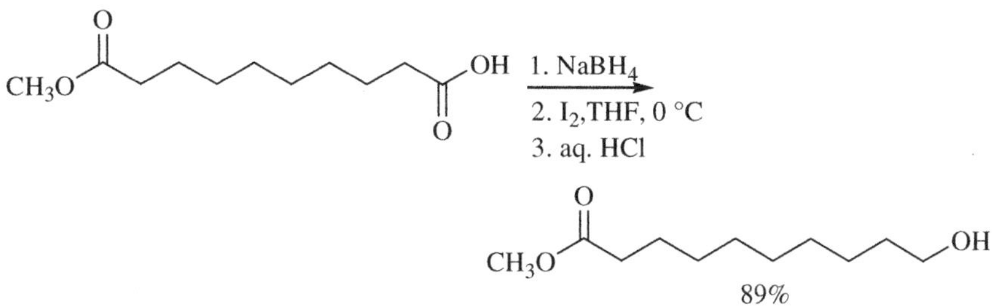

(6.42)

acids in the presence of esters [71]. Aldehydes and ketones are reduced with sodium borohydride or hydrogen over platinum.

Alcohols can be prepared by substitution reactions and addition reactions. Alkali hydroxides will convert appropriate alkyl chlorides, bromides, iodides, and sulfonates to alcohols. Oxymercuration of alkenes gives Markovnikov alcohols, and hydroboration followed by oxidation gives anti-Markovnikov alcohols. Hindered boranes like 9-borabicyclononane (9-BBN) can be used to improve selectivity. Commercially available pyridine borane can be activated with iodine to provide good yields and functional group tolerance (Eq. 6.43) [72].

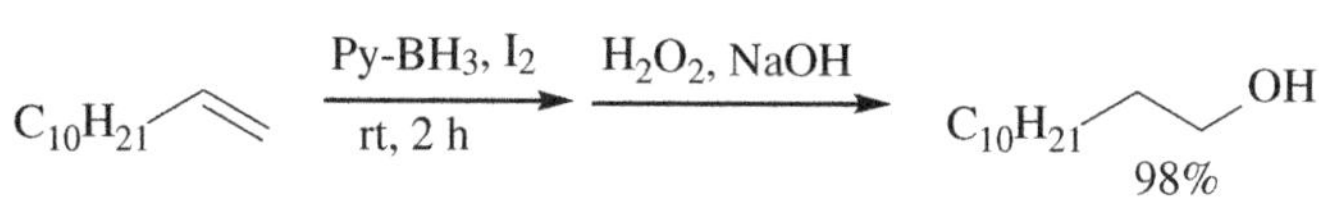

(6.43)

6.4 ETHERS

Like alcohols, ethers are commonly prepared by nucleophilic substitution or by addition to alkenes. In the Williamson method, an alkoxide will displace a halide or sulfate group from a primary carbon, with elimination being a significant problem in more substituted halides. Potassium hydride can be used to generate the ether even for particularly bulky examples, as shown in Equation 6.44 [73]. In the example shown in Equation 6.45, a silver oxide catalyst allowed for mild conditions [74].

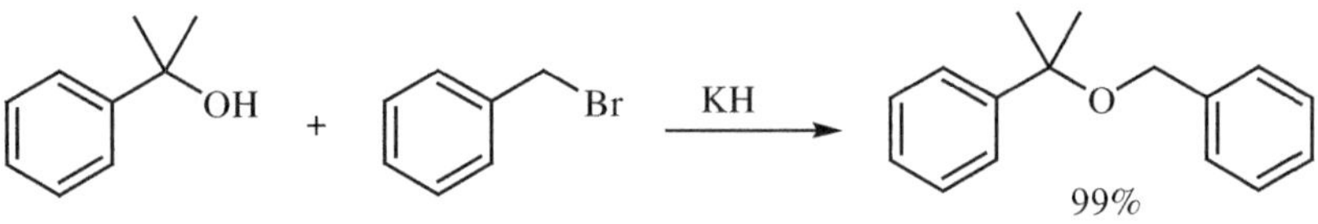

(6.44)

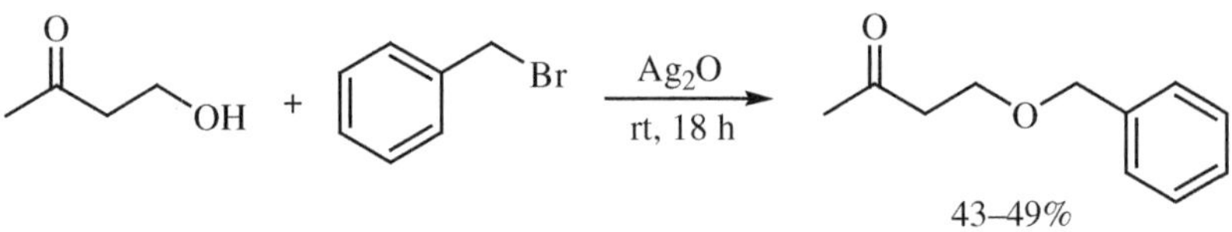

(6.45)

Overall, the addition of alcohols to alkenes is accomplished by alkoxymercuration followed by reduction, as shown in Equation 6.46 [75]. The process gives high-yielding, fast reactions with overall Markovnikov addition. Iron trichloride is a successful catalyst for these reactions (Eq. 6.47) [76].

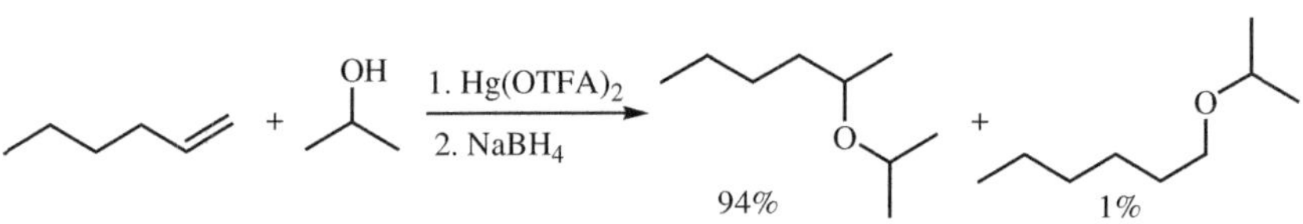

(6.46)

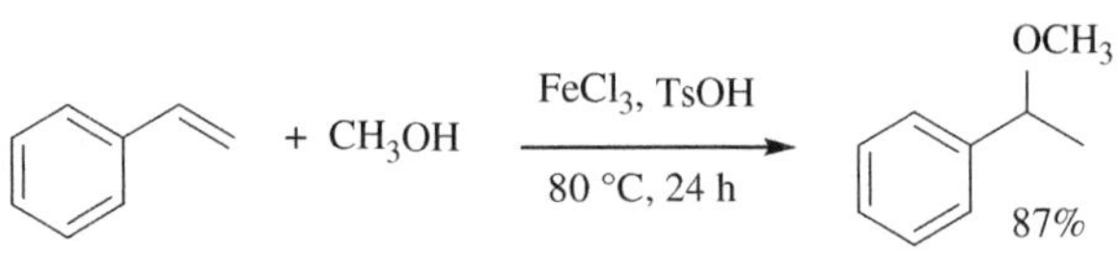

(6.47)

Silyl ethers are commonly used as temporary protection for alcohol groups [77–79]. Trimethylsilyl ethers are unstable toward mild nucleophiles such as methanol, especially in the presence of acid, but *tert*-butyldimethylsilyl ethers are stable toward a variety of mildly basic, reducing, and oxidizing conditions. They are commonly prepared from primary and secondary alcohols by treatment with *tert*-butyldimethylchlorosilane (TBSCl) and an amine in tetrahydrofuran (THF). The chlorosilane and a catalytic amount of 4-(*N,N*-dimethylamino)pyridine (DMAP) can be used to selectively silylate primary alcohols in the presence of secondary alcohols (Eq. 6.48) [80].

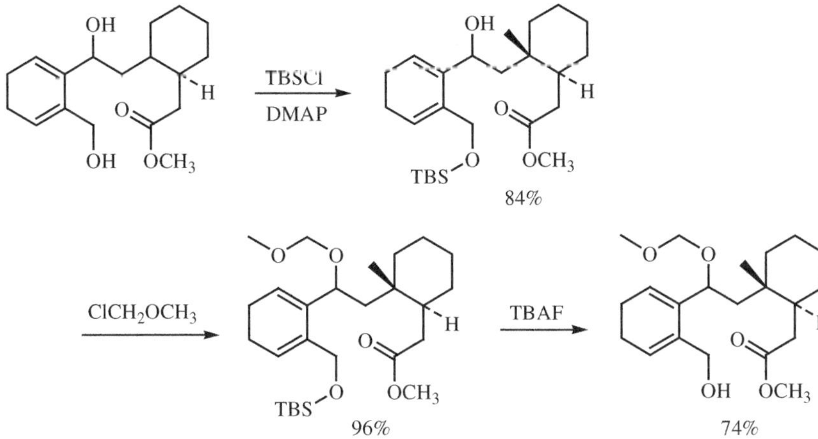

(6.48)

In this example, protection at the primary alcohol was used temporarily to allow acetal protection at the secondary alcohol. The silyl protecting group was then removed selectively to free the alcohol using tetrabutylammonium fluoride in THF, taking advantage of the high affinity of fluoride for silicon. Other reagents include dilute hydrofluoric acid (HF) or acetic acid [27].

6.5 ALKYL HALIDES

Chlorine or bromine may be incorporated by substitution for hydrogen under free radical conditions. This is particularly useful when the substrate is a highly symmetric compound or contains a site especially prone to free radical formation. Otherwise, complex mixtures of isomers are obtained.

Anhydrous hydrogen chloride, bromide, or iodide will add to alkenes by a carbocationic mechanism to give Markovnikov products. The products may have rearranged structures if the first intermediate carbocation can improve in stability by a 1,2-hydride or alkyl shift. Addition to α,β-unsaturated carbonyl compounds affords the β-halo compounds (Eq. 6.49) [81], which are prone to elimination and are thus often isolated

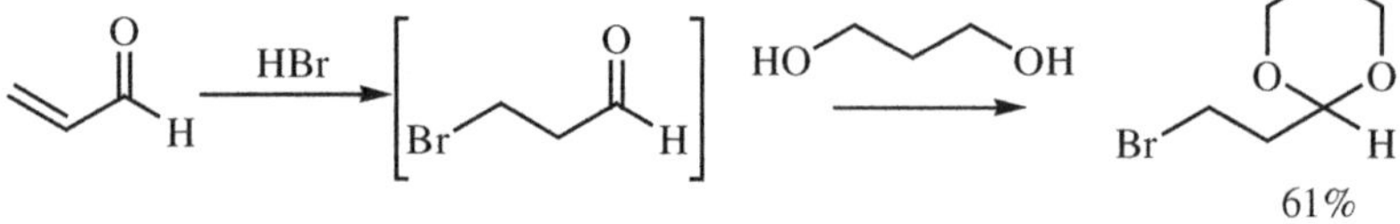

(6.49)

as acetals. Aqueous hydrogen halides may be used with alkenes by heating in the presence of a PTC (Section 9.6.1) such as hexadecyltributylphosphonium bromide (Eq. 6.50) [82]. Only Markovnikov products are formed, even with hydrobromic acid and added peroxides.

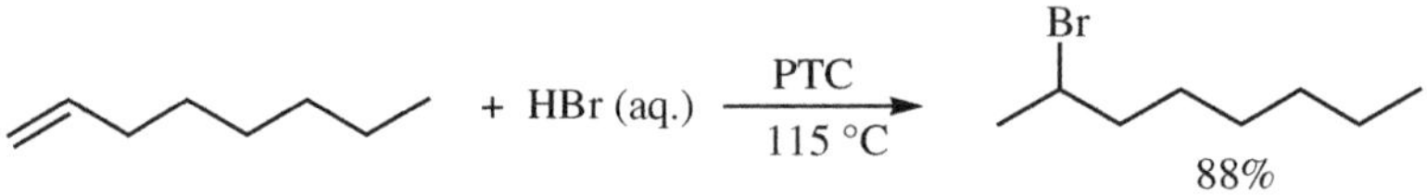

(6.50)

6.5.1 Alkyl Chlorides and Alkyl Bromides

Halogenation of an alcohol may be performed stereoselectively under very mild conditions using 2,4,6-trichloro[1,3,5]triazine and DMF. The reaction occurs with an inversion of configuration (Eq. 6.51) [83]. Adding sodium bromide to the reaction mixture in Equation 6.51 can provide the alkyl bromide in high yield.

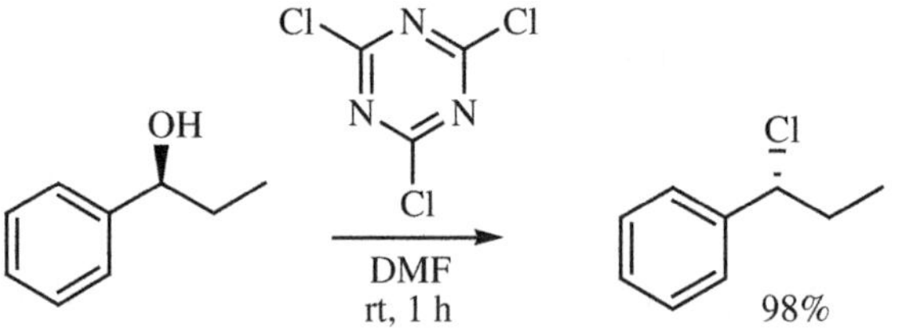

(6.51)

Iminium salts (Vilsmeier reagents) can be readily prepared from DMF and phthaloyl chloride. Stirring in dioxane solution for 3 hours at 40°C results in the precipitation of the product in 78% yield [84]. Because of the hygroscopic nature of the product, it has also been used *in situ* (Eq. 6.52). The Vilsmeier reagent can be used to convert alcohols to alkyl chlorides, or aldehydes to dichlorides [84].

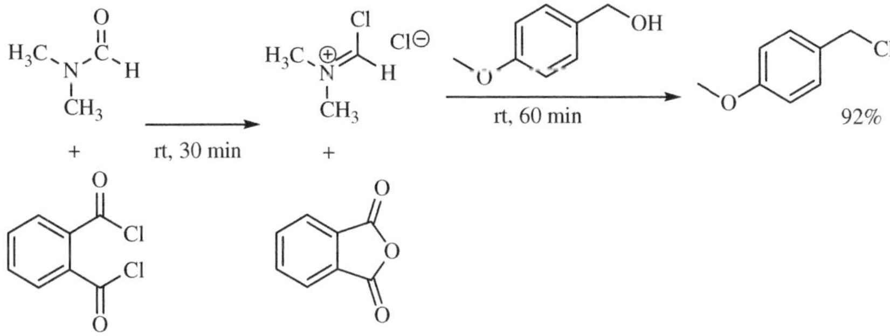

(6.52)

The use of a catalytic amount of triphenylphosphine oxide and oxalyl chloride in the presence of lithium bromide allows for the conversion of alcohols to alkyl bromides (Eq. 6.53). The use of oxalyl chloride alone allows chlorination. Both reactions occur under mild conditions [85].

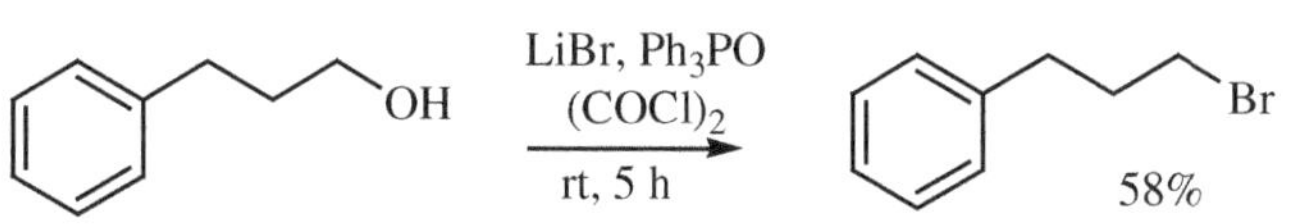

(6.53)

Reductive bromination of carboxylic acids to alkyl halides has been accomplished using 1,1,3,3-tetramethyldisiloxane (TMDS) and trimethylbromosilane (TMBS) in the presence of an indium tribromide catalyst. The reaction occurs in high yield in the presence of a variety of functionalities [86].

Allylic hydrogens may be replaced with bromine using *N*-bromosuccinimide (NBS) and a free radical initiator [87, 88]. This is a free radical mechanism, and two structural isomers may result from the attachment at either end of the conjugated radical system. Benzylic hydrogens may be replaced similarly with NBS.

Bromodimethylsulfonium bromide (BDMS) can be used to form α-bromoesters from β-ketoesters (Eq. 6.54) [89]. The same reagent serves to brominate 1,3-diketones in the α position.

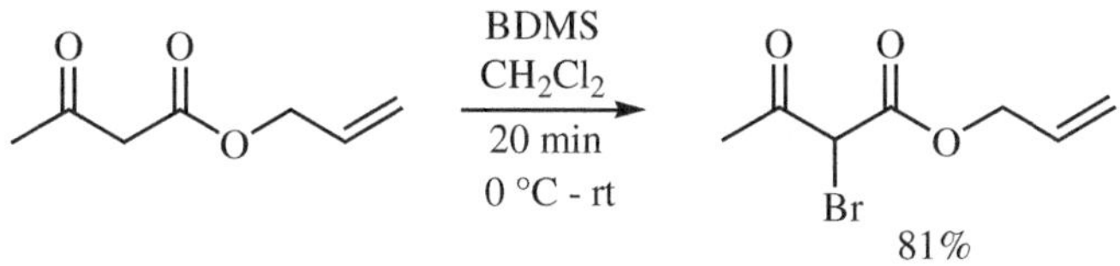

(6.54)

6.5.2 Alkyl Iodides

Primary, secondary, and tertiary alkyl iodides may be prepared from alcohols by treatment with chlorotrimethylsilane in the presence of sodium iodide, without rearrangement, in yields ranging from 78% to 98% (Eq. 6.55) [90].

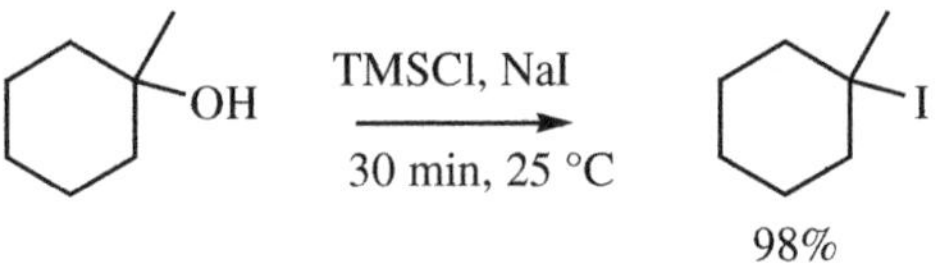

(6.55)

Primary and secondary alcohols may be warmed with commercially available *N,N*-dimethyl-*N*-(methylsulfanylmethylene)ammonium iodide to give high yields of alkyl iodides in the presence of a wide variety of functionalities [91]. Primary alcohols can be iodinated selectively in the presence of secondary alcohols with this procedure. A simple and efficient procedure calls for cerium trichloride and sodium iodide in refluxing acetonitrile to convert primary and secondary alcohols to the corresponding iodides in 69–93% yields [92].

Ketone enol silyl ethers or acetates are readily converted to α-iodoketones in high yield by treatment with a molar equivalent of iodine and copper (II) nitrate in acetonitrile (Eq. 6.56) [93].

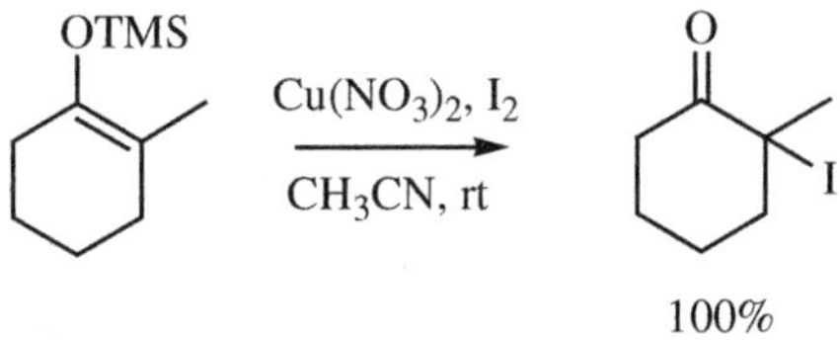

(6.56)

6.5.3 Alkyl Fluorides

Primary, secondary, and tertiary alcohols can be converted directly to fluorides, usually in high yield, by treatment with a mixture of perfluoro-1-butanesulfonyl fluoride (PBSF), triethylamine tris(hydrogen fluoride), and triethylamine (Eq. 6.57) [94]. This combination of reagents tolerates other functionalities and the base present neutralizes HF as it forms.

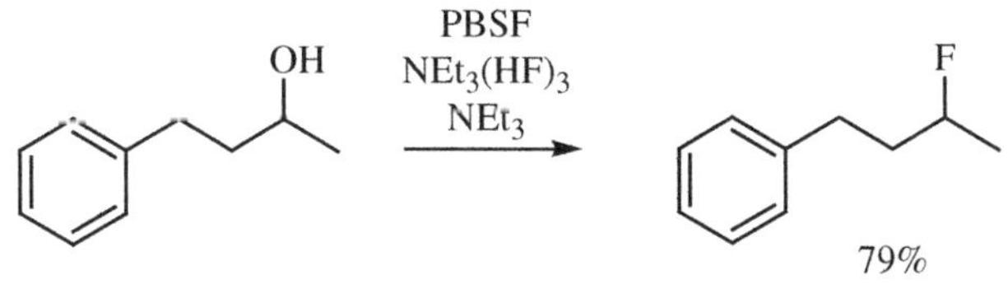

(6.57)

Fluorination of primary, secondary, tertiary, allylic, and benzylic alcohols can be accomplished with diethylaminosulfur trifluoride (DAST), but the reagent reacts violently with moisture and air and is thermally unstable, decomposing explosively above 50°C [95]. An alternative reagent, diethylaminodifluorosulfinium tetrafluoroborate (XtalFluor-E), is more thermally stable and provides similar yields (Eq. 6.58) [96].

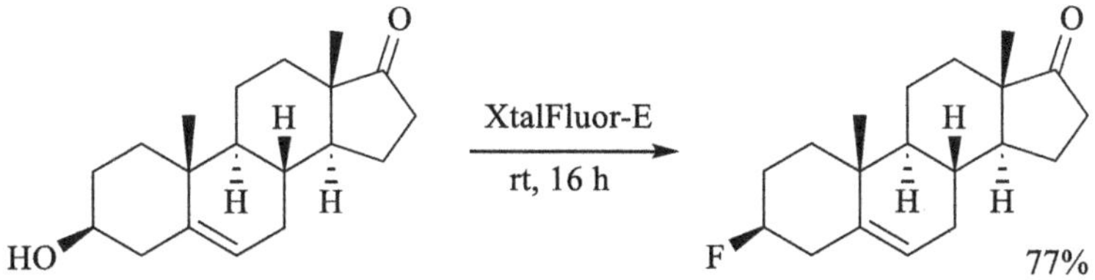

(6.58)

A variety of nucleophilic substitutions, including fluorination, can be accomplished in ionic liquids (Section 9.5) such as 1-*n*-butyl-3-methylimidazolium tetrafluoroborate ([bmim][BF_4]) [97]. Fluoride can replace the hydroxy group on alcohols using triphenylphosphorus and 1,2-difluoroethane as the halogen source (Eq. 6.59) [98]. Chloride, bromide, and iodide can also be added using the appropriate 1,2-dihaloethane. The gem-dihalide can be formed from the aldehyde.

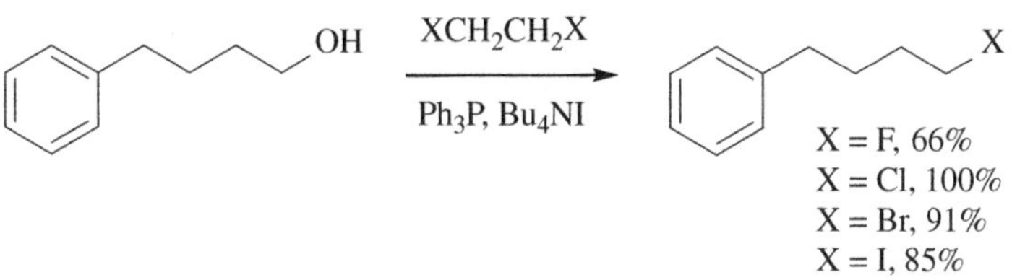

(6.59)

1-Chloromethyl-4-fluoro-1,4-diazoniabicyclo[2.2.2]octane bis(tetrafluoroborate) (Selectfluor or F-TEDA-BF_4) is a water-soluble fluorinating agent with a wide range of applicability [99]. It is possible to accomplish

α-fluorination of the enol acetate in water with a surfactant to assist with the solubility of the organic component (Eq. 6.60) [100].

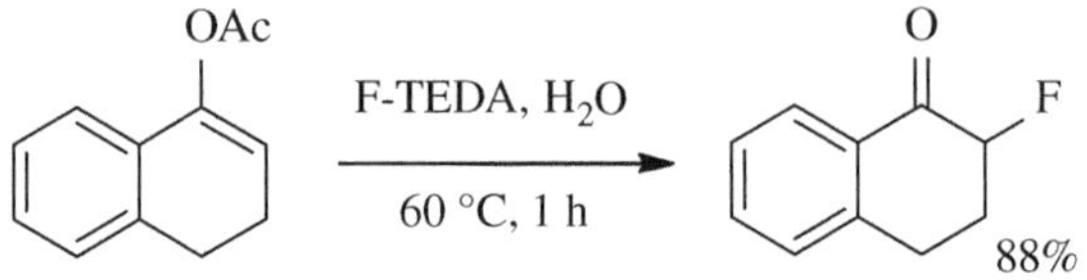

(6.60)

6.6 AMINES

Amines are commonly prepared by treating primary or secondary halides or sulfonates with excess ammonia. The primary amine produced may compete with ammonia to give some secondary and even tertiary amine products, so an excess is used to minimize this. A secondary or tertiary amine may be prepared by treating a primary or secondary amine with an alkyl halide or sulfonate in the same way. Primary amines can be made cleanly using potassium phthalimide instead of ammonia, and a number of improvements to the original procedure have been suggested. Osby et al. treated the phthalimide with sodium borohydride and propanol followed by acetic acid to isolate the primary amine in high yield [101].

Reductive amination of aldehydes and ketones gives primary, secondary, and tertiary amines. Sodium triacetoxyborohydride (STAB-H) [102] or platinum-catalyzed hydrogenations are used, as shown in Equations 6.61 [103] and 6.62 [104].

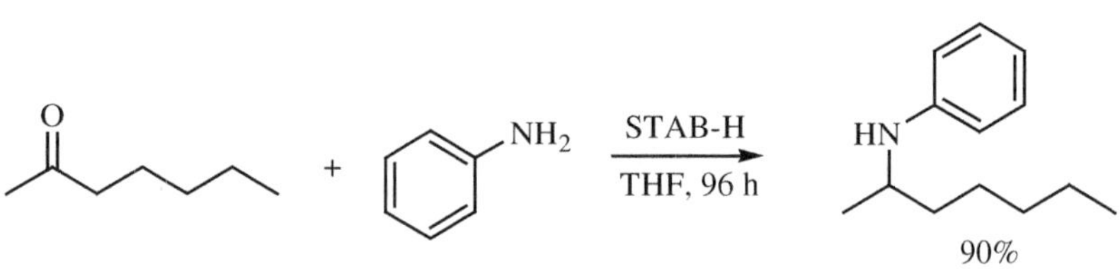

(6.61)

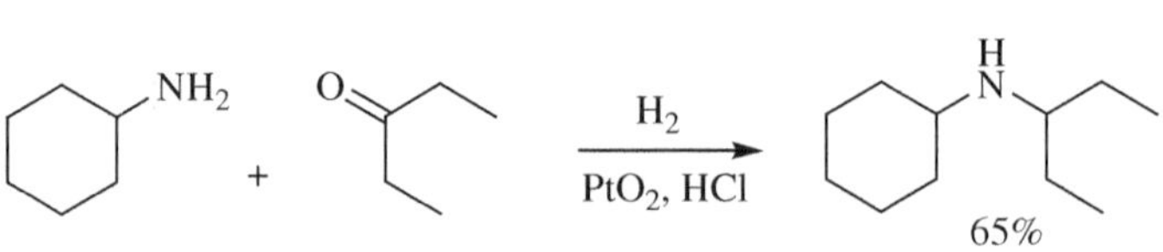

(6.62)

Tertiary alkyl groups cannot be attached to nitrogen by the preceding reactions, and so carbocation methods are used. In the Ritter reaction, an alcohol is converted to a carbocation that attacks the nitrogen of the cyanide ion to give, after hydration, a hydrolyzable amide (Eq. 6.63) [105].

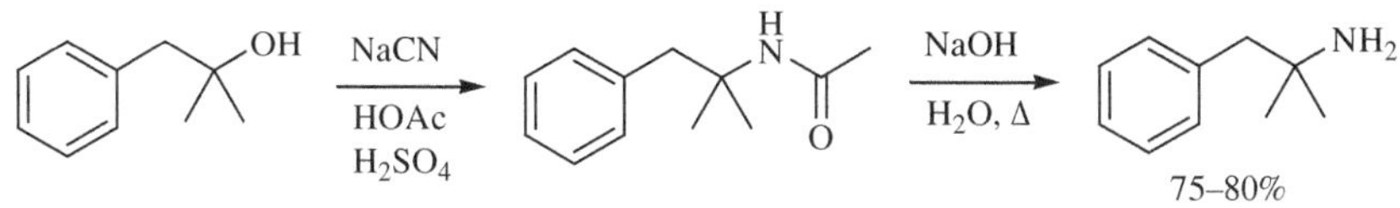

(6.63)

Other nitrogen-containing functional groups may be reduced to give amines. Nitriles may be reduced to primary amines with diisopropylaminoborane [106]. Catalytic hydrogenation can successfully produce amines from amides or nitriles with a careful choice of catalyst [107]. A number of reducing agents such as lithium aluminum hydride or the milder sodium *bis*(2-methoxyethoxy)aluminum hydride (Red-Al®) will reduce amides to secondary or tertiary amines and will also reduce nitriles, oximes, and nitroalkanes to primary amines [108]. Diisobutylaluminum borohydride, formed by combining sodium borohydride and diisobutylaluminum chloride at room temperature for an hour, shows a wide range of these and other reduction reactions under mild conditions [109].

6.7 ISOCYANATES

Isocyanates have been prepared by treating amine hydrochlorides with the highly toxic phosgene in a hot solvent. Far friendlier synthesis of this functional group is accomplished from the isonitrile [110] using DMSO as the oxidant (Eq. 6.64) [111].

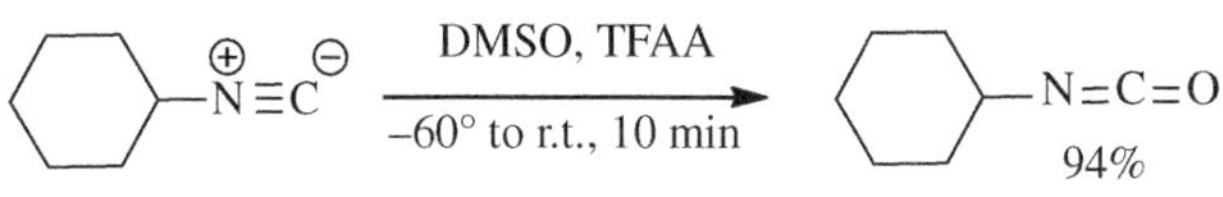

(6.64)

6.8 ALKENES

Alkenes may be made from saturated compounds by various β-elimination reactions. A vicinal dihalide may be dehalogenated by sodium iodide or activated zinc to give a double bond specifically between the carbons that bore the halogens. One practical use of such a process is the inversion of the configuration of alkenes. The anti-addition of chlorine followed by net syn elimination using sodium iodide does so, as shown in Equation 6.65 [112]. Anti-elimination may be obtained with activated zinc and acetic acid in DMF [113].

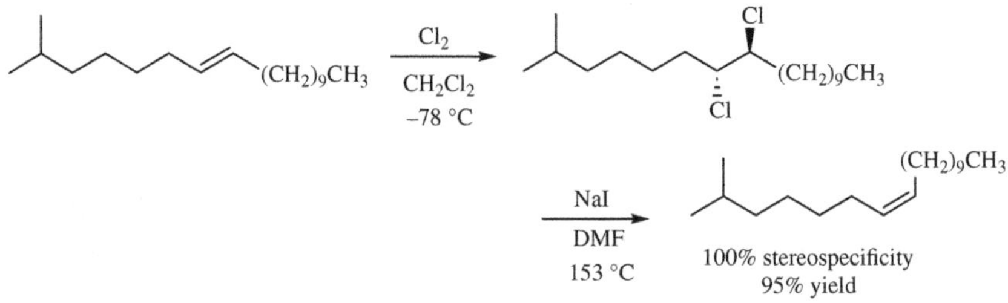

(6.65)

Alkenes are more commonly formed by the elimination of a leaving group and a hydrogen atom. β-Dehydrohalogenation is usually done with strong bases. If substitution competes, sterically hindered, less nucleophilic bases such as potassium *tert*-butoxide are chosen. Here, regiospecificity is a problem because there is usually a choice of β hydrogens. Anti-elimination is usual where conformationally allowed. Among anti-possibilities, conjugated products predominate over the nonconjugated, and more substituted alkenes predominate over less substituted ones. If the alkene will be conjugated, less potent bases may be used, especially when there are other base-sensitive functional groups in the molecule. Triphosgene in dimethylaminopyridine (DMAP) can dehydrate tertiary alcohols with regiocontrol obtained by varying the conditions of the reaction (Eq. 6.66) [114].

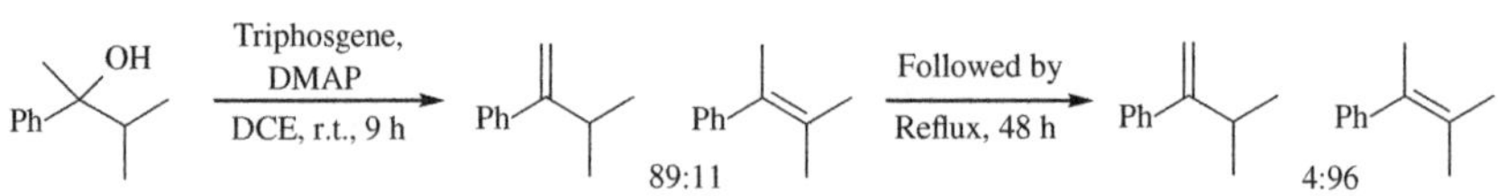

(6.66)

Alcohols undergo elimination by treatment with acid. Rearrangements are likely because carbocations are intermediates (Eq. 6.67) [115].

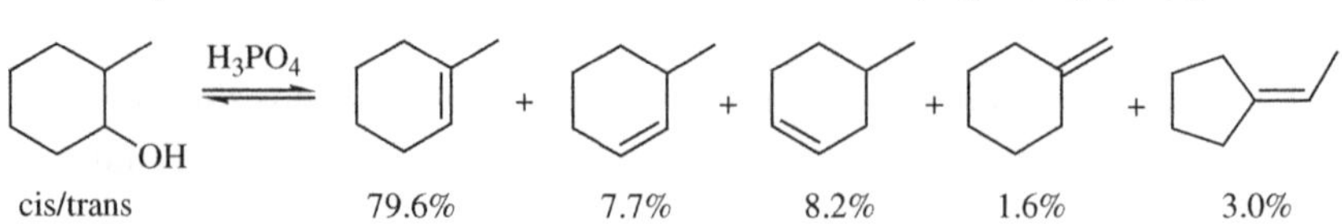

(6.67)

Selenides eliminate readily without a base. They are generally prepared from enolate anions by reaction with diphenyl diselenide or phenylselenyl bromide to give phenyl selenides. The phenyl selenides are oxidized with sodium periodate, hydrogen peroxide, or peracids to the selenoxides, which eliminate even at room temperature to afford the α,β-unsaturated ketones and esters [116].

One-step synthesis of α, β-unsaturated carbonyls is possible via *o*-iodoxybenzoic acid (IBX) [117]. The mild conditions that are typical of this reaction are demonstrated in Equation 6.68 [118].

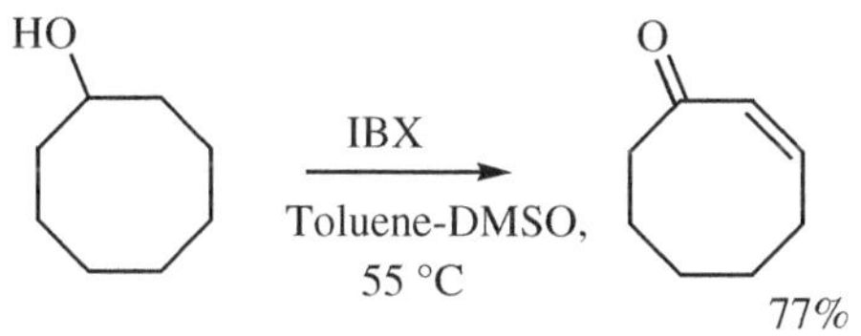

(6.68)

Ketones may be reductively eliminated via their tosylhydrazones [119], as illustrated in Equation 6.69 [120]. If there is a choice, the least substituted alkene will predominate. Trisubstituted alkenes require lithium diisopropylamide (LDA) in place of the alkyllithium [121].

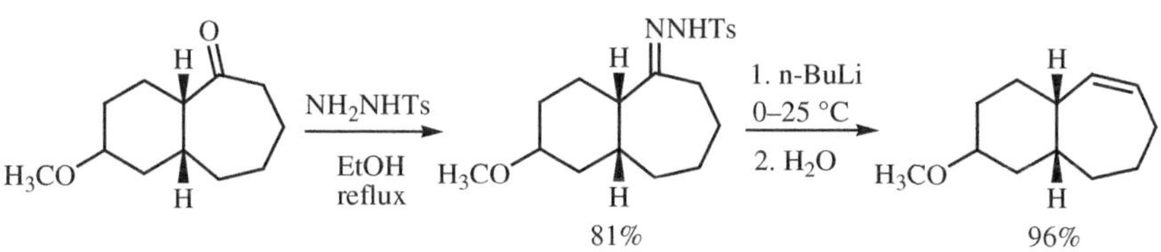

(6.69)

cis-Alkenes may be prepared by partial hydrogenation of appropriate alkynes. P-2 nickel, a hydrogenation catalyst prepared by stirring nickel (II) acetate with sodium borohydride in ethanol, can give high selectivity for the cis-alkene [122]. A convenient method for the reduction of an alkyne to an alkene uses a copper catalyst, as shown in Equation 6.70 [123].

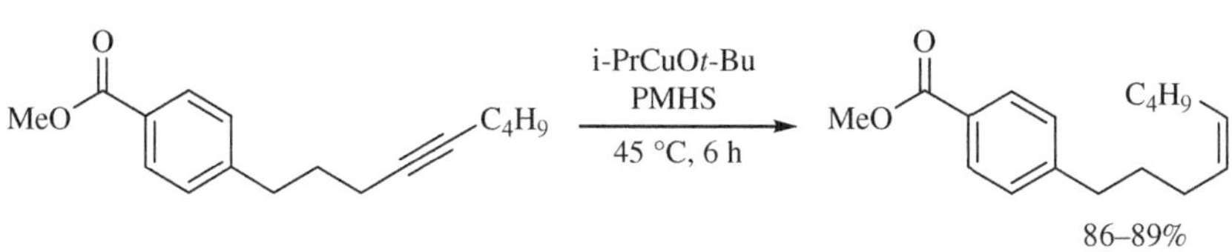

(6.70)

trans-Alkenes are available via dissolving metal reductions of alkynes. Using the catalyst shown in Equation 6.71, the Trost group has reduced alkynes to *trans* alkenes in the presence of a wide variety of functionalities [124]. Carbanions will also add to certain alkynes to give alkenes (Eq. 7.11).

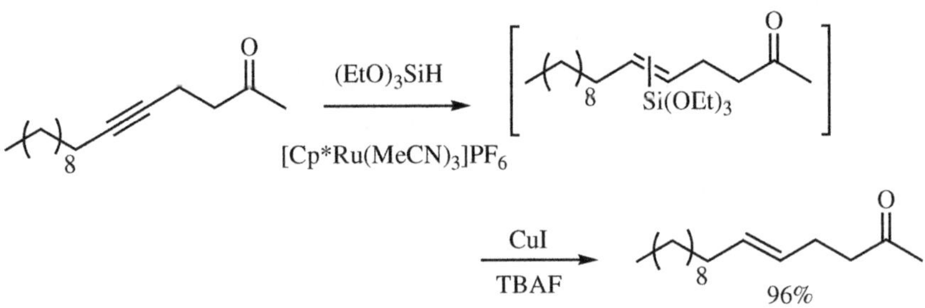

(6.71)

6.9 REDUCTIVE REMOVAL OF FUNCTIONALITY

It is sometimes necessary to reduce a functionalized site to a CH_3 or CH_2 group. Starting from the acid level of oxidation, this is usually done in two stages. For example, an ester may be reduced to an alcohol with $LiAlH_4$, the alcohol is converted to a tosylate, and then it is reduced again to a CH_3 group. Ketones and aldehydes can be reduced completely via the tosylhydrazone using catecholborane followed by decomposition of the hydroboration intermediate, as shown in Equation 6.71 [125].

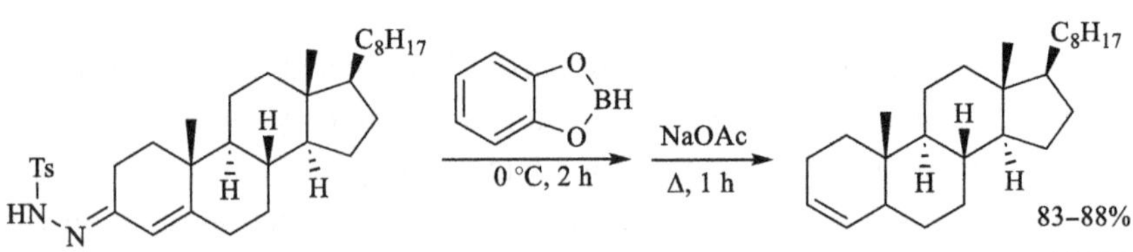

(6.72)

A modification of the Clemmensen reduction allows the removal of an aliphatic ketone under acidic but otherwise mild conditions (Eq. 6.73) [126].

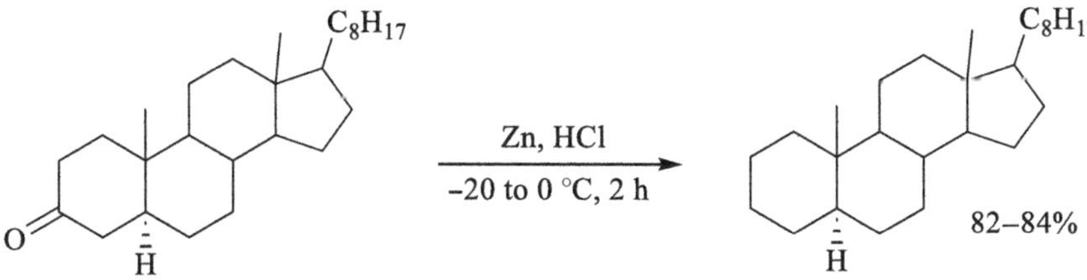

(6.73)

The reduction of alkyl halides has been important in many syntheses. Sodium cyanoborohydride in HMPA will reduce alkyl iodides, bromides, and tosylates selectively in the presence of ester, amide, nitro, chloro, cyano, alkene, epoxide, and aldehyde groups [127]. Tris(trimethylsilyl) silane can be used to remove, sometimes selectively, a variety of functionalities including halogens, alcohols, and carboxylates [128].

RESOURCES

1. The Organic Chemistry Portal allows searching of synthetic methods by the desired bond. https://www.organic-chemistry.org/synthesis/
2. The same website lists examples of named reactions. https://www.organic-chemistry.org/namedreactions/
3. Organic Syntheses provides references to a wide variety of detailed synthetic procedures, all of which have been independently verified. https://www.orgsyn.org/Default.aspx

PROBLEMS

Show how you would prepare each of the following products from the given starting materials. Where more than one step is required, show each step distinctly. Use the literature references as necessary. Questions marked with an asterisk are answered in Appendix C.

6.1*

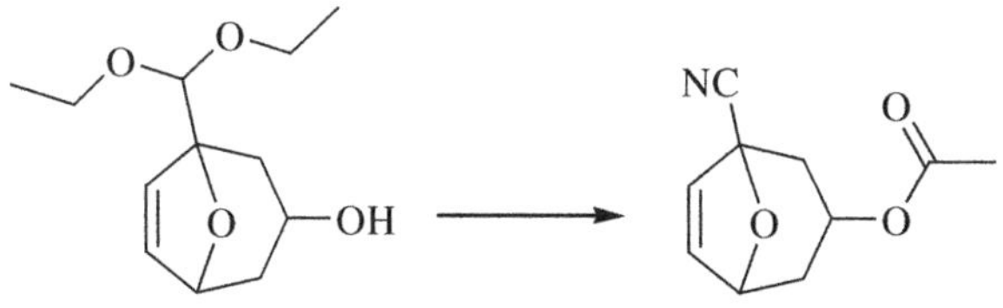

Ref. [129]

6.2

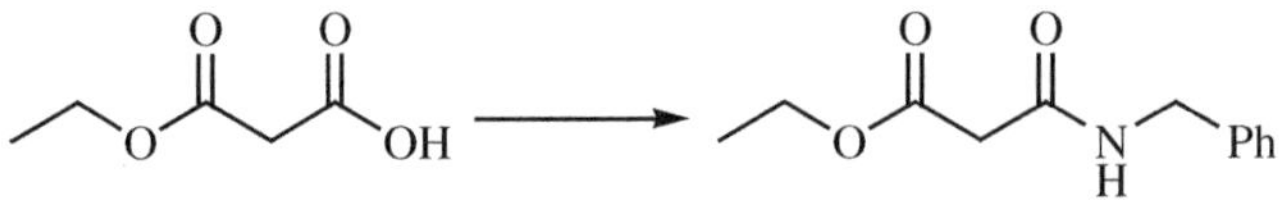

Ref. [130]

6.3

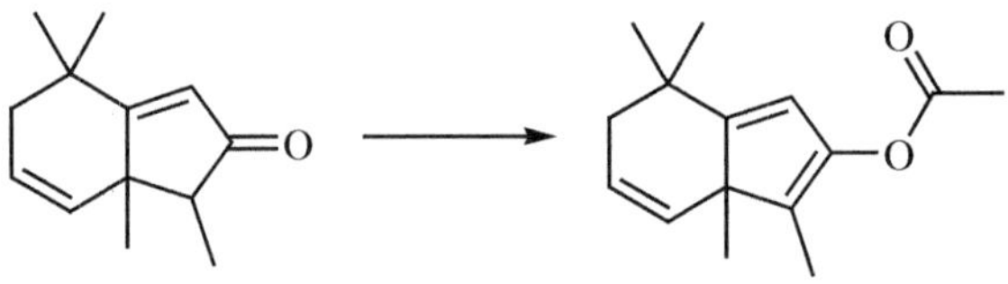

Ref. [131]

6.4

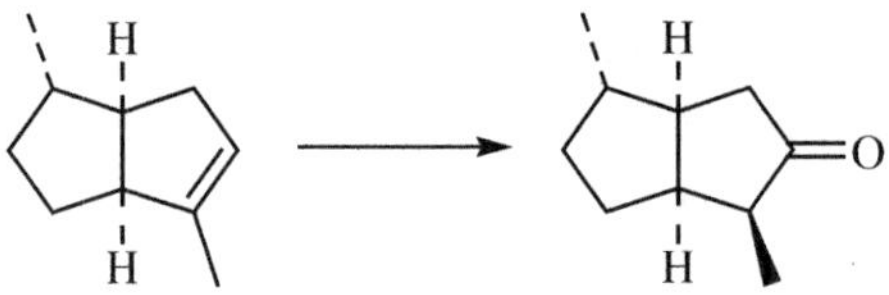

Ref. [132]

6.5*

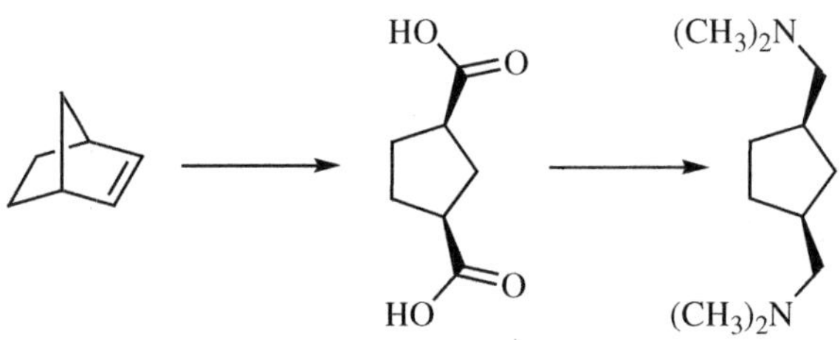

Ref. [133]

6.6*

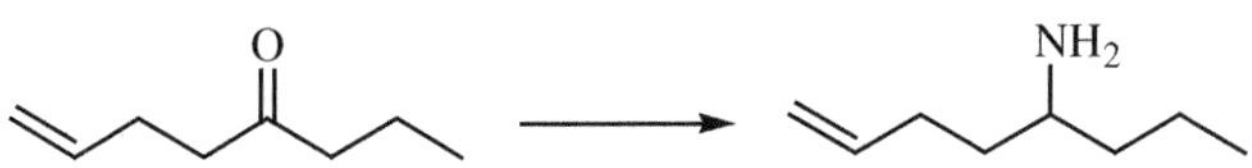

Ref. [134]

6.7

Ref. [135]

6.8

Ref. [136]

6.9*

Ref. [137]

6.10*

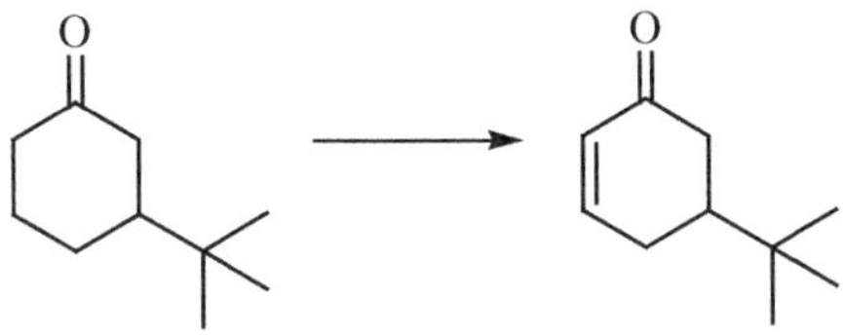

Ref. [138]

6.11

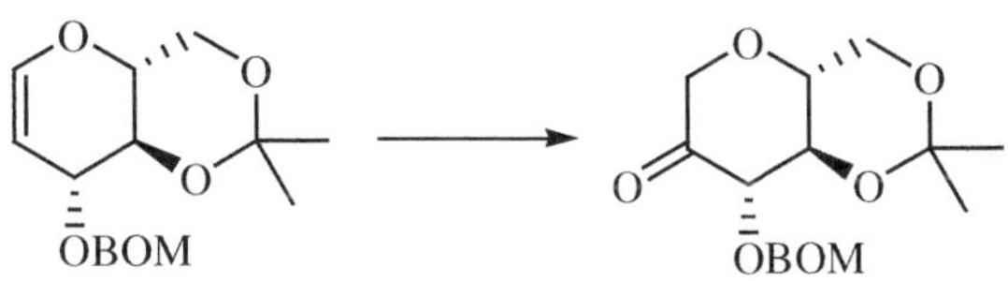

Ref. [139]

6.12

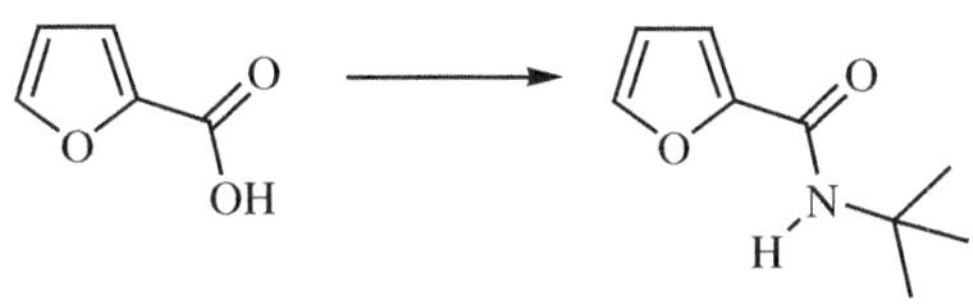

Ref. [140]

6.13

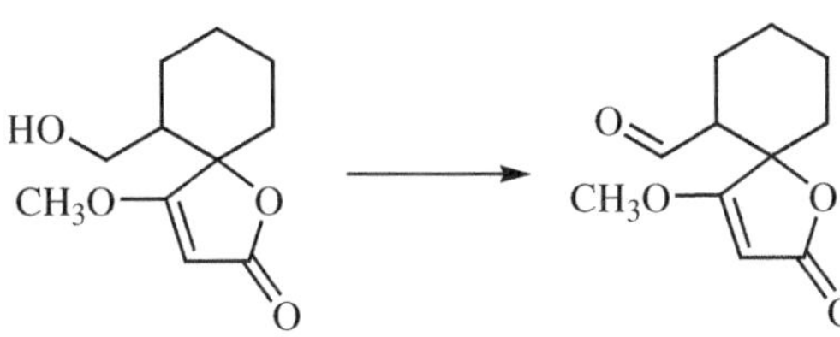

Ref. [141]

6.14

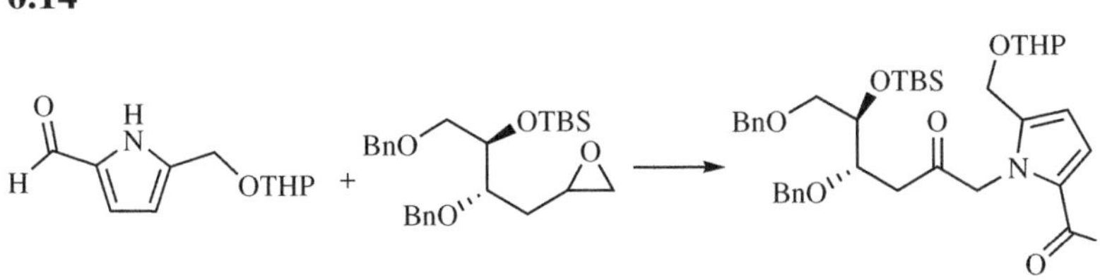

Ref. [142]

6.15*

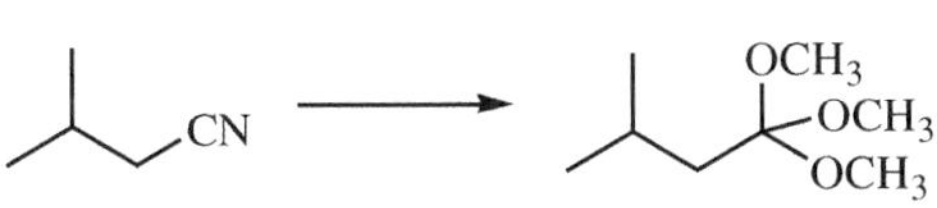

Ref. [143]

6.16

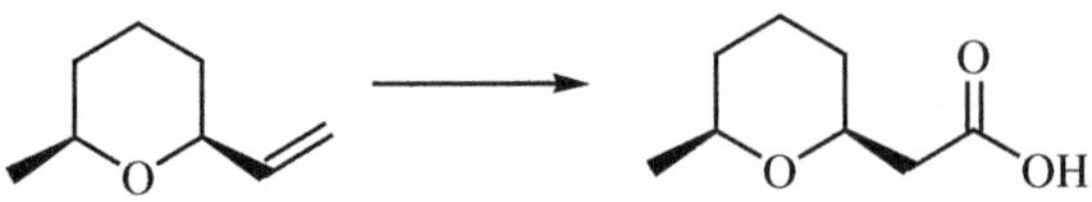

Ref. [144]

6.17*

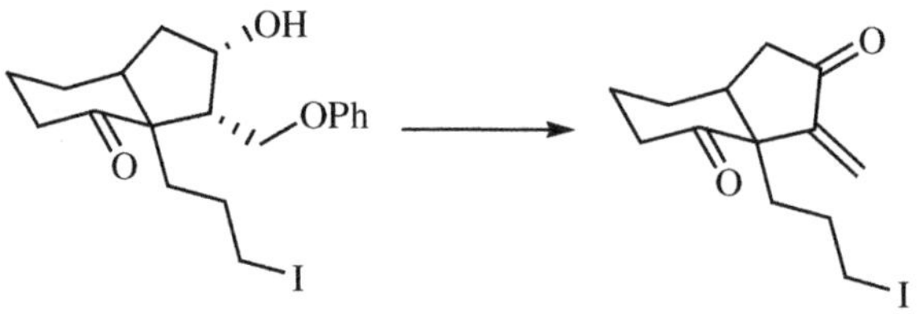

Ref. [145]

6.18*

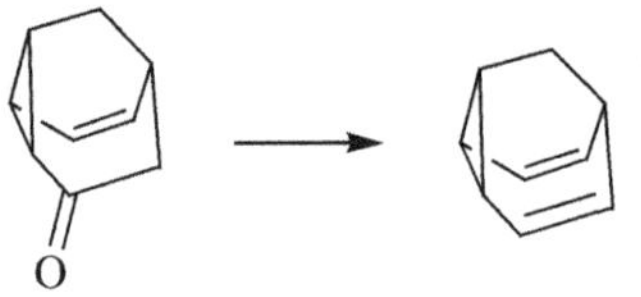

Ref. [146]

6.19

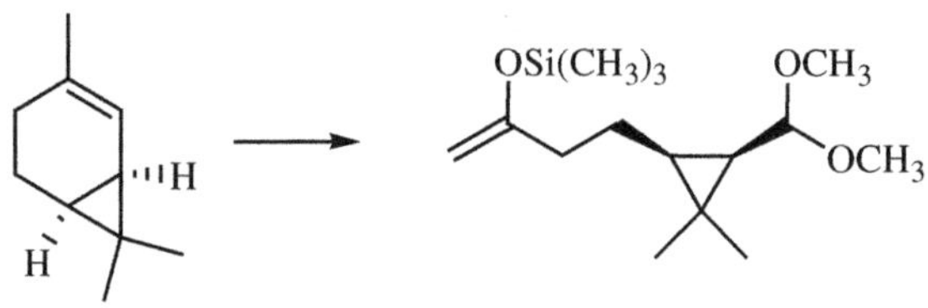

Ref. [147]

6.20

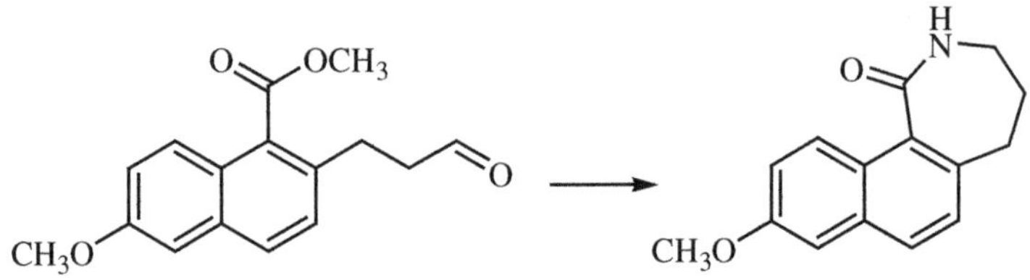

Ref. [148]

6.21

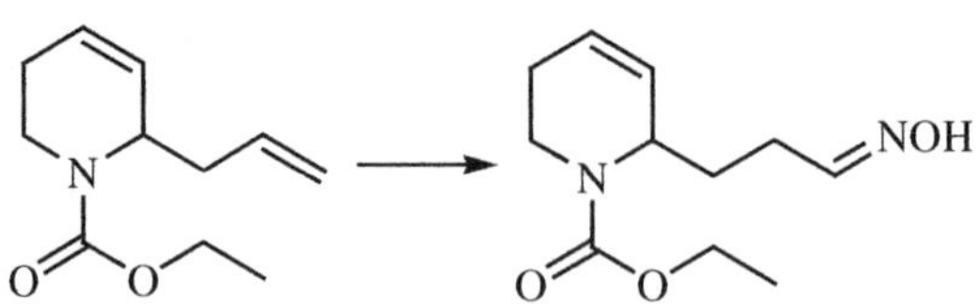

Ref. [149]

6.22*

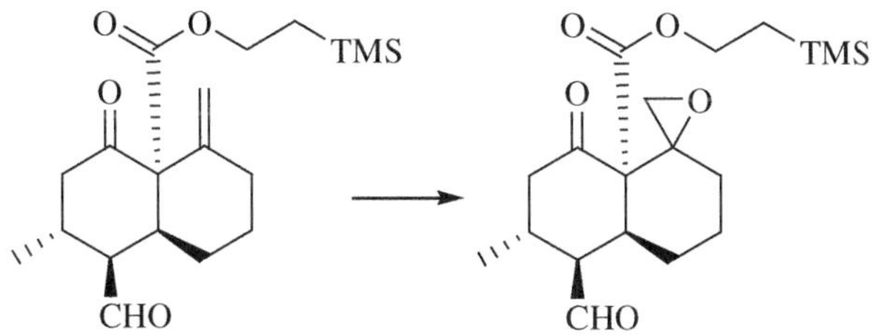

Ref. [150]

6.23

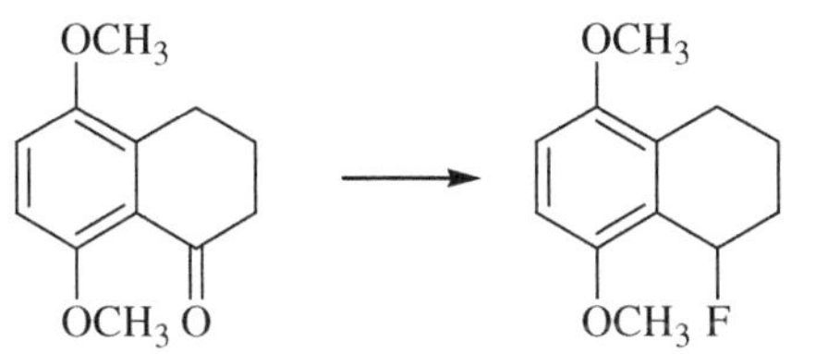

Ref. [151]

6.24

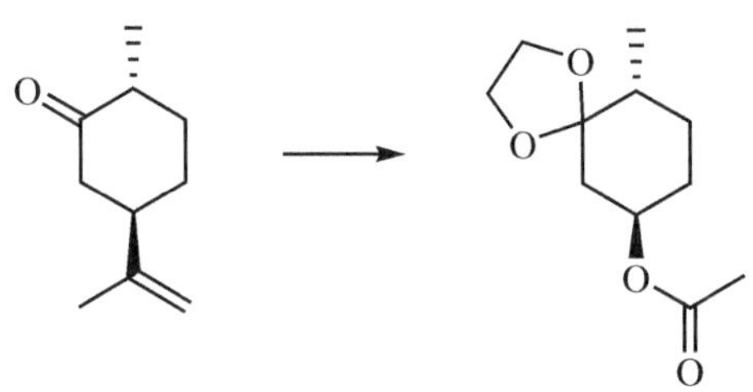

Ref. [152]

6.25*

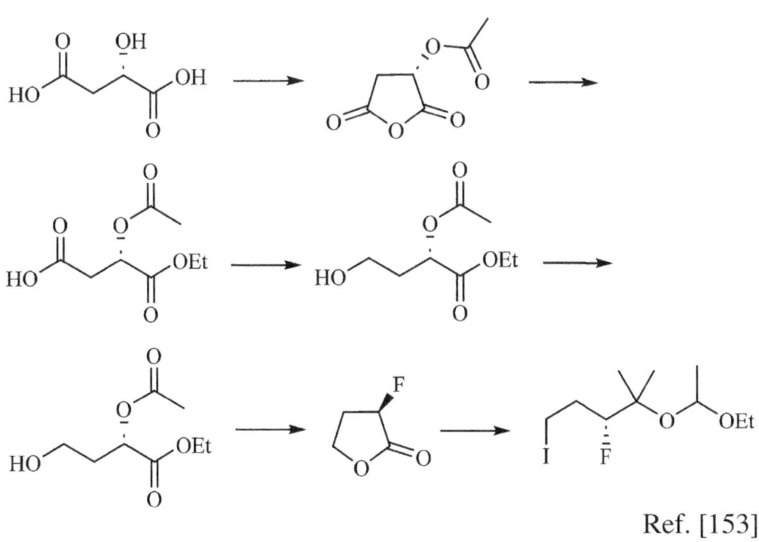

Ref. [153]

6.26

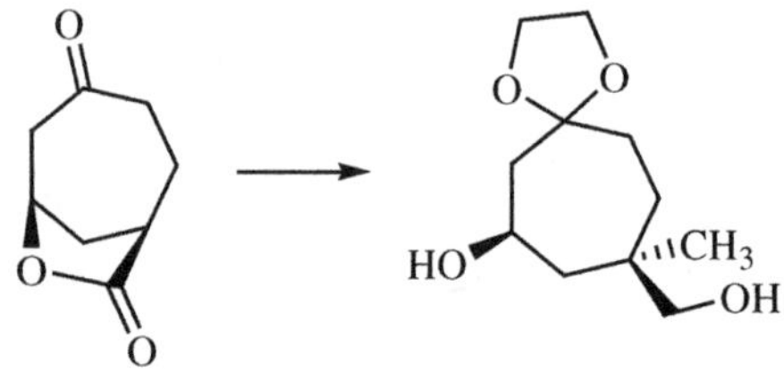

Ref. [154]

6.27

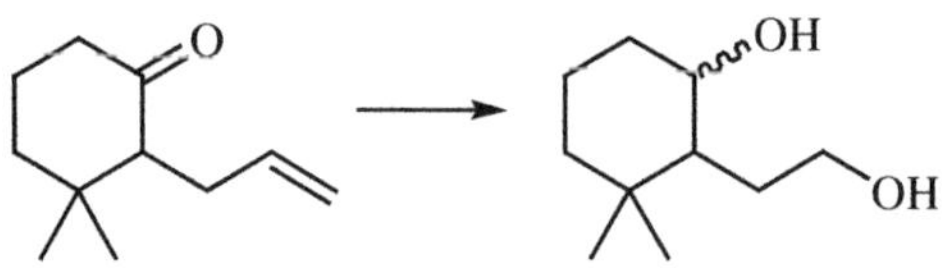

Ref. [155]

6.28*

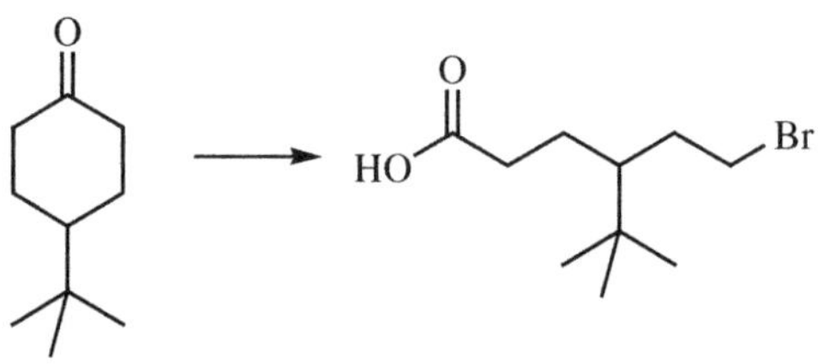

Ref. [156]

6.29*

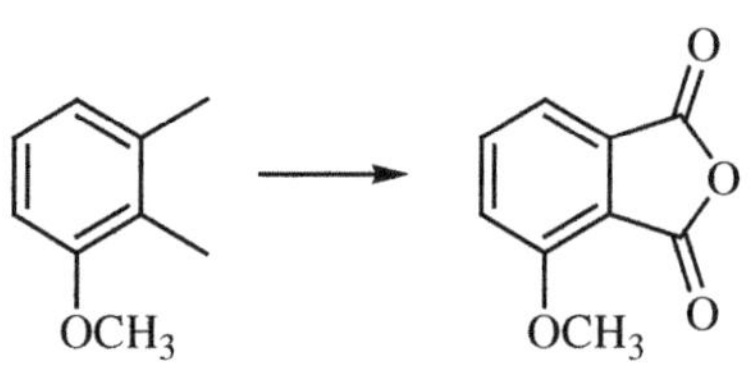

Ref. [157]

6.30

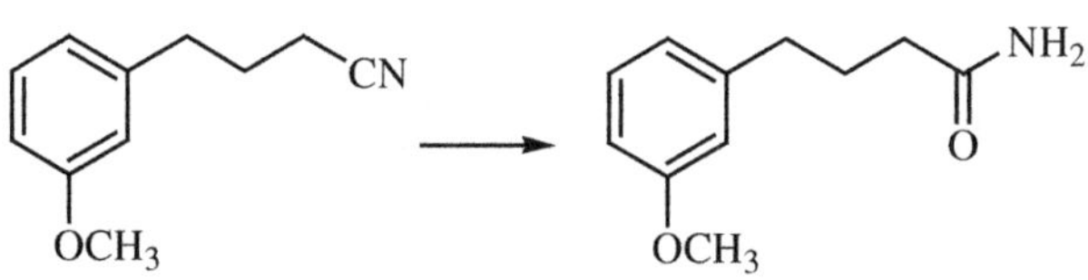

Ref. [158]

6.31

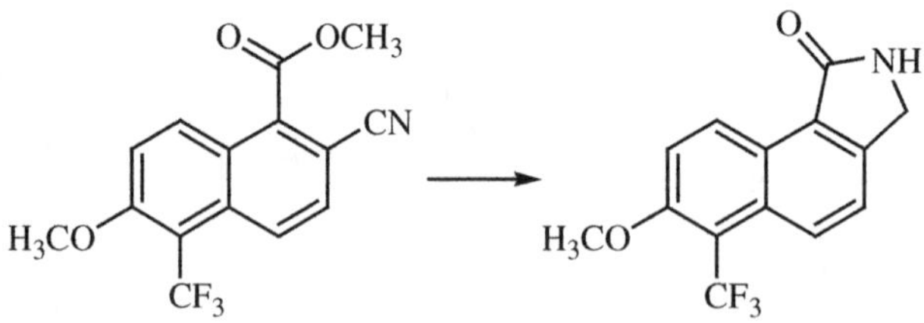

Ref. [148]

6.32*

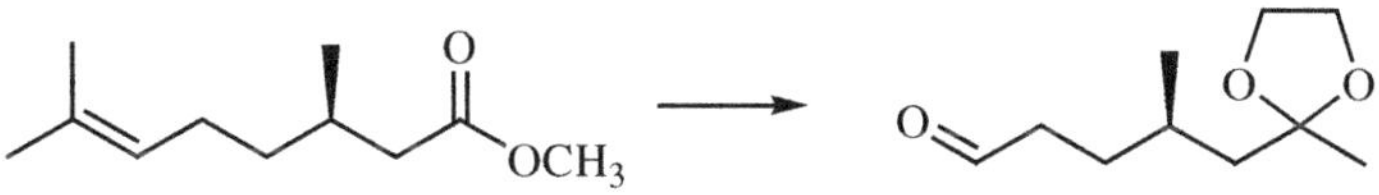

Ref. [159]

6.33

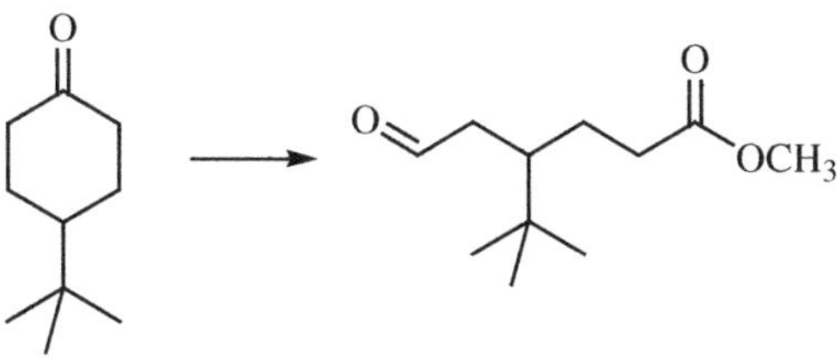

Ref. [160]

6.34*

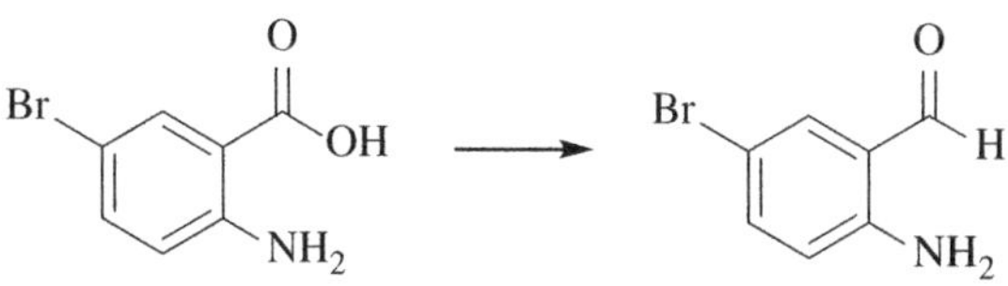

Ref. [161]

6.35

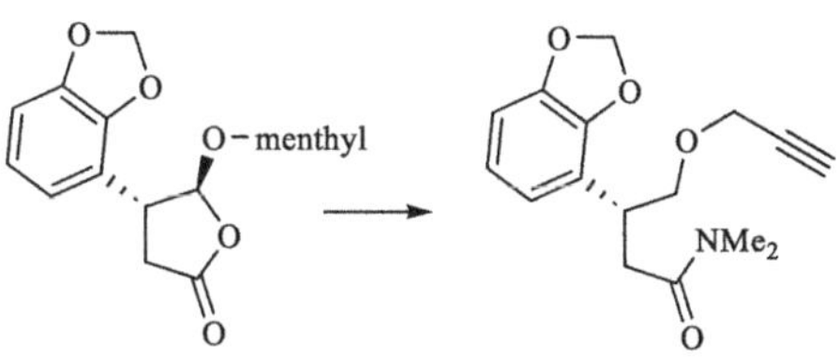

Ref. [162]

6.36*

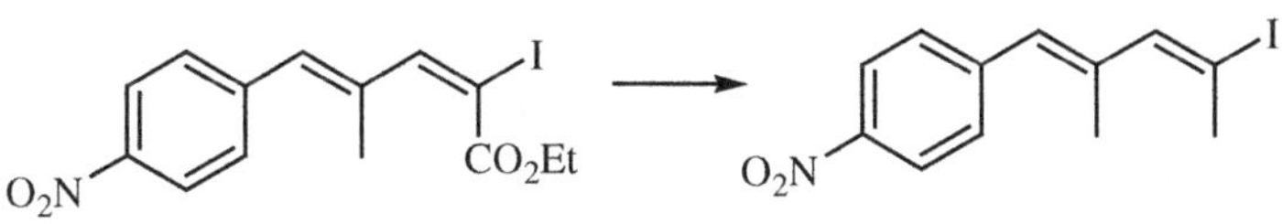

Ref. [163]

6.37

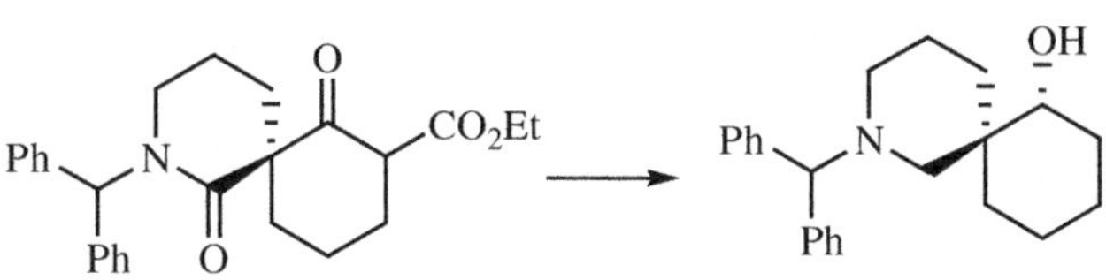

Ref. [164]

6.38

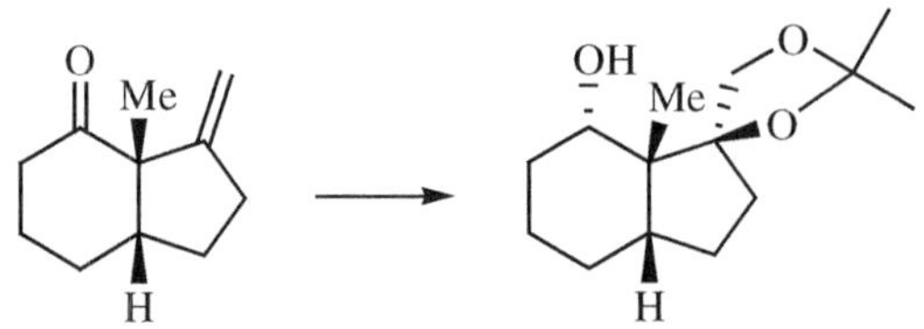

Ref. [165]

REFERENCES

1. Patai, S.; Rappoport, Z.; et al., Eds. *Patai's Chemistry of Functional Groups*; Wiley-Interscience: New York, 1964-present, multiple volumes; available online by subscription.
2. Ferguson, L.; Wims, A. *J. Org. Chem.* **1960**, *25*, 668–670.
3. Treley, W. F.; Marvel, C. S. *Org. Synth.* **1947**, *27*, 86–87.
4. Moorthy, J. N.; Parida, K. N. *J. Org. Chem.* **2014**, *79*, 11431–11439.
5. Reed, S. M.; Hutchison, J. E. *J. Chem. Educ.* **2000**, *77*, 1627–1629.
6. Corey, E. J.; Schmidt, G. *Tetrahedron Lett.* **1979**, *20*, 399–402.
7. Nakamura, K.; Baker, T. J.; Goodman, M. *Org. Lett.* **2000**, *2*, 2967–2970.
8. Kiyotsuka, Y.; Katayama, Y.; Acharya, H. P.; Hyodo, T.; Kobayashi, Y. *J. Org. Chem.* **2009**, *74*, 1939–1951.
9. Tan, W.-Y.; Lu, Y.; Zhao, J.-F.; Chem, W.; Zhang, H. *Org. Lett* **2021**, *23*, 6648–6653.
10. Travis, B. R.; Sivakumar, M.; Hollist, G. O.; Borhan, B. *Org. Lett.* **2003**, *5*, 1031–1034.
11. Pearl, I. A. *Org. Synth.* **1950**, *30*, 101–105.
12. Rich, R. H.; Bartlett, P. A. *J. Org. Chem.* **1996**, *61*, 3916–3919.
13. Nugent, W. A.; McKinney, R. J. *J. Org. Chem.* **1985**, *50*, 5370–5372.
14. Degenhardt, C. R. *J. Org. Chem.* **1980**, *45*, 2763–2766.
15. Burgstahler, A. W.; Weigel, L. O.; Bell, W. J.; Rust, M. K. *J. Org. Chem.* **1975**, *40*, 3456–3458.
16. Svec, R. L.; Hergenrother, P. J. *Angew. Chem. Int. Ed.* **2020**, *59*, 1857–1862.
17. Jones, G.; Raphael, R. A.; Wright, S. *J. Chem. Soc. Perkin Trans.* **1974**, *1*, 1676–1683.
18. Bertolinin, V.; Appiani, R.; Pallavicni, M.; Bolchi, C. *J. Org. Chem.* **2021**, *86*, 15712–15716.
19. Okamoto, N.; Miwa, Y.; Minami, H.; Takeda, K.; Yanada, R. *J. Org. Chem.* **2011**, *76*, 9133–9138.

20. Guthrie, J. L.; Rabjohn, N. *Org. Synth.* **1957**, *37*, 50–51.
21. Moorthy, J. N.; Singhal, N. *J. Org. Chem.* **2005**, *70*, 1926–1929.
22. Lee, J.; Kim, M.; Chang, S.; Lee, H. Y. *Org. Lett.* **2009**, *11*, 5598–5601.
23. Dey, C.; Kündig, E. P. *Org. Synth.* **2014**, *91*, 221–232.
24. Adams, R.; Ulich, L. H. *J. Am. Chem. Soc.* **1920**, *42*, 599–611.
25. Scattolin, T.; Deckers, K.; Schoenebeck, F. *Org. Lett.* **2017**, *19*, 5740–5748.
26. Hurd, C. D.; Christ, R.; Thomas, C. L. *J. Am. Chem. Soc.* **1933**, *55*, 2589–2592.
27. Krimen, L. I. *Org. Synth.* **1970**, *50*, 1–2.
28. Matheson, M.; Pasqua, A. E.; Sewell, A. L.; Marquez, R. *Org. Synth.* **2013**, *90*, 358–366.
29. Mowry, D. T. *Chem. Rev.* **1948**, *42*, 189–283.
30. Kent, R. E.; McElvain, S. M. *Org. Synth.* **1945**, *25*, 61–64.
31. Krynitsky, J. A.; Carhart, H. W. *Org. Synth.* **1952**, *32*, 65–67.
32. Ding, R.; Liu, Y.; Han, M.; Jiao, W.; Tian, H.; Sun, B. *J. Org. Chem.* **2018**, *83*, 12939–12944.
33. Quinn, D. J.; Haun, G. J.; Moura-Letts, G. *Tetrahedron Lett.* **2016**, 3844–3847.
34. Disney, N.; Smyth, M.; Wharry, S.; Moody, T. S.; Baumann, M. *React. Chem. Eng.* **2024**, *9*, 349–354.
35. Casy, G.; Patterson, J. W.; Taylor, R. J. K. *Org. Synth.* **1989**, *67*, 193–201.
36. DeWolfe, R. H. *Synthesis* **1974**, 153–172.
37. Rose, N. G. W.; Blaskovich, M. A.; Evindar, G.; Wilkinson, S.; Luo, Y.; Fishlock, D.; Reid, C.; Lajoie, G. A. *Org. Synth.* **2002**, *79*, 216–227.
38. Rachlin, A. I.; Gurien, H.; Wagner, D. P. *Org. Synth.* **1971**, *51*, 8–10.
39. Fukuyama, T.; Tokuyama, H. *Aldrichimica Acta* **2004**, *37*, 87–96.
40. Takayama, H.; Fujiwara, R.; Kasai, Y.; Kitajima, M.; Aimi, N. *Org. Lett.* **2003**, *5*, 2967–2970.
41. Evans, D. A.; Johnson, J. S. *J. Org. Chem.* **1997**, *62*, 786–787.
42. Marshall, J. A.; Crooks, S. L.; DeHoff, B. S. *J. Org. Chem.* **1988**, *53*, 1616–1623.
43. Marshall, J. A.; Yanik, M. M.; Adams, N. D.; Ellis, K. C.; Chobanian, H. R. *Org. Synth.* **2005**, *81*, 157–170.
44. Webb, D.; Jamison, T. F. *Org. Lett.* **2011**, *14*, 568–571 Britton, J.; Jamison, T. F. *Nat. Protoc.* **2017**, *12*, 2423–2446.
45. Huang, J. H.; Lou, J. D.; Zhu, L. H.; Ping, L. *Molecules* **2005**, *10*, 794–797.

46. Mancuso, A. J.; Huang, S. L.; Swern, D. *J. Org. Chem.* **1978**, *43*, 2480–2482.
47. Hecker, S. J.; Heathcock, C. H. *J. Org. Chem.* **1985**, *50*, 5159–5166.
48. Boeckman, R. K. Jr.; Shao, P.; Mullins, J. J. *Org. Synth.* **2000**, *77*, 141–152 Dess, D. B.; Martin, J. C. *J. Am. Chem. Soc.* **1991**, *113*, 7277–7287.
49. Ansari, I. A.; Gree, R. *Org. Lett.* **2002**, *4*, 1507–1509.
50. Maas, D. D.; Blagg, M.; Wiemer, D. F. *J. Org. Chem.* **1984**, *49*, 853–856.
51. Willand-Charnley, R.; Fisher, T. J.; Johnson, B. M.; Dussault, P. H. *Org. Lett.* **2012**, *14*, 2242–2245.
52. Meinwald, J.; Crandall, J.; Hymans, W. E. *Org. Synth.* **1965**, *45*, 77–79.
53. Stevens, R. V.; Chapman, K. T.; Weller, H. N. *J. Org. Chem.* **1980**, *45*, 2030–2032 Mohrig, J. R.; Nienhuis, D. M.; Linck, C. F.; Van Zoeren, C.; Fox, B. G.; Mahaffy, P. G. *J. Chem. Educ.* **1985**, *62*, 519–521.
54. Ma, X.; Herzon, S. B. *J. Org. Chem.* **2016**, *81*, 8673–8695.
55. Layer, R. W. *Chem. Rev.* **1963**, *63*, 489–510.
56. Pearce, G. T.; Gore, W. E.; Silverstein, R. M. *J. Org. Chem.* **1976**, *41*, 2797–2803.
57. Whitesell, J. K.; Whitesell, M. A. *Synthesis* **1983**, 517–536.
58. Hunig, S.; Lucke, E.; Brenninger, W. *Org. Synth.* **1961**, *41*, 65–66.
59. Nickerson, D. M.; Angeles, V. V.; Mattson, A. E. *Org. Lett.* **2013**, *15*, 5000–5003.
60. Dong, J.-L.; Yu, L.-S.-H.; Xie, J.-W. *ACS Omega* **2018**, *3*, 4974–4985.
61. Gopinath, R.; Haque, S. J.; Patel, B. K. *J. Org. Chem.* **2002**, *67*, 5842–5845.
62. Čorić, I.; Vellalath, S.; List, B. *J. Am. Chem. Soc.* **2010**, *132*, 8536–8537.
63. Nicolaou, K. C.; Wang, J.; Tang, Y.; Botta, L. *J. Am. Chem. Soc.* **2010**, *132*, 11350–11363.
64. Millar, J. G.; Oehlschlager, A. C.; Wong, J. W. *J. Org. Chem.* **1983**, *48*, 4404–4407.
65. Sabitha, G.; Babu, R. S.; Rajkumar, M.; Srividya, R.; Yadav, J. S. *Org. Lett.* **2001**, *3*, 1149–1151.
66. Gassman, P. G.; Burns, S. J. *J. Org. Chem.* **1988**, *53*, 5574–5576.
67. Takai, K.; Kataoka, Y.; Miyai, J.; Okazoe, T.; Oshima, K.; Utimoto, K. *Org. Synth.* **1996**, *73*, 73–84.
68. Okimoto, Y.; Sakaguchi, S.; Ishii, Y. *J. Am. Chem. Soc.* **2002**, *124*, 1590–1591 and references therein; Hirabayashi, T.; Sakaguchi, S.; Ishii, Y. Org. *Synth.* **2005**, *82*, 55–58.

69. House, H. O.; Czuba, L. J.; Gall, M.; Hugh, D.; Olmstead, H. D. *J. Org. Chem.* **1969**, *34*, 2324–2336.
70. Kanth, J. V. B.; Periasamy, M. *J. Org. Chem.* **1991**, *56*, 5964–5965.
71. Frye, S. V.; Eliel, E. L. *J. Org. Chem.* **1985**, *50*, 3402–3404.
72. Clay, J. M.; Vedejs, E. *J. Am. Chem. Soc.* **2005**, *127*, 5766–5767.
73. Huang, H.; Nelson, C. G.; Taber, D. F. *Tetrahedron Lett.* **2010**, *51*, 3545–3546.
74. Tanabe, M.; Peters, R. H. *Org. Synth.* **1981**, *60*, 92–100.
75. Brown, H. C.; Kurek, J. T.; Rei, M. H.; Thompson, K. L. *J. Org. Chem.* **1984**, *49*, 2551–2557.
76. Ke, F.; Li, Z.; Xiang, H.; Zhou, X. *Tetrahedron Lett.* **2011**, *52*, 318–320.
77. Bartoszewicz, A.; Kalck, M.; Nilsson, J.; Hiresova, R.; Stawinski, J. *Synlett.* **2008**, *1*, 37–40.
78. Lalonde, M.; Chan, T. H. *Synthesis* **1985**, 817–845.
79. Corey, E. J.; Venkateswarlu, A. *J. Am. Chem. Soc.* **1972**, *94*, 6190–6191.
80. Funk, R. L.; Daily, W. J.; Parvez, M. *J. Org. Chem.* **1988**, *53*, 4141–4143.
81. Roman, B. I.; De Kimpe, N.; Stevens, C. V. *Chem. Rev.* **2010**, *110*, 5914–5988.
82. Landini, D.; Rolla, F. *J. Org. Chem.* **1980**, *45*, 3527–3529.
83. De Luca, L.; Giacomelli, G.; Porcheddu, A. *Org. Lett* **2002**, *4*, 553–555.
84. Kimura, Y.; Matsuura, D.; Hanawa, T.; Kobayashi, Y. *Tetrahedron Lett.* **2012**, *53*, 1116–1118.
85. Denton, R. M.; An, J.; Adeniran, B.; Blake, A. J.; Lewis, W.; Poulton, A. M. *J. Org. Chem.* **2011**, *76*, 6749–6767.
86. Moriya, T.; Yoneda, S.; Kawana, K.; Ikeda, R.; Konakahara, T.; Sakai, N. *Org. Lett.* **2012**, *14*, 4842–4845.
87. Djerassi, C. *Chem. Rev.* **1948**, *43*, 271–317.
88. Oda, M.; Kawase, T.; Kurata, H. *Org. Synth.* **1996**, *73*, 240–245.
89. Khan, A. T.; Ali, M. A.; Goswami, P.; Choudhury, L. H. *J. Org. Chem.* **2006**, *71*, 8961–8963.
90. Olah, G. A.; Narang, S. C.; Gupta, B. G. B.; Malhotra, R. *J. Org. Chem.* **1979**, *44*, 1247–1251.
91. Ellwood, A. R.; Porter, M. J. *J. Org. Chem.* **2009**, *74*, 7982–7985.
92. Di Deo, M.; Marcantoni, E.; Torregiani, E.; Bartoli, G.; Bellucci, M. C.; Bosco, M.; Sambri, L. *J. Org. Chem.* **2000**, *65*, 2830–2833.
93. Dalla Cort, A. *J. Org. Chem.* **1991**, *56*, 6708–6709.

94. Yin, J.; Zarkowsky, D. S.; Thomas, D. W.; Zhao, M. M.; Huffman, M. A. *Org. Lett.* **2004**, *6*, 1465–1468.
95. Middleton, W. J.; Bingham, E. M. *Org Synth.* **1977**, *57*, 50–52.
96. L'Heureux, A.; Beaulieu, F.; Bennett, C.; Bill, D. R.; Clayton, S.; LaFlamme, F.; Mirmehrabi, M.; Tadayon, S.; Tovell, D.; Couturier, M. *J. Org. Chem.* **2010**, *75*, 3401–3411.
97. Kim, D. W.; Song, C. E.; Chi, D. Y. *J. Org. Chem.* **2003**, *68*, 4281–4285.
98. Chen, J.; Lin, J.-H.; Xiao, J.-C. *Org. Lett.* **2018**, *20*, 3061–3064.
99. Nyffeler, P. T.; Duron, S. G.; Burkart, M. D.; Vincent, S. P.; Wong, C.-H. *Angew. Chem. Int. Ed.* **2005**, *44*, 192–212.
100. Stavber, G.; Zupan, M.; Jareb, M.; Stavber, S. *Org. Lett.* **2004**, *6*, 4973–4976.
101. Osby, J. O.; Martin, M. G.; Ganem, B. *Tetrahedron Lett.* **1984**, *25*, 2093–2096.
102. Abdel-Magid, A. F.; Mehrman, S. J. *Org. Process Res. Dev.* **2006**, *10*, 971–1031.
103. Abdel-Magid, A. F.; Carson, K. G.; Harris, B. D.; Maryanoff, C. A.; Shah, R. D. *J. Org. Chem.* **1996**, *61*, 3849–3862.
104. Stowell, J. C.; Padegimas, S. J. *J. Org. Chem.* **1974**, *39*, 2448–2449.
105. Ritter, J. J.; Kalish, J. *Org. Synth.* **1964**, *44*, 44–46.
106. Haddenham, D.; Pasumansky, L.; DeSoto, J.; Eagon, S.; Singaram, B. *J. Org. Chem.* **2009**, *74*, 1964–1970.
107. Werkmeister, S.; Junge, K.; Beller, M. *Org. Process Res. Dev.* **2014**, *18*, 289–302.
108. Cha, J. S. *Bull. Korean Chem. Soc.* **2011**, *32*, 1808–1846.
109. Amberchan, G.; Snelling, R. A.; Moya, E.; Landi, M.; Lutz, K.; Gatihi, R.; Singaram, B. *J. Org. Chem.* **2021**, *86*, 6207–6227.
110. Wang, X.; Wang, Q.-G.; Luo, Q.-L. *Synthesis* **2015**, *47*, 49–54.
111. Le, H. V.; Ganem, B. *Org. Lett.* **2011**, *13*, 2584–2585.
112. Sonnet, P. E.; Oliver, J. E. *J. Org. Chem.* **1976**, *41*, 3284–3286.
113. Sonnet, P. E.; Oliver, J. E. *J. Org. Chem.* **1976**, *41*, 3279–3283.
114. Ganiu, M. O.; Cleveland, A. H.; Paul, J. L.; Kartika, R. *Org. Lett.* **2019**, *21*, 5611–5615.
115. Friesen, J. B.; Schretzman, R. *J. Chem. Educ.* **2011**, *88*, 1141–1147.
116. Liotta, D. *Acc. Chem. Res.* **1984**, *17*, 28–34.
117. Frigerio, M.; Santagostino, M.; Sputore, S. *J. Org. Chem.* **1999**, *64*, 4537–4538.

118. Nicolaou, K. C.; Zhong, Y.-L.; Baran, P. S. *J. Am. Chem. Soc.* **2000**, *122*, 7596–7597.
119. Shapiro, R. H. *Org. React.* **1976**, *23*, 405–507.
120. Gleiter, R.; Muller, G. *J. Org. Chem.* **1988**, *53*, 3912–3917.
121. Kolonko, K. J.; Shapiro, R. H. *J. Org. Chem.* **1978**, *43*, 1404–1408.
122. Brown, C. A.; Ahuja, V. K. *J. Org. Chem.* **1973**, *38*, 2226–2230.
123. Cox, N.; Dang, H.; Whittaker, A. M.; Lalic, G. *Org. Synth.* **2016**, *93*, 385–400.
124. Trost, B. M.; Ball, Z. T.; Jöge, T. *J. Am. Chem. Soc.* **2002**, *124*, 7922–7923.
125. Kabalka, G. W.; Hutchins, R.; Natale, N. R.; Yang, D. T. C.; Broach, V. *Org Synth.* **1979**, *59*, 42–48.
126. Yamamura, S.; Toda, M.; Hirata, Y. *Org. Synth.* **1973**, *53*, 86–89.
127. Lane, C. F. *Aldrichchimica Acta* **1975**, *8*, 3–10.
128. Chatgilialoglu, C.; Ferreri, C.; Landais, Y.; Timokhin, V. I. *Chem. Rev.* **2018**, *118*, 6516–6572.
129. La Belle, B. E.; Knudsen, M. J.; Olmstead, M. M.; Hope, H.; Yanuck, M. D.; Schore, N. E. *J. Org. Chem.* **1985**, *50*, 5215–5222.
130. Moyer, M. P.; Feldman, P. L.; Rapoport, H. *J. Org. Chem.* **1985**, *50*, 5223–5230.
131. Ishihara, M.; Tsuneya, T.; Shiota, H.; Shiga, M.; Nakatsu, K. *J. Org. Chem.* **1986**, *51*, 491–495.
132. Mathews, R. S.; Whitesell, J. K. *J. Org. Chem.* **1975**, *40*, 3312–3313.
133. Wiberg, K. B.; Bailey, W. F.; Jason, M. E. *J. Org. Chem.* **1976**, *41*, 2711–2714.
134. Harding, K. E.; Burks, S. R. *J. Org. Chem.* **1984**, *49*, 40–44.
135. Kogura, T.; Eliel, E. L. *J. Org. Chem.* **1984**, *49*, 576–578.
136. Bright, Z. R.; Luyeye, C. R.; Morton, A. S. M.; Sedenko, M.; Landolt, R. G.; Bronzi, M. J.; Bohovic, K. M.; Gonser, M. W. A.; Lapainis, T. E.; Hendrickson, W. H. *J. Org. Chem.* **2005**, *70*, 684–687.
137. Unelius, C. R.; Bohman, B.; Lorenzo, M. G.; Troger, A.; Franke, S.; Francke, W. *Org. Lett.* **2010**, *12*, 5601–5603.
138. Gorthey, L. A.; Vairamani, M.; Djerassi, C. *J. Org. Chem.* **1985**, *50*, 4173–4182.
139. Wu, W.; Xiao, M.; Wang, J.; Li, Y.; Xie, Z. *Org. Lett.* **2012**, *14*, 1624–1627.
140. Carpenter, A. J.; Chadwick, D. J. *J. Org. Chem.* **1985**, *50*, 4362–4368.
141. Takeda, K.; Shibata, Y.; Sagawa, Y.; Urahata, M.; Funaki, K.; Hori, K.; Sasahara, H.; Yoshii, E. *J. Org. Chem.* **1985**, *50*, 4673–4681.

142. Sudhakar, G.; Kadam, V. D.; Bayya, S.; Pranitha, G.; Jagadeesh, B. *Org. Lett.* **2011**, *13*, 5452–5455.

143. Wiberg, K.; Martin, E. J.; Squires, R. R. *J. Org. Chem.* **1985**, *50*, 4717–4720.

144. Kim, Y.; Mundy, B. P. *J. Org. Chem.* **1982**, *47*, 3556–3557.

145. Canham, S. M.; France, D. J.; Overman, L. E. *J. Org. Chem.* **2013**, *78*, 9–34.

146. Henkel, J. G.; Hane, J. T. *J. Org. Chem.* **1983**, *48*, 3858–3859.

147. Taylor, M. D.; Minaskanian, G.; Winzenberg, K. N.; Santone, P.; Smith, A. B., III *J. Org. Chem.* **1982**, *47*, 3960–3964.

148. Wrobel, J.; Dietrich, A.; Gorham, B. J.; Sestanj, K. *J. Org. Chem.* **1990**, *55*, 2694–2702.

149. Kozikowski, A. P.; Park, P. *J. Org. Chem.* **1990**, *55*, 4668–4682.

150. Liu, X.; Lee, C.-S. *Org. Lett.* **2012**, *14*, 2886–2889.

151. Chen, C.-P.; Swenton, J. S. *J. Org. Chem.* **1985**, *50*, 4569–4576.

152. Solladie, G.; Hutt, J. *J. Org. Chem.* **1987**, *52*, 3560–3566.

153. Shiuey, S. J.; Partridge, J. J.; Uskokovic, M. R. *J. Org. Chem.* **1988**, *53*, 1040–1046.

154. Baldwin, J. E.; Broline, B. M. *J. Org. Chem.* **1982**, *47*, 1385–1391.

155. Magatti, C. V.; Kaminski, J. J.; Rothberg, I. *J. Org. Chem.* **1991**, *56*, 3102–3108.

156. Paquette, L. A.; Ra, C. S. *J. Org. Chem.* **1988**, *53*, 4978–4985.

157. Hansen, D. W. Jr.; Pappo, R.; Garland, R. B. *J. Org. Chem.* **1988**, *53*, 4244–4253.

158. Venit, J. J.; DiPerro, M.; Magnus, P. *J. Org. Chem.* **1989**, *54*, 4298–4301.

159. Chiang, Y. C. P.; Yang, S. S.; Heck, J. V.; Chabala, J. C.; Chang, M. N. *J. Org. Chem.* **1989**, *54*, 5708–5712.

160. Beckwith, A. L. J.; Zimmerman, J. *J. Org. Chem.* **1991**, *56*, 5791–5796.

161. Hoover, J. M.; Stahl, S. S. *Org. Synth.* **2013**, *90*, 240–250.

162. Ogura, A.; Yamada, K.; Yokoshima, S.; Fukuyama, T. *Org. Lett.* **2012**, *14*, 1632–1635.

163. Lim, H. N.; Parker, K. A. *J. Am. Chem. Soc.* **2011**, *133*, 20149–20151.

164. Park, Y.; Lee, Y. J.; Hong, S.; Lee, M.; Park, H.-G. *Org. Lett.* **2012**, *14*, 852–854.

165. Lee, H.-Y.; Sha, C.-K. *J. Org. Chem.* **2012**, *77*, 598–605.

7

CARBON–CARBON BOND FORMATION

For two carbons to be mutually attractive and join together, they usually begin with opposite-charge polarizations. The nucleophilic carbon has available electrons and is seeking a plus charge or nucleus, and the electrophilic carbon carries a partial or full positive charge. In a basic solution, the nucleophile carries a negative charge because it is bonded to, or associated ionically with, a more electropositive element, that is, a metal. This connection may be direct or may include intervening π-conjugated atoms. The electrophile has a partially positive carbon because of a dipolar bond with a more electronegative atom. In an acid solution, the nucleophile is an alkene or an arene with its projecting π electrons, with or without polarization (but not a carbanion because acid solutions simply protonate them), and the electrophile is a complexed or free carbocation. Examples of each are shown in Table 7.1.

Some carbon–carbon bond-forming reactions occur with little or no charge polarization. Examples include the Diels–Alder, Cope, and Claisen

Understanding Organic Chemistry: Bridging Fundamental and Advanced Courses, Fourth Edition. Ann M. Fabirkiewicz and John C. Stowell.

Companion website: www.wiley.com/go/fabirkiewiczorganic4e

TABLE 7.1 Examples of Nucleophilic and Electrophilic Carbon–Carbon Bond-Forming Agents

Nucleophiles	Electrophiles
CH_3MgBr	CH_3Br
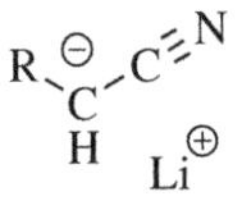	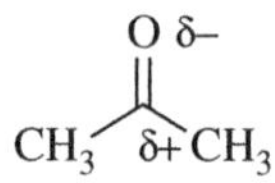
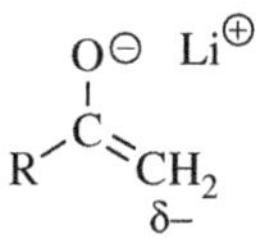	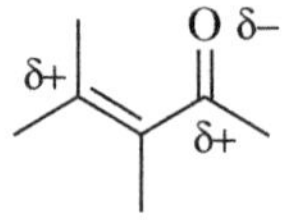
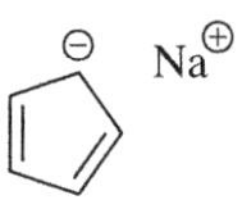	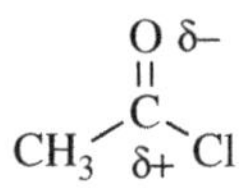
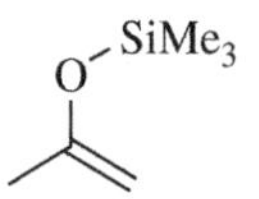	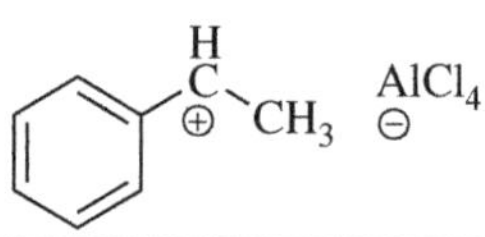

reactions, where concerted bond reorganization occurs (Chapter 5). Others involve transient neutral but electrophilic intermediates, carbenes, and arynes.

Example applications of these in the formation of single and double bonds are shown in the following sections and in Chapter 8 [1, 2].

7.1 CARBON–CARBON SINGLE-BOND FORMATION

7.1.1 Reactions in Basic Solutions

Generation of carbanions for carbon–carbon bond formation is accomplished by several methods. The more electropositive metals, in the metallic state, are strong reducing agents, that is, electron donors. They will react with most organohalides, reducing them to halide ions and negatively charged, strongly basic carbon, as in the synthesis of n-butyl lithium from 1-bromobutane and lithium metal. Carbon will have a complete octet of electrons at the expense of the metal. The new carbon–metal

TABLE 7.2 Some Bases Suitable for Preparing Carbanions

Base	Et_3N	K_2CO_3	NaOH	$NaOC_2H_5$	KO*t*-Bu	LiN(*i*-Pr)$_2$[a]	*t*-BuLi[a]
pK_a [6] (H_2O)	10.7	10.3	15.7		17		
pK_a [7] (DMSO)	18		31.4	29.8	32.2	36	53

[a]The basicity of alkyllithium reagents is even higher in the presence of tetramethylethylenediamine (TMEDA).

bond is substantially ionic, and the reagents are referred to as *carbanions*, with the metals being essentially cations [3]. Magnesium [4], lithium [5], zinc, and sometimes sodium and potassium are used in this way. These carbanions are powerful nucleophiles toward most electrophiles. In some cases, the reactivity is lowered by exchanging less electropositive metal cations (such as copper, mercury, or cadmium) for those initially used in order to obtain selectivity on polyfunctional electrophiles. This direct use of reducing metals on organohalides is commonly the source of simple alkyl, aryl, and vinyl carbanions, and ester enolates.

Another routine method of preparing carbanions begins, not with organohalides but with reagents where carbon carries a hydrogen of sufficient acidity to be removed by a stronger base. A selection of bases for this purpose is given in Table 7.2. The weakest bases are used to generate small equilibrium concentrations of carbanions that can be sufficient for significant carbon–carbon bond formation. Removal of the α-hydrogen of diethyl malonate by sodium ethoxide will give a near-complete formation of a carbanion that is resonance delocalized to two oxygens. Ketone and ester enolates are usually prepared using lithium diisopropylamide (LDA) (Eq. 7.1) [8]. This stronger base has little nucleophilicity because of the high steric hindrance of the isopropyl groups. LDA is strong enough to remove both the acid and α-protons from carboxylic acids, giving a dianion that has high reactivity toward electrophiles at the α position (Eq. 7.2) [9].

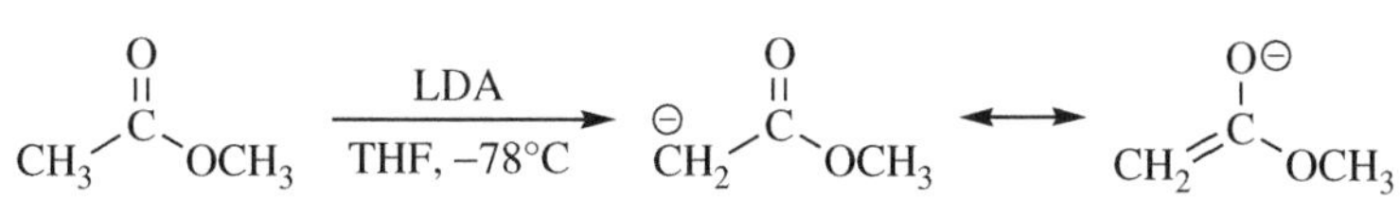

(7.1)

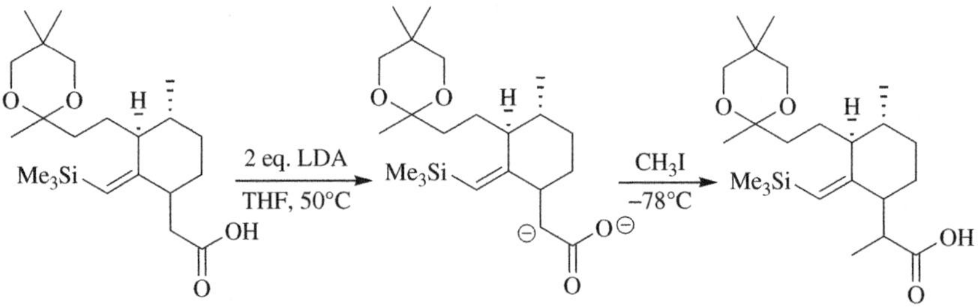

(7.2)

Directed *ortho*-metalation reactions use alkyl lithium bases in TMEDA to generate a carbanion ortho to a heteroatom-containing site. Amide and O-carbamate groups, which serve to coordinate the alkyl lithium, are typical directing groups. Addition of an electrophile completes the reaction [10]. A variation on this reaction uses lithium 2,2,6,6-tetramethylpiperidide (LiTMP) as the base (Eq. 7.3) [11].

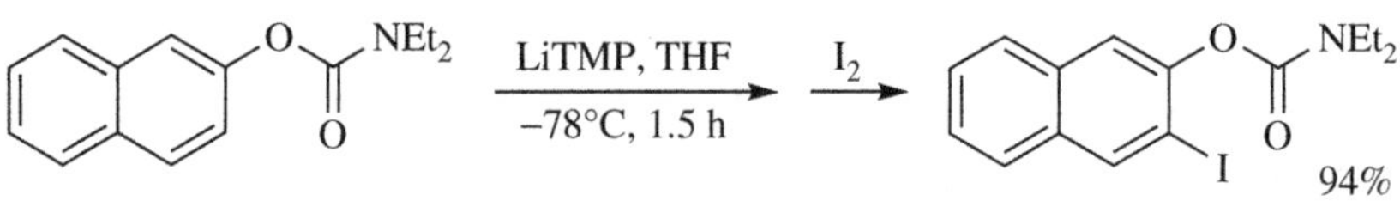

(7.3)

Metal halogen exchange reactions can also be used to prepare carbanions. An aryl or vinyl halide may be treated with an alkyllithium, which gives the aryl or vinyllithium and the alkyl halide (Eq. 7.4) [12]. Vinyl bromide can be converted to vinyllithium by treatment with *tert*-butyllithium in ether at −78°C [13].

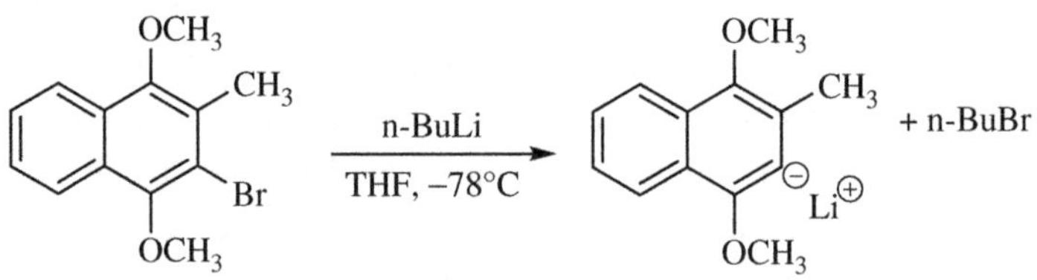

(7.4)

The carbanions from any of these methods are usually combined with electrophiles immediately after they are prepared or even during their preparation. The four common kinds of carbon–carbon bond-forming electrophiles give alkylation, acylation, addition, and conjugate addition.

Alkylations occur in the reaction of carbanions with simple primary or secondary alkyl chlorides, bromides, iodides, or tosylates. For the reaction in Equation 7.5, a weak base was used to generate a small concentration of the resonance delocalized enolate anion [14]. In this case, the electrophile, methyl iodide, is present from the initial stage. The least hindered face of the carbanion attacked to give the diastereomer shown.

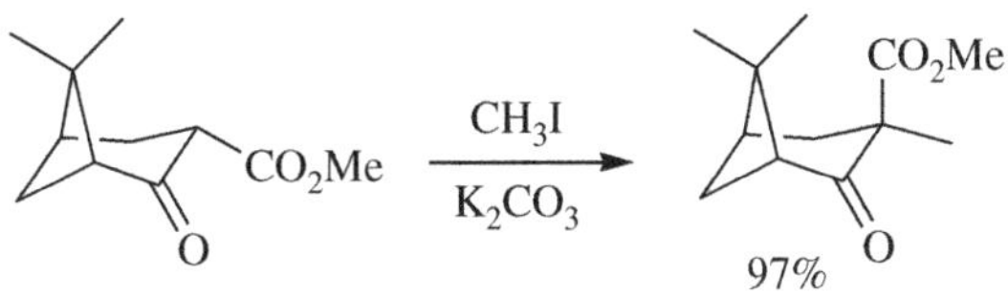

(7.5)

Acylation reactions occur between carbanions and acid chlorides, anhydrides, or esters, forming ketones. An intramolecular example is shown in Equation 7.6 [15]. Treatment with a base gives the lactone enolate, which is then acylated by the methyl ester. Alkyl- or aryllithium and magnesium reagents tend to react twice with these acylating agents, leading to tertiary alcohols. Aromatic ketones may be made in high yield from acid chlorides by treating the Grignard reagent with bis[2-(*N,N*-dimethylamino)ethyl] ether, a complexing agent, which mediates its reactivity [16]. The Fukuyama coupling reaction allows acylation as well, using the thioester as the starting electrophile (Eq. 7.7) [17]. The less reactive nitriles or tertiary amides will acylate Grignard reagents (Eq. 7.8) [18].

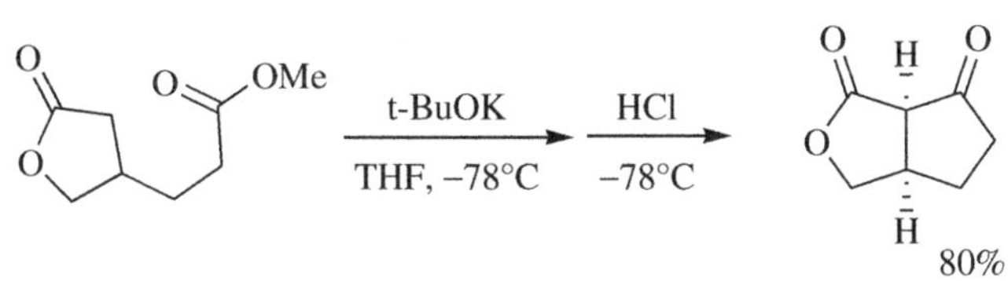

(7.6)

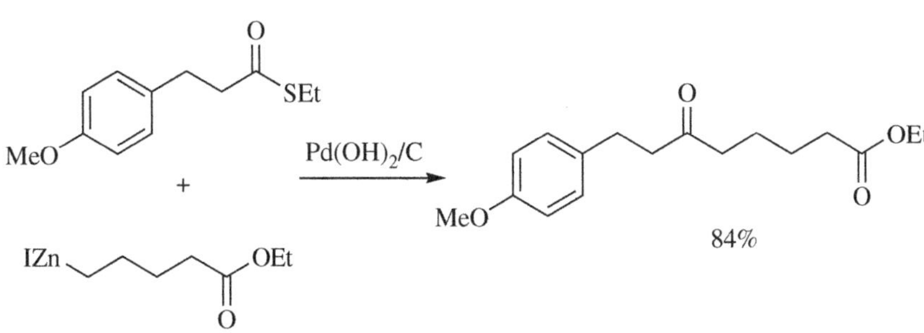

(7.7)

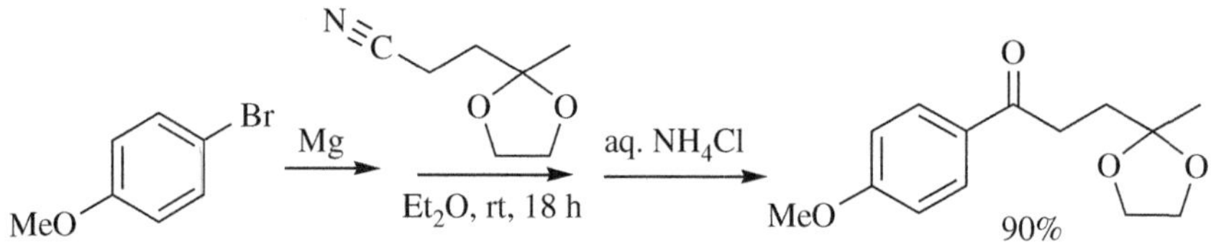

(7.8)

Addition reactions occur between carbanions and ketones or aldehydes to give alcohols. In Equation 7.9, two equivalents of LDA were used to form the dianion, which was added to an aldehyde as shown [19].

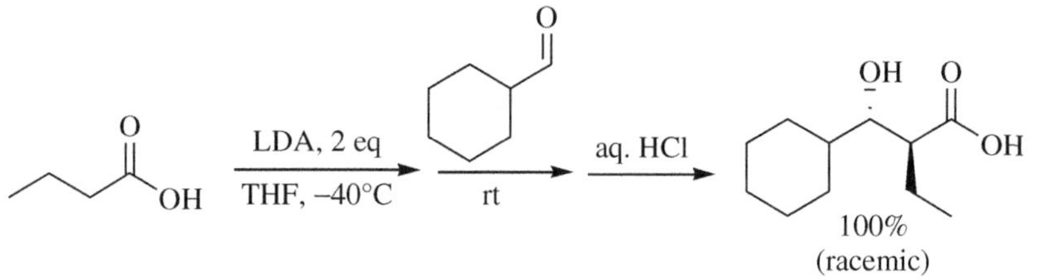

(7.9)

Conjugate addition reactions, also known as Michael additions, occur between α,β-unsaturated carbonyl compounds and highly resonance-stabilized carbanions such as those from 1,3-dicarbonyl compounds. These nucleophiles will add to the β position of an α,β-unsaturated carbonyl compound. Subsequent protonation will give an overall addition to the α,β unsaturation (Eq. 7.10) [20]. More reactive, less stabilized nucleophiles such as Grignard reagents, alkyllithium reagents, and simple enolates from esters, nitriles, amides, or carboxylic *O*,α-dianions usually do not give conjugate addition but bond at the carbonyl carbon.

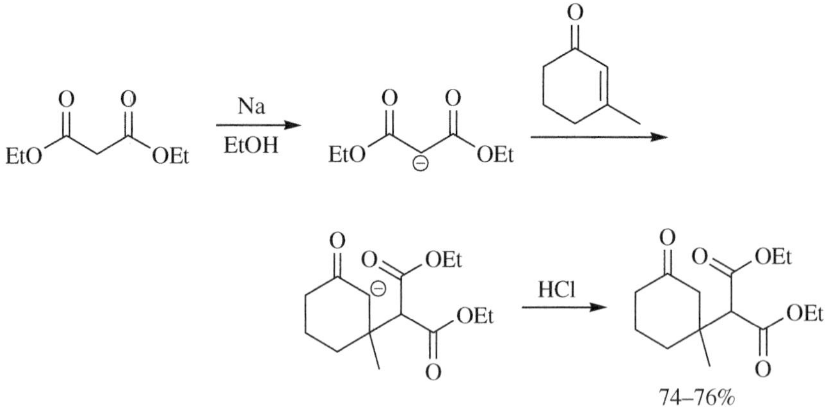

(7.10)

α,β-Acetylenic esters and acetals also undergo conjugate addition. If an organocuprate is added at low temperature and the resulting anion protonated also at low temperature, the process is a stereoselective syn addition of R and H, as shown in Equation 7.11 [21]. This amounts to a stereoselective synthesis of trisubstituted alkenes. If instead of protonating, the anion is alkylated, even tetrasubstituted alkenes are available stereospecifically. The requisite acetylenic acetals are readily made from the anions of the 1-alkynes plus methyl orthoformate. Esters may be used similarly, but the intermediate anion is less reactive [22].

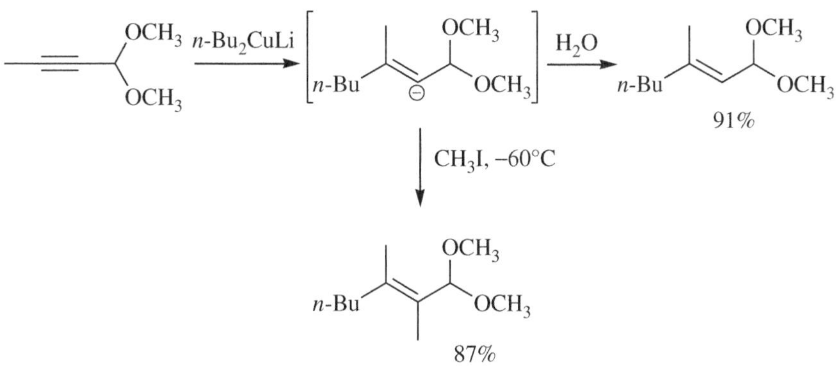

(7.11)

Organocuprates add at the β position of α,β-unsaturated aldehydes, ketones, esters, and *N,N*-dialkylamides using trimethylchlorosilane as an activator [23]. The example shown in Equation 7.12 uses transmetallation of the initially formed organozinc to generate an organocuprate that contains an ester moiety. The enolate anion formed by addition to the β carbon is trapped with trimethylchlorosilane, which is then removed with tetrabutylammonium fluoride [24].

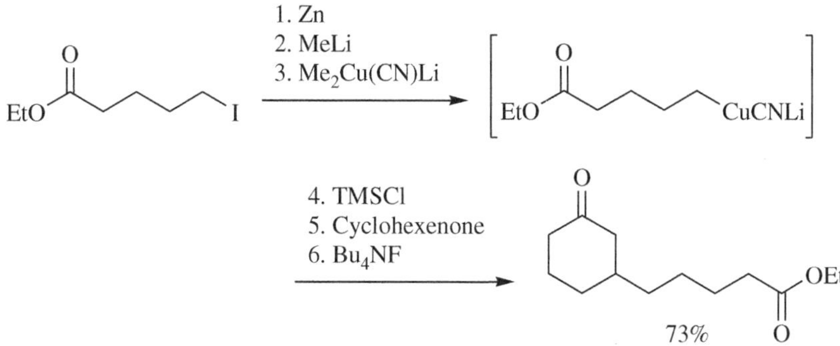

(7.12)

The cyanide ion is a source of nucleophilic carbon and is only weakly basic. It is thus available as a salt that can be used in water solutions. It can be alkylated most readily by nucleophilic substitution under phase-transfer conditions using a quaternary ammonium catalyst (Eq. 7.13, Section 9.6.1) [25]. The alkylations also proceed well under homogeneous conditions in dimethylsulfoxide (DMSO).

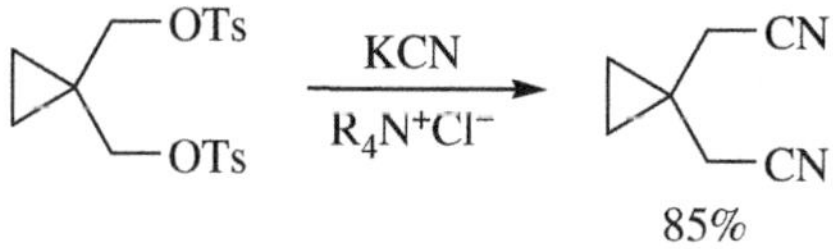

(7.13)

Asymmetric ketones can present a problem in specificity. Under basic conditions, they may react with two or more molecules of the electrophile to give a mixture of products. Furthermore, asymmetric ketones may present a choice of two enolate sites so that control is necessary to direct to the desired one. There are several possible solutions to this problem. One solution is to incorporate a temporary group on one enolate site to render that site more acidic so that the electrophile will react there. The familiar β-ketoester reactions (acetoacetic ester synthesis) are widely used. Another possibility is to first convert the ketone to an imine (Section 6.2.3) or a dimethylhydrazone, and the enolate of that derivative can be reacted with electrophiles [26]. Stereospecificity of the addition is obtained by forming a derivative with (*S*)-1-amino-2-methoxymethylpyrrolidine (SAMP), as shown in Equation 7.14 [27]. Without derivatization, alkylation of unsymmetric ketones will occur mostly at the more substituted enolate site under reversible deprotonating conditions. Using a base such as LDA will give alkylation primarily at the least substituted enolate.

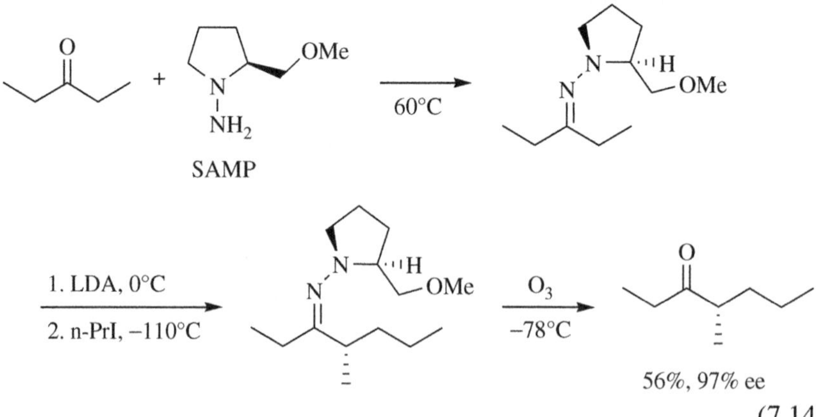

(7.14)

Aldehyde enolates present another problem. They tend to give self-condensation before an electrophile can be added. This may be solved again by the use of imine enolates or *N,N*-dimethylhydrazones, which are themselves of low electrophilicity and allow for good crossed aldol condensations and alkylations. For example, the *tert*-butyl imine of propanal was converted to the enolate with LDA and used in a crossed aldol condensation (Eq. 7.15) [28].

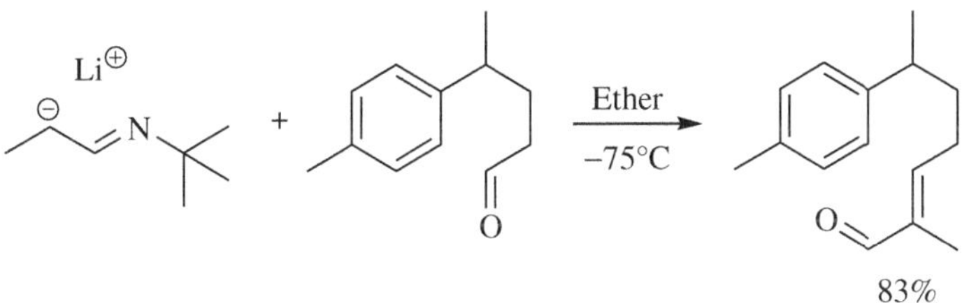

(7.15)

7.1.2 Reactions in Acidic Solutions

Strong acids produce carbocations from a variety of functional molecules. Protonation of alcohols, epoxides, carbonyl compounds, and alkenes does so. Lewis acids such as anhydrous aluminum chloride can combine with these substrates and can also remove halide ions from carbon to give carbocations. Diazotization of primary amines in acid solution is another source.

Carbocations are transient intermediates, generated in the presence of alkene or arene nucleophiles to give carbon–carbon bond formation. This gives a new carbocation requiring a second step, which may be deprotonation, bonding to an oxygen or a halide ion, loss of a silyl group, or abstraction of a hydride, to complete the reaction. Carbocations may react with alkenes and dienes to give new carbocations, which may do likewise repetitively to give polymers [29]. Some varieties of synthetic rubber are produced in this way. Although σ-bonding electrons are more tightly held than π electrons, they will react with carbocations particularly when they are arranged close by for intramolecular (rearrangement) processes. Migration of an atom or a group with the σ-bonding pair from an adjacent carbon to the initial carbocationic site will occur if the new carbocation has greater stability (delocalization) from electron-donating alkyl groups, adjacent nonbonding electron pairs, or resonance to allylic sites, or if strain energy is released.

The example in Equation 7.16 shows the formation of a carbocation from an epoxide, a reaction with an alkene, and finally aromatic substitution on a furan [30].

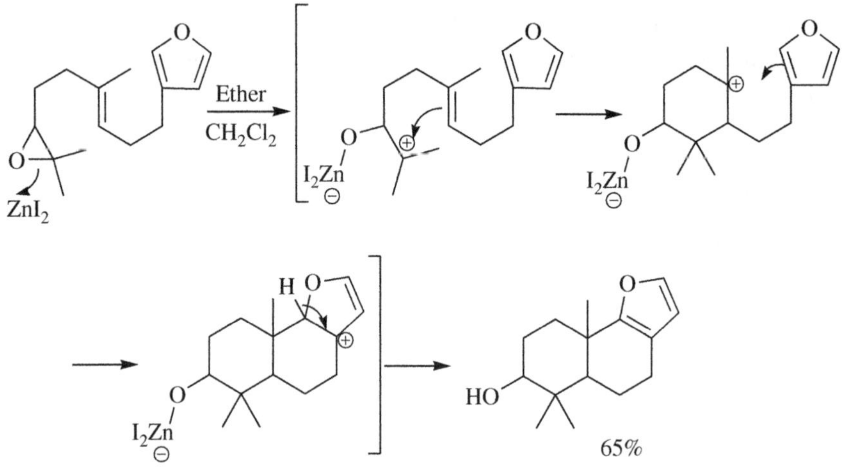

(7.16)

Trimethylsilyl enol ethers are formed readily from ketones (Eqs. 6.40 and 6.41) and are especially valuable nucleophiles toward a variety of electrophilic carbon species [31]. After the attachment of the electrophilic carbon, the trimethylsilyl group is readily removed to afford a ketone. Alkylation is achieved with the use of a Lewis acid catalyst (Eq. 7.17) [32]. The reaction is analogous to the alkylation of a ketone enolate anion, but with some advantages. Here, a specific enol ether (Section 6.2.5) can be used, restricting the alkylation to one site and giving no dialkylation (which sometimes competes in enolate anion alkylation). It also allows for the attachment of tertiary alkyl groups and others that would have given mostly elimination in basic solutions.

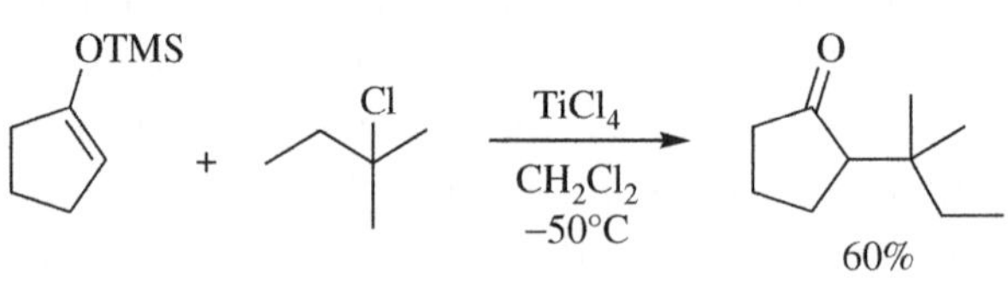

(7.17)

Some aldol reactions can be carried out in acid. In this case, the nucleophile is an enol and the electrophile is the protonated carbonyl group.

Equation 7.18 shows the cyclization of a keto aldehyde [33]. The acidic conditions generally give dehydration of the aldol.

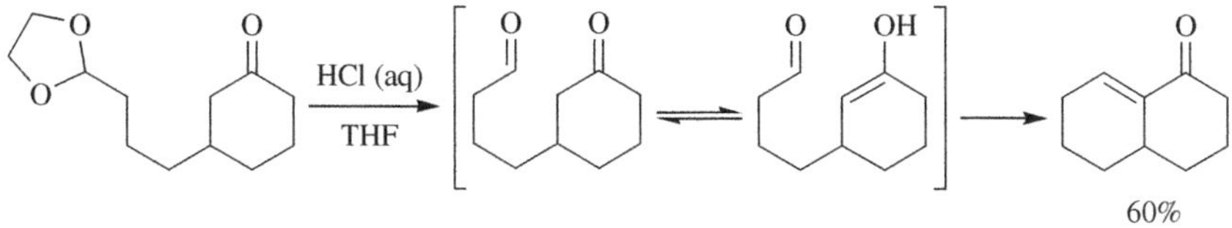

(7.18)

Use of an acetal as an electrophile with a silyl enol ether, as shown in Equation 7.19, demonstrates the equivalent of an aldol reaction [34].

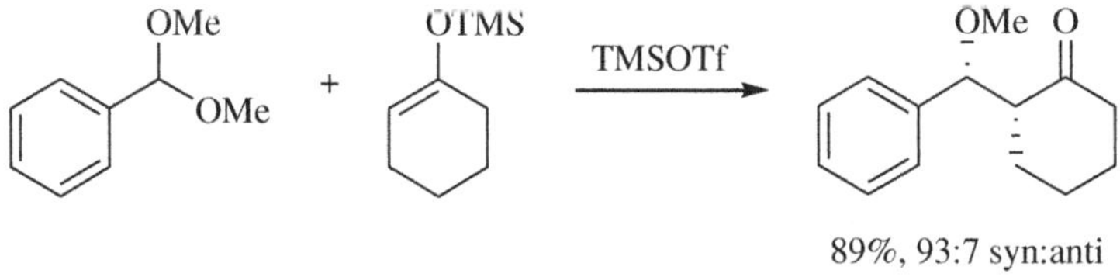

(7.19)

Crossed aldol condensations [35] between dissimilar ketones may be carried out under Lewis acid conditions using the silyl enol ether of that ketone intended as the nucleophile. This affords the aldols without dehydration or polycondensation (Eq. 7.20) [36].

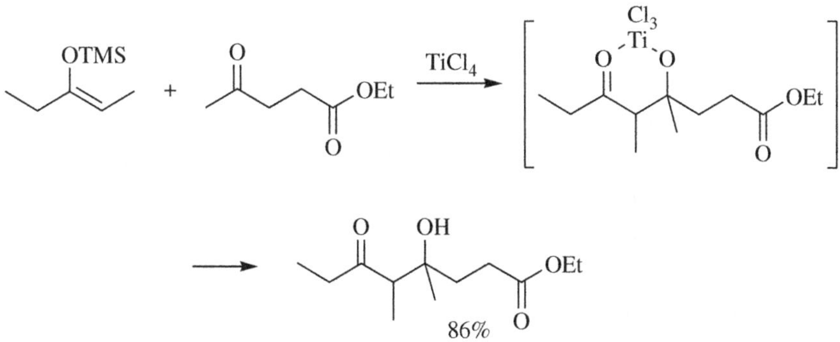

(7.20)

Simple ketones and 1,3-diketones give conjugate addition in an acidic solution, as shown in Equation 7.21 [37]. Here, too, the silyl enol ethers and $TiCl_4$ may be used (Eq. 7.22) [38].

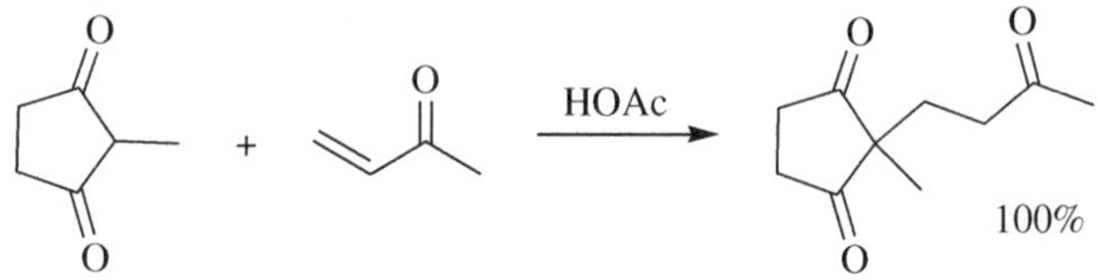

(7.21)

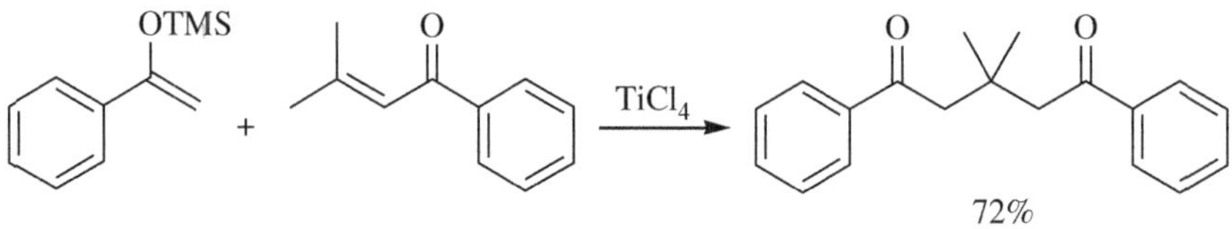

(7.22)

7.1.3 Organometallic Coupling Reactions

Organometallic compounds are a source of strongly nucleophilic carbon, which can react in a coupling reaction with an alkyl halide as the electrophile to form a carbon–carbon bond. One of the earliest examples of this sort of coupling reaction is the Wurtz reaction, which uses sodium metal to form the nucleophile (Eq. 7.23) [39]. The mixture of products obtained is a significant disadvantage of this early reaction.

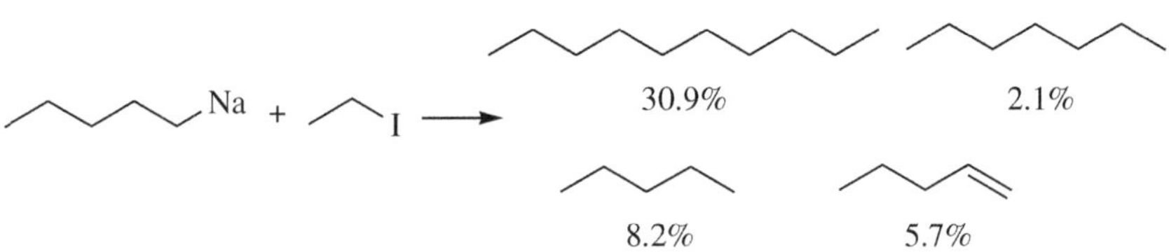

(7.23)

Negishi and related coupling reactions [40] are synthetically valuable because of their broad scope and chemo- and regioselectivity [41]. A variety of mostly platinum catalysts have been used to accomplish the transmetallation reaction that is part of the mechanism. Homogeneous palladium catalysts increase reaction efficiency, but heterogenous catalysts are easier to separate from the products, simplifying product isolation and catalyst reuse. Ligand choice for the palladium atom catalyst can improve efficiency and selectivity. These reactions are important in pharmaceutical chemistry owing to their tolerance for heterocyclic substrates. Recent developments include working toward greener conditions and moving away from expensive transition metal catalysts [42].

The Suzuki–Miyaura reaction uses boron in the organometallic reagent, as shown in Equation 7.24 [43]. The palladium catalyst is commercially available. The Negishi reaction uses organozinc reagents in the cross-coupling reaction, as shown in Equation 7.25 [44]. An example of the Kumada reaction, which uses a Grignard reagent to mediate the cross-coupling, is shown in Equation 7.26. The catalyst is a phosphine–nickel complex [45], though other metals have been used. A green example of a cross-coupling designed for classroom use is shown in Equation 7.27 [46].

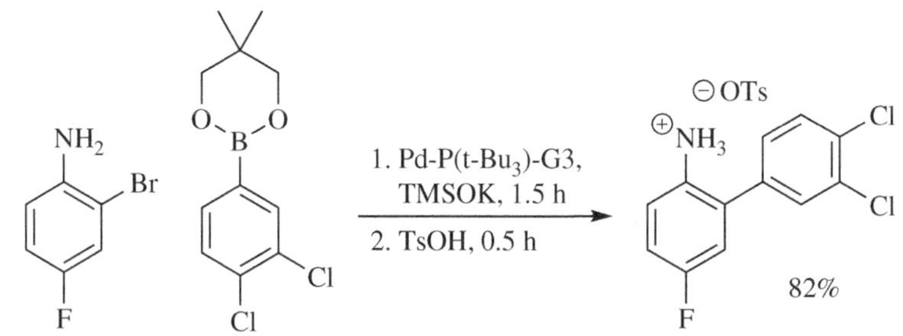

(7.24)

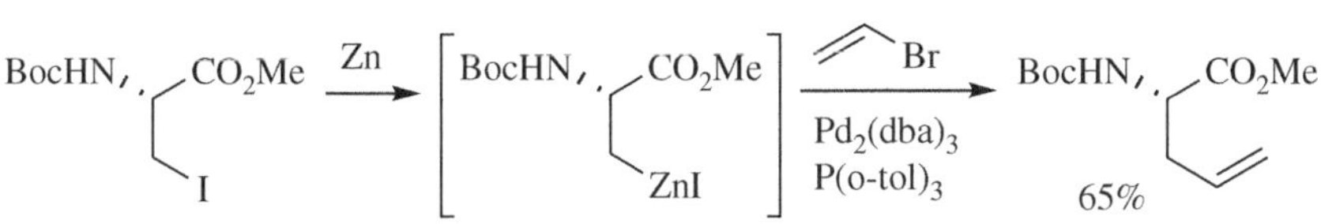

(7.25)

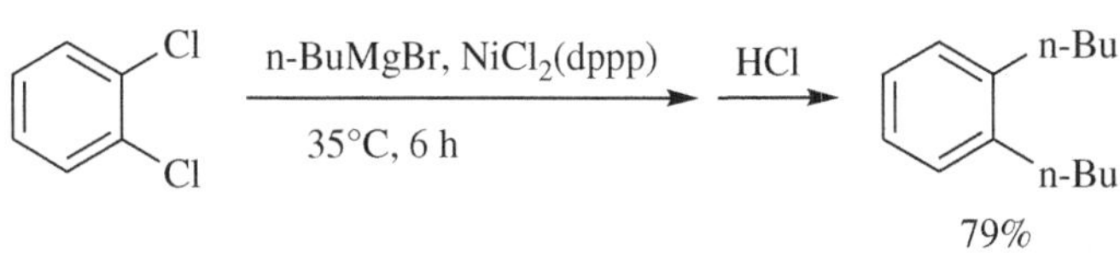

(7.26)

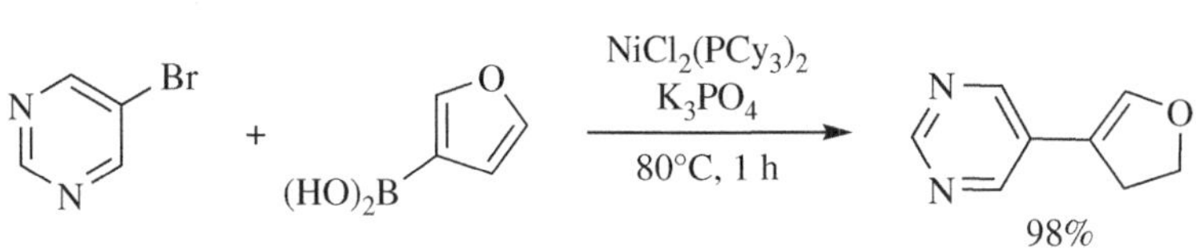

(7.27)

Coupling reactions have been achieved with a variety of aromatic and heteroaromatic rings using a thiomethyl group and an organozinc reagent. A variety of functionalities is tolerated in the reaction (Eq. 7.28) [47].

SPhos, 2-(2′,6′-dimethoxybiphenyl)dicyclohexylphosphine, is a catalyst for the reaction [48].

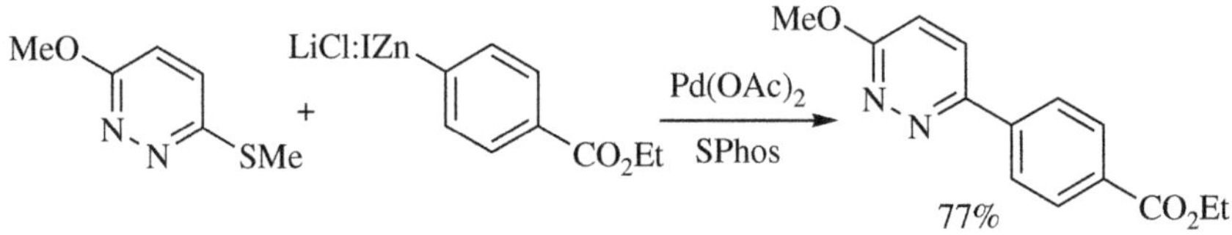

(7.28)

7.2 CARBON–CARBON DOUBLE-BOND FORMATION

Conjugated carbon–carbon double bonds can be synthesized using an aldol or similar reaction, followed by the dehydration of the initial product to form the α,β-unsaturated carbonyl compound. In Equation 7.29, the aldol product is formed using an azo analog of the enolate anion, which prevents self-condensation of the aldehyde. An acid-catalyzed dehydration forms the double bond [49].

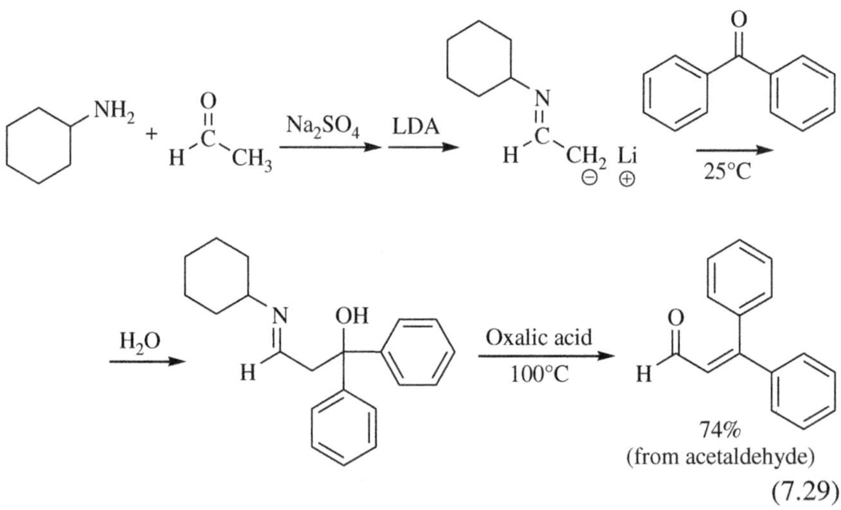

(7.29)

Carbonyl olefination reactions allow two carbons to come together joined by a double bond. The electrophilic side comes from an aldehyde or ketone. If the nucleophilic side is an alkylmagnesium or an alkyllithium, the secondary or tertiary alcohol intermediate may be dehydrated, but in many cases, there is a choice of sites for the new double bond and it may arise elsewhere than the newly joined carbons. The double bond can be formed specifically if the nucleophilic carbon bears a group with high oxygen affinity that will leave with the oxygen. Phosphorus and silicon are excellent in this role, as demonstrated in the following several examples.

The Peterson reaction is the result of elimination of a β-hydroxysilane that selectively forms the *Z*-alkene under acidic conditions, and the *E*-alkene under basic conditions (Eq. 7.30) [50]. This selectivity is a consequence of the mechanism, in which the acid catalyst protonates the alcohol leaving group and anti-elimination occurs. Basic conditions result in a syn elimination.

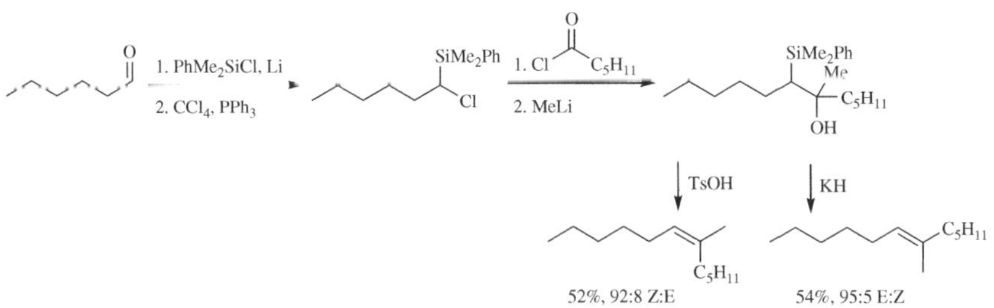

(7.30)

Phosphorus has been used to a far greater extent for specific alkene synthesis [51]. Alkyl chlorides and bromides may be treated with triphenylphosphine to give quaternary salts. The positively charged phosphorus allows the removal of a proton from the carbon alpha to the phosphorus to generate an ylide. Although the ylide carries no net charge, the substantial dipole gives high nucleophilic reactivity toward aldehydes and ketones to give an intermediate 1,2-oxaphosphetane that cleaves to the alkene and triphenylphosphine oxide. This is known as the *Wittig reaction* (Eq. 7.31) [52]. The triphenylphosphine oxide is nonvolatile and somewhat organic soluble and can be a nuisance to remove. Generally, the reaction favors the (*Z*)-alkene when the ylide contains only alkyl groups.

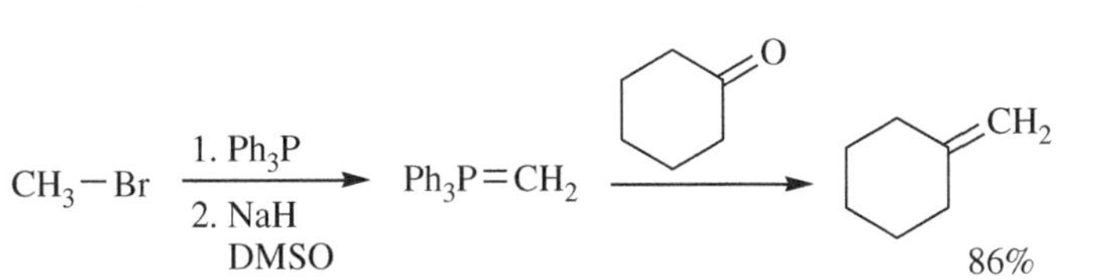

(7.31)

The use of a sulfonyl-substituted imine in place of the aldehyde has allowed for the preferential formation of either the (*E*)- or (*Z*)-alkene, depending on the specific sulfonyl substitution (Eq. 7.32) [53].

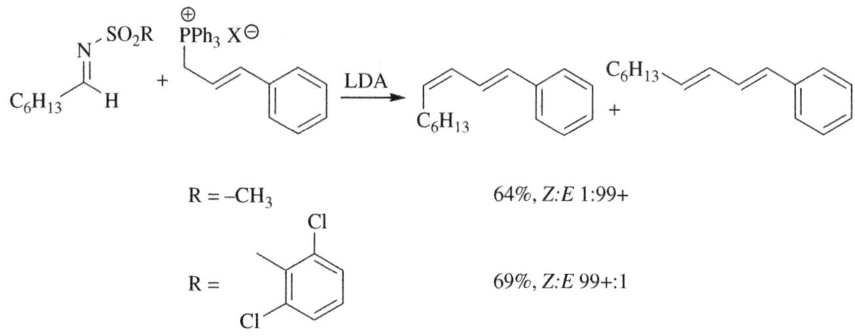

(7.32)

The Wadsworth–Emmons modification [54] of the Wittig reaction uses the phosphonate esters to form α,β-unsaturated esters, in which the (*E*)-alkene is favored (Eq. 7.33) [55]. *In situ* alkylation is also possible (Eq. 7.34) [54]. The phosphonoesters are prepared by treating the α-bromoesters with triethyl phosphite (Arbuzov reaction) [56]. The diethyl phosphate by-product is water-soluble and easily removed.

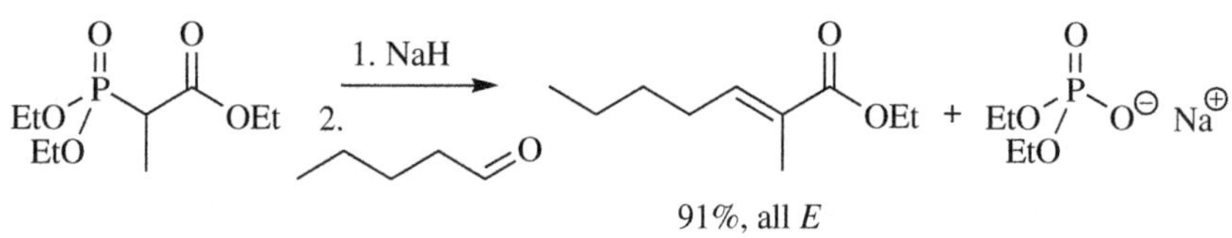

(7.33)

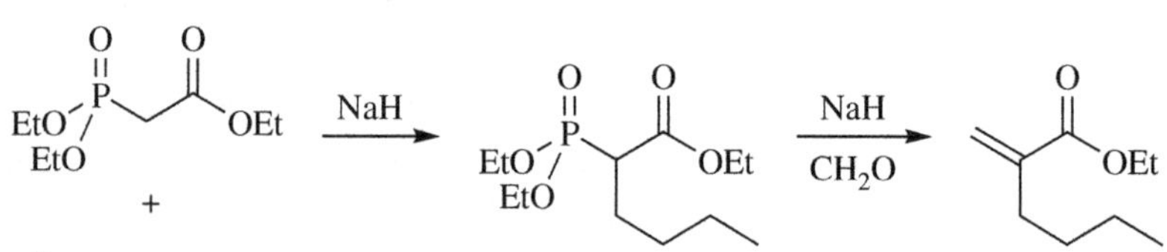

(7.34)

In the Still–Gennari modification, a potassium base is used with crown ether complexation of the cation and the (*Z*)-product is favored. Even higher *Z* selectivity is obtained using trifluoroethyl phosphonoesters (Eq. 7.35) [57].

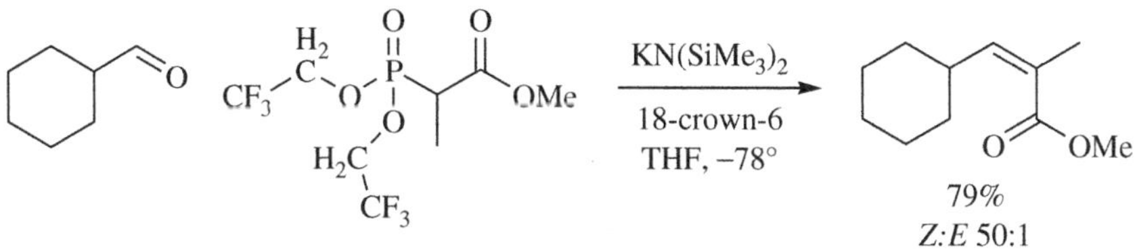

(7.35)

The Heck reaction is used industrially to form octyl 4-methoxycinnamate, a commercial sunscreen ingredient. The catalyst in this case is the heterogeneous Pd on carbon (Eq. 7.36) [58].

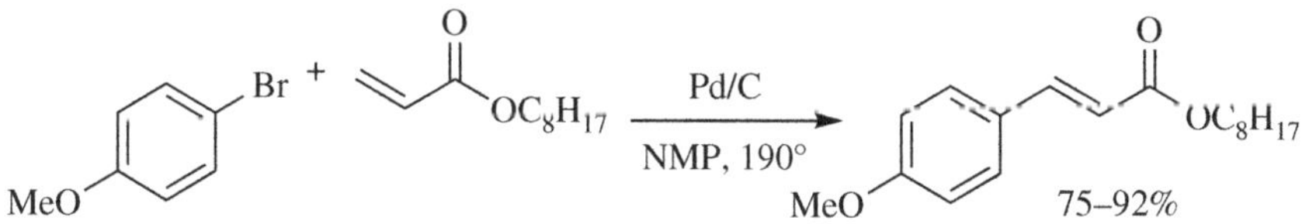

(7.36)

7.3 MULTIBOND PROCESSES

Many reactions result in a nearly simultaneous formation of a pair of σ bonds. In some cases, a carbene is a transient intermediate. Carbene, $:CH_2$, is electron-deficient; it lacks two electrons for a complete octet. Although there is no net charge and little or no dipole, it is highly electrophilic and will attack both π and σ electrons to form pairs of new bonds. The lack of specificity in this high reactivity renders $:CH_2$ of little synthetic value, but selective cyclopropanation of alkenes may be accomplished using Simmons–Smith conditions (Eq. 7.37) [59], with a variety of methods available [60].

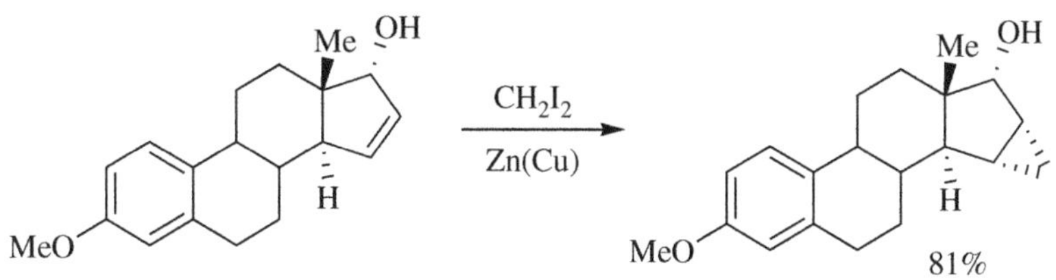

(7.37)

Cyclopropanation of alkenes can also be accomplished via other transient electrophilic intermediates that are possibly metal complexes of carbenes. Copper, rhodium, ruthenium, cobalt, or palladium catalyze the decomposition of diazocarbonyl compounds [61], which, in the presence of alkenes, gives cyclopropyl derivatives. The intramolecular example shown in Equation 7.37 uses rhodium bis(1-adamantate) dimer as the catalyst [62].

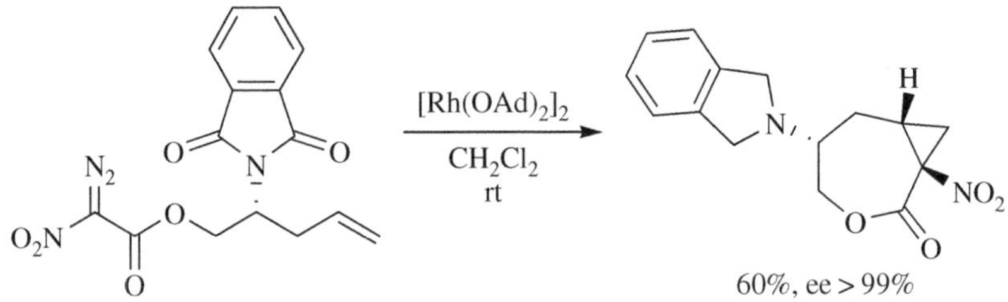

(7.38)

Concerted reactions are commonly used to join carbons. For example, the Diels–Alder reaction is the formation of a cyclohexene from a diene and an alkene. Usually, the alkene is rendered electrophilic by conjugation with a carbonyl group, and the diene may be rendered nucleophilic by electron-donating substituents. In the case shown in Equation 7.39 the alkene is further electron-depleted by association with a Lewis acid [63], a common technique for accelerating Diels–Alder reactions. In some cases, the alkene is nucleophilic and the diene is electrophilic, as shown in Equation 7.40 [64]. Examples of this sort are called *reverse-electron-demand* Diels–Alder reactions. It is important to point out here that the concerted reactions differ from earlier examples in that no carbanion or cation intermediate is involved, and in many cases, electrophilic and nucleophilic factors are not present, as in the very favorable dimerization of cyclopentadiene. These reactions are covered in more detail in Chapter 5.

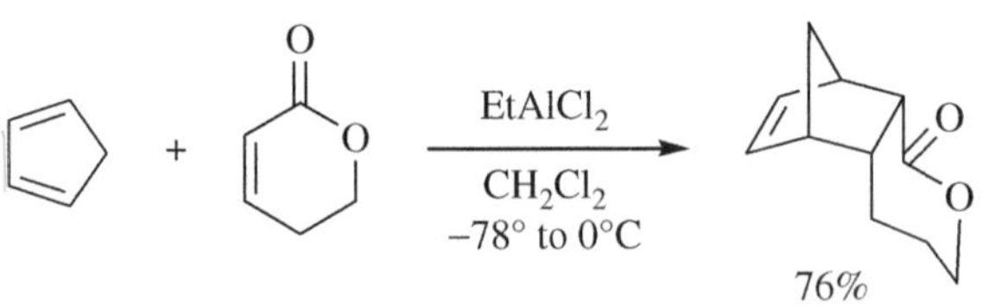

(7.39)

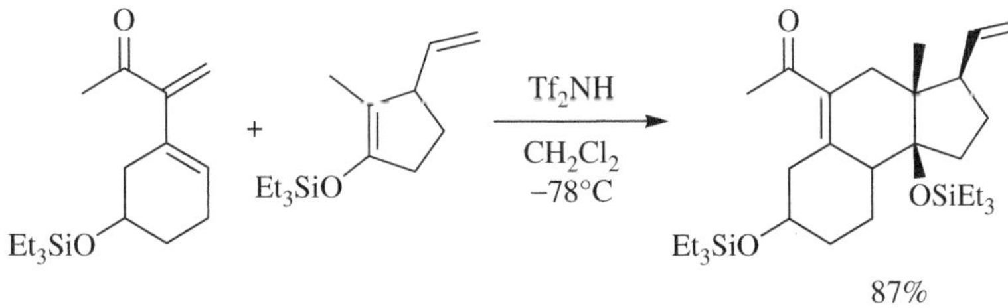

(7.40)

Photocatalysts have been developed to allow 1,2 additions to a double bond. These reactions have a wide tolerance for functionality. In the example shown in Equation 7.41, sodium decatungstate (NaDT) and nickel 4,4′-di-t-butyl-2,2′-dipyridyl bromide (Ni(dtbbpy)Br_2) catalyze electron transfer reactions initiated by light-emitting diode (LED) irradiation at 390 nm for 24 hours. This activates the C–H bond of the alkane to add to the alkene, which then adds to the carbonyl component in an oxidative addition [65].

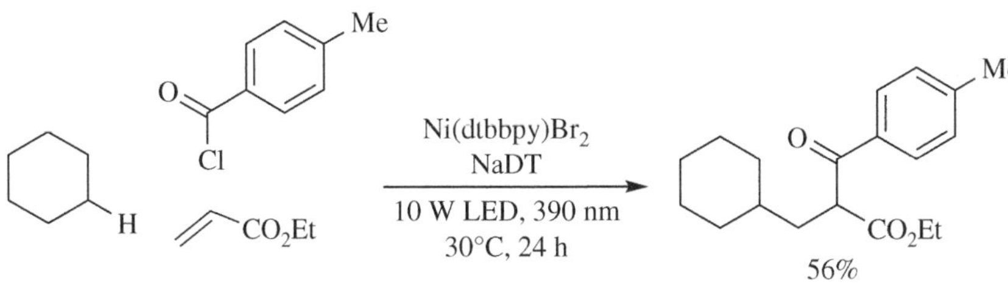

(7.41)

RESOURCES

1. The Organic Syntheses website includes the full contents of the publication of the same title. https://www.orgsyn.org/ (Organic Syntheses: A Publication of Reliable Methods for the Preparation of Organic Compounds, R. L. Danheiser, Editor in Chief, CambridgeSoft.)
2. The Organic Chemistry Portal allows searching by bond for many synthetic methods. The site is maintained by Dr. Douglass Taber. https://www.organic-chemistry.org/synthesis/

PROBLEMS

Show how you would prepare each of the following products from the given starting materials. Where more than one step is required, show each step distinctly. Problems marked with an asterisk are answered in Appendix C.

7.1*

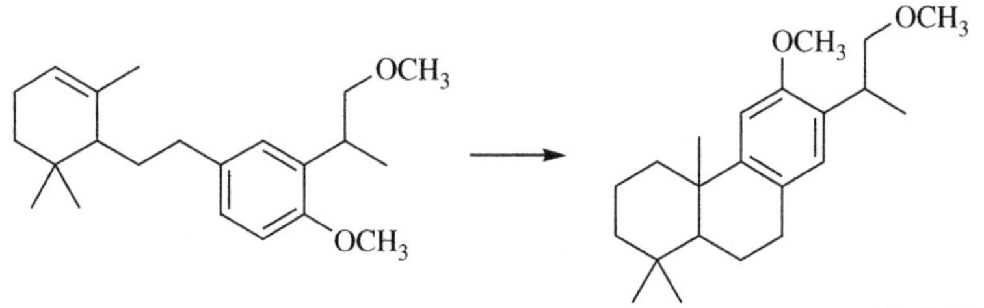

Ref. [66]

7.2*

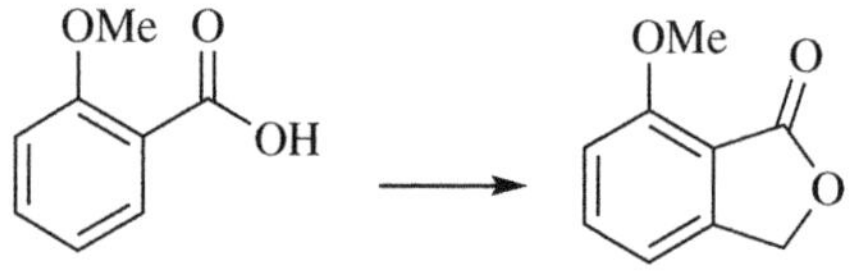

Ref. [67]

7.3

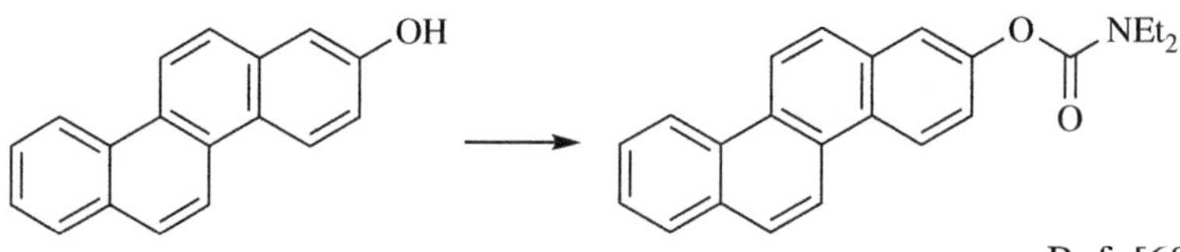

Ref. [68]

7.4*

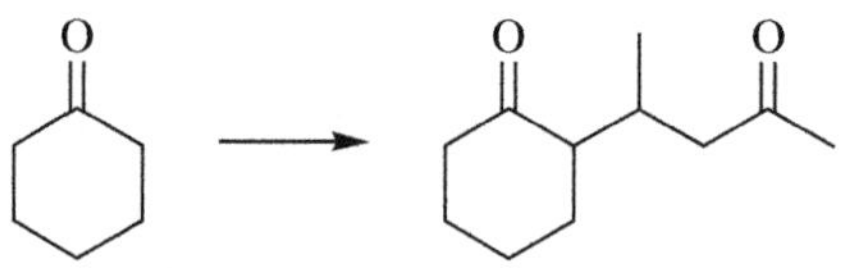

Ref. [69]

7.5

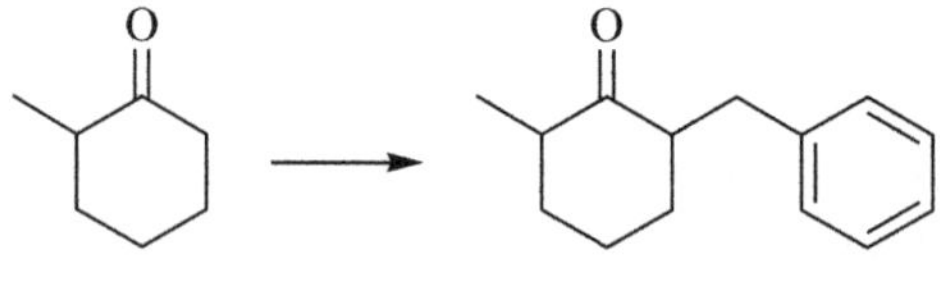

Ref. [70]

7.6

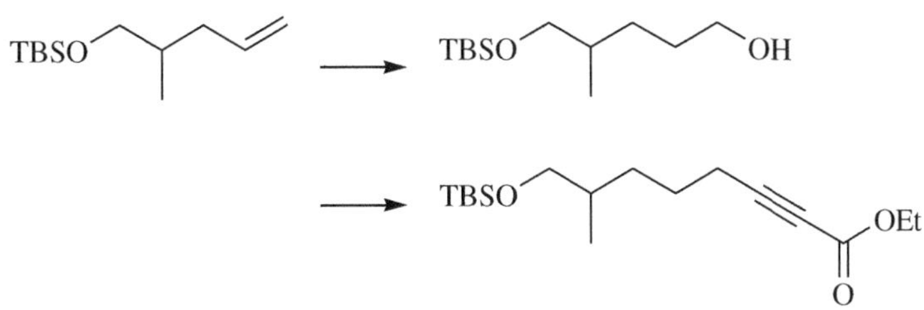

Ref. [71]

7.7*

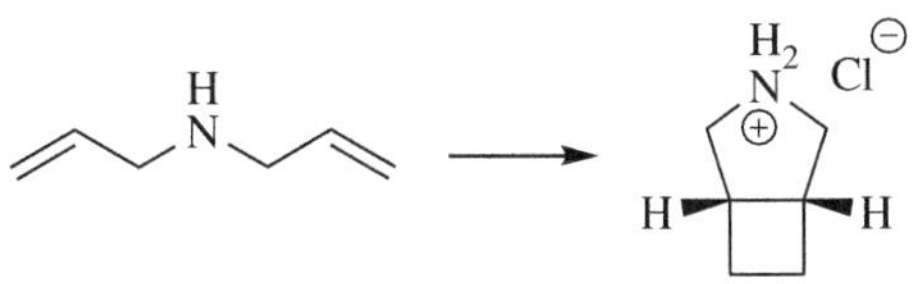

Ref. [72]

7.8

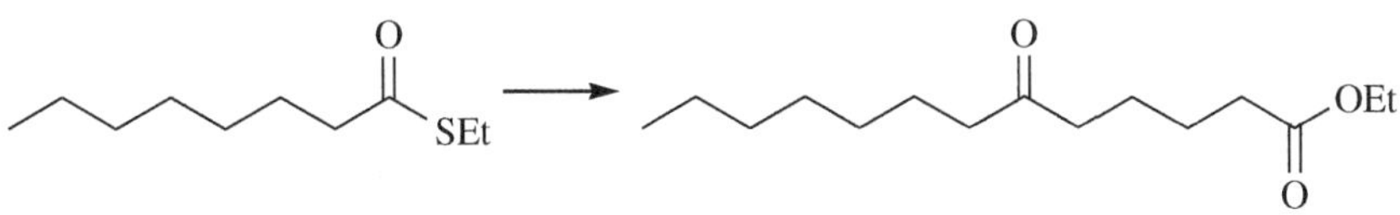

Ref. [73]

7.9*

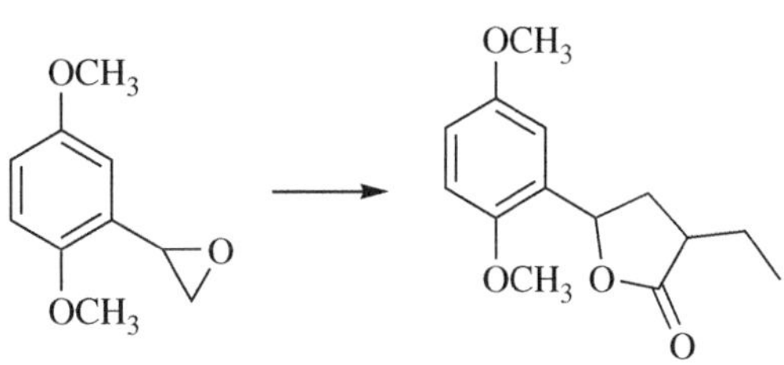

Ref. [74]

7.10*

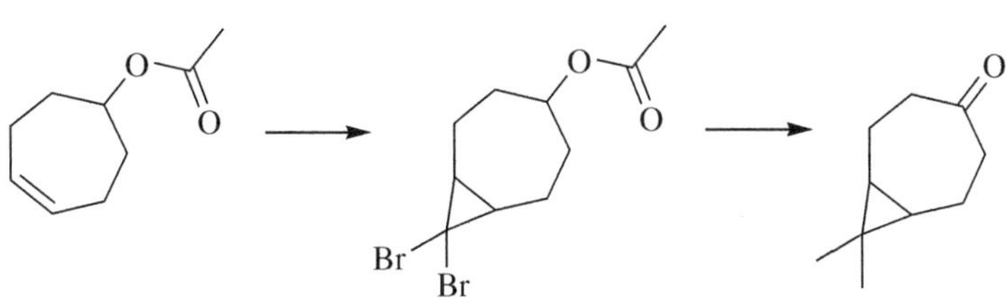

Ref. [75]

7.11

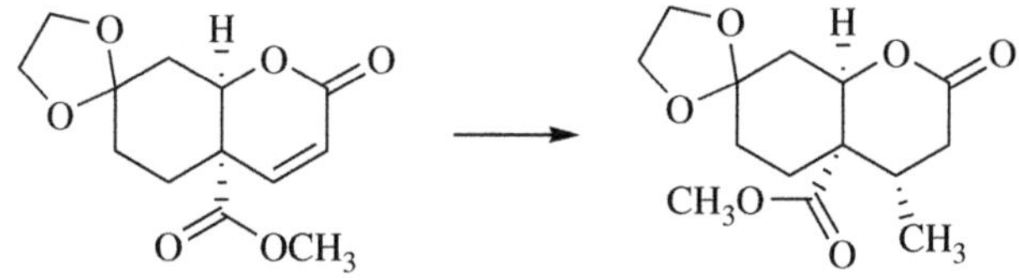

Ref. [76]

7.12

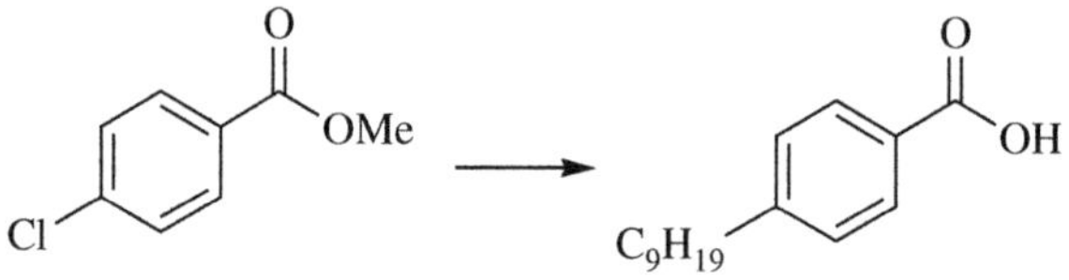

Ref. [77]

7.13*

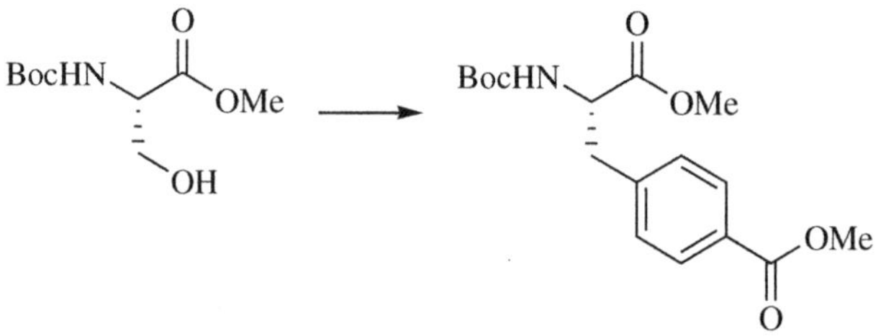

Ref. [78]

7.14

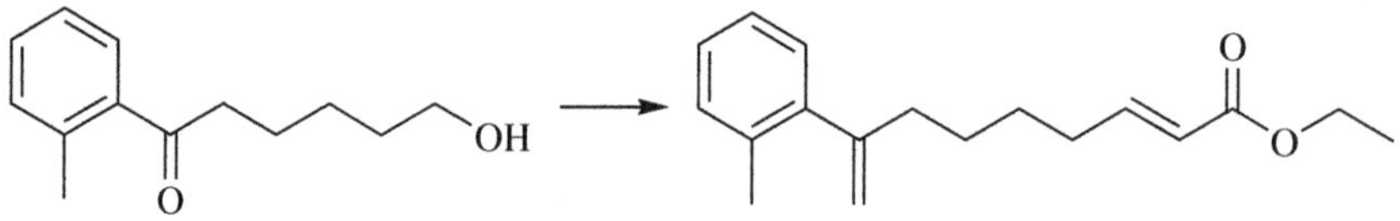

Ref. [79]

7.15

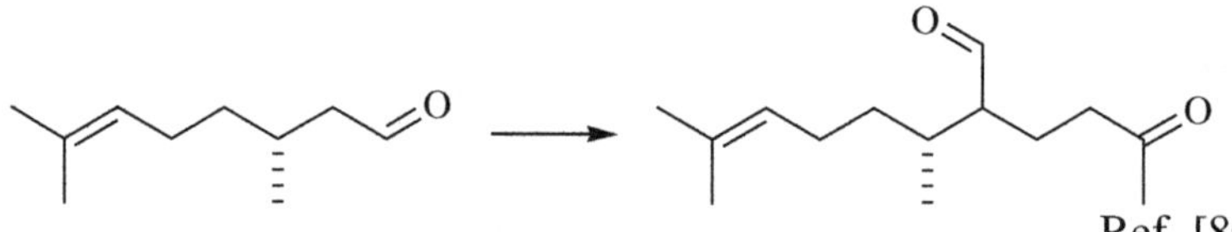

Ref. [80]

7.16

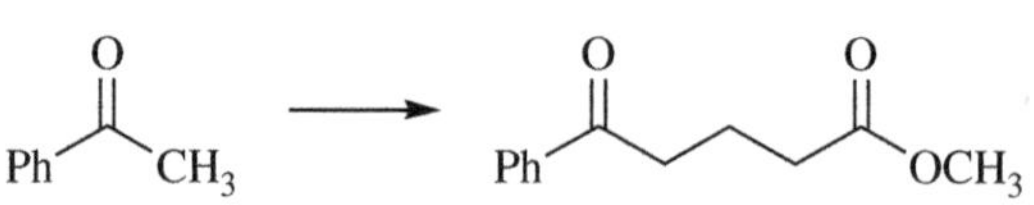

Ref. [81]

7.17*

Ref. [82]

7.18*

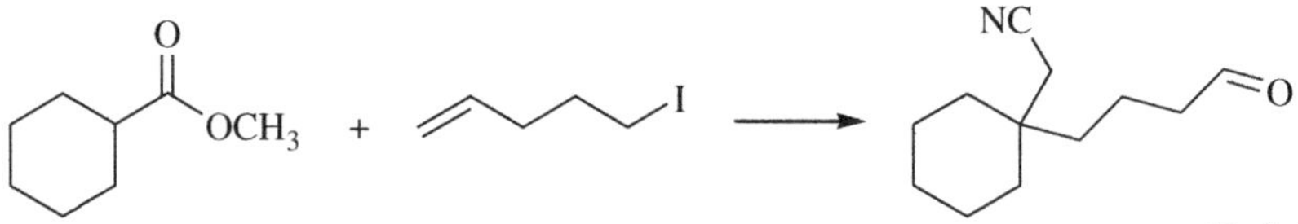

Ref. [83]

7.19

Ref. [84]

7.20

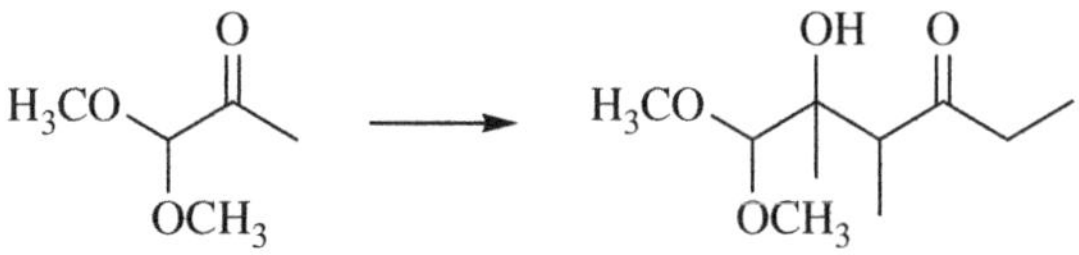

Ref. [85]

7.21

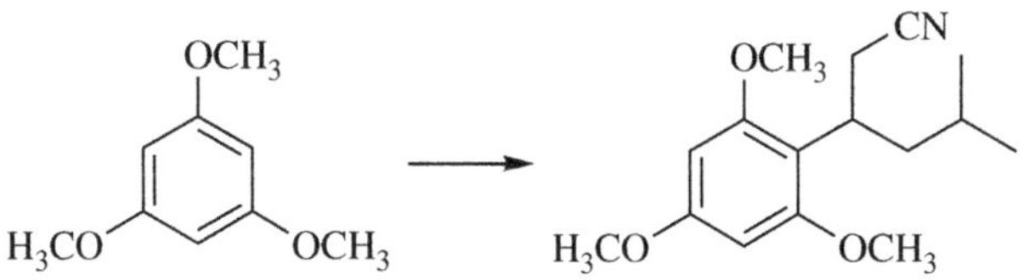

Ref. [86]

7.22

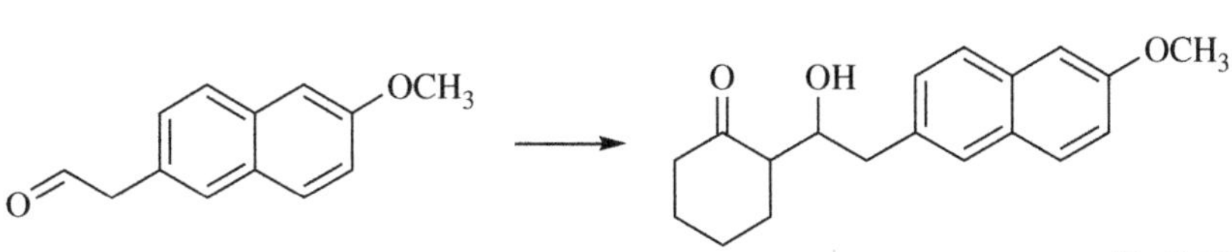

Ref. [87]

7.23*

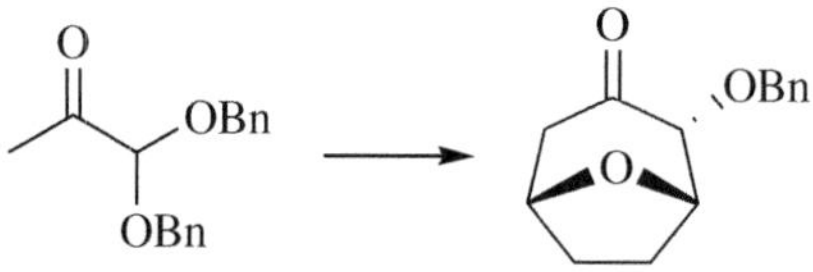

Ref. [88]

7.24

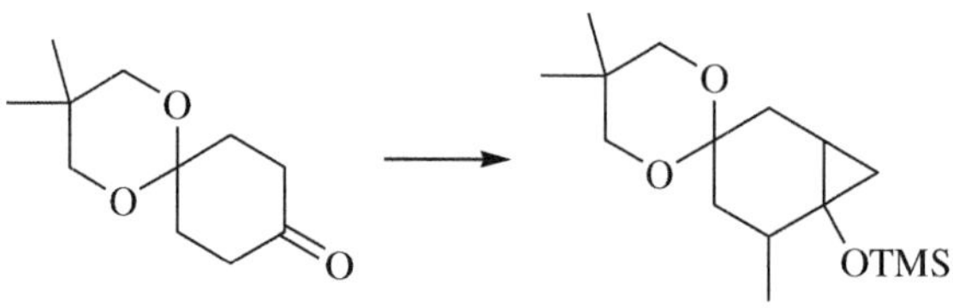

Ref. [89]

7.25*

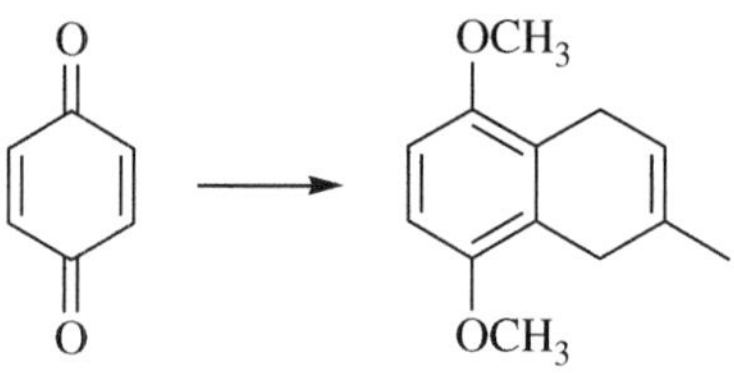

Ref. [90]

7.26

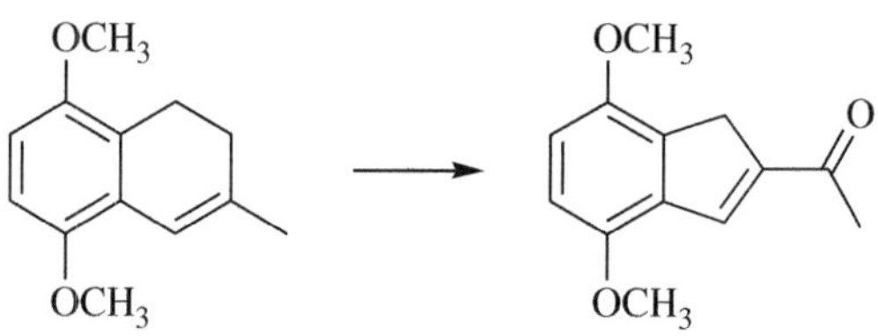

Ref. [90]

7.27

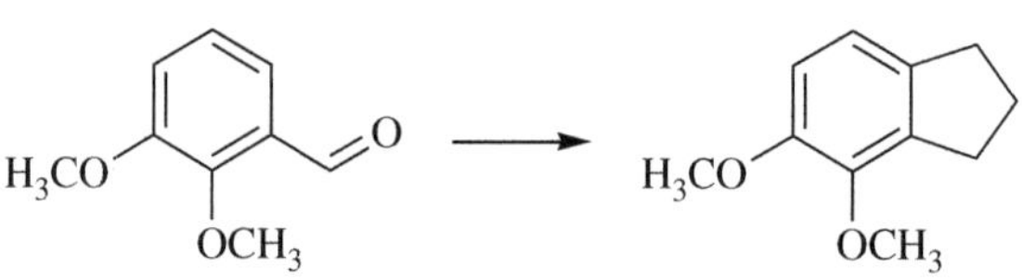

Ref. [91]

7.28

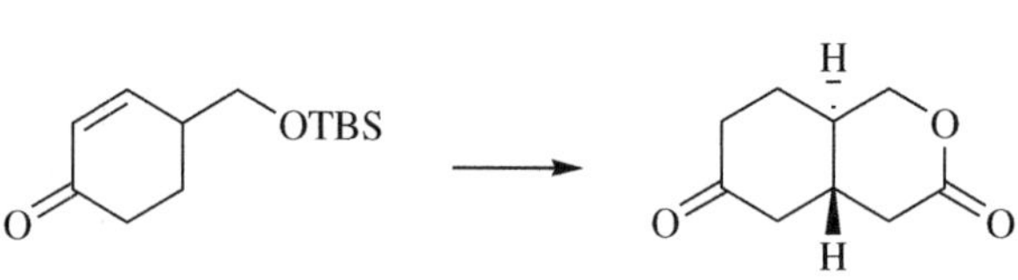

Ref. [92]

7.29*

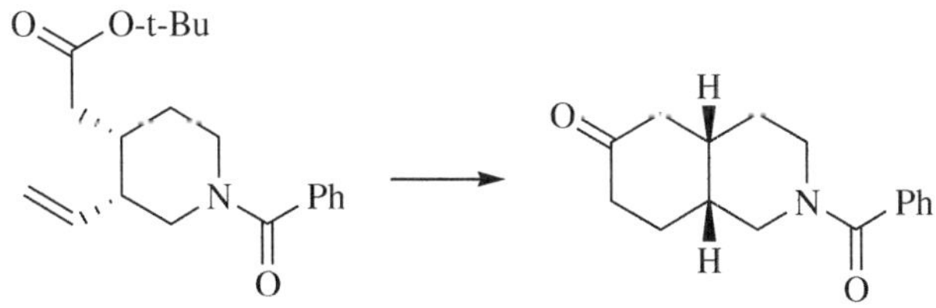

Ref. [93]

7.30

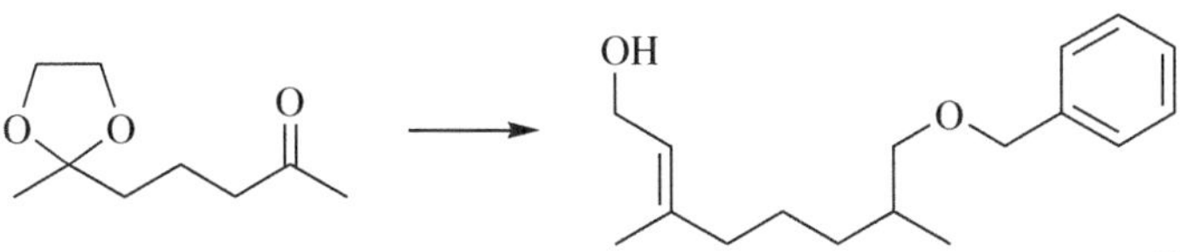

Ref. [94]

7.31

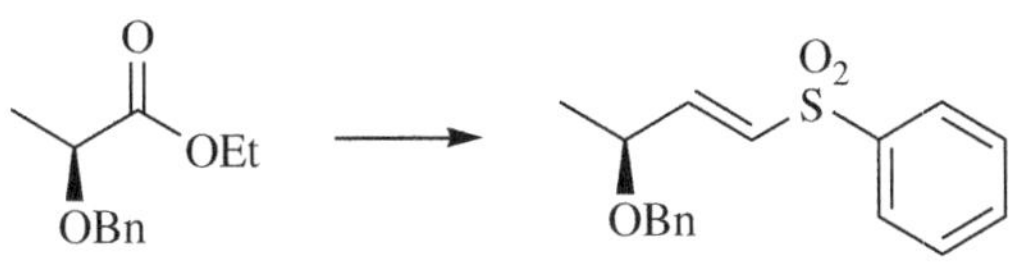

Ref. [95]

7.32

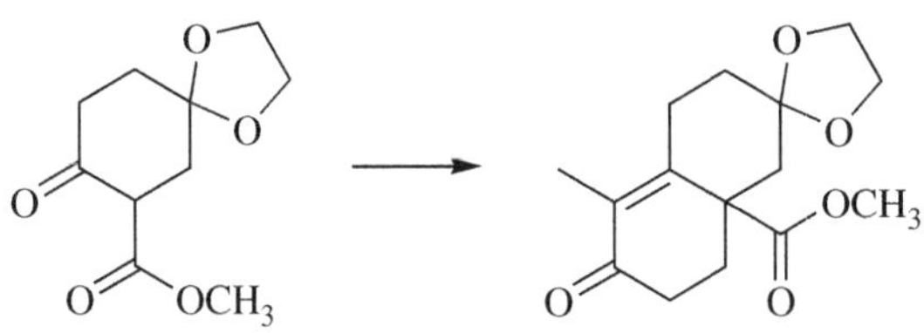

Ref. [96]

7.33

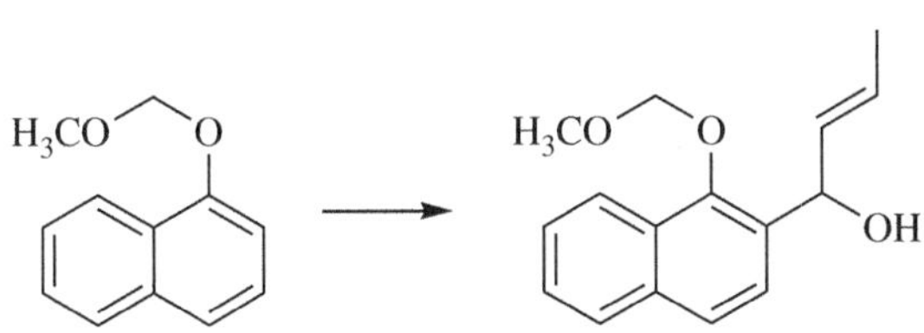

Ref. [97]

7.34*

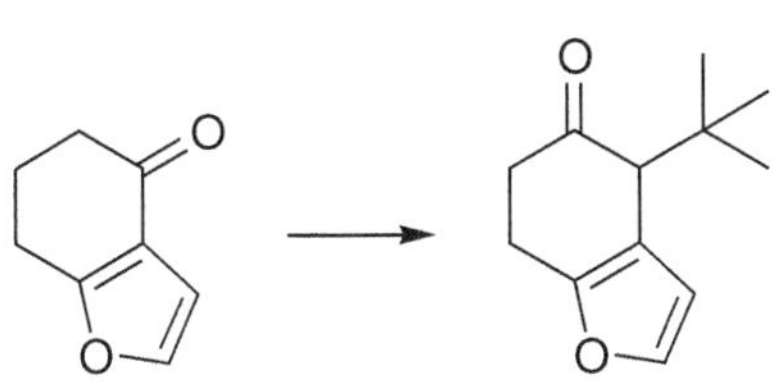

Ref. [98]

7.35 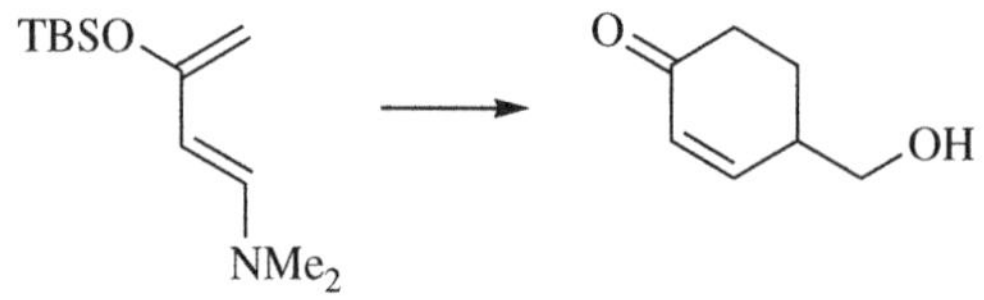 Ref. [99]

REFERENCES

1. Smith, M. B. *March's Advanced Organic Chemistry: Reactions, Mechanisms, and Structure*, 8th ed.; John Wiley & Sons, Inc: Hoboken, 2019.
2. Augustine, R. L., Ed. *Carbon–Carbon Bond Formation*, Volume *1*; Marcel Dekker: New York, 1979.
3. Stowell, J. C. *Carbanions in Organic Synthesis*; Wiley-Interscience: New York, 1979.
4. Rappoport, Z.; Marek, I., Eds. *The Chemistry of Organomagnesium Compounds*; John Wiley & Sons, Inc: New York, 2008.
5. Wakefield, B. J. *Organolithium Methods*; Academic Press: New York, 1988.
6. Ripin, D. H.; Evans, D. A. pK_a's of Inorganic and Oxo-Acids. http://ccc.chem.pitt.edu/wipf/MechOMs/evans_pKa_table.pdf (accessed January 16, 2025).
7. Reich, H. Bordwell pK_a Table. https://organicchemistrydata.org/hansreich/resources/pka/ (accessed January 16, 2025).
8. Taber, D. F.; Amedio, J. C. Jr.; Raman, K. *J. Org. Chem.* **1988**, *53*, 2984–2990.
9. Zhou, W. -S.; Xu, X. -X. *Acc. Chem. Res.* **1994**, *27*, 211–216.
10. Snieckus, V. *Chem Rev.* **1990**, *90*, 879–933.
11. Groom, K.; Hussain, S. M. S.; Morin, J.; Nilewski, C.; Rantanen, T.; Snieckus, V. *Org. Lett.* **2014**, *16*, 2378–2381.
12. Ghera, E.; Ben-David, Y. *J. Org. Chem.* **1988**, *53*, 2972–2979.
13. Hecker, S. J.; Heathcock, C. H. *J. Org. Chem.* **1985**, *50*, 5159–5166.
14. Inokuchi, T.; Asanuma, G.; Torii, S. *J. Org. Chem.* **1982**, *47*, 4622–4626.
15. Boeckman, R. K. Jr.; Naegley, P. C.; Arthur, S. D. *J. Org. Chem.* **1980**, *45*, 752–754.
16. Wang, X.-J.; Zhang, L.; Sun, X.; Xu, Y.; Krishnamurthy, D.; Senanayake, C. H. *Org. Lett.* **2005**, *7*, 5593–5595.
17. Mori, Y.; Seki, M. *J. Org. Chem.* **2003**, *68*, 1571–1574.
18. Martin, S. F.; Puckette, T. A.; Colapret, J. A. *J. Org. Chem.* **1979**, *44*, 3391–3396.

19. Black, T. H.; DuBay, W. J.; III; Tully, P. S. *J. Org. Chem.* **1988**, *53*, 5922–5927.
20. Poupart, M.-A.; Lassalle, G.; Paquette, L. A. *Org. Synth.* **1990**, *69*, 173–179.
21. Alexakis, A.; Commercon, A.; Coulentianos, C.; Normant, J. F. *Tetrahedron* **1984**, *40*, 715–731.
22. Corey, E. J.; Katzenellenbogen, J. A. *J. Am. Chem. Soc.* **1969**, *91*, 1851–1852.
23. Yoshikai, N.; Nakamura, E. *Chem. Rev.* **2012**, *112*, 2339–2372.
24. Lipshutz, B. H.; Wood, M. R.; Tirado, R. *Org. Synth.* **1999**, *76*, 252–262.
25. Foos, J.; Steel, F.; Rizvi, S. Q. A.; Fraenkel, G. *J. Org. Chem.* **1979**, *44*, 2522–2529.
26. Lazny, R.; Nodzewska, A. *Chem. Rev.* **2010**, *110*, 1386–1434.
27. Enders, D.; Kipphardt, H.; Fey, P. *Org. Synth.* **1987**, *65*, 183–202.
28. Büchi, G.; Wüest, H. *J. Org. Chem.* **1969**, *34*, 1122–1123.
29. Aoshima, S.; Kanaoka, S. *Chem. Rev.* **2009**, *109*, 5245–5287.
30. Tanis, S. P.; Herrinton, P. M. *J. Org. Chem.* **1983**, *48*, 4572–4580.
31. Gawronski, J.; Wascinska, N.; Gajewy, J. *Chem. Rev.* **2008**, *108*, 5227–5252.
32. Reetz, M. T.; Chatziiosifidis, I.; Hübner, F.; Heimbach, H. *Org. Synth.* **1984**, *62*, 95–100.
33. Abbott, R. E.; Spencer, T. A. *J. Org. Chem.* **1980**, *45*, 5398–5399.
34. Dilman, A. D.; Ioffe, S. L. *Chem. Rev.* **2003**, *103*, 733–772.
35. Mukaiyama, T. *Org. React.* **1982**, *28*, 203–331.
36. Banno, K. *Bull. Chem. Soc. Jpn.* **1976**, *49*, 2284–2291.
37. Hajos, Z. G.; Parrish, D. R. *Org. Synth.* **1985**, *63*, 26–36.
38. Narasaka, K. *Org. Synth.* **1987**, *65*, 12–16.
39. Morton, A. A.; Davidson, J. B.; Hakan, B. L. *J. Am. Chem. Soc.* **1942**, *64*, 2242–2247.
40. Negishi, E.; King, A. O.; Okukado, N. *J. Org. Chem.* **1977**, *42*, 1821–1823.
41. Jana, R.; Pathak, T. P.; Sigman, M. S. *Chem. Rev.* **2011**, *111*, 1417–1492.
42. Sun, C.-L.; Shi, Z.-J. *Chen. Rev.* **2014**, *114*, 9219–9280.
43. Delaney, C. P.; Heyboer, E. M.; Denmark, S. E. *Org. Synth.* **2020**, *97*, 245–261.
44. Atmuri, N. D.; Lubell, W. D. *Org Synth.* **2015**, *92*, 103–116.
45. Kumada, M.; Tamao, K.; Sumitani, K. *Org. Synth.* **1978**, *58*, 127–133.
46. Hie, L.; Chang, J. J.; Garg, N. K. *J. Chem. Educ.* **2015**, *92*, 571–574. Hie, L.; Garg, N. K. *Org. Synth.* **2016**, *93*, 306–318.

47. Metzger, A.; Melzig, L.; Despotopoulou, C.; Knochel, P. *Org. Lett.* **2009**, *11*, 4228–4231.
48. Barder, T. E.; Walker, S. D.; Martinelli, J. R.; Buchwald, S. L. *J. Am. Chem. Soc.* **2005**, *127*, 4685–4696.
49. Wittig, G.; Hesse, A. *Org. Synth.* **1970**, *50*, 66–71.
50. Barrett, A. G. M.; Flygare, J. A.; Hill, J. M.; Wallace, E. M. *Org. Synth.* **1996**, *73*, 50–59.
51. Maryanoff, B. E.; Reitz, A. D. *Chem. Rev.* **1989**, *89*, 863–927.
52. Greenwald, R.; Chaykovsky, M.; Corey, E. J. *J. Org. Chem.* **1963**, *28*, 1128–1129.
53. Dong, D. J.; Li, H. H.; Tian, S. K. *J. Am. Chem. Soc.* **2010**, *132*, 5018–5020.
54. Wadsworth, W. S.; Emmons, W. D. *J. Am. Chem. Soc.* **1961**, *83*, 1733–1738.
55. White, J. D.; Takabe, K.; Prisbylla, M. P. *J. Org. Chem.* **1985**, *50*, 5233–5244.
56. Bhattacharya, A. K.; Thyagarajan, G. *Chem. Rev.* **1981**, *81*, 415–430.
57. Still, W. C.; Gennari, C. *Tetrahedron Lett.* **1983**, *24*, 4405–4408.
58. Yin, L.; Liebscher, J. *Chem. Rev.* **2007**, *107*, 133–173. de Vries, J. G. *Can. J. Chem.* **2001**, *79*, 1086–1092. Eisenstadt, A.; Keren, Y. Process for the Preparation of Octyl Methoxy Cinnamate. U.S. Patent 5,187,303, February 16, 1993.
59. Clive, D. L. J.; Daigneault, S. *J. Org. Chem.* **1991**, *56*, 3801–3814.
60. Lebel, H.; Marcoux, J. F.; Molinaro, C.; Charette, A. B. *Chem. Rev.* **2003**, *103*, 977–1050.
61. Doyle, M. P.; Forbes, D. C. *Chem. Rev.* **1998**, *98*, 911–936.
62. Vanier, S. F.; Larouche, G.; Wurz, R. P.; Charette, A. B. *Org. Lett.* **2010**, *12*, 672–675.
63. Ikeda, T.; Yue, S.; Hutchinson, C. R. *J. Org. Chem.* **1985**, *50*, 5193–5199.
64. Jung, M. E.; Yoo, D. *Org. Lett.* **2011**, *13*, 2698–2701.
65. Wang, D.; Ackermann, L. *Chem. Sci.* **2022**, *13*, 7256–7263.
66. Matsumoto, T.; Imai, S.; Miuchi, S.; Sugibayashi, H. *Bull. Chem. Soc. Jpn.* **1985**, *58*, 340–345.
67. Wang, X.; DeSilva, S. O.; Reed, J. N.; Billadeau, R.; Griffen, E. J.; Chan, A.; Sniekus, V. *Org Synth.* **1995**, *72*, 163–172.
68. Kancherla, S.; Lorentzen, M.; Snieckus, V.; Jørgensen, K. B. *J. Org. Chem.* **2018**, *83*, 3590–3598.
69. Huffman, J. W.; Potnis, S. M.; Satish, A. V. *J. Org. Chem.* **1985**, *50*, 4266–4270.
70. Gall, M.; House, H. O. *Org. Synth.* **1972**, *52*, 39–52.

71. Walba, D. M.; Stoudt, G. S. *J. Org. Chem.* **1983**, *48*, 5404–5406.
72. Mansson, C. M. F.; Burns, N. Z. *Org. Synth.* **2024**, *101*, 410–422.
73. Mori, Y.; Seki, M. *Org. Synth.* **2007**, *84*, 285–294.
74. Coburn, C. E.; Anderson, D. K.; Swenton, J. S. *J. Org. Chem.* **1983**, *48*, 1455–1461.
75. Taylor, M. D.; Minaskanian, G.; Winzenberg, K. N.; Santone, P.; Smith, A. B. III. *J. Org. Chem.* **1982**, *47*, 3960–3964.
76. White, J. D.; Matsui, T.; Thomas, J. A. *J. Org. Chem.* **1981**, *46*, 3376–3378.
77. Fürstner, A.; Leitner, A.; Seidel, G. *Org. Synth.* **2005**, *81*, 33–41.
78. Jackson, R. F. W.; Perez-Gonzalez, M. *Org. Synth.* **2005**, *81*, 77–88.
79. Hornback, J. M.; Barrows, R. D. *J. Org. Chem.* **1983**, *48*, 90–98.
80. Hagiwara, H.; Ono, H.; Hoshi, T. *Org. Synth.* **2003**, *80*, 195–199.
81. Narasaka, K.; Soai, K.; Aikawa, Y.; Mukaiyama, T. *Bull. Chem. Soc. Jpn.* **1976**, *49*, 779–783.
82. Bailey, W. F.; Luderer, M. R.; Mealy, M. J.; Punzalan, E. R. *Org. Synth.* **2005**, *81*, 121–133.
83. Callahan, J. F.; Newlander, K. A.; Bryan, H. G.; Huffman, W. F.; Moore, M. L.; Yim, N. C. F. *J. Org. Chem.* **1988**, *53*, 1527–1530.
84. Kraus, G. A.; Hon, Y. S.; Sy, J.; Raggon, J. *J. Org. Chem.* **1988**, *53*, 1397–1400.
85. Tanabe, Y.; Ohno, N. *J. Org. Chem.* **1988**, *53*, 1560–1563.
86. Lal, K.; Zarate, E. A.; Youngs, W. J.; Salomon, R. G. *J. Org. Chem.* **1988**, *53*, 3673–3680.
87. Pataki, J.; Di Raddo, P.; Harvey, R. G. *J. Org. Chem.* **1989**, *54*, 840–844.
88. Vidal-Pascual, M.; Martinez-Lamenca, C.; Hoffmann, H. M. R. *Org. Synth.* **2006**, *83*, 61–69.
89. Piers, E.; Friesen, R. W. *J. Org. Chem.* **1986**, *51*, 3405–3406.
90. Flynn, G. A.; Vaal, M. J.; Stewart, K. T.; Wenstrup, D. L.; Beight, D. W.; Bohme, E. H. *J. Org. Chem.* **1984**, *49*, 2252–2258.
91. Jardon, P. W.; Vickery, E. H.; Pahler, L. F.; Pourahmady, N.; Mains, G. J.; Eisenbraun, E. J. *J. Org. Chem.* **1984**, *49*, 2130–2135.
92. Godleski, S. A.; Villhauer, E. B. *J. Org. Chem.* **1984**, *49*, 2246–2252.
93. Hutchison, D. R.; Khau, V. V.; Martinelli, M. J.; Nayyar, N. K.; Peterson, B. C.; Sullivan, K. A. *Org. Synth.* **1998**, *75*, 223–228.
94. Hajos, Z. G.; Wachter, M. P.; Werblood, H. M.; Adams, R. E. *J. Org. Chem.* **1984**, *49*, 2600–2608.
95. Enders, D.; von Berg, S.; Jandeleit, B. *Org. Synth.* **2002**, *78*, 177–188.

96. Pariza, R. P.; Fuchs, P. L. *J. Org. Chem.* **1983**, *48*, 2306–2308.
97. Harvey, R. G.; Hahn, J. T.; Bukowska, M.; Jackson, H. *J. Org. Chem.* **1990**, *55*, 6161–6166.
98. Zambias, R. A.; Caldwell, C. G.; Kopka, I. E.; Hammond, M. L. *J. Org. Chem.* **1988**, *53*, 4135–4137.
99. Kozmin, S. A.; He, S.; Rawal, V. H. *Org. Synth.* **2002**, *78*, 160–168.

8

PLANNING MULTISTEP SYNTHESES

The challenge in synthesis is to devise a set of reactions that will convert inexpensive, readily available materials into complex, valuable products. Ordinarily, this is not an obvious following of a roadmap, but rather a complex puzzle requiring much strategy. This chapter gives samples of the planning process with actual syntheses of relatively simple cases. Enough of the procedure is provided to enable you to analyze and devise syntheses for many molecules. A more extensive treatment is given in an excellent book by Warren [1].

8.1 RETROSYNTHETIC ANALYSIS

You will become familiar with the sorts of compounds that are available as you read and study the chemical literature, but the actual process of synthesis begins at the end; that is, you must study the desired structure and work *backward*. What penultimate intermediate would be readily

Understanding Organic Chemistry: Bridging Fundamental and Advanced Courses, Fourth Edition. Ann M. Fabirkiewicz and John C. Stowell.

Companion website: www.wiley.com/go/fabirkiewiczorganic4e

converted to that product, and then what before that? This process is called *retrosynthetic analysis* [2], and each backward step is indicated by a double-shafted arrow (⇒). A backward scheme is drawn, and then a forward process is developed with actual reagents, indicated with ordinary arrows. In more complicated syntheses, you will need to look ahead toward steps in the middle of the process, but still, a backward approach is most practical.

The steps include functional group interconversions as given in Chapter 6 and carbon backbone construction as illustrated in Chapter 7. Viewed as the disassembly of the product, you should first disconnect the parts that are joined by functional groups; for example, esters should be separated into acid and alcohol parts. The carbon–carbon bonds should be disconnected at or near functional groups and at branch points in the backbone. There are often a great many choices of dividing points and starting materials. For example, jasmone and dihydrojasmone have been made by hundreds of routes [3]. In selecting among choices, the number of steps should be minimized, cheaper starting materials selected, and high-yield reactions favored, and the scheme should converge instead of following a long linear sequence of steps. Sometimes, a closely related molecule will be available, requiring a minimum of construction effort, as in making other steroids from diosgenin [4].

8.2 DISCONNECTION AT A FUNCTIONAL GROUP OR A BRANCH POINT

Carbon–carbon bonds are frequently built using carbonyl compounds. A carbonyl group normally confers a pattern of alternating potential electrophilic or nucleophilic reactivity along a carbon chain, as shown in Structure **8.1** [5]. The electrophilic character exists in the carbonyl compounds themselves, continuing along the chain as far as conjugating *p* orbitals are present to transmit it (**8.2**). The nucleophilic character at the

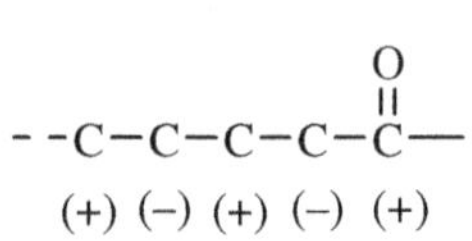

8.1

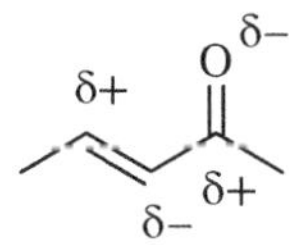

8.2

α carbon exists in the derived enolate (**8.3**) or enol (**8.4**) form. The charges at these sites can be propagated through a conjugated π system and serve to attract another carbon reagent of opposite charge and give a new bond.

8.3 **8.4**

As we consider disconnection along the chain, we can decide whether the reactive site backed by the carbonyl group will be nucleophilic or electrophilic. Three different choices are taken in the following examples to illustrate the rationale.

Disconnection at the α carbon gives fragments with a (+) on the carbonyl, requiring the incoming reagent to be nucleophilic. Compound **8.5** is an intermediate in the synthesis of hemlock alkaloids. The carbamate functional group is made from a chlorocarbonate and the amine; therefore, this disconnection is the first retro step (Scheme 8.1). Ketones are commonly used to orient polarity in synthesis, so we will take the amine back to the

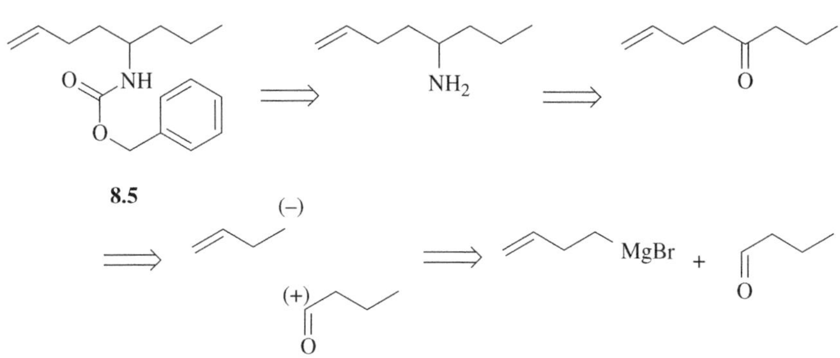

SCHEME 8.1 Retrosynthesis of 4-[(carbobenzyloxy)amino]-7-octene.

ketone. The key carbon–carbon bond-forming step in this synthesis forms the alpha bond. We can now write a forward scheme with commercially available starting materials (Eq. 8.1) [6].

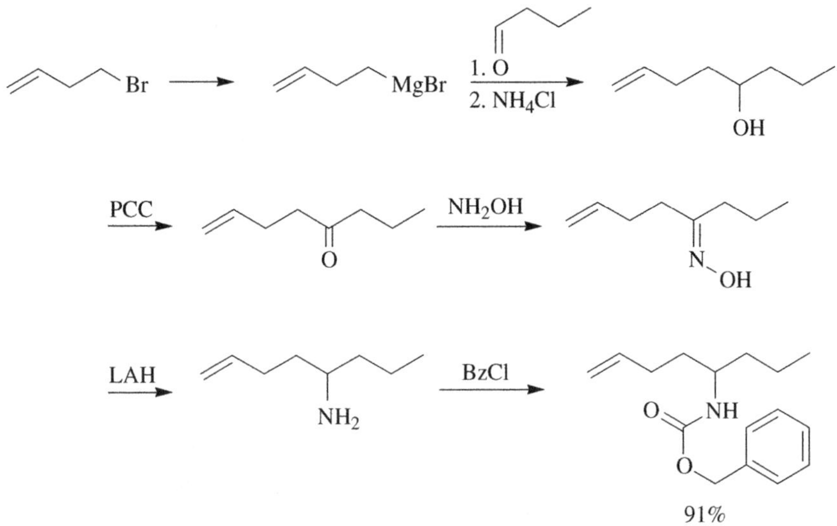

(8.1)

The starting point, 4-bromo-1-butene, is commercially available or can be prepared from the alcohol, which may, in turn, be prepared from vinylmagnesium chloride plus ethylene oxide. The Grignard reagent may be treated with an acylating agent, or as these authors chose, an aldehyde followed by an oxidizing agent. In this example, the amine functional group suggested a carbonyl intermediate. This same retro step should be suggested by many other functional groups, including alcohols and halides.

Disconnection between the α and β carbons can be used to synthesize compound **8.6**, on both sides of the carbonyl (Scheme 8.2). In this

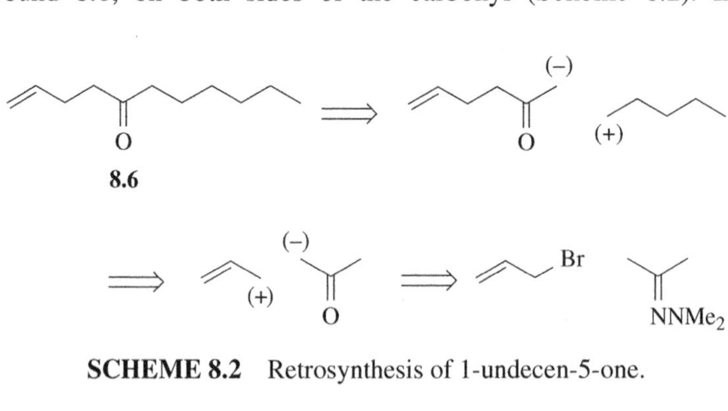

SCHEME 8.2 Retrosynthesis of 1-undecen-5-one.

disconnection, the carbonyl fragment is the nucleophile, specifically the enolate of a hydrazone derivative (Section 7.1.1). The synthesis is shown in Equation 8.2 [7].

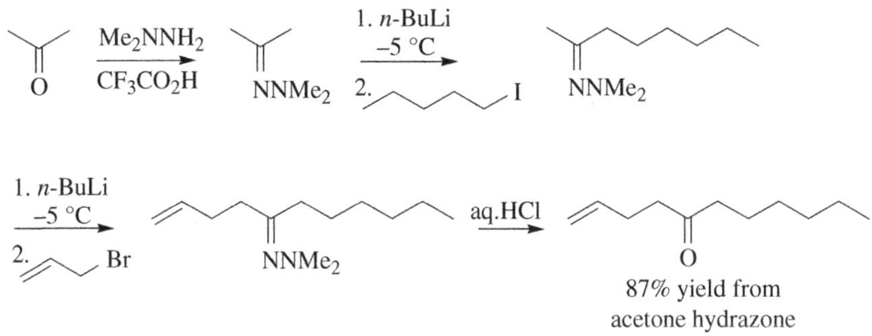

(8.2)

Disconnection one atom further along the chain, that is between the β and γ carbons, requires a conjugate addition as in the synthesis of **8.7** (Scheme 8.3). The actual reaction is illustrated in Equation 8.3 [8]. Lithium di-*n*-butylcuprate is usually used in two- to fivefold excess, but the use of dicyclohexylphosphide as an auxiliary ligand to the cuprate requires only one equivalent of the reagent.

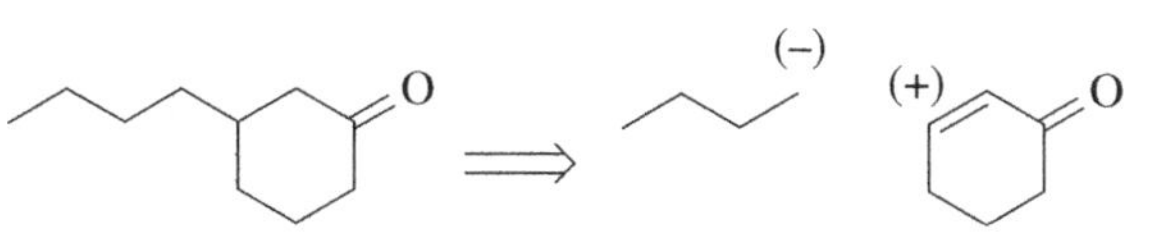

8.7

SCHEME 8.3 Retrosynthesis of 3-butylcyclohexanone.

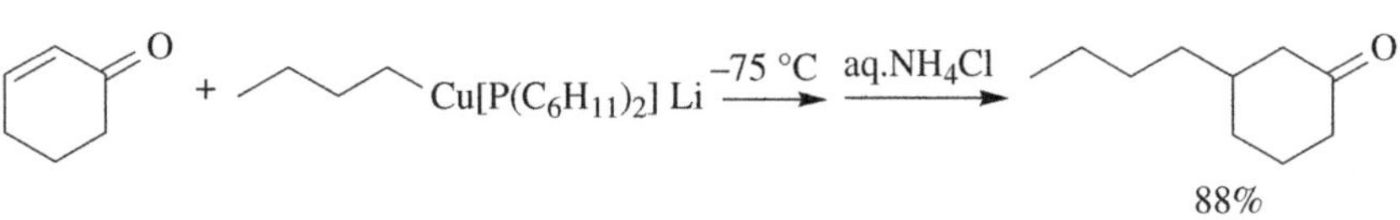

(8.3)

Which of the three disconnections we select depends on other structural features of the particular product and on the availability of materials. In the synthesis of **8.7**, the other two disconnection choices would have required more steps and difunctional intermediate compounds because the compound is cyclic. Compounds **8.5** and **8.6** could be made by any of the three methods.

8.3 COOPERATION FOR DIFUNCTIONALITY

Molecules that contain two functional groups at particular distances apart are assembled considering the electronic influence of both groups together. As with the monofunctional compounds, we can consider carbonyl compounds to be primary intermediates and examine their influence on the fragments from disconnection between the groups. If the synthetic target is substituted 1,3 (or 1,5), then the polarities required are supported by those induced by the presence of the carbonyl group. When the synthetic target has 1,4 functionality, the normal electrophilic or nucleophilic sites in carbonyl compounds do not give the easiest routes to products. For these situations, we substitute another compound that allows the opposite reactivity but can subsequently be converted to the carbonyl compound. Reagents with this reversed or "abnormal" reactivity are designated by the German term *umpolung* [9]. Several such reagents are illustrated in Table 8.1.

Considering first a 1,3-difunctional chain, the retrosynthetic analysis suggests cleavage adjacent to the carbonyl, and the normal polarity influence of the carbonyl provides the charges shown. Since they are opposite, they will attract each other, and this would be a favorable approach to the

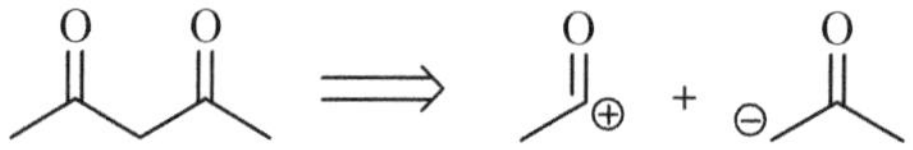

synthesis of 1,3-difunctional compounds. To be specific, consider the synthesis of **8.8**. This hydroxyester may be disconnected between the α and β carbons to give appropriate fragments (Scheme 8.4).

The actual reagents could be the ester enolate in the form of a Reformatsky reagent and the ketone. Alternatively, β-hydroxycarbonyl compounds are often dehydrated; thus, the α,β- or β,γ-unsaturated compounds should be approached in planning by first rehydrating. The initial adduct could also be reduced or oxidized or further converted to other functional groups; therefore, Compounds **8.9** and **8.10** would also be approached with the same intermediates and disconnection in mind (Scheme 8.5). The actual synthesis of **8.10** was carried out as shown in Equation 8.4 [18]. The dehydration gives the endocyclic double bond in five-membered or smaller rings.

TABLE 8.1 Electrophilic and Nucleophilic Umpolung Reagents

Reagent	Equivalent	Bond Formed	Example [Reference]
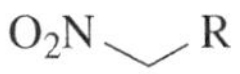	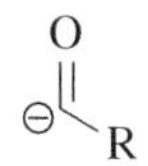	α	Equation 8.6 [10]
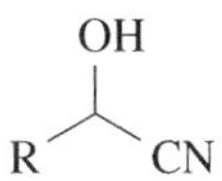	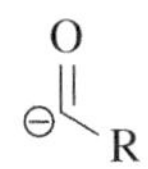	α	Equation 8.7 [11]
	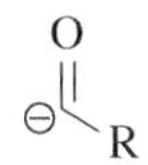	α	[12]
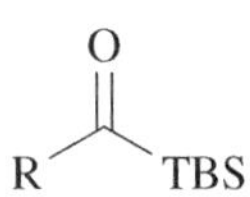	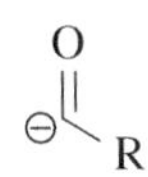	α	Equation 8.10 [13]
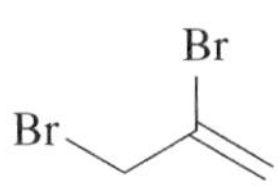	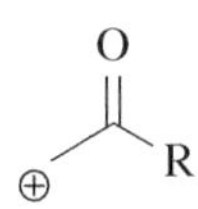	α,β	Equation 8.8 [14]
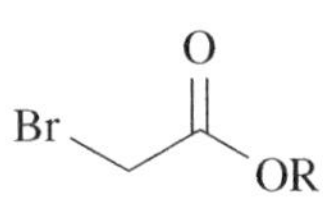	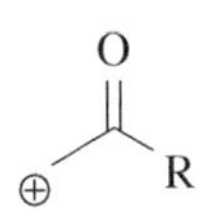	α,β	Equation 8.9 [15]
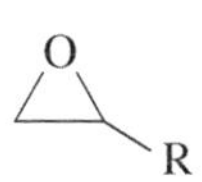	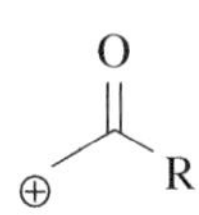	α,β	Equation 8.26 [16]
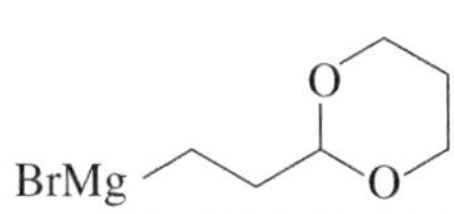	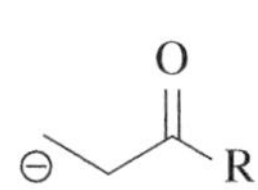	β,γ	Equation 8.16 [17]

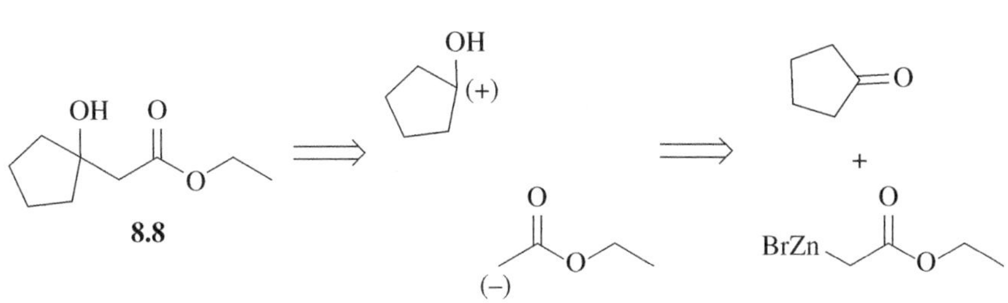

SCHEME 8.4 Retrosynthesis of ethyl (1-hydroxycyclopentyl)acetate.

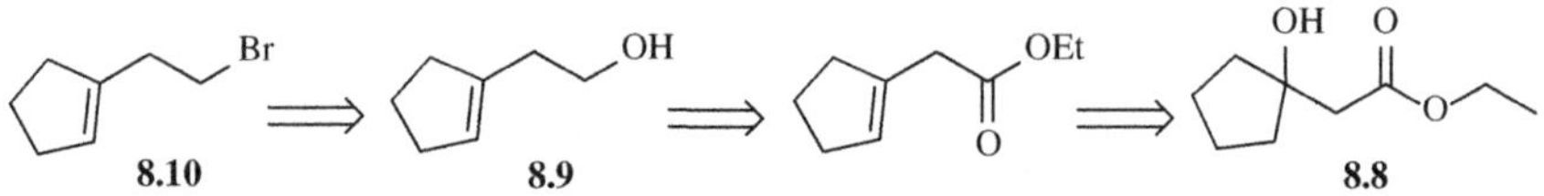

SCHEME 8.5 Retrosynthesis of 1-(2-bromoethyl)cyclopentene.

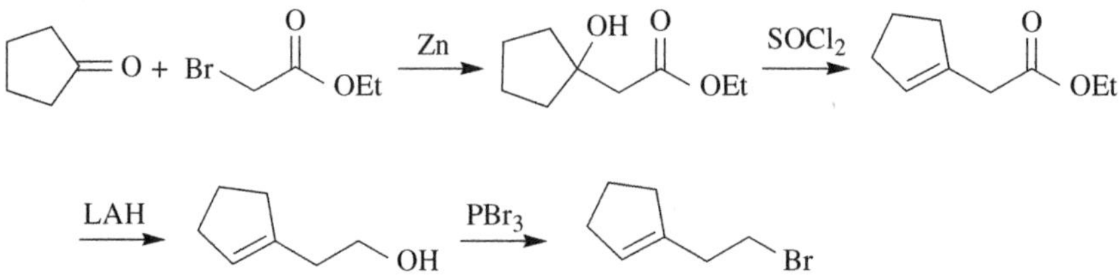

(8.4)

Bromoalcohol **8.11** is also 1,3-difunctional. Maximum simplification would result from disconnecting the C3 alcohol fragment, but first, a carbonyl function should probably be considered. This gives a β-hydroxyester, so again the Reformatsky reaction is suggested (Scheme 8.6). Equation 8.5 shows the sequence used [19] to prepare **8.11**.

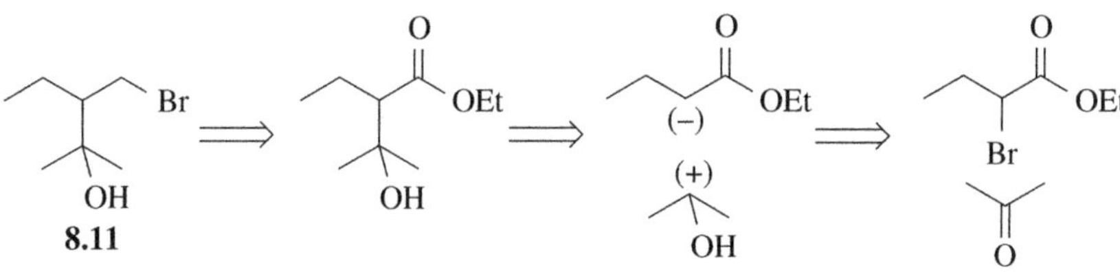

SCHEME 8.6 Retrosynthesis of 3-bromo-3-ethyl-2-methyl-2-butanol.

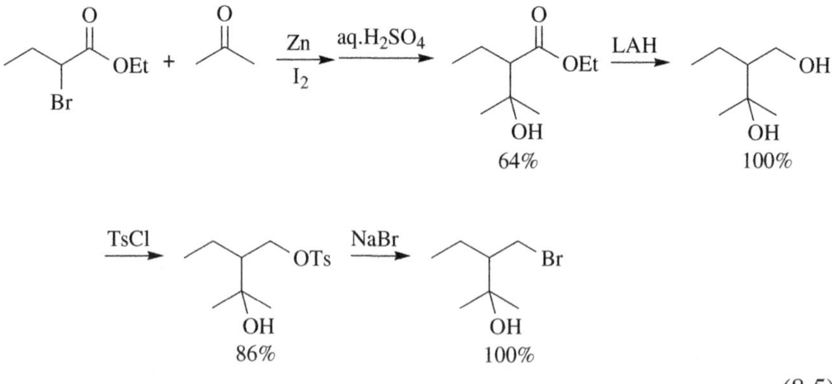

(8.5)

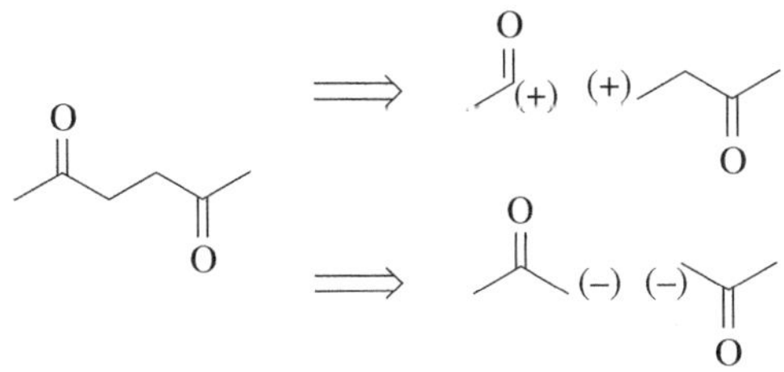

SCHEME 8.7 Retrosynthesis of 2,5-heptanedione.

Turning next to a 1,4-difunctional molecule, 2,5-heptanedione, we find that neither of the two likely disconnections give mutually attracting fragments (Scheme 8.7). It is therefore necessary to use an umpolung (reverse polarity) reagent in place of one of the usual components. A nitro group on a carbon facilitates the removal of a proton from that carbon, giving a nucleophilic nitronate anion suitable for the 1,2 disconnection. Later, the nitro group and carbon may be converted to a carbonyl group (which itself would have been electrophilic). 2,5-Heptanedione was prepared in this way (Eq. 8.6) [10].

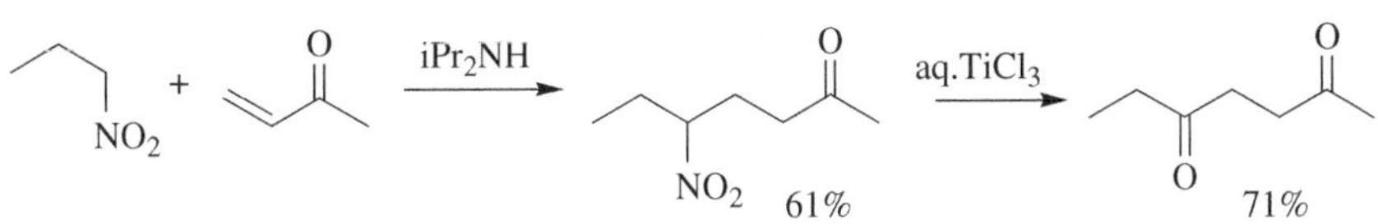

(8.6)

An abnormal polarity carbonyl α disconnection step is shown for Compound **8.12** in Scheme 8.8. The carbonyl carbon is not nucleophilic, but an umpolung reagent will allow the carbon to behave as a nucleophile for the carbon–carbon bond-forming reaction.

The use of cyanide on the aldehyde forms the cyanohydrin in which the anion is stabilized by the nitrile. Attack on the β carbon of acrylonitrile

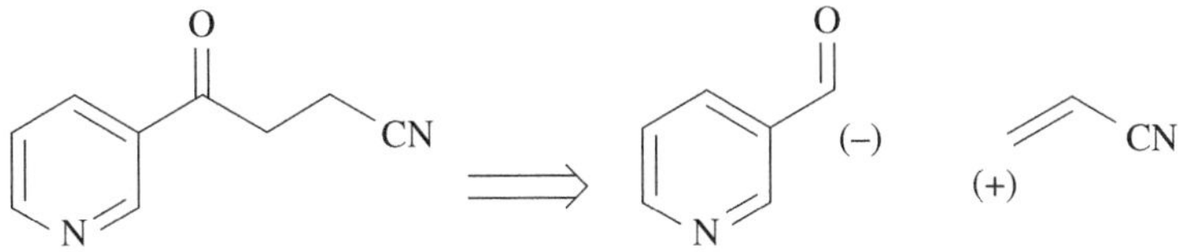

8.12

SCHEME 8.8 Retrosynthesis of 4-oxo-4-(3-pyridyl)butyronitrile.

forms the new bond, as shown in Equation 8.7 [11].

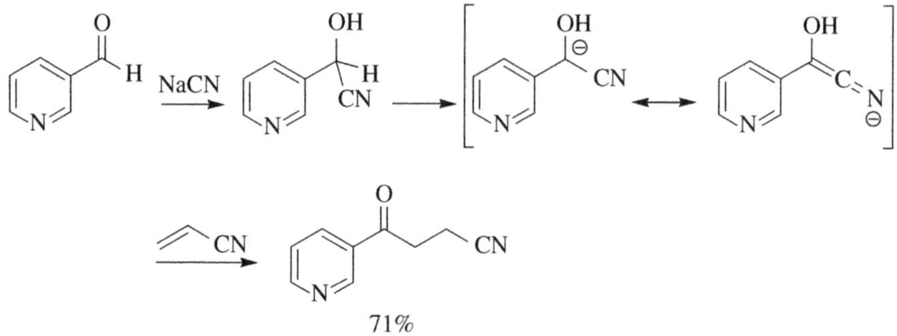

(8.7)

Disconnecting one bond farther from the carbonyl, the α,β abnormal case is provided indirectly by the reaction of Grignard reagents with epoxides. This is followed by oxidation if the carbonyl group is the desired functionality. The β,γ disconnection with abnormal polarity is readily arranged if the α carbon is sp^3-hybridized and insulates the carbonyl group from the nucleophilic β carbon.

The 2,3 disconnection of 1,4-difunctional molecules is also of value. The diketone in Scheme 8.9 disconnects to a pair of mutually unattractive enolate ions, but one may be replaced with an umpolung reagent. 2,3-Dibromopropene is equivalent to acetone with an electrophilic α carbon. The actual steps are shown in Equation 8.8. The initial ketone enolate was stirred at room temperature to equilibrate to the more stable more substituted enolate. The vinyl bromide intermediate was hydrolyzed to the ketone using mercury (II) acetate in formic acid [14].

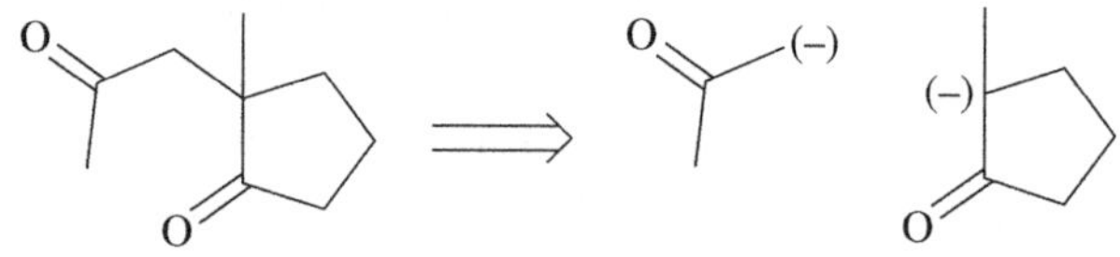

SCHEME 8.9 Retrosynthesis of 2-methyl-2-(2-oxopropyl)cyclopentanone.

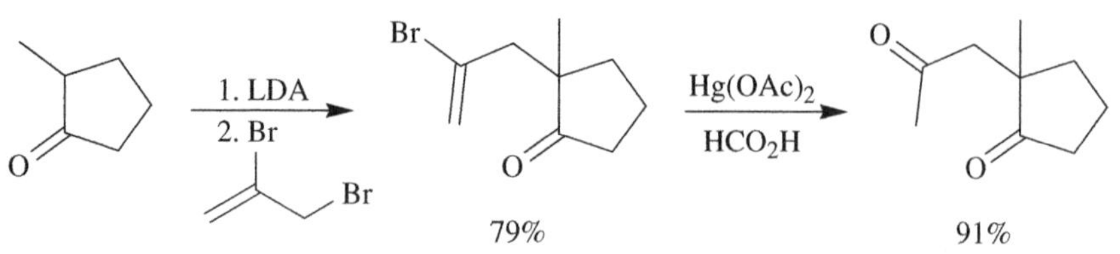

(8.8)

The retrosynthetic analysis of γ-ketoester, **8.13**, yields a pair of mutually unattractive enolate ions, stemming from the 1,4 arrangement of functionality. One enolate may be rendered electrophilic by placing a bromine on the α carbon (Scheme 8.10), creating an umpolung reagent. The retrosynthetic analysis of intermediate **8.14** indicates a 1,5 difunctional molecule, which is supported by similar principles already developed for 1,3-difunctional cases, that is, the polarity induced by the carbonyl group supports bond formation and an umpolung reagent is not necessary (Scheme 8.10).

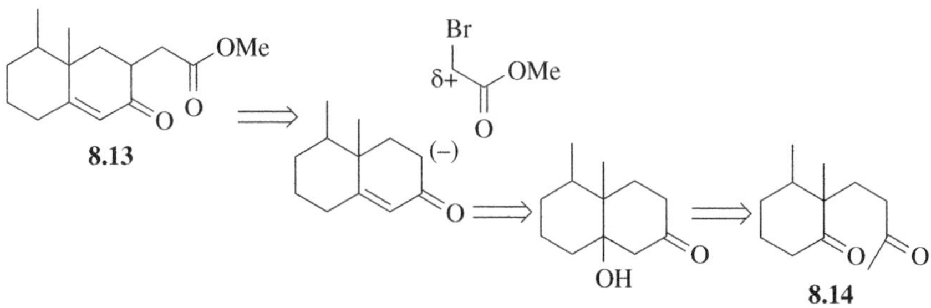

SCHEME 8.10 Retrosynthesis of methyl (1,2,6,7,8,8a hexahydro-8,8a-dimethyl-3(4*H*)-oxonaphthyl)acetate.

The actual synthesis of **8.13** and **8.14** was accomplished as shown in Equation 8.9. We expect 2,3-dimethylcyclohexanone to give an enolate or enol preferentially at C2 to form **8.14**. Closure of the six-membered ring is accomplished with sodium methoxide. Formation of the enolate via lithium diisopropylamide (LDA) and alkylation with methyl bromoacetate provides **8.13** [15].

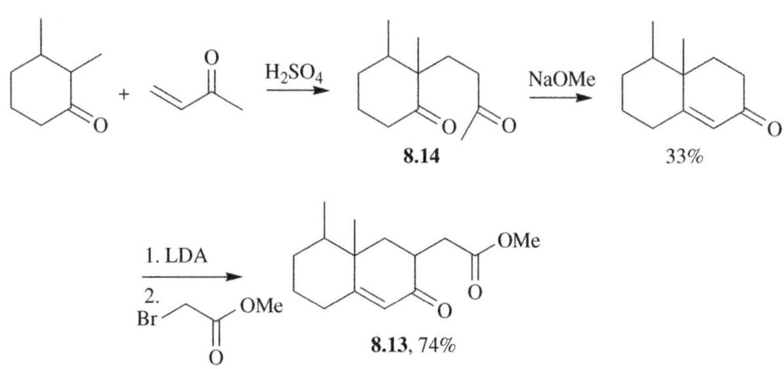

(8.9)

A variety of umpolung reagents have been developed (Table 8.1), and stereocontrol of such reactions is possible as well. Dithioacetals are readily converted to the corresponding carbonyl compound and can be treated with *n*-butyl lithium to form a nucleophilic acyl anion equivalent [12]. The silyl benzoin reaction generates α-hydroxyketones with complete regiocontrol [20]. Stereocontrol in a conjugate addition using an umpolung reagent has also been accomplished [21].

Acyl silanes are umpolung reagents, allowing the carbonyl group to take on nucleophilic character due to the favorable formation of an O—Si bond as silicon migrates to oxygen (Brook rearrangement) [22]. The related silyl glyoxylates [23] can be used to add both a nucleophilic vinyl Grignard and an electrophilic aldehyde to the carbonyl carbon (Eq. 8.10) [13]. This reaction has been adapted in a key step in the synthesis of trachyspic acid, using the lithium enolate as the nucleophile and an α,β-unsaturated carbonyl compound as the electrophile adding to the silyl glyoxylate (Eq. 8.11) [13].

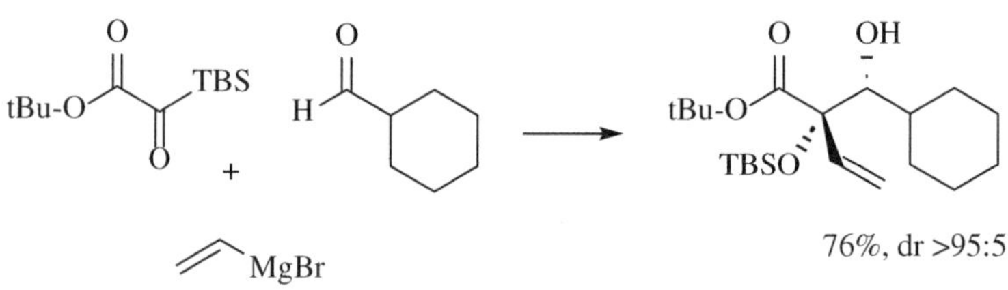

(8.10)

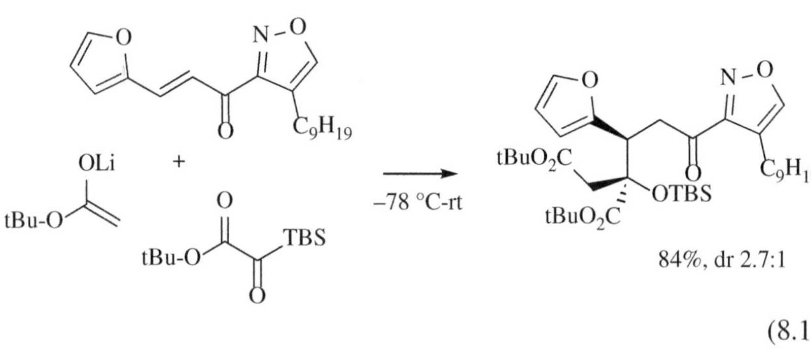

(8.11)

We can use these retrosynthetic choices in some longer sequences. *exo*-Brevicomin (**8.15**) is a pheromone from the Western pine beetle. Examining the functionality, we see a carbon attached to two oxygens, that is, an acetal derivative of a ketone. A Wittig reaction was used to

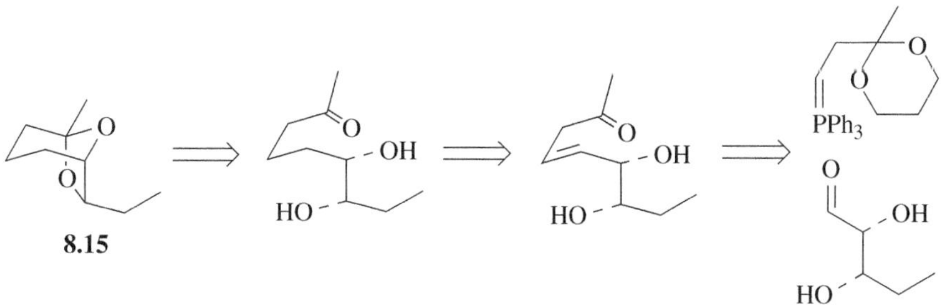

SCHEME 8.11 Retrosynthesis of *exo*-brevicomin.

bring the two functional groups together (Scheme 8.11). A *threo*-diol is needed.

The Whitesides group [24] chose to use transketolase (TK), a commercially available enzyme isolated from yeast, to affect the kinetic resolution (Section 3.4) of the racemic starting material and form the required stereoisomer (Eq. 8.12).

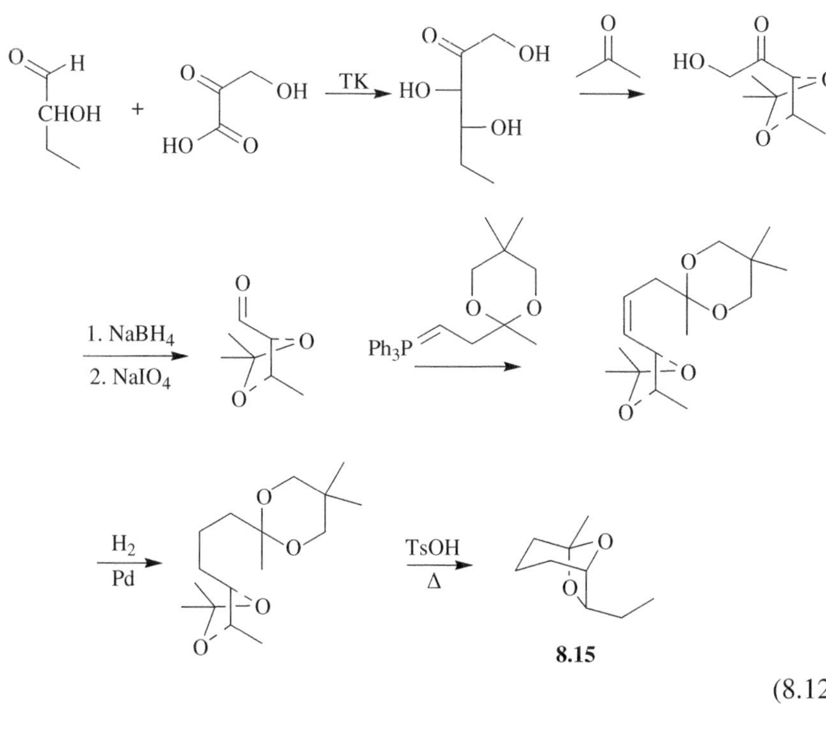

(8.12)

The lactone 8.16 is a 1,5-difunctional molecule at a lower oxidation state than **8.14** and is accessible by the same disconnection rationale (Scheme 8.12). First disconnect the functionality and then the carbon chain at the branch point (C3—C4 bond). The butyraldehyde enolate called for here is not practical because of self-condensation (Section 7.1.1); therefore, a nonelectrophilic derivative of the aldehyde is used instead.

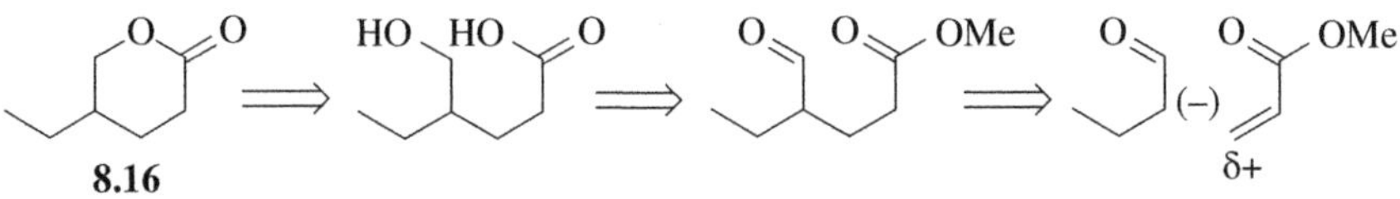

8.16

SCHEME 8.12 Retrosynthesis of 5-ethyltetrahydro-2*H*-pyran-2-one.

The piperidine enamine is sufficiently nucleophilic for the Michael addition and can be hydrolyzed readily after that step (Eq. 8.13) [25, 26].

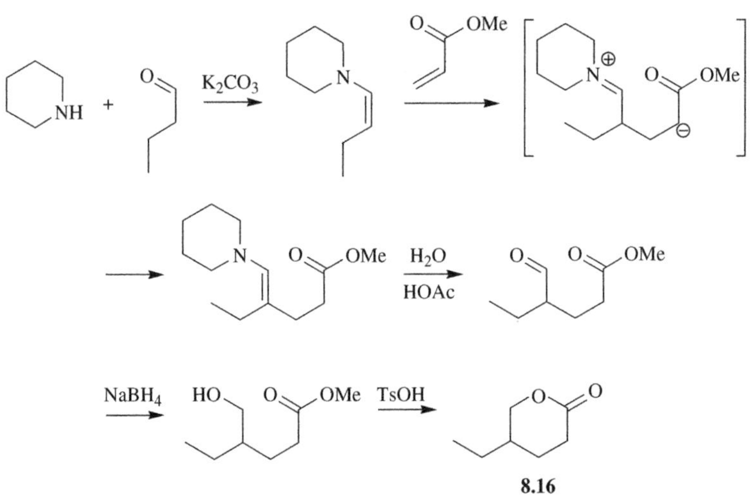

8.16

(8.13)

1,6-Difunctional molecules are not often formed by connecting intervening carbons. The ready availability of cyclohexenes and cyclohexanones allows for oxidative ring opening to give the 1,6-functionality spacing, as exemplified in Equation 8.14 [27]. In Equation 8.15, enantiocontrol of the Baeyer–Villiger reaction (Eqs. 6.8 and 6.9) is demonstrated using an enzymatic reagent, PAMO-P3, a mutant strain phenylacetone monooxygenase (Section 3.4) [28]. The lactone can be hydrolyzed to introduce 1,6 functionality. Some 1,5-difunctional molecules are also formed by the oxidative opening of five-membered ring compounds.

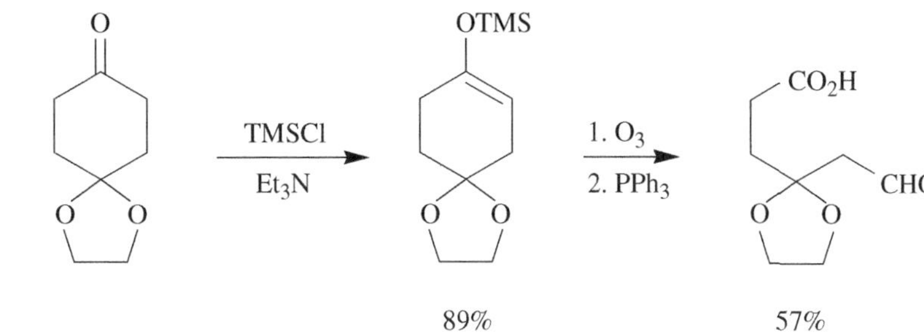

(8.14)

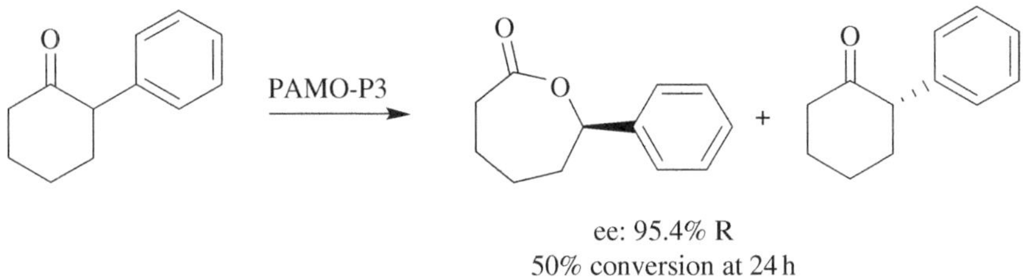

(8.15)

Difunctional molecules are sometimes assembled, ignoring the polarizing possibilities of one of the functional groups. The nucleophilic character can be developed at a site on a chain where it is insulated from a carbonyl group by intervening sp^3 carbons. The carbonyl group will need protection as the acetal. Equation 8.16 shows the preparation of 1,4 functionality this way [17].

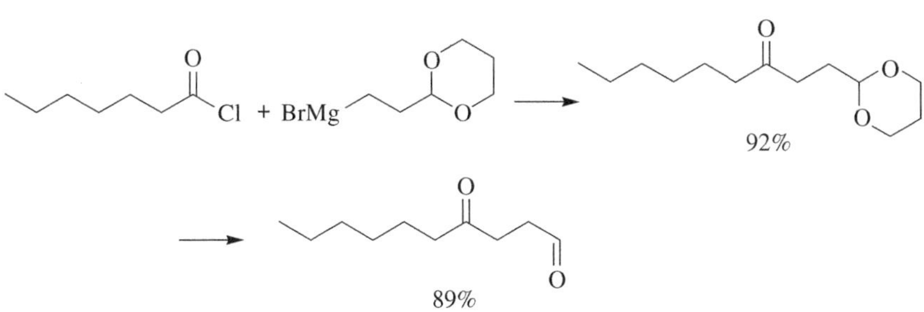

(8.16)

8.4 RING CLOSURE

Many difunctional molecules in this chapter and in Chapter 7 react to give rings. A competition may exist between inter- and intramolecular reactions. In most cases, the formation of three-, five-, and six-membered rings is more favorable than polymerization because the intermolecular process requires a bimolecular collision, while cyclization requires only conformational alignment. This becomes less probable for rings larger than six; therefore, high dilution is used to lessen the frequency of intermolecular collisions. The acyloin condensation using chlorotrimethylsilane as a trapping agent is efficient for ring closures ranging from 4 to 14 members (Eq. 8.17) [29].

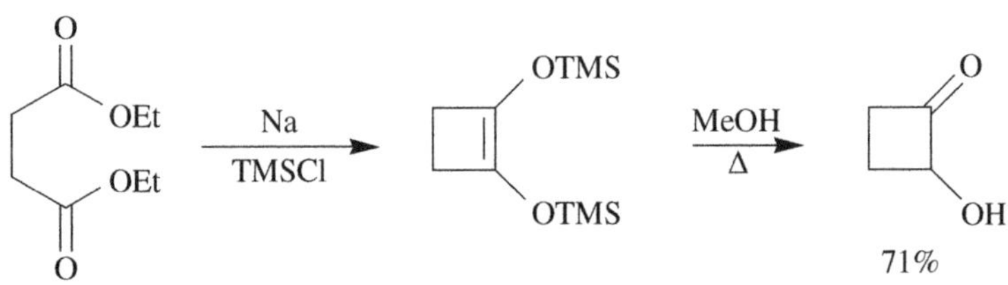

(8.17)

In some cases, there is a competition within a molecule for closure to rings of different sizes. If the closure is an irreversible reaction such as alkylation of an enolate, kinetic control may allow three-membered ring closure to predominate over five-membered, as shown in Equation 8.18 [30]. The electrophilic carbon closer to the nucleophilic site is found first, in spite of the incorporation of ring strain. With reversible reactions, the thermodynamic product predominates. That is, the low-strain five- or six-membered rings will form to the exclusion of three- or four-membered alternatives (Scheme 8.13). In Claisen and aldol cyclizations, five- and six-membered rings are not in competition with each other. Seven-membered rings may sometimes be closed readily without high dilution (Eq. 8.19) [31].

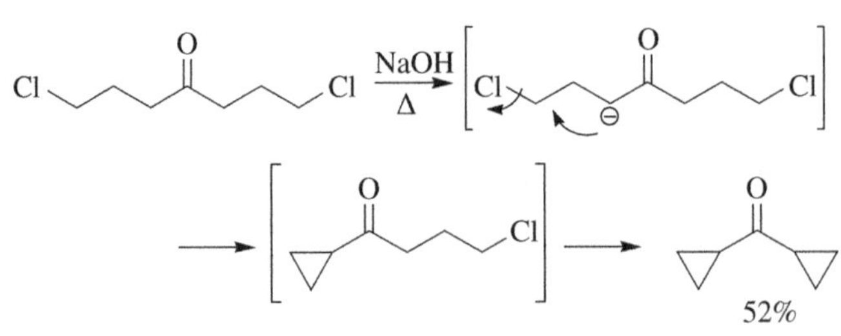

(8.18)

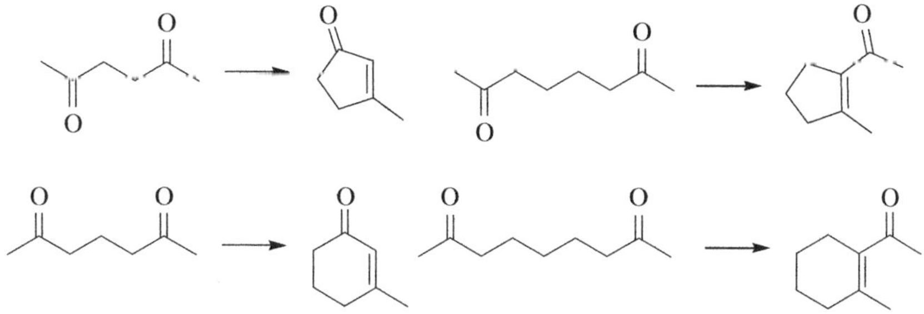

SCHEME 8.13 Products expected from aldol cyclizations of 1,4- to 1,7-diketones.

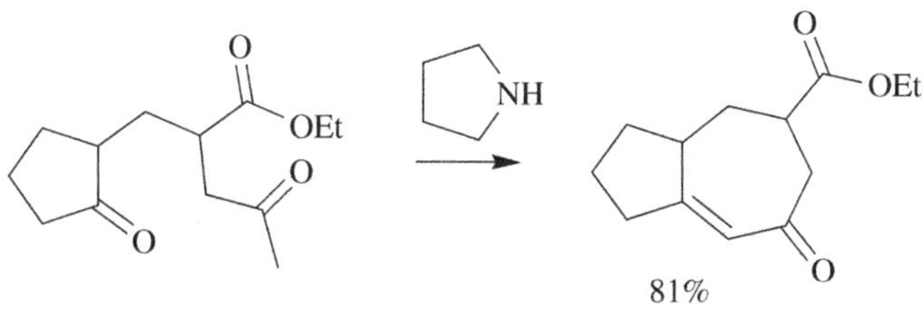

(8.19)

Heterocyclic rings show the same size preferences. Halohydrins give epoxides under basic, irreversible conditions. Hemiacetals, acetals, and lactones under reversible conditions favor five- or six-membered rings, where conformational flexibility permits, over intermolecular polymerization.

Click chemistry has been developed to predictably produce specific chemical bonds in high yield. A common example is the copper-catalyzed azide-alkyne cycloaddition (CuAAC) which forms 1,2,3-triazoles from azides and alkynes. While the synthesis of azides typically requires the use of toxic and hazardous materials, recent developments allow for the formation of the azide from a primary amine using fluorosulfuryl azide (FSO_2N_3). The Sharpless group performed this reaction in well plates to generate over 1200 triazoles, with 80% of the reactions occurring in 70% or greater yield (Eq. 8.20) [32]. A copper catalyst forms the 1,5-disubstituted ring, which is an important building block in medicinal chemistry [33]. The modular nature of this reaction allows a wide range of possible triazoles.

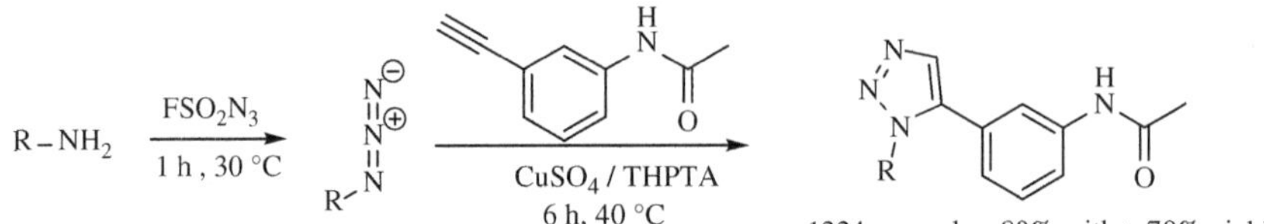

1224 examples, 80% with > 70% yield

(8.20)

The closure of four-membered rings requires special methods. Three reactions are frequently used: the acyloin condensation (Eq. 8.17), photochemical cycloaddition (Eq. 5.7), and thermal ketene cycloadditions.

Alkenes and alkynes will cycloadd photochemically to other alkene molecules, especially those conjugated to carbonyl groups, to give cyclobutanes or cyclobutenes [34]. The molecules are raised to an excited electronic state, sometimes via a radiation-absorbing sensitizer compound, combine to form the ring, and descend to the electronic ground state. In doubly unsymmetric cases, the regio- and stereochemistry can be complex and dependent on conditions. Nevertheless, many are synthetically useful. Two examples are shown in Equations 8.21 [35] and 8.22 [36].

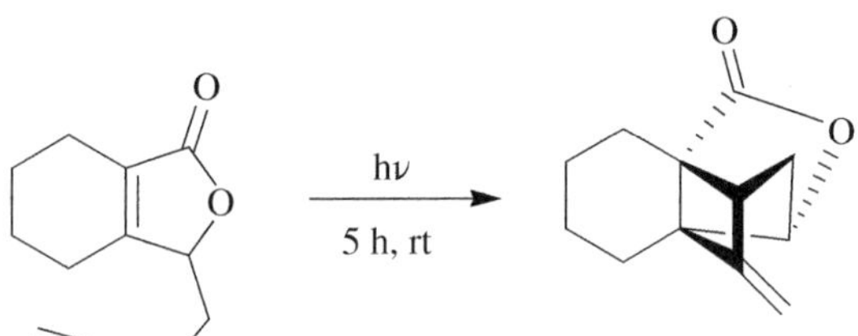

70%

(8.21)

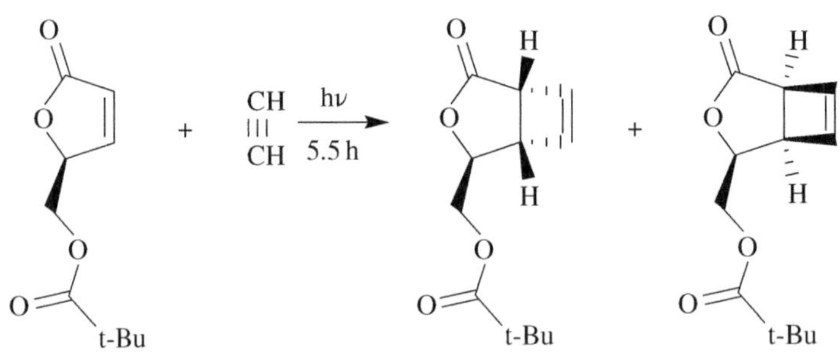

Anti Syn

53%, anti:syn 70:30

(8.22)

Ketenes add thermally to alkenes to give cyclobutanones. Dichloroketene is readily generated *in situ* from trichloroacetyl chloride and copper-activated zinc metal. In Equation 8.23, 1-hexyne was treated with dichloroketene to give the four-membered ring. The chlorine atoms were then removed reductively with zinc dust [37].

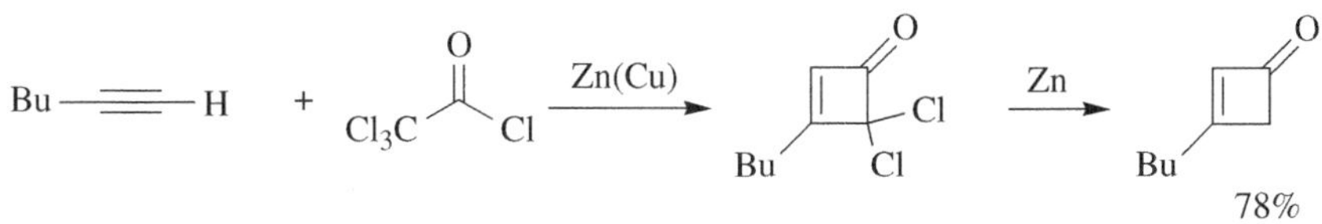

(8.23)

A catalytic 2 + 2 cycloaddition using trifluoromethanesulfonimide has been reported [38]. The *trans* product shown in Equation 8.24 is the kinetic product, and thermodynamic conditions lead to the more stable *cis*-isomer.

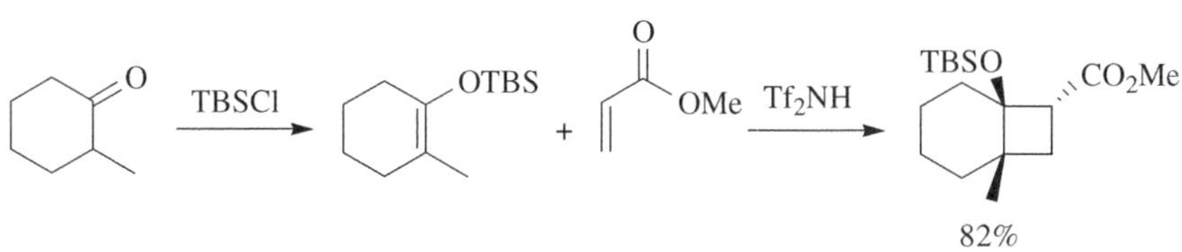

(8.24)

Benzocyclobutenones are useful synthetic intermediates that are readily available using the reaction shown in Equation 8.25 [39]. This approach tolerates a variety of functionalities with good yields.

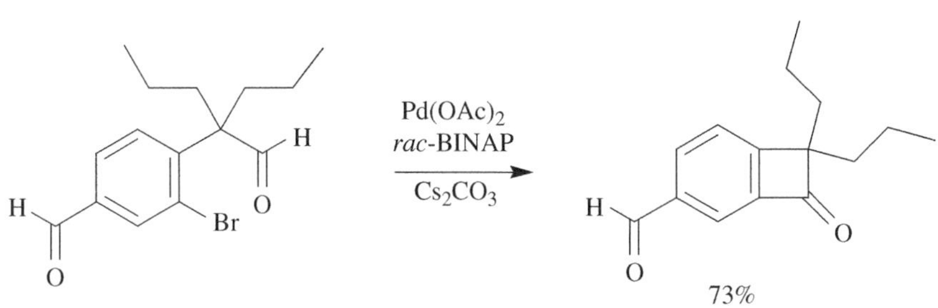

(8.25)

8.5 ACETYLIDE ALKYLATION AND ADDITION

A disconnection between a double or triple bond and an allylic carbon should suggest acetylide chemistry, while a disconnection between the double-bonded carbons should suggest the Wittig and allied reactions (Section 7.2). Acetylide ions are useful for linking carbon chains, particularly where a double bond is desired with stereoselectivity. Acetylene and 1-alkynes may be deprotonated with strong bases such as LDA and then treated with alkyl halides or carbonyl compounds. Preformed lithium acetylide complexed with ethylenediamine is available as a dry powder [40].

In Scheme 8.14, retrosynthetic analysis suggests targeting the carbon alpha to the carbonyl. Using the epoxide as the source of oxygen allows a way around the polarity influence of the carbonyl group, which would be expected to induce a negative charge at the alpha carbon. The actual synthesis is shown in Equation 8.26 [16].

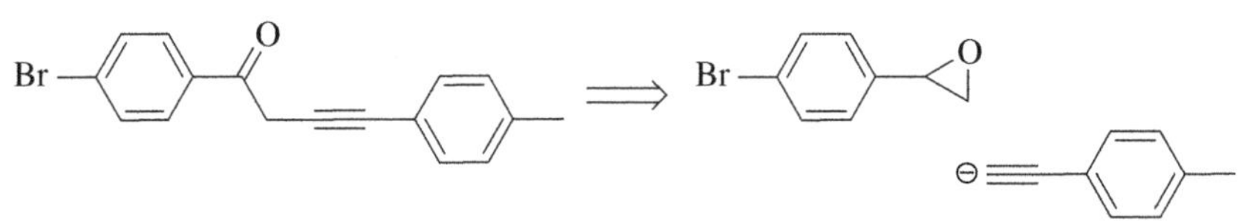

SCHEME 8.14 Retrosynthesis of 1-(4-bromophenyl)-4-(4-methylphenyl) but-3-yn-1-one.

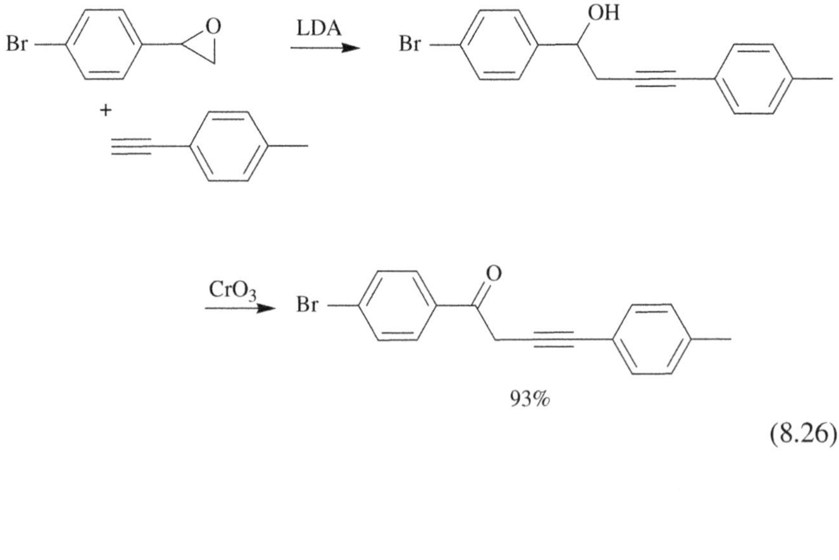

(8.26)

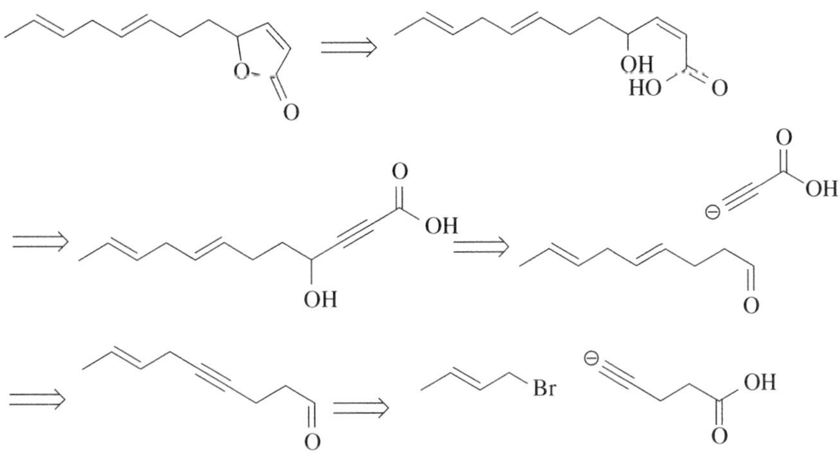

SCHEME 8.15 Retrosynthesis of *cis,trans,trans*-dodeca-2,7,10-trienoic acid γ-lactone.

The lactone in Scheme 8.15 was constructed by using acetylide chemistry at two sites. The disconnections follow the principles given earlier, that is, open the lactone, disconnect beside the alcohol, and so forth. The synthesis devised by Jakubowski et al. is shown in Equation 8.27 [41]. Two equivalents of ethylmagnesium bromide were used in order to deprotonate both the carboxylic acid and the alkyne. The reactivity of dianions is generally greater at the last site of proton removal. Lithium in liquid ammonia with ethanol gave the *trans* alkene, and several steps later, hydrogenation was used to prepare the *cis* double bond for ring closure.

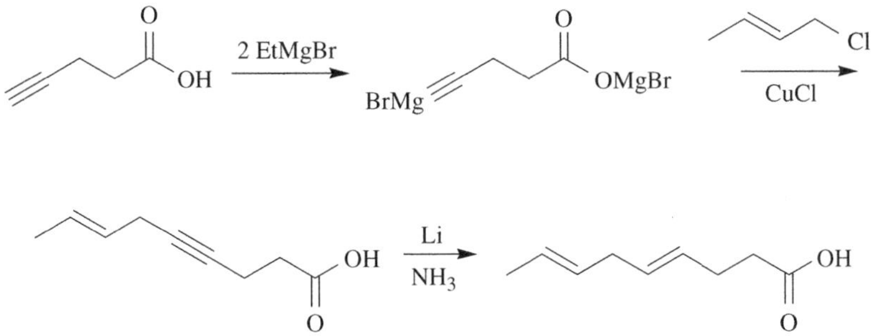

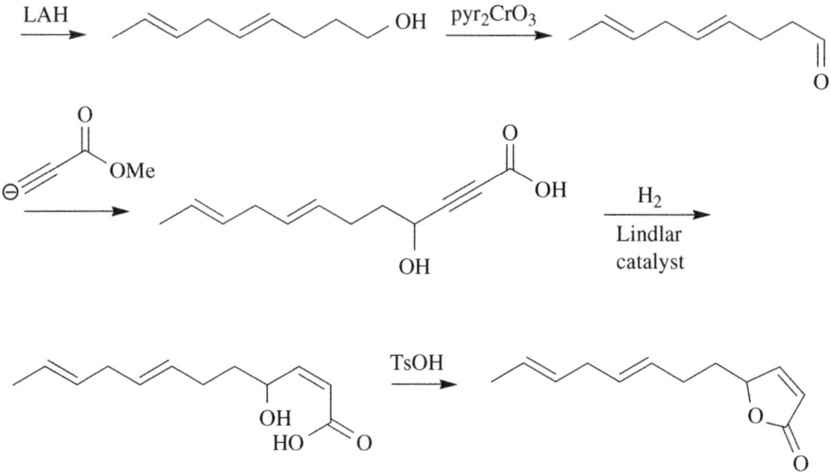

(8.27)

8.6 THE DIELS–ALDER REACTION

In the Diels–Alder reaction, a diene and a dienophile combine through a cyclic transition state to give a six-membered ring (Section 5.3.1). Good reactivity is found when the reacting double bond of the dienophile is electron-poor owing to conjugation with one or more electron-withdrawing groups, such as esters or nitriles. Additionally, the diene is more reactive with electron-donating groups attached. Thus, in planning a synthesis, one should select an appropriately conjugated dienophile even when this may require a subsequent reduction of a carbonyl group. Examples include propenoates, propynoates, maleates, and dimethyl acetylenedicarboxylate. Other dienophiles are reactive because they contain ring strain that is partly relieved on reacting. This factor contributes to the reactivity of cyclopropene, cyclobutadiene, and cyclopentadiene. The very unstable transient benzyne is also a reactive dienophile. There are other Diels–Alder combinations in which the diene is electron-poor and the dienophile is electron-rich. These less common cases are called *reverse-electron-demand reactions* (Eq. 7.40).

The product of a Diels–Alder reaction is generally a cyclohexene; thus, finding that feature in a structure suggests a Diels–Alder disconnection (Scheme 8.16). The cyclohexene may be part of a bicyclic or fused-ring structure, or it may be a cyclohexadiene as when an activated acetylene is the dienophile. Furthermore, if the ring is hydrogenated,

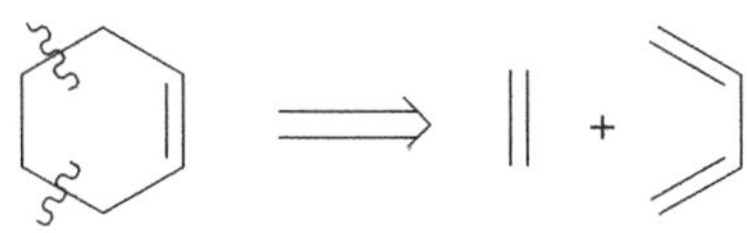

SCHEME 8.16 The Diels–Alder disconnection.

aromatized, or modified with new functionality, the simple presence of a six-membered ring may be a sufficient reason to propose a Diels–Alder reaction step. The Diels–Alder and retro-Diels–Alder reactions are used to form 2-cyclohexene-1,4-dione from 1,4-benzoquinone, as shown in Equation 8.28 [42].

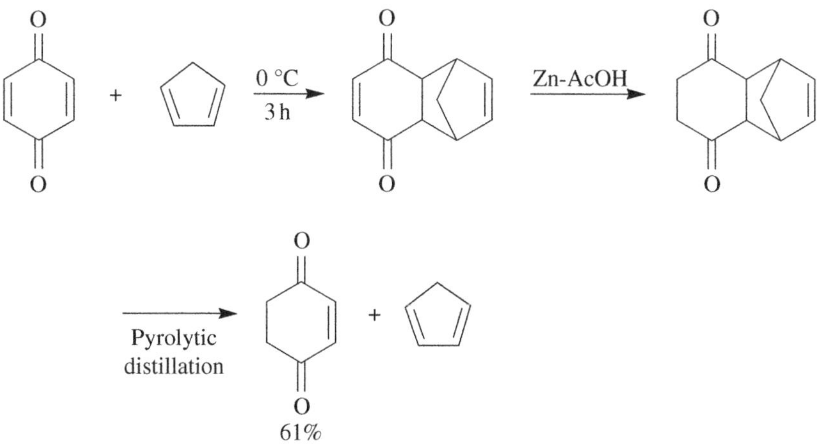

(8.28)

Although four stereogenic centers are formed in many Diels–Alder reactions, often only one or two pairs of enantiomers are formed in appreciable amounts. Lewis acid catalysts are known to accelerate Diels–Alder reactions and also serve to greatly affect the regiochemistry, as shown in Equation 8.29 [43].

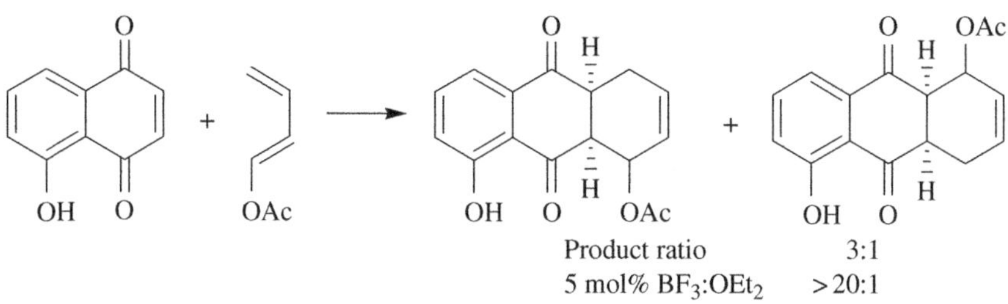

(8.29)

The Diels–Alder reaction is stereospecific. The diene and dienophile approach suprafacially with respect to both reactants to form the new σ bonds, while the original geometry is minimally shifted. Because of this, groups that are cis in the dienophile remain cis in the cyclohexene. The stereochemical relationships in the diene are maintained as well, which limits the number of possible regio- and stereoisomers. The endo/exo ratio can be controlled, often by the use of a catalyst. An example is shown in Equation 8.30 [44]. Chloroaluminate *N*-1-butylpyridinium chloride ($AlCl_3$:BPC) is an ionic liquid which has been used as a substitute for water in these reactions.

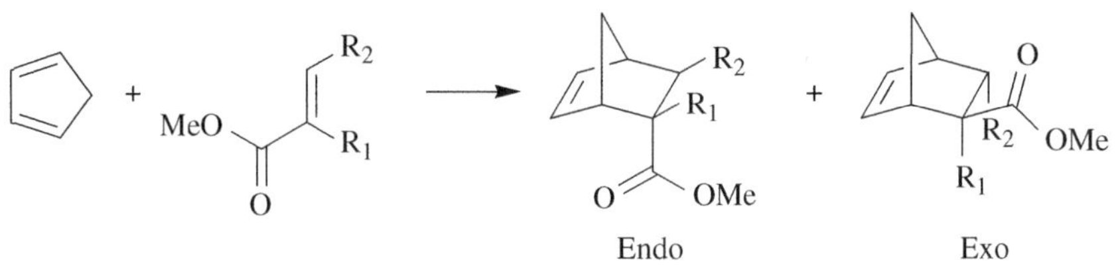

(8.30)

Substitution	Solvent	Endo:exo
R_1=CH_3, R_2=H	Ethanol	35:65
	45% $AlCl_3$:BPC	26:74
	60% $AlCl_3$:BPC	75:25
R_1=H, R_2=CH_3	Ethanol	62:38
	45% $AlCl_3$:BPC	48:52
	60% $AlCl_3$:BPC	86:14

Narrowing Diels–Alder reaction products to one enantiomer requires a chiral influence either as part of the dienophile [45], as shown in Equation 8.31 [46], or as part of a Lewis acid catalyst [47], as shown in Equation 8.32 [48]. In many Diels–Alder reactions, steric hindrance or intramolecular restrictions limit the number of isomers. Altogether, stereospecificity, regiospecificity, and endo/exo control make most Diels–Alder reactions quite practical.

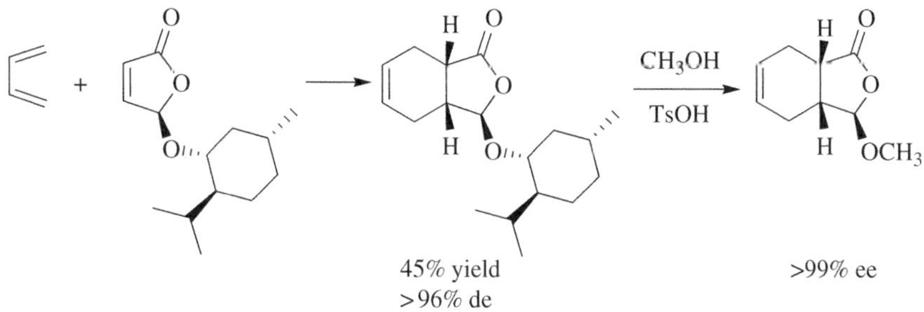

(8.31)

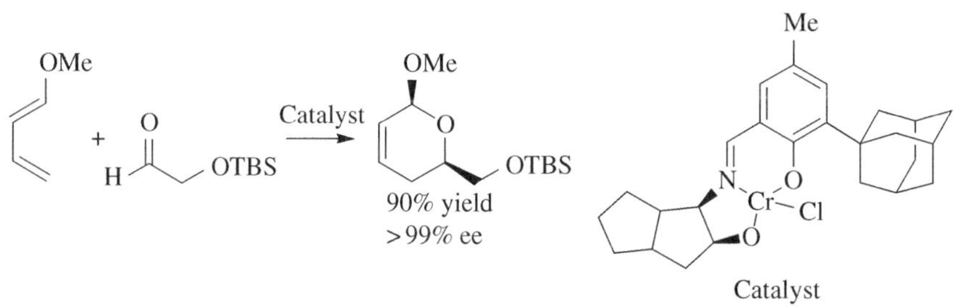

(8.32)

Compound **8.17** is a simple case with no stereochemistry, but it is not a cyclohexene. The enol tautomer would be a cyclohexene, and with this idea, we can disconnect as shown in Scheme 8.17. Allyl alcohol is not reactive as a dienophile, but acrolein has the activating carbonyl group, which can be reduced later. The silyl enol ether of 1-buten-3-one is a good diene (Eq. 8.33) [49].

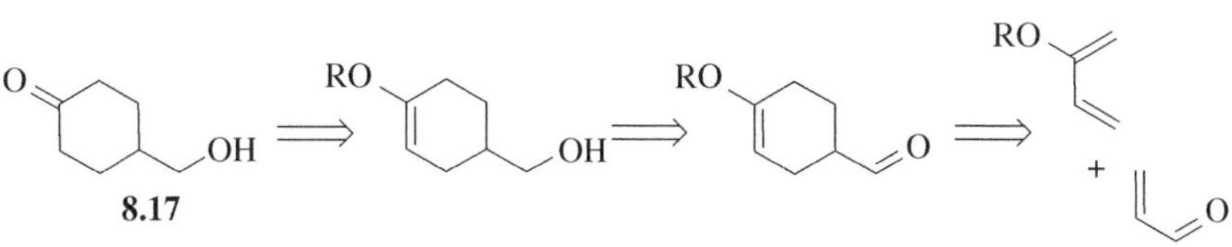

SCHEME 8.17 Retrosynthesis of 4-(1-hydroxymethyl)cyclohexanone.

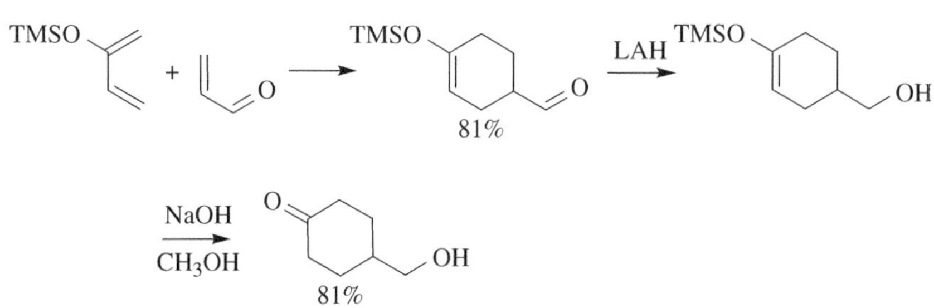

(8.33)

The Robinson annulation is also convenient for forming six-membered rings combining a Michael addition to an α,β unsaturated ketone with a ring closing Aldol condensation. Michael condensation of methyl vinyl ketone with the enolate of 2-methylcyclohexane-1,3-dione in a one-pot reaction catalyzed by *S*-proline results in the formation of a conjugated 2-decalone with stereocontrol (Eq. 8.34). The product is an important intermediate in steroid synthesis [50].

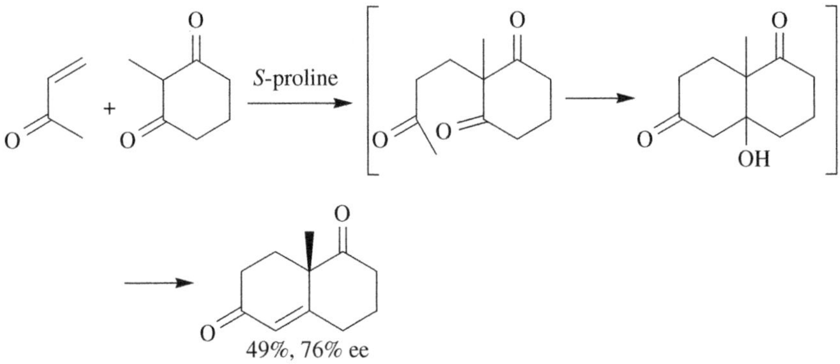

(8.34)

8.7 THE CLAISEN REARRANGEMENT

In the Claisen rearrangement, an allyl enol ether undergoes a [3,3] sigmatropic shift to form a γ,δ-unsaturated carbonyl compound [51]. The ether is derived from an enol and an allyl alcohol (Scheme 8.18). Carbon-based [3,3] sigmatropic reactions (Cope rearrangements) are discussed in Section 5.3.3.

In this reaction, an allyl alcohol rearranges as it attaches to the carbon alpha to a carbonyl group. A ketal exchange reaction [52] can be used to

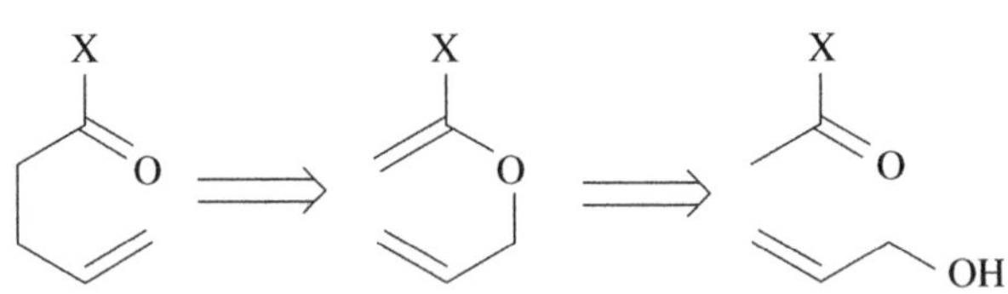

SCHEME 8.18 Retrosynthetic analysis of the Claisen rearrangement.

form the allyl enol ether, as illustrated in Equation 8.35 [53]. The related Carroll rearrangement uses an allylic ester to produce a β-keto ester. In the example shown in Equation 8.36, conditions are mild and reactions high-yielding [54]. The β-keto ester is decarboxylated in the final step.

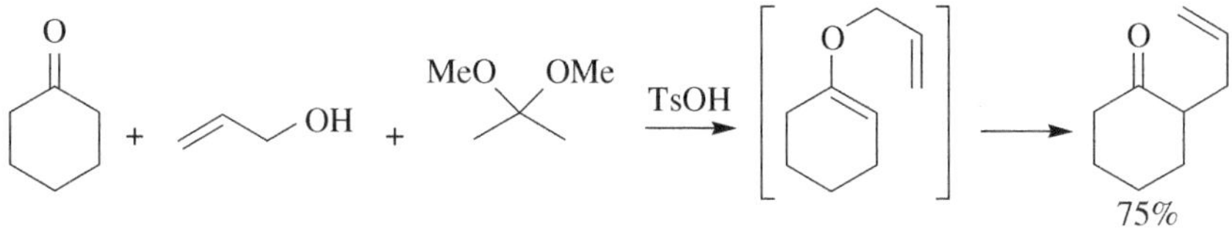

(8.35)

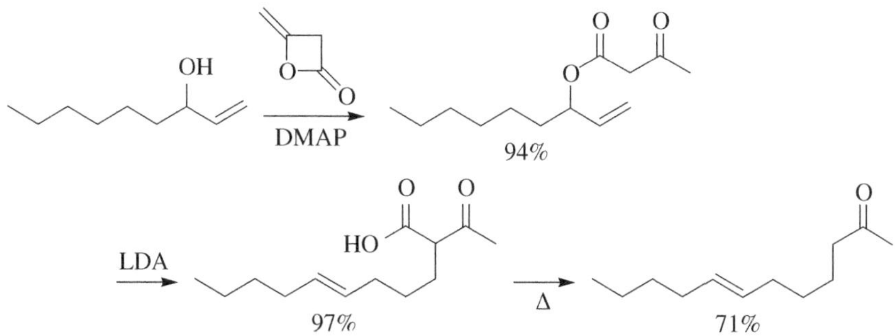

(8.36)

An aza-Claisen rearrangement can be used to form allylic amines, which are important synthetic intermediates. In this reaction (Eq. 8.37), the allylic alcohol is used to form an allylic N-tosylcarbamate. Following the rearrangement, which requires *N,N*-diisopropylethylamine to form the imidate that prevents the unwanted oxo-rearrangement, a gold catalyst serves to decarboxylate the carbamate [55]. Chirality is maintained if the starting alcohol is chiral.

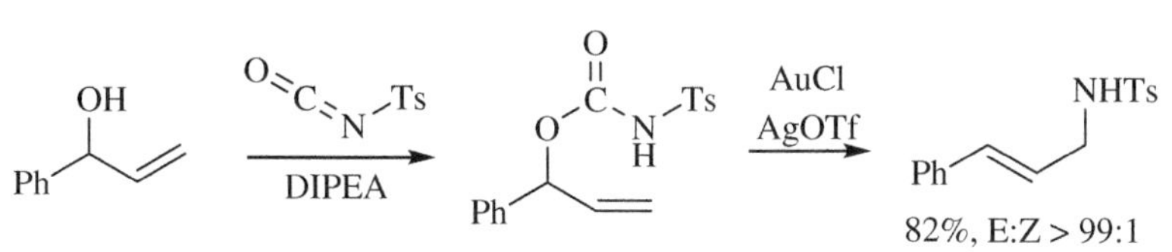

(8.37)

It is possible to begin with an allyl ester and prepare the silyl enol derivative for rearrangement, as shown in Equation 8.38. This reaction often proceeds through a chair-like transition state, although a boat-like transition state may be observed under particular electronic and steric demands [56]. The example in Equation 8.38 uses the expected chair conformation as well as the steric demands of the starting material to achieve excellent stereocontrol and high yield [57].

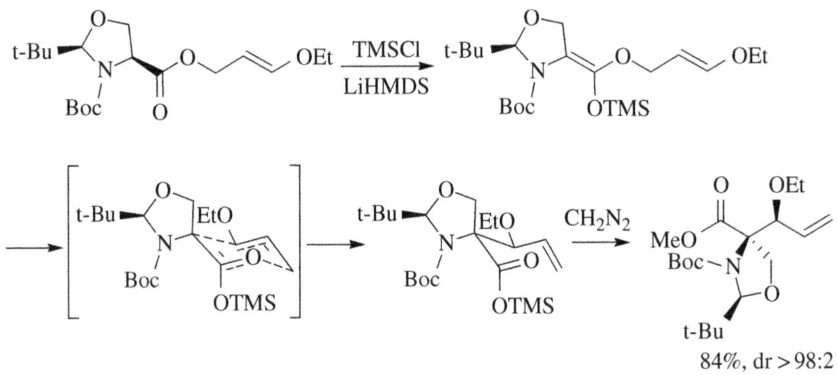

(8.38)

In retrosynthetic analysis, we recognize the need for the Claisen rearrangement when we see a γ,δ-unsaturated carbonyl compound. Compound **8.18** contains an unsaturation in the γ,δ position relative to the position of the bromine. Consider a carbonyl compound as a likely intermediate and then write a retro-Claisen rearrangement (Scheme 8.19). The ketene acetal could come from ethyl orthoacetate. Continue with a disconnection at the

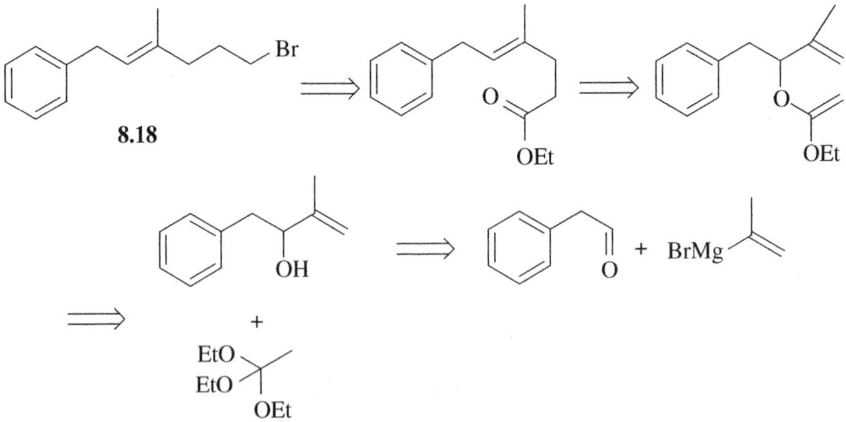

SCHEME 8.19 Retrosynthesis of *trans*-6-bromo-3-methyl-1-phenylhexene.

alcohol. The actual synthesis is shown in Equation 8.39 [58].

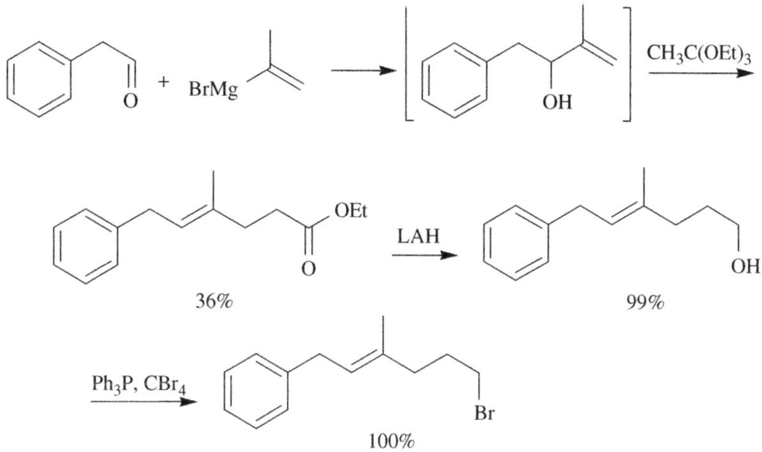

(8.39)

Tetrapyrroles find a wide distribution in nature [59], leading to an interest in them as synthetic targets. Compound **8.19** is an intermediate in the synthesis of tetrapyrroles. The retrosynthetic analysis of **8.19** suggests

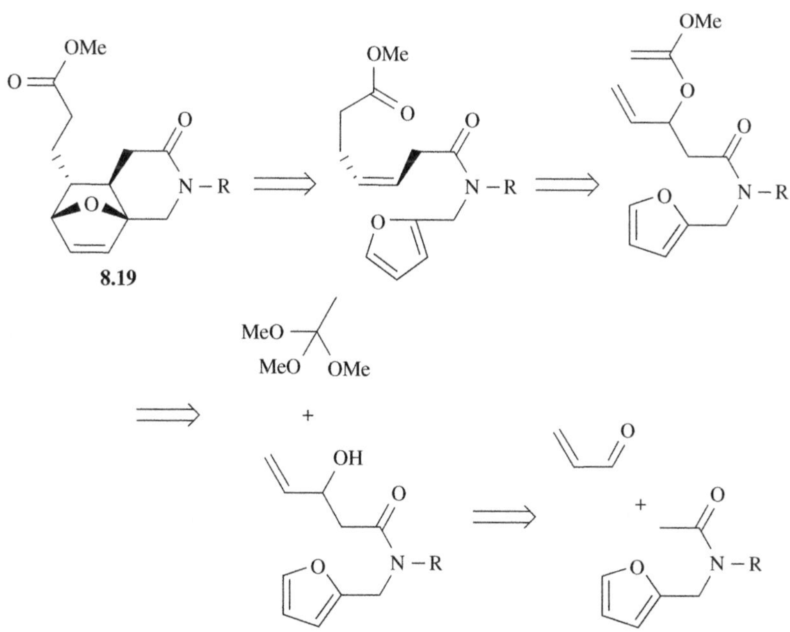

SCHEME 8.20 Retrosynthesis of 3-[4-oxo-11-oxa-3-azatricyclo[6.2.1.0] undec-9-en-7-yl]propionic acid methyl ester.

first a Diels–Alder reaction to form the six-membered ring. Further analysis suggests a Claisen rearrangement to form the starting material for the Diels–Alder reaction (Scheme 8.20).

The actual synthesis, shown in Equation 8.40, uses the dibenzosuberyl group (R-) to protect the nitrogen, which greatly increases the yield in the Diels–Alder reaction, from 14% when the R- group is benzyl. The Claisen rearrangement and Diels–Alder reaction occur in one step [60].

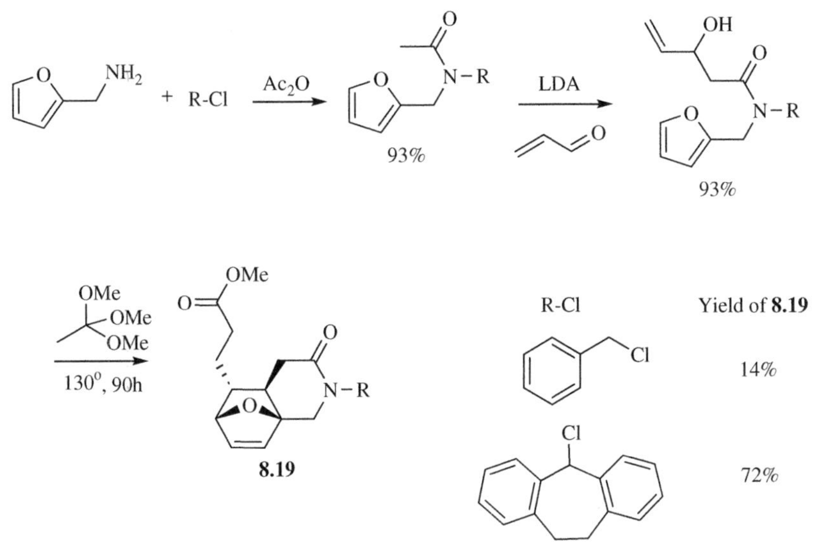

(8.40)

8.8 ASYMMETRIC SYNTHESIS

Asymmetric synthesis, or enantioselective synthesis, is characterized by reactions that produce one enantiomer or diastereomer in favor of others. This selectivity can be difficult to achieve but is critical in many applications of organic chemistry, particularly in the pharmaceutical industry. The origin of chiral selectivity in a reaction or sequence comes from chiral catalysts, chiral auxiliaries like protecting groups, and enzymatic catalysis (as discussed in Section 3.5), or separation techniques can be used to isolate the desired product (as discussed in Section 3.4).

In a recent synthesis of bedaquiline [61], used in the treatment of multidrug-resistant tuberculosis, these concepts were combined to improve the synthesis and isolation of the desired enantiomer in a greener process. The chiral base bis(1-phenylethyl)amine was used in place of LDA to allow a diastereoselective ratio of 90:10 in the final condensation step. Chiral supercritical fluid chromatography (SFC) [62] is comparable to high-performance liquid chromatography (HPLC) but uses supercritical CO_2 as the mobile phase and provides ultrafast and excellent separations, resulting in greener conditions for the purification step. SFC gave the desired 1*R*, 2*S* enantiomer in 13% overall yield with >99% ee (Eq. 8.41).

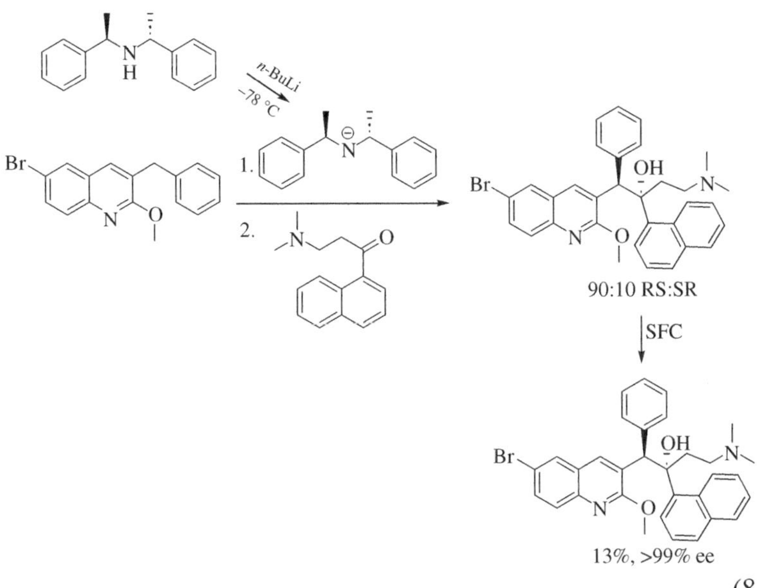

(8.41)

8.9 SYNTHETIC STRATEGIES

At the beginning of this chapter, it was noted that there are hundreds of reported syntheses of jasmone and dehydrojasmone [3]. This chapter and the previous two have presented a number of reactions by which chemical bonds can be formed. As we consider the ways in which an entire molecule can be assembled, it is important to note that there are many possible pathways that can provide access to a target molecule, and the development of these methodologies advances organic chemistry as a whole. A reaction developed to solve a synthetic challenge for one molecule provides inspiration or methodology for the next. The art of our science then is to see these challenges both within the box of historical context and outside the box of what has been done to look for new approaches.

Jasmone, **8.20**, is a volatile component of the fragrant oil isolated from jasmine flowers and is thus an important compound in the fragrance industry. In plants, it plays an important role in defense and development [63]. Methodologies developed for this molecule also have applications in the field of prostaglandin synthesis. Three syntheses have been chosen as examples.

A key step in the Grieco synthesis [64] (Eq. 8.42) forms the *cis* double bond in a Wittig reaction with a hemiacetal.

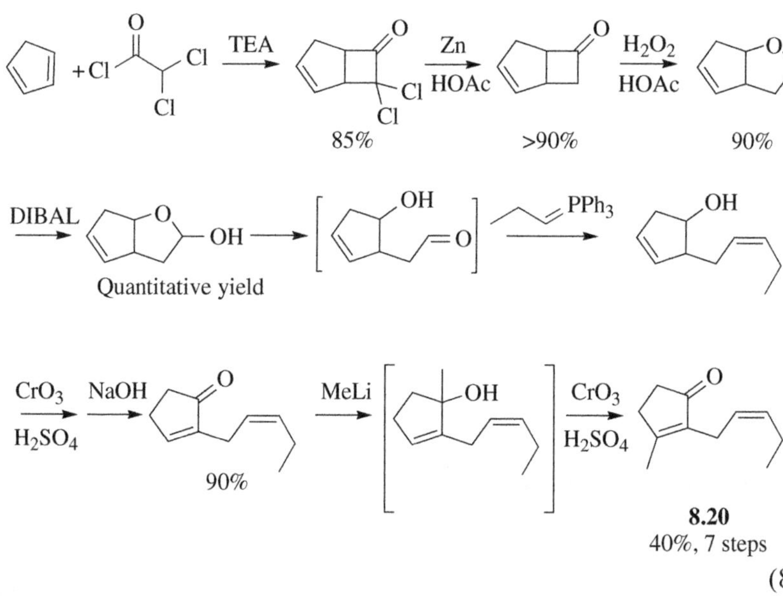

8.20
40%, 7 steps

(8.42)

The McCurry synthesis [65] uses a kinetic retro-Aldol reaction to form the five-membered ring (Eq. 8.43).

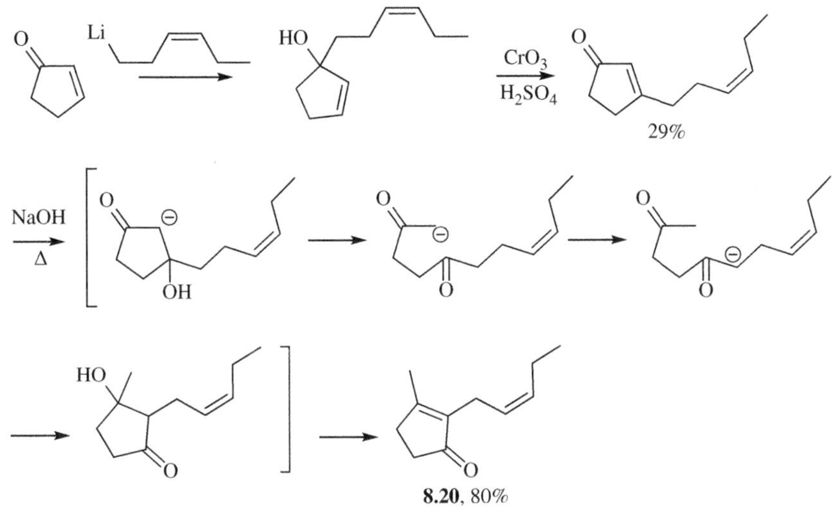

(8.43)

A titanium catalyst is used to control a crossed Claisen reaction in the Tanabe synthesis (Eq. 8.44) [66].

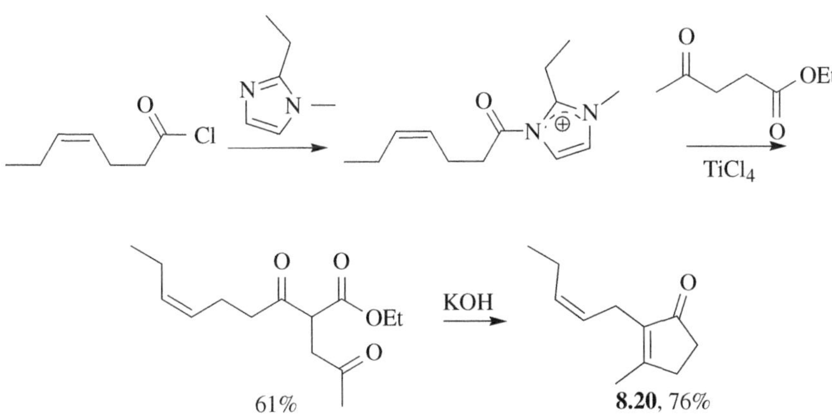

(8.44)

8.10 FINAL NOTE

In longer syntheses, it is not always routine to apply certain key steps. There are highly novel creations that few people would bring together. To pick one example among a great many, would Scheme 8.21 seem obvious [67]? On the other hand, bringing together reactions for a complex scheme from a list larger than a person could bring to mind can now be done by computer. Programs have been written, backed by large data collections, that use retrosynthetic analysis and molecular templates as well as artificial intelligence to suggest reaction schemes for the synthesis of complex molecules [68–71]. Rapid developments in this field prompt regular research to keep up with changes.

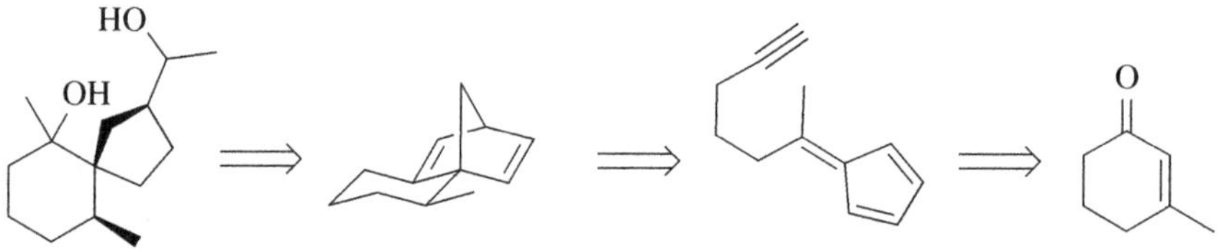

SCHEME 8.21 Retrosynthesis of 2-(1-hydroxyethyl)-6,10-dimethylspiro[4.5]decan-6-ol.

RESOURCES

1. Name-reaction.com lists name reactions alphabetically along with mechanisms and literature citations. https://www.name-reaction.com/list.
2. Similarly, the Organic Synthesis Archive lists named reactions and provides mechanisms and literature citations as well. https://www.synarchive.com/named-reactions.
3. The Organic Synthesis Archive allows for searching of total syntheses by molecule, by author, by date, and by carbon count, including references to the original literature. https://www.synarchive.com.
4. Professor Hans Reich maintains an extensive total synthesis archive, including literature references, searchable by reagent, keyword, and year. https://www.chem.wisc.edu/areas/reich/syntheses/syntheses.htm.

5. Professor Dennis Cao developed a database of organic reactions with answer keys. The reactions are sorted by mechanism. http://www.realorganicchemistry.com.

PROBLEMS

Show how you would synthesize each of the following compounds from simple, readily available materials. Answers are provided in Appendix C for questions marked with an asterisk.

8.1* Ref. [72]

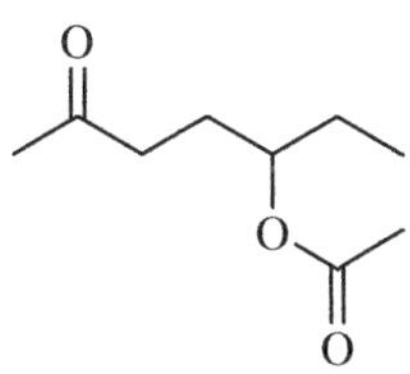

8.2* Ref. [73]

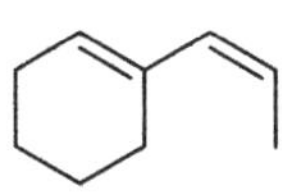

Stereoselectively

8.3 Ref. [74]

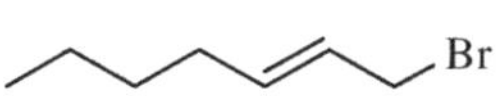

8.4 Ref. [75]

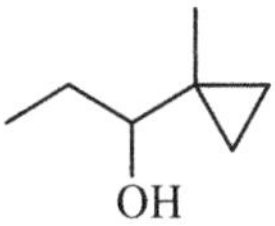

8.5* Ref. [76]

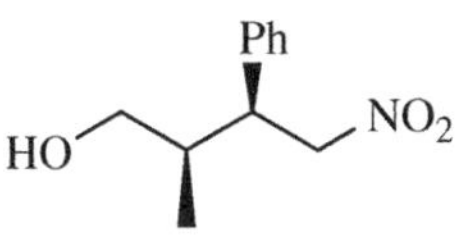

8.6 Ref. [77]

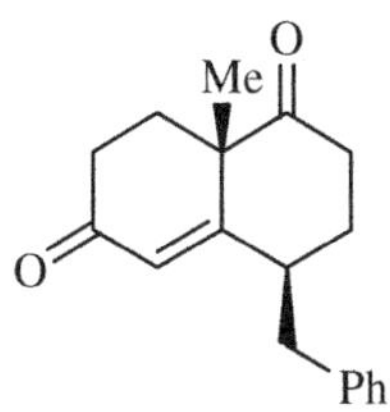

8.7* Ref. [78]

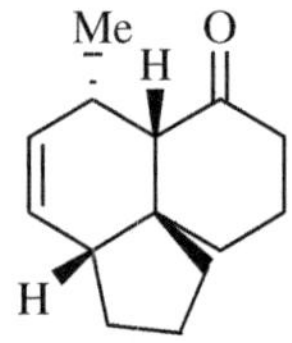

8.8 Ref. [79]

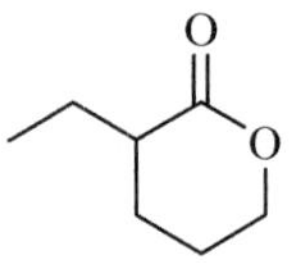

8.9 Ref. [80]

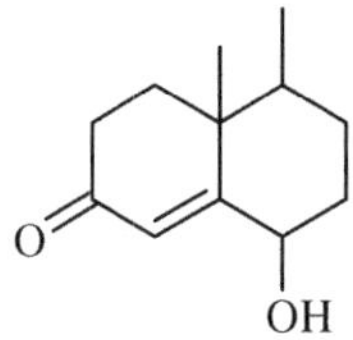

8.10* Ref. [81]

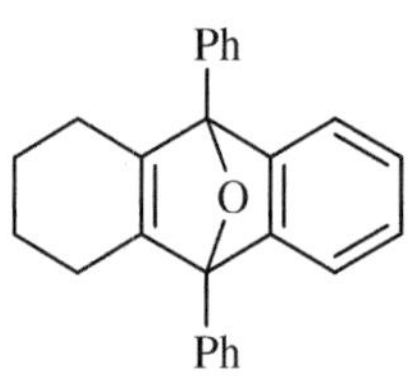

8.11 Ref. [82]

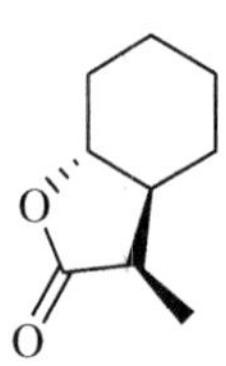

8.12* Ref. [83]

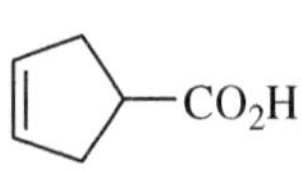

8.13 Ref. [84]

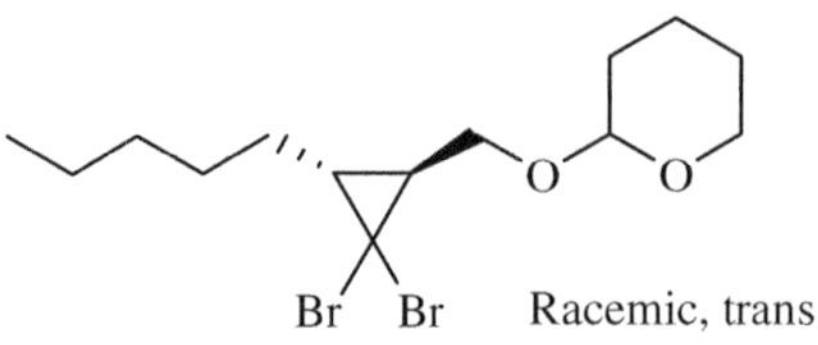

8.14* Ref. [85]

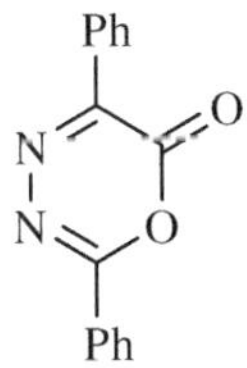

8.15 Ref. [86]

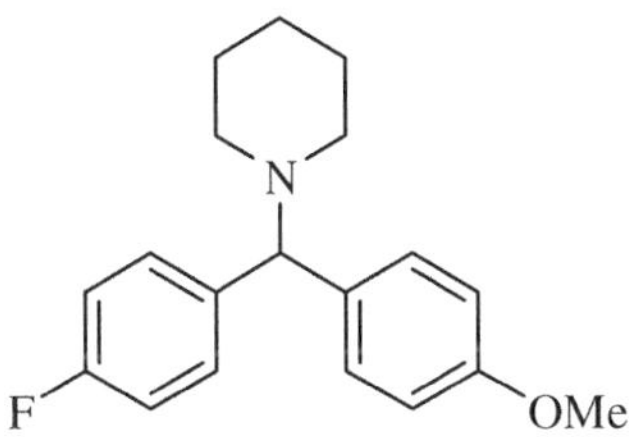

8.16 Ref. [87]

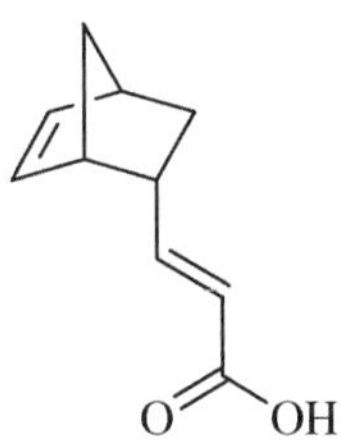

8.17 Ref. [88]

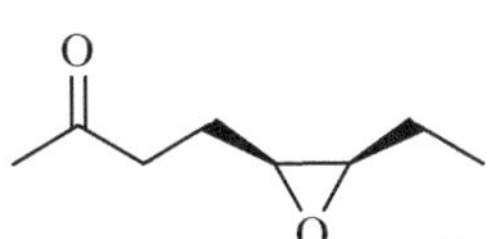

Racemic, cis

8.18 Ref. [89]

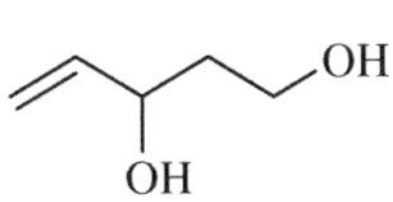

8.19* Ref. [90]

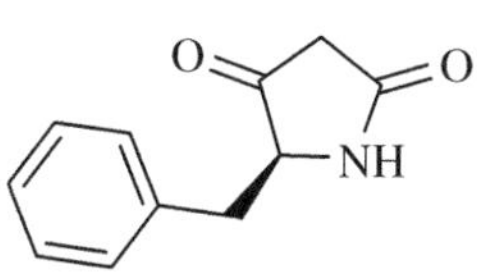

8.20 Ref. [91]

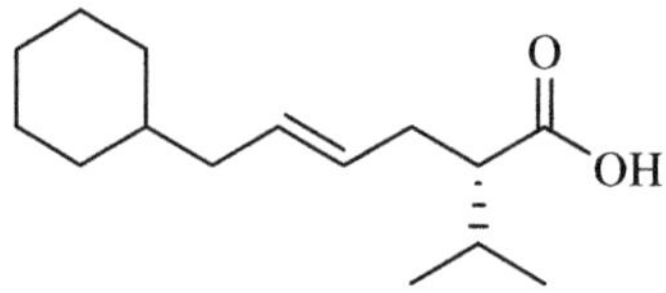

8.21 Ref. [55]

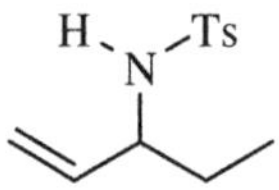

8.22 Ref. [92]

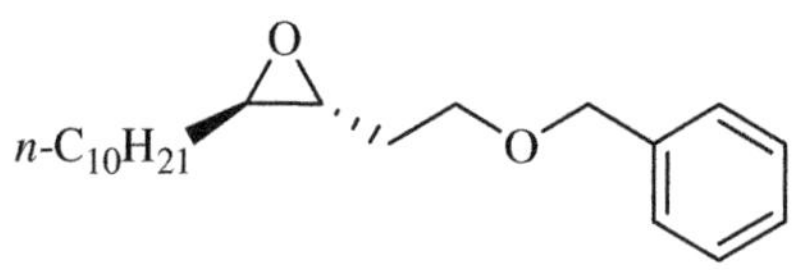

8.23 Ref. [93]

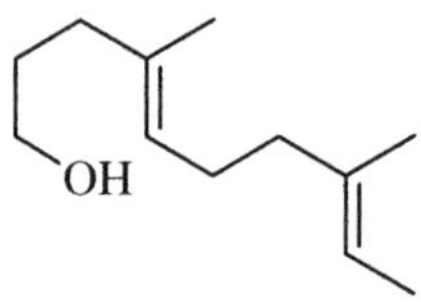

8.24 Ref. [94]

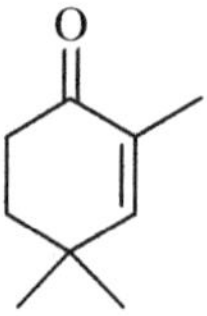

8.25 Ref. [95]

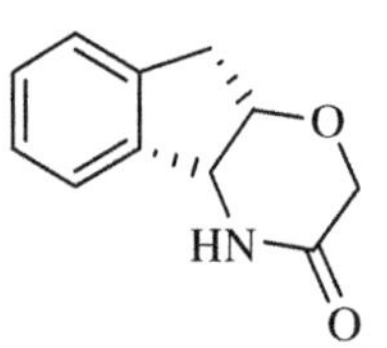

8.26 Ref. [96]

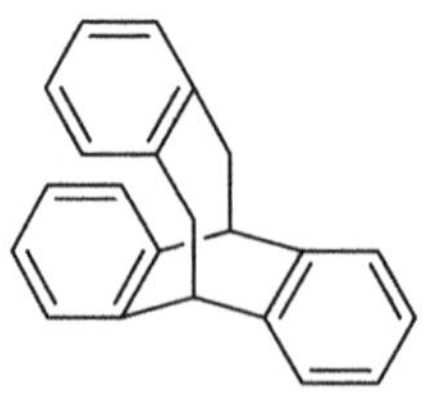

8.27* 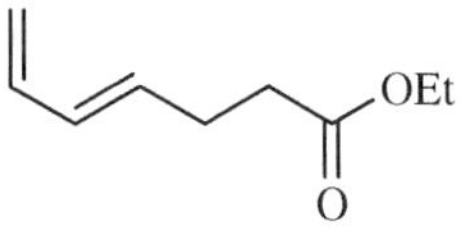 Ref. [97]

8.28 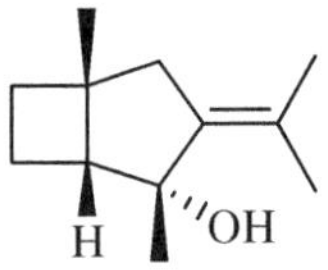 Ref. [98]

8.29 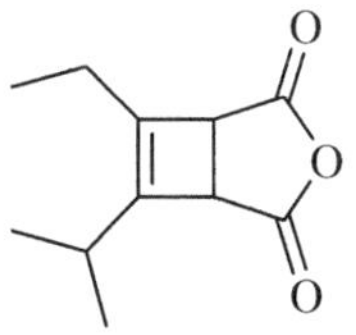 Ref. [99]

8.30 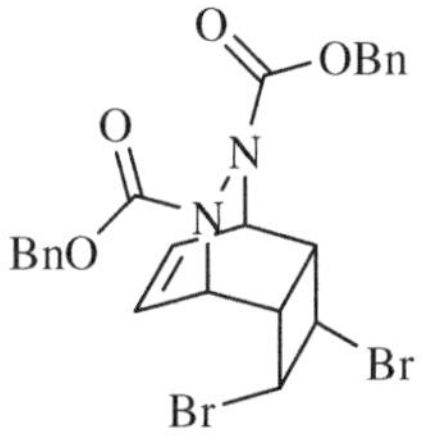 Ref. [100]

8.31* 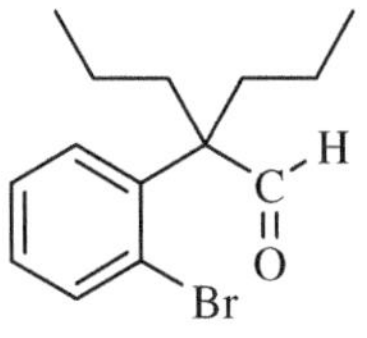 Ref. [101]

8.32 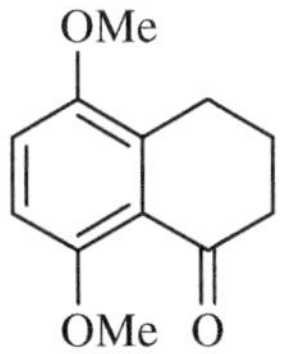 Ref. [102]

8.33 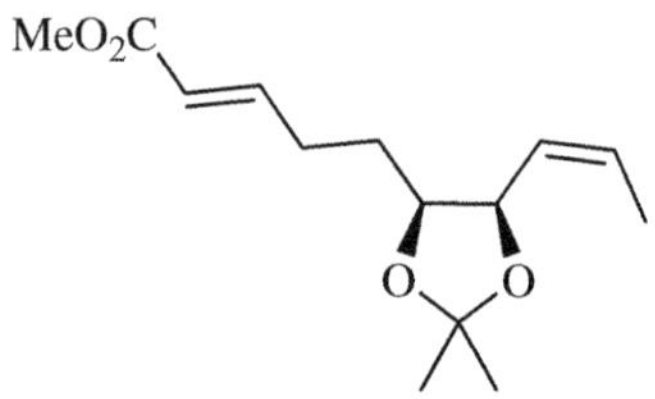 Ref. [103]

8.34 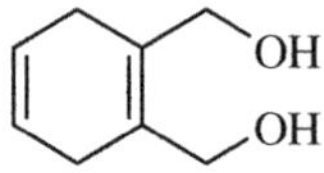 Ref. [104]

8.35* 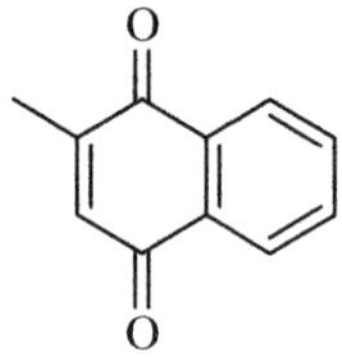 Ref. [105]

8.36*  Ref. [106]

8.37* 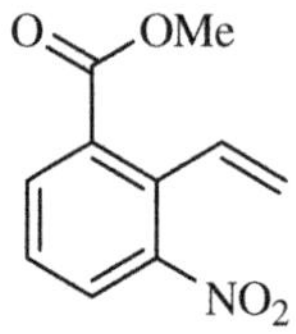 Ref. [107]

REFERENCES

1. Warren, S. *Organic Synthesis: The Disconnection Approach*, 2nd ed.; John Wiley & Sons, Inc: New York, 2008.
2. Van Veller, B. *J. Chem. Educ.* **2021**, *98*, 2726–2729.
3. Fortineau, A.-D. *J. Chem. Educ.* **2004**, *81*, 45–50; Crombie, L.; Harper, S. H. *J. Chem. Soc.* **1952**, 869–875; Tanaka, K.; Fu, G. C. *J. Am. Chem. Soc.* **2001**, *123*, 11492–11493; and many other papers.

4. Marker, R. E.; Tsukamoto, T.; Turner, D. L. *J. Am. Chem. Soc.* **1940**, *62*, 2525–2532; American Chemical Society International Historic Chemical Landmarks. The “Marker Degradation” and Creation of the Mexican Steroid Hormone Industry 1938–1945. https://www.acs.org/education/whatischemistry/landmarks/progesteronesynthesis.html (accessed January 23, 2025).
5. Evans, D. A.; Andrews, G. C. *Acc. Chem. Res.* **1974**, *7*, 147–155.
6. Harding, K. E.; Burks, S. R. *J. Org. Chem.* **1984**, *49*, 40–44.
7. Yama, hita, M.; Matsumiya, K.; Tanabe, M.; Suemitsu, R. *Bull. Chem. Soc. Jpn.* **1985**, *58*, 407–408.
8. Bertz, S. H.; Dabbagh, G. *J. Org. Chem.* **1984**, *49*, 1119–1122.
9. Seebach, D. *Angew. Chem., Int. Ed.* **1979**, *18*, 239–258; *Umpoled Synthons: A Survey of Sources and Uses in Synthesis*; Hase, T. A., Ed.; John Wiley & Sons, Inc: New York, 1987.
10. McMurry, J. E.; Melton, J. *Org. Synth.* **1977**, *56*, 36–39.
11. Stetter, H.; Kuhlmann, H.; Lorenz, G. *Org. Synth.* **1979**, *59*, 53–57.
12. Seebach, D.; Corey, E. J. *J. Org. Chem.* **1975**, *40*, 231–237; Jung, N.; Grässle, S.; Lütjohann, D. S.; Bräse, S. *Org. Lett.* **2014**, *16*, 1036–1039; Stütz, P.; Stadler, P. A. *Org. Synth.* **1977**, *56*, 8–14.
13. Boyce, G. R.; Greszler, S. N.; Johnson, J. S.; Linghu, X.; Malinowski, J. T.; Nicewicz, D. A.; Satterfield, A. D.; Schmitt, D. C.; Steward, K. M. *J. Org. Chem.* **2012**, *77*, 4503–4515.
14. Welch, S. C.; Chayabunjonglerd, S. *J. Am. Chem. Soc.* **1979**, *101*, 6768–6769.
15. Zoretic, P. A.; Ferrari, J. L.; Bhakta, C.; Barcelos, F.; Branchard, B. *J. Org. Chem.* **1982**, *47*, 1327–1329.
16. Sniady, A.; Morreale, M. S.; Dembinski, R. *Org. Synth.* **2007**, *84*, 199–208.
17. Stowell, J. C. *Chem. Rev.* **1984**, *84*, 409–435.
18. Ruppert, J. F.; White, J. D. *J. Am. Chem. Soc.* **1981**, *103*, 1808–1813.
19. Yamagiwa, S.; Kosugi, H.; Uda, H. *Bull. Chem. Soc. Jpn.* **1978**, *51*, 3011–3015.
20. Linghu, X.; Bausch, C. C.; Johnson, J. S. *J. Am. Chem. Soc.* **2005**, *127*, 1833–1840.
21. Fernández, M.; Uria, U.; Vicario, J. L.; Reyes, E.; Carrillo, L. *J. Am. Chem. Soc.* **2012**, *134*, 11872–11875.
22. Brook, A. G. *Acc. Chem. Res.* **1974**, *7*, 77–84.
23. Nicewicz, D. A.; Breteche, G.; Johnson, J. S. *Org. Synth.* **2008**, *85*, 278–286.

24. Myles, D. C.; Andrulis, P. J. III; Whitesides, G. M. *Tetrahedron Lett.* **1991**, *32*, 4835–4838.

25. Kuehne, M. E.; Kirkeno, C. L.; Matsko, T. H.; Bohnert, J. C. *J. Org. Chem.* **1980**, *45*, 3259–3265.

26. Stork, G.; Brizzollara, A.; Landesman, H. K.; Smuszkovicz, J.; Terrel, R. *J. Am. Chem. Soc.* **1963**, *85*, 207–222.

27. Nguyen, H.; Oh, H. S.; Henry-Riyad, H.; Sepulveda, D.; Romo, D. *Org. Synth.* **2011**, *88*, 121–137.

28. Schulz, F.; Leca, F.; Hollmann, F.; Reetz, M. T. *Beilstein J. Org. Chem.* **2005**, *1*, 10.

29. Bloomfield, J. J.; Nelke, J. M. *Org. Synth.* **1977**, *57*, 1–7.

30. Curtis, O. E. Jr.; Sandri, J. M.; Crocker, R. E.; Hart, H. *Org. Synth.* **1958**, *38*, 19–21.

31. Kumar, V. T. R.; Swaminathan, S.; Rajagopalan, K. *J. Org. Chem.* **1985**, *50*, 5867–5869.

32. Meng, G.; Guo, T.; Ma, T.; Zhang, J.; Shen, Y.; Sharpless, K. B.; Dong, J. *Nature* **2019**, *574*, 86–89.

33. Bonandi, E.; Christodoulou, M. S.; Fumagalli, G.; Perdicchia, D.; Rastelli, G.; Passarella, D. *Drug Discov. Today* **2017**, *22*, 1572–1581.

34. Hoffmann, N. *Chem. Rev.* **2008**, *108*, 1052–1103.

35. Hue, B. T. B.; Dijkink, J.; Kuiper, S.; Larson, K. K.; Guziec, F. S. Jr.; Goubitz, K.; Fraanje, J.; Schenk, H.; van Maarseveena, J. H.; Hiemstra, H. *Org. Biomol. Chem.* **2003**, *1*, 4364–4366.

36. Alibés, R.; de March, P.; Figueredo, M.; Font, J.; Fu, X.; Racamonde, M.; Álvarez-Larena, A.; Piniella, J. F. *J. Org. Chem.* **2003**, *68*, 1283–1289.

37. Danheiser, R. L.; Savariar, S.; Cha, D. D. *Org. Synth.* **1990**, *68*, 32–40.

38. Takasu, K.; Ishii, T.; Inanaga, K.; Ihara, M. *Org. Synth.* **2006**, *83*, 193–199.

39. Martin, R.; Flores-Gaspar, A. *Org. Synth.* **2012**, *89*, 159–169.

40. Bassan, E. M.; Baxter, C. A.; Beutner, G. L.; Emerson, K. M.; Fleitz, F. J.; Johnson, S.; Keen, S.; Kim, M. M.; Kuethe, J. T.; Leonard, W. R.; Mullens, P. R.; Muzzio, D. J.; Roberge, C.; Yasuda, N. *Org. Process Res. Dev.* **2012**, *16*, 87–95.

41. Jakubowski, A. A.; Guziec, F. S. Jr.; Sugiura, M.; Tam, C. C.; Tishler, M.; Omura, S. *J. Org. Chem.* **1982**, *47*, 1221–1228.

42. Oda, M.; Kawase, T.; Okada, T.; Enomoto, T. *Org. Synth.* **1996**, *73*, 253–261.

43. Trost, B. M.; Ippen, J.; Vladuchick, W. C. *J. Am. Chem. Soc.* **1977**, *99*, 8116–8118.
44. Kumar, A.; Pawar, S. S. *J. Org. Chem.* **2004**, *69*, 1419–1420.
45. Oppolzer, W. *Angew. Chem., Int. Ed.* **1984**, *23*, 876–889.
46. Feringa, B. L.; de Jong, J. C. *J. Org. Chem.* **1988**, *53*, 1125–1127.
47. Kagan, H. B.; Riant, O. *Chem. Rev.* **1992**, *92*, 1007–1019.
48. Chavez, D. E.; Jacobsen, E. N. *Org. Synth.* **2005**, *82*, 34–42.
49. Yin, T.-K.; Lee, J. G.; Borden, W. T. *J. Org. Chem.* **1985**, *50*, 531–534.
50. Bui, T.; Barbas, C. F. III. *Tetrahedron Lett.* **2000**, *41*, 6951–6954.
51. Martin Castro, A. M. *Chem. Rev.* **2004**, *104*, 2939–3002.
52. Lorette, N. B.; Howard, W. L. *J. Org. Chem* **1960**, *25*, 521–525.
53. Crandall, J. K.; Magaha, H. S.; Henderson, M. A.; Widener, R. K.; Thark, G. A. *J. Org. Chem.* **1982**, *47*, 5372–5380.
54. Wilson, S. R.; Augelli, C. E. *Org. Synth.* **1990**, *68*, 210–219.
55. Xing, D.; Yang, D. *Org. Synth.* **2012**, *89*, 450–459 Xing, D.; Yang, D. *Org Lett*. **2010**, *12*, 1068–1071.
56. Ireland, R. E.; Wipf, P.; Xiang, J. N. *J. Org. Chem.* **1991**, *56*, 3572–3582.
57. Fairhurst, N. W. G.; Mahon, M. F.; Munday, R. H.; Carbery, D. R. *Org. Lett.* **2012**, *14*, 756–759.
58. Guthrie, A. E.; Semple, J. E.; Joulie, M. M. *J. Org. Chem.* **1982**, *47*, 2369–2376.
59. Heinemann, I. U.; Jahn, M.; Jahn, D. *Arch. Biochem. Biophys.* **2008**, *474*, 238–251.
60. Jacobi, P. A.; Li, Y. *J. Am. Chem. Soc.* **2001**, *123*, 9307–9312.
61. Lubanyana, H.; Arvidsson, R. I.; Govender, T.; Kruger, H. G.; Naicker, T. *ACS Omega* **2020**, *5*, 3607–3611.
62. Barhate, C. L.; Wahab, M. F.; Tognarelli, D. J.; Berger, T. A. *Anal. Chem.* **2016**, *88*, 8664–8672.
63. Yan, Y.; Borrego, E.; Kolomiets, M. V. Jasmonate Biosynthesis, Perception and Function in Plant Development and Stress Responses. In R. Valenzuela Baez; Ed., *Lipid Metabolism*; InTech, 2013. doi: 10.5772/52675. https://www.intechopen.com/chapters/42122 (accessed December 31, 2024)
64. Grieco, P. A. *J. Org. Chem.* **1972**, *37*, 2363–2364.
65. McCurry, P. M. Jr.; Singh, R. K. *J. Org. Chem.* **1974**, *39*, 2317–2319.
66. Misaki, T.; Nagase, R.; Matsumoto, K.; Tanabe, Y. *J. Am. Chem. Soc.* **2005**, *127*, 2854–2855.

67. Nyström, J.-E.; Helquist, P. *J. Org. Chem.* **1989**, *54*, 4695–4698.
68. Ishida, S.; Terayama, K.; Kojima, R.; Takasu, K.; Okuno, Y. *J. Chem. Inf. Model.* **2022**, *62*, 1357–1367.
69. Coley, C. W.; Rogers, L.; Green, W. H.; Jensen, K. F. *ACS Cent. Sci.* **2017**, *3*, 1237–1245.
70. Chen, J. H.; Baldi, P. *J. Chem. Inf. Model.* **2009**, *49*, 2034–2043.
71. Gao, H.; Struble, T. J.; Coley, C. W.; Wang, Y.; Green, W. H.; Jensen, K. F. *ACS Cent. Sci.* **2018**, *4*, 1465–1476.
72. Johnston, B. D.; Oehlschlager, A. C. *J. Org. Chem.* **1986**, *51*, 760–763.
73. Inman, W. D.; Sanchez, K. A. J.; Chaidez, M. A.; Paulson, D. R. *J. Org. Chem.* **1989**, *54*, 4872–4881.
74. Watanabe, Y.; Iida, H.; Kibayashi, C. *J. Org. Chem.* **1989**, *54*, 4090–4097.
75. Bartlett, P. A.; Mori, I.; Bose, J. A. *J. Org. Chem.* **1989**, *54*, 3236–3239.
76. Hayashi, Y. *Org. Synth.* **2022**, *99*, 68–78.
77. Peng, F.; Dai, M.; Angeles, A. R.; Danishefsky, S. *J. Chem. Sci.* **2012**, *3*, 3076–3080.
78. Ross, A. G.; Li, X.; Danishefsky, S. J. *J. Am. Chem. Soc.* **2012**, *134*, 16080–16084.
79. Edwards, M. P.; Ley, S. V.; Lister, S. G.; Palmer, B. D.; Williams, D. J. *J. Org. Chem.* **1984**, *49*, 3503–3516.
80. Bian, M.; Wang, Z.; Xiong, X.; Sun, Y.; Matera, C.; Nicolaou, K. C.; Li, A. J. *Am. Chem. Soc.* **2012**, *134*, 8078–8081; Diaz, S.; Gonzalez, A.; Bradshaw, B.; Cuesta, J.; Bonjoch, J. *J. Org. Chem.* **2005**, *70*, 3749–3752.
81. Inoue, K.; Inoue, K.; Mori, A.; Okano, K. *Org. Synth.* **2021**, *98*, 509–530.
82. Jensen, B. L.; Malkawi, A.; McGowan, V. *J. Chem. Educ.* **2000**, *77*, 1474–1476; Jefford, C. W.; Li, Y.; Wang, Y. *Org. Synth.* **1993**, *71*, 207–213.
83. Depres, J.-P.; Greene, A. E. *Org. Synth.* **1998**, *75*, 195–200.
84. Porter, N. A.; Ziegler, C. B. Jr.; Khouri, F. F.; Roberts, D. H. *J. Org. Chem.* **1985**, *50*, 2252–2258.
85. Kelleghan, A. V.; Spence, K. A.; Garg, N. K. *Org. Synth.* **2020**, *97*, 189–206.
86. Le Gall, E.; Martens, T. *Org. Synth.* **2012**, *89*, 283–293.
87. Polniaszek, R. P.; Stevens, R. V. *J. Org. Chem.* **1986**, *51*, 3023–3027.
88. Mulholland, R. L. Jr.; Chamberlin, A. R. *J. Org. Chem.* **1988**, *53*, 1082–1085.
89. Reitz, A. B.; Nortey, S. O.; Maryanoff, B. E.; Liotta, D.; Monahan, R. III. *J. Org. Chem.* **1987**, *52*, 4191–4202.

90. Lambert, K. M.; Medley, A. W.; Jackson, A. C.; Markham, L. E.; Wood, J. L. *Org. Synth.* **2019**, *96*, 528–585.
91. Herold, P.; Duthaler, R.; Rihs, G.; Angst, C. *J. Org. Chem.* **1989**, *54*, 1178–1185.
92. Flippin, L. A.; Jalai-Araghi, K.; Brown, P. A.; Burmeister, H. R.; Vesonder, R. F. *J. Org. Chem.* **1989**, *54*, 3006–3007.
93. Ho, N.-H.; le Noble, W. J. *J. Org. Chem.* **1989**, *54*, 2018–2021.
94. Schuster, D. I.; Rao, J. M. *J. Org. Chem.* **1981**, *46*, 1515–1521.
95. Vora, H. U.; Lathrop, S. P.; Reynolds, N. T.; Kerr, M. S.; de Alaniz, J. R.; Rovis, T. *Org. Synth.* **2010**, *87*, 350–361.
96. Sadana, A. K.; Saini, R. K.; Billups, W. E. *Chem. Rev.* **2003**, *103*, 1539–1602.
97. Parker, K. A.; Iqbal, T. *J. Org. Chem.* **1982**, *47*, 337–342.
98. Cargill, R. L.; Wright, B. W. *J. Org. Chem.* **1975**, *40*, 120–122.
99. Smith, A. B. III; Boschelli, D. *J. Org. Chem.* **1983**, *48*, 1217–1226.
100. Mascitti, V.; Corey, E. J. *J. Am. Chem. Soc.* **2004**, *126*, 15664–15665.
101. Martin, R.; Flores-Gaspar, A. *Org. Synth.* **2012**, *89*, 450–459.
102. Chen, C.-P.; Swenton, J. S. *J. Org. Chem.* **1985**, *50*, 4569–4576.
103. Calo, F.; Richardson, J.; Barrett, A. G. M. *Org. Lett.* **2009**, *11*, 4910–4913; Gypser, A.; Peterek, M.; Scharf, H.-F. J. *Chem. Soc., Perkin Trans.* **1997**, *1*, 1013–1016.
104. Angus, R. O. Jr.; Johnson, R. P. *J. Org. Chem.* **1983**, *48*, 273–276.
105. Budwitz, J.; Newton, C. G. *Org. Synth.* **2023**, *100*, 159–185.
106. Mikami, K.; Shimizu, M. *Chem. Rev.* **1992**, *92*, 1021–1050.
107. Söderberg, B. C. *Org. Synth.* **2011**, *88*, 291–295.

9

PHYSICAL INFLUENCES ON REACTIONS

The experimental section of a journal article is the core of factual observation. It remains of value even when the explanations and theories are revised. This is the part of the article that is consulted in detail when you want to prepare a reported compound or a closely related one. From introductory texts, it is easy to get the impression that simply mixing starting materials gives the product. Some cases are as easy as that, but many require particular conditions or techniques to stimulate or control the process.[1]

Reaction conditions are chosen for efficiency and selectivity. The reactants may be capable of giving many products, but with appropriate handling, the desired one may predominate. Appropriate techniques may make the reaction easy to conduct, without temperature extremes or lengthy reaction times. Many reactions are kinetically controlled, that is, given sufficient energy and opportunity, the reactants will proceed

[1]There are many safety considerations in the design of experiments, and this should be done under the supervision of experienced personnel.

Understanding Organic Chemistry: Bridging Fundamental and Advanced Courses,
Fourth Edition. Ann M. Fabirkiewicz and John C. Stowell.

Companion website: www.wiley.com/go/fabirkiewiczorganic4e

to products. Other reactions are thermodynamically controlled and will proceed readily in either direction until an equilibrium amount of reactants and products are present. In the latter, progress toward products is made by using an excess of a reactant or removing a product from the reaction phase by distillation or crystallization.

The larger part of the effort is usually the separation and purification of the products, but these procedures are found in laboratory texts [1] and will not be covered here.

9.1 UNIMOLECULAR REACTIONS

Many eliminations and rearrangements require only heat to proceed, and the simplest procedure is heating with no solvent. Heating dicyclopentadiene at 200–210°C gives the retro-Diels–Alder product cyclopentadiene, which escapes as a gas and is condensed [2]. Heating a chlorooxazolidinone neat gives HCl elimination directly (Eq. 9.1) [3].

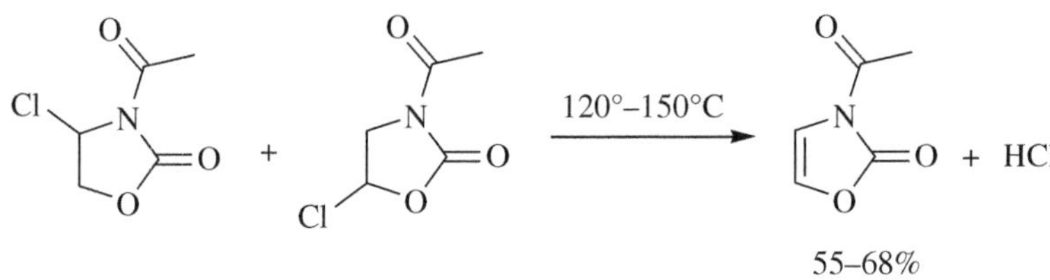

(9.1)

More often, reactions are carried out in a solvent, which serves several purposes. A solvent with an appropriate boiling point may be heated at reflux to provide a constant temperature for the reaction. If a reaction is particularly exothermal, the solvent can be a heat sink to dissipate the exotherm rather than allowing the reactant to rise to a precipitously high temperature. If there is a competing bimolecular reaction between molecules of the reactant, the solvent is a diluent that lessens the collisions and favors the unimolecular reaction.

An alternative that avoids the expense of a solvent and its removal is flash vacuum pyrolysis [4]. This avoids bimolecular reactions because in the low density of a gas (especially under vacuum), molecular collisions are relatively infrequent. Short reaction times at extremely high temperatures may be used where warranted. For example, 2-acetoxy-1,4-dioxane is heated at 425°C for 25 minutes to produce 1,4-dioxene (Eq. 9.2) [5].

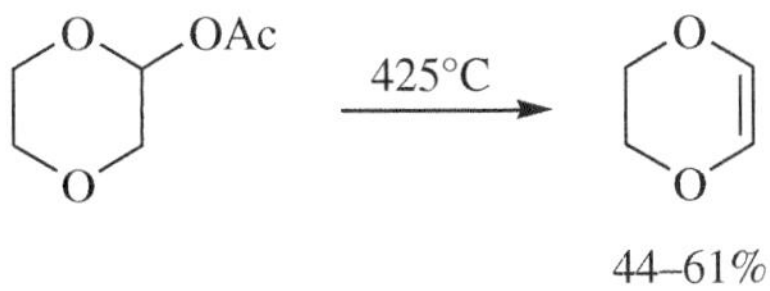

(9.2)

9.2 HOMOGENEOUS TWO-COMPONENT REACTIONS

Efficient interaction of two or more components occurs if they are combined in a single liquid phase. The Diels–Alder reaction between (2*E*,4*E*)-2,4-hexadien-1-ol and maleic anhydride occurs without solvent as the two components melt together over approximately 3 minutes [6]. This reaction can be performed safely at a 5-mmole scale in a 50-mL beaker, but doubling the reaction scale or performing the reaction in a smaller vessel causes an extremely exothermic reaction due to the lack of solvent as a heat sink.

If one or more of the reactants remain solid at the desired temperature, the reaction zone is often limited to the surface of the solid which may soon be blocked by a layer of reaction product. A solvent that dissolves both reactants can alleviate this problem and also aid in temperature control. More solvent will give less frequent collisions between reactants and slow reaction progress. This may be desirable for moderation of very fast reactions or minimized for slower ones. In some cases, one reactant may be used in excess as the solvent to maintain the high collision frequency. It is also possible to carry out solvent-free reactions in the solid phase using ball milling, a process in which chemically inert balls are added to the reaction mixture. Tumbling the reaction vessel allows the balls to grind and mix the solid reactants. In Equation 9.3, the balls are first milled with the reagents, then additional balls are added with the alcohol. In this way, benzylic alcohols are converted in high yield to aldehydes or ketones, with no evidence of overoxidation, a common problem in the synthesis of aldehydes [7]. A similar solvent-free procedure for generating ylides has been developed using a mortar and pestle to replace the ball mill [8].

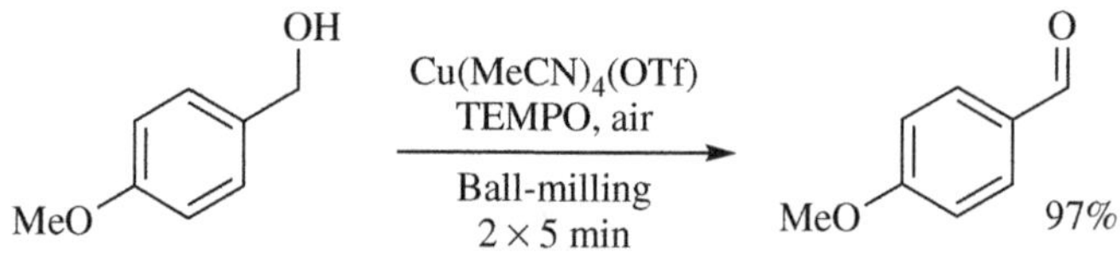

(9.3)

It is common to control fast reactions by adding one of the reactants over a period of time to a solution of the other(s). Under these conditions, the added component does not become diluted by all the solvent in the reaction flask, and locally high concentrations exist. This is usually of no consequence, but with especially reactive reagents, competing reactions may occur in these concentrated zones. For example, when the reaction in Equation 9.4 is carried out by adding the 1-aminobenzotriazole (ABT) to a solution of lead tetraacetate and 4-phenyloxazole, the locally high concentration of the benzyne intermediate allows it to dimerize to biphenylene, and the Diels–Alder product is obtained in low yield. This is corrected by adding the ABT and lead tetraacetate simultaneously on opposite sides of the stirred solution. In this way, there is no locally high concentration of benzyne and the Diels–Alder adduct is obtained in quantitative yield [9]. It is interesting to note that if the product is heated in benzene, a retro-Diels–Alder reaction occurs affording isobenzofuran (a potent diene for other Diels–Alder reactions) along with benzonitrile.

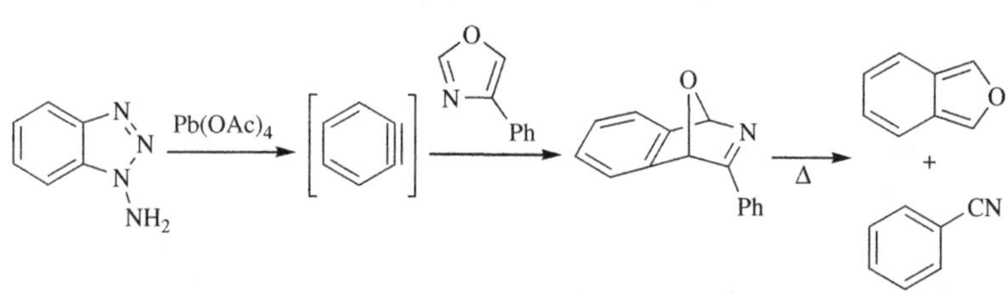

(9.4)

9.3 TEMPERATURE EFFECTS

The temperature selected for a reaction is a compromise, arrived at by experimentation using similar cases as guides. A lower temperature will give a longer reaction time, while a higher temperature will bring on competing reactions that will lessen the yield of the desired product and complicate the purification process.

A series of experiments were used to choose 1,2-dichloroethane as the optimum solvent for the reaction shown in Equation 9.5. The optimum temperature and time were determined similarly, and the yield was optimized in this solvent at 85°C [10].

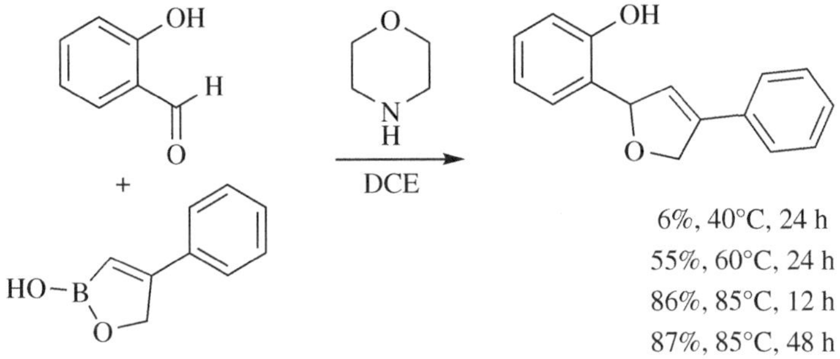

(9.5)

The reaction of LDA with carboxylic acids to afford the *O*,α-dianions proceeds rapidly at −40°C, but if the carbon chain is very long, the reaction is very slow. Eicosanoic acid gives only the monoanion at room temperature, but heating at 50°C for several hours gives the dianion [11]. Similar effects are found in other reactions of long-chain molecules.

Microwave-assisted organic synthesis (MAOS) allows considerable rate acceleration due to the increased heating efficiency, which often improves yields and minimizes side reactions [12]. As an example, the multicomponent reaction in Equation 9.6 is accomplished in 30 minutes, and products can be determined by adjusting the reaction solvents [13].

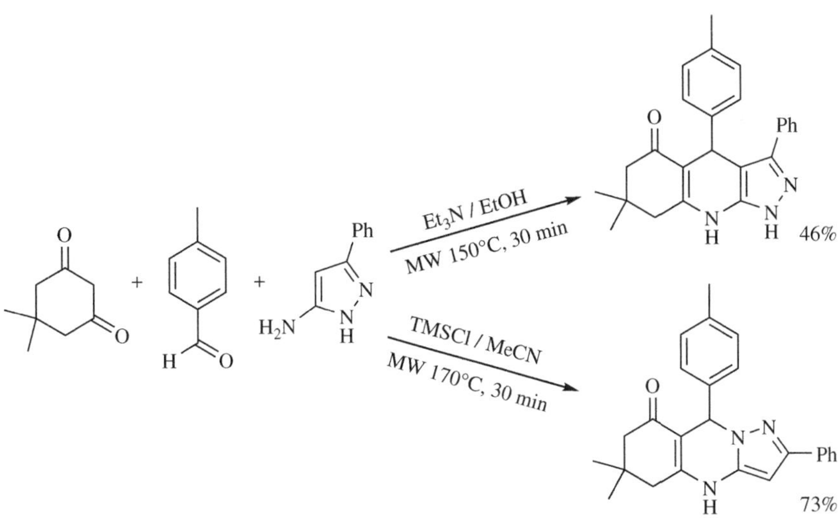

(9.6)

9.4 PRESSURE EFFECTS

Laboratory reactors are available from many suppliers for applying pressures of up to 20 kbar (19,700 atm, 2 GPa) to reaction solutions in volumes of 1–50 mL. Those reactions that have negative activation volumes ($\Delta V^{\ddagger}$), that is, those in which the transition state including solvation occupies a smaller volume than the starting molecules, will be considerably accelerated by very high pressures [14–16]. Negative $\Delta V^{\ddagger}$ values are common among quaternization of amines and phosphines, hydrolyses, esterifications, Claisen and Cope rearrangements, nucleophilic substitutions, and Diels–Alder reactions.

Very high pressure is especially valuable where heat alone leads to undesired reaction products. For example, in the reaction of citraconic acid with furan (Eq. 9.7), heat favors the retro-Diels–Alder reaction as the product is thermodynamically less stable than the starting material, but under the conditions indicated, a high yield was obtained [17].

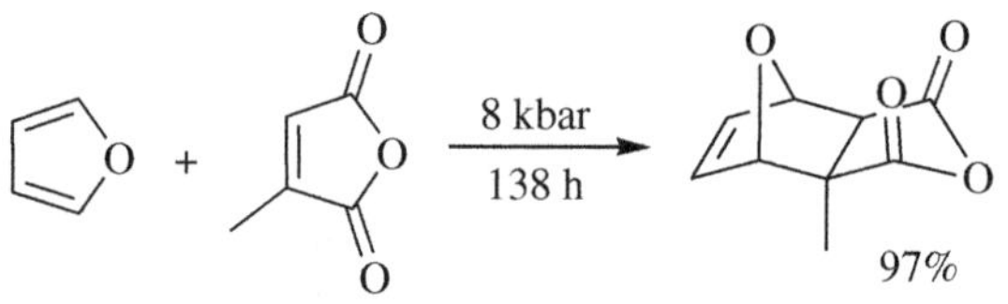

(9.7)

While high-pressure reactions typically require specialized equipment, a Bayliss–Hillman reaction between *p*-bromobenzaldehyde and methyl acrylate has been induced by the pressure created as water freezes (2 kbar). The reaction is performed in a test tube placed in a water-filled sealed autoclave in a household refrigerator at −20°C. An 86% yield was obtained in this manner compared to a 65% yield at 1 atm [18].

Esters may be hydrolyzed under mildly basic conditions at room temperature at 10 kbar. This avoids the potential side reactions, such as epimerization, dehydration, and migration of double bonds, which occur in hot acidic or basic conditions at 1 atm on sensitive compounds like the one in Equation 9.8 [19].

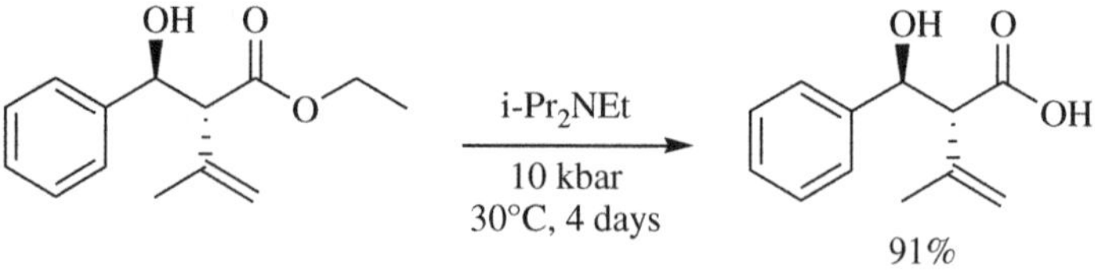

(9.8)

9.5 SOLVENT EFFECTS

It is frequently necessary to treat an organic compound with an ionic reagent. Ionic reagents are not appreciably soluble in nonpolar organic solvents; therefore, polar solvents are used [20]. The ions are solvated by coordination with the oppositely charged end of the solvent dipole or by specific hydrogen bonding. Hydrogen bonding (protic) solvents give higher rates for S_N1 reactions compared to aprotic solvents because they aid the departure of anionic leaving groups by hydrogen bonding with them. The smaller the anion, the more concentrated the charge and the more tightly it will be solvated by protic solvents. Thus, in methanol, the order of nucleophilicity of halides toward iodomethane is $I^- > Br^- > Cl^- > F^-$ [21].

The more frequently used S_N2 reactions are generally not aided by these protic solvents because the solvents cluster around the nucleophile anion, rendering it less reactive. S_N2 reactions are greatly aided by polar aprotic solvents such as dimethylsulfoxide (DMSO), dimethylformamide (DMF), hexamethylphosphoramide (HMPA), tetramethylurea, or 1,3-dimethyl-2-imidazolidinone. In these solvents, the positive end of the dipole is relatively encumbered, while the negative end is exposed and available for association with cations. The anions are thus not well solvated and exceptionally reactive. For example, NaCN in DMSO reacts with primary and secondary alkyl chlorides to give nitriles in 0.5–2 hours, while 1–4 days are required in aqueous alcohol [22, 23]. In polar aprotic solvents, the relative nucleophilicity of anions follows charge density and is the reverse of that found in protic solvents. The displacement on *n*-butyl tosylate in DMSO gives the order $F^- > Cl^- > Br^- > I^-$ [24].

Ionic liquids are essentially low-melting salts, and they have found increasing use as solvents in organic synthesis as "green" alternatives to common organic solvents [25, 26]. Because these salts are nonvolatile, the risk of exposure to the environment and the experimenter is lessened, but this does not imply a lack of toxicity and appropriate care is warranted. The *anti*-addition of bromine to an alkene is typically carried out in carbon tetrachloride, a solvent of known human toxicity that is also ozone-depleting with a very long environmental half-life. Replacing this solvent with the ionic liquid 1-butyl-3-methylimidazolium tetrafluroborate [27], [bmim][BF_4], allows for the halogenation to occur in 92% yield with greater than 99% of the *anti*-addition product [28].

An empirical scale of solvent polarity has been developed on the basis of shifts of the ultraviolet (UV)–visible absorption maximum of a pyridinophenylate indicator [29], which should be useful for predicting solvent effects on rates.

9.6 BIPHASIC REACTIONS

Inorganic reagents are frequently insoluble in solvents that dissolve organic materials, presenting a challenge in getting those reagents in sufficient contact with the organic substrates to generate good yields. Some inorganic reagents are soluble in water and are used with an organic solution with vigorous stirring to promote a reaction at the interface between the water and organic phases. Some reagents will remain in the solid phase. The frequency of successful collision on a surface is far less than in the bulk of a homogeneous solution. The problem of incompatible solubilities can be overcome by several means, as discussed below.

9.6.1 Phase-Transfer Catalysis

Frequently, incompatible solubility is overcome by adding a small amount of a tetraalkylammonium or phosphonium salt or a crown ether. This is called *phase-transfer catalysis* [30–32]. The term phase-transfer catalysis (PTC) is applied to both solid–liquid reactions and liquid–liquid cases.

Quaternary cations (Q+) with sufficiently large alkyl groups have an affinity for organic solvents and will carry reactive anions with them into solution in the organic layer. These anions are particularly reactive because they carry only a small hydration shell. Some stirring is still necessary because the quaternary salts are used in catalytic amounts and must repeatedly exchange product anions for reactant anions at the phase boundary (Fig. 9.1).

Alkyl halides and sulfonates will undergo nucleophilic substitution by these reactive inorganic anions or carbanions in the organic layer. Chloride or bromide ions cannot displace iodide or tosylate ions efficiently because the iodide and tosylate anions are relatively lipophilic and remain in the organic phase, which shifts the equilibrium toward starting materials. Organochlorides can be converted to bromides or iodides and

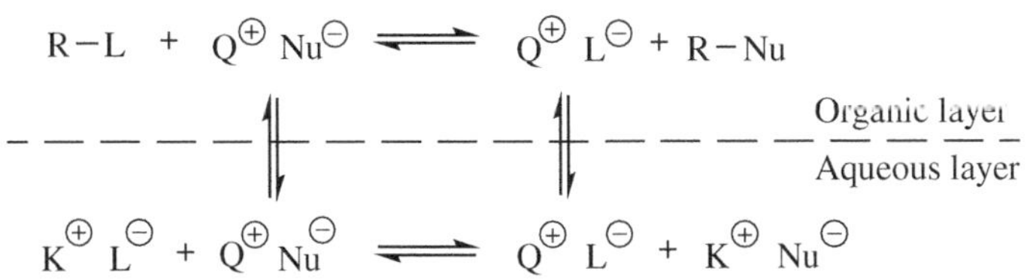

FIGURE 9.1 The equilibria present in a phase-transfer reaction. Starks [33]/American Chemical Society.

organobromides can be converted to iodides in good yield. Mesylates are good leaving groups with low lipophilicity and are displaceable by all halides (Eq. 9.9) [34]. Other nucleophilic ions that are not themselves good leaving groups, such as cyanide, phenoxides, carboxylates, carbanions, alkoxides, or sulfinates, can give high yields of substitution products [35].

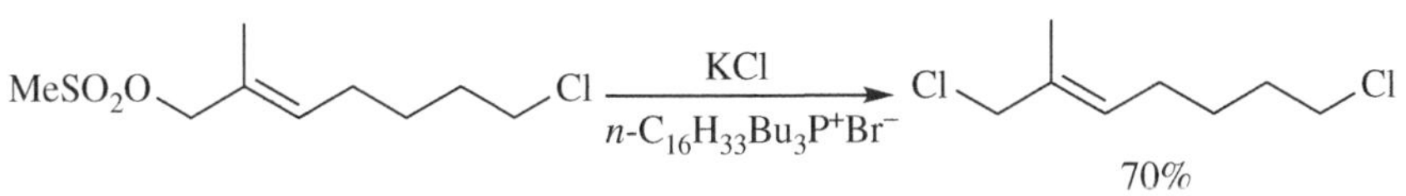

(9.9)

Aliquat 336 (Starks' catalyst, trioctylmethyl ammonium chloride) [33] has been widely used in PTC. A green procedure to form adipic acid from cyclohexene takes advantage of this catalyst in combination with sodium tungstate, both of which are recovered for reuse in subsequent reactions (Eq. 9.10) [36].

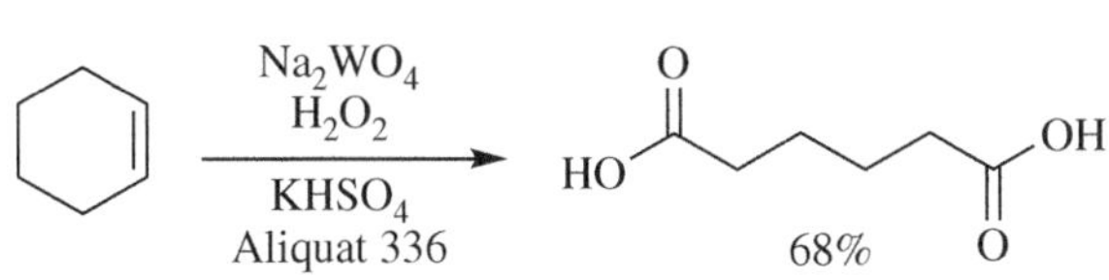

(9.10)

It is possible to omit the solvent altogether. Sterically hindered mesitoic esters are slow to form and slow to hydrolyze under ordinary conditions. When the hydrolysis is carried out with water and hydrocarbon solvents, yields are low. Using powdered KOH and Aliquat 336, hydrolysis can be accomplished, with the organic reactants and products serving as the liquid phase (Eq. 9.11) [37].

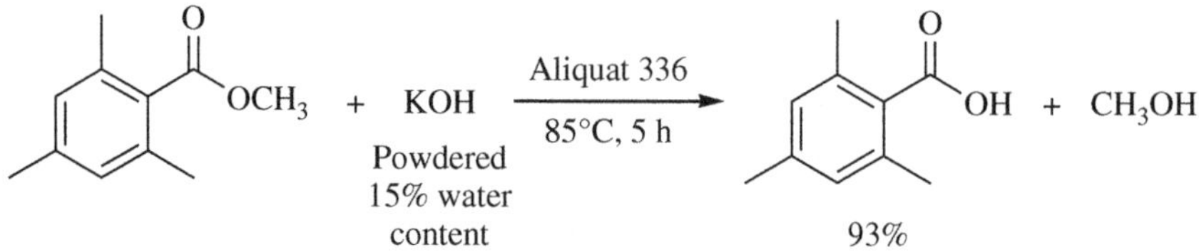

(9.11)

The use of chiral quaternary ammonium ions allows for stereocontrol in phase-transfer reactions [38], as shown in Equation 9.12 [39]. The catalyst, (*S*)-4,4-dibutyl-2,6-bis(3,4,5-trifluorophenyl)-4,5-dihydro-3*H*-dinaphtho [2,1-c:1′,2′-*e*]azepinium bromide, (*S*)-1, is commercially available.

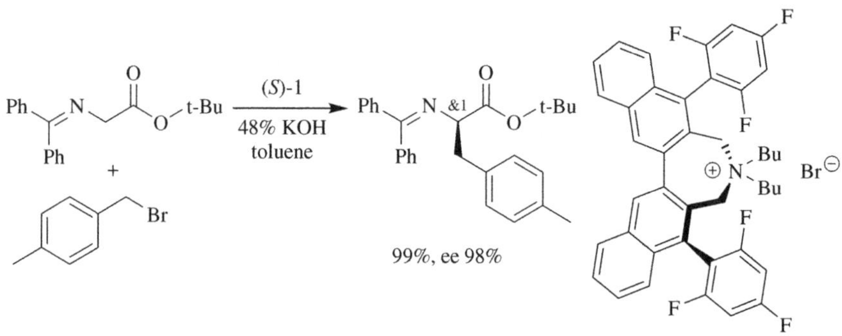

(*S*)-1

(9.12)

Alkali metal salts become soluble in low-polarity organic solvents when those cations are coordinated by close-fitting cyclic ethers, giving them an organic-compatible exterior [40]. Dicyclohexyl-18-crown-6 is particularly suitable for potassium, and the combination of it and potassium permanganate has a high solubility in benzene [41].

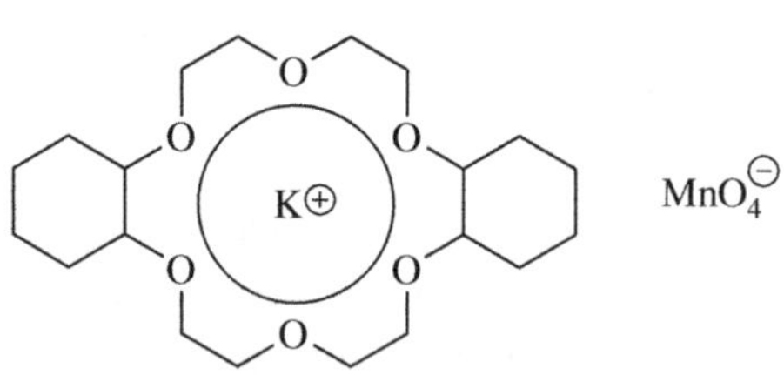

Substitution reactions may be carried out in high yield using 18-crown-6 with potassium salts. In Equation 9.13, potassium acetate reacts efficiently in a substitution reaction in the presence of 18-crown-6. Without the crown ether, this reaction shows no product after 48 hours. Interestingly, the same reaction proceeds in 95% yield when an ionic liquid is used instead of the crown ether, but the ionic liquid has the advantage of much lower toxicity [42].

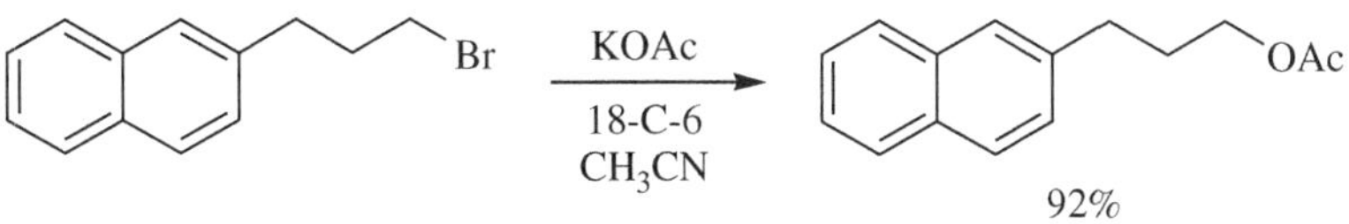

(9.13)

9.6.2 Increasing Solubility

An obvious solution to insoluble inorganic reagents is to use a more organic soluble salt of the same reactive anion. For example, although sodium borohydride has very low solubility in dichloromethane, ether, or tetrahydrofuran (THF), tetrabutylammonium borohydride has high solubility in dichloromethane. This salt is useful for the reduction of aldehydes and ketones that are not soluble in hydroxylic solvents or that react with those solvents to give hydrates, acetals, or ethers [43]. Potassium permanganate is insoluble in organic solvents, and when used in water, it decomposes, requiring a large excess. Tetrabutylammonium permanganate is easily prepared and may be used in stoichiometric amounts in pyridine solution at room temperature to give oxidation products quickly and in high yield [44].

9.6.3 Increasing Surface Area

In the preceding cases, the intent is to bring at least small amounts of the reagent anions into the organic solution for reaction. In contrast to these, solid metals must undergo oxidation at their surfaces, and obtaining convenient rates of reaction depends on selecting appropriately fine particles. Zinc metal is available as lumps (mossy), granular particles, and as dust. Even finer zinc may be prepared by reducing anhydrous zinc chloride with lithium naphthalenide [45]. This method is useful for the preparation of a number of active metals, known as Rieke metals. These are very reactive and usually pyrophoric in air [46].

Magnesium in the form of lathe turnings is sufficient for most Grignard reactions, but for some temperature-sensitive or -reluctant cases, activation is needed. Stirring magnesium turnings under a nitrogen atmosphere with a Teflon bar for 15 hours gives smaller dark-gray particles, which react promptly with allylic and benzylic chlorides in ether at 0°C to give clear solutions of the Grignard reagents in high yield, with no coupling products. With untreated turnings, the slower-reacting magnesium allows for a competing reaction of the halide with the Grignard reagent as it is formed, leading to appreciable amounts of coupled hydrocarbon and precipitated $MgCl_2$ [47]. A fine powder from potassium metal reduction of $MgCl_2$ is even more reactive [48].

Sodium is available as cast ingots that can be cut into small pieces with a knife. When more surface area is needed, it can be finely pulverized by stirring it in molten form in refluxing toluene, and then cooling. This was done for the process in Equation 9.14 [49], where the cooled toluene was replaced with ether once the sodium was prepared. Trimethylchlorosilane is slow to react with carbanions; therefore, it can be present in a reaction mixture where it will capture oxyanions as they are formed.

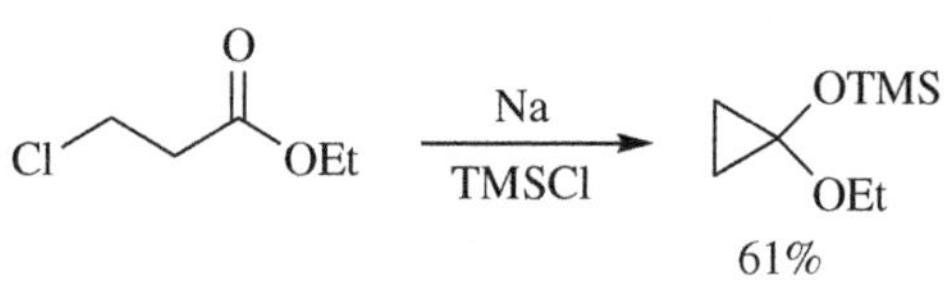

(9.14)

9.6.4 Flow Chemistry

Flow chemistry combines many of the advantages of the techniques above. Reagents are combined in channels and reactions conducted in a continuous stream, rather than a flask. This technique allows control over temperature, pressure, mixing rates, and multiphasic reactions in order to optimize yields [50].

Choosing one of many examples, it is possible to use flow techniques to accomplish the controlled oxidation of alcohols to aldehydes, ketones, or methyl esters, depending on reaction conditions. The alcohol is dissolved in ethyl acetate and a catalytic amount of tetrabutylammonium bromide (TBAB). In a second stream, sodium hypochlorite is added, with the concentration varied depending on the desired product. The two streams are combined and passed through a perfluoroalkoxy (PFA) tube reactor at room temperature for 5–30 minutes. Mostly high yields of ketones, methyl esters, and aldehydes are obtained under these conditions (Eqs. 9.15 and 9.16) [51].

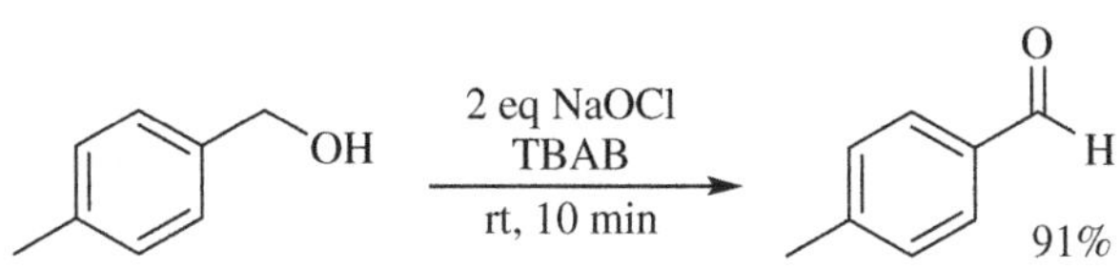

(9.15)

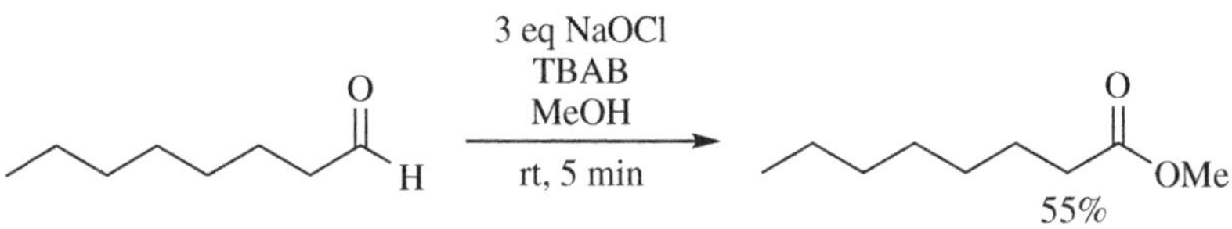

(9.16)

9.6.5 Ultrasound

Metals can be made to react dramatically faster using ultrasound (sonication) [52–54]. A reaction flask may be partly immersed in an ultrasonic bath, or an ultrasonic probe can be inserted directly into a reaction mixture. The oscillator is typically 60–250 W at 20–60 kHz.

Organometallic reagents may be formed by ultrasonic irradiation [55]. The Reformatsky reaction, in the presence of iodine, proceeds in high yield in minutes at 25–30°C with ultrasonic erosion of ordinary zinc dust in dioxane (Eq. 9.17) [56]. The same reaction under conventional conditions proceeds over several hours to give a 61% yield.

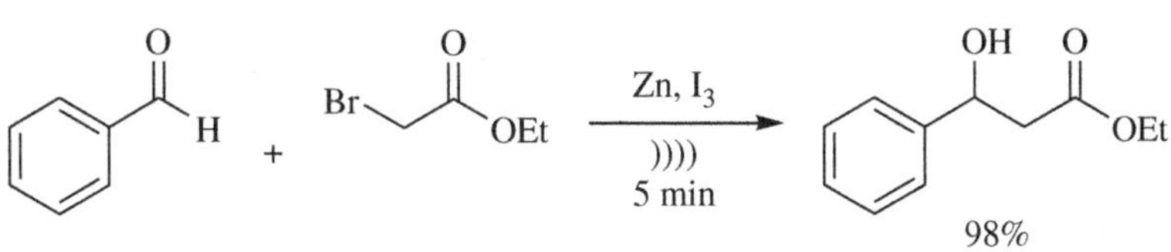

(9.17)

Sonication of Ni powder prior to its use as a hydrogenation catalyst greatly increases the rate of that reaction, though this effect is likely due at least in part to the sonication pre-treatment cleaning the oxide coating off the surface of the Ni catalyst [57].

Dichlorocyclopropanes are important synthetic intermediates that are formed from dichlorocarbene. Sonication can be used to produce dichlorocarbene from carbon tetrachloride and magnesium, and form the cyclopropane in one pot in high yields. An example is shown in Equation 9.18 [58].

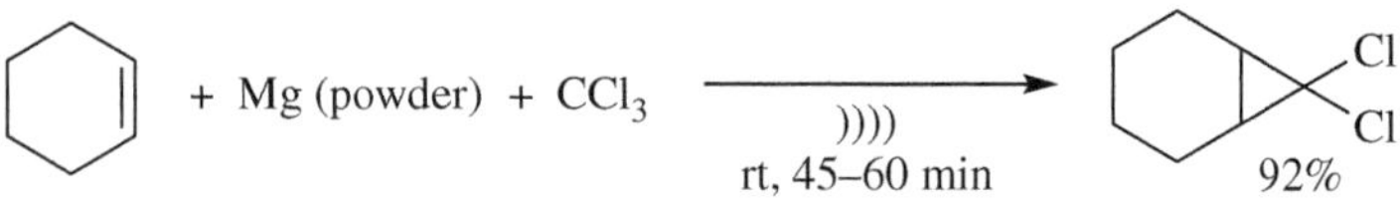

(9.18)

The solvent-free domino reaction in Equation 9.19 results in a 94% yield when heated at 80°C for 15 minutes. A similar 92% yield is obtained at room temperature in 20 minutes with sonication, resulting in a green improvement in reaction conditions by avoiding the need for heat [59].

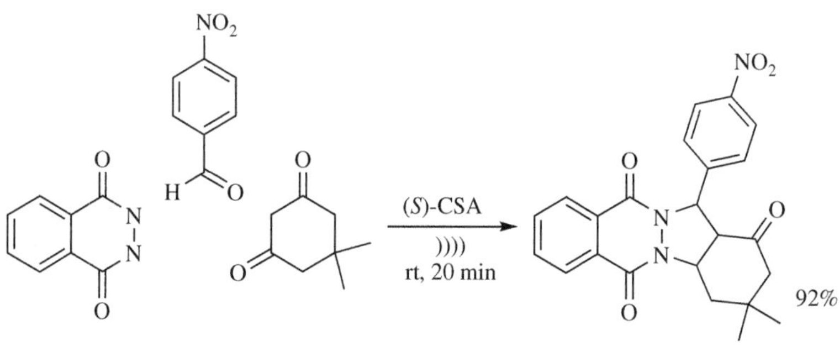

(9.19)

9.7 REACTIONS ON CHEMICAL SUPPORTS

Numerous reactions that are inefficient in solution will proceed at lower temperatures and with higher selectivity in the adsorbed state on porous inorganic solids [60, 61]. Just as polar solvents have substantial effects on reactions in homogeneous solutions, the very polar adsorbents can affect reactants. It is also possible to adsorb or chemically bond the reagent or reactant to a polymeric support [62–64]. These methods have a significant advantage in practice, as purification generally involves only filtration or other straightforward separation, and it may be possible to recover the catalyst.

Ozone, because of its very low solubility in ordinary organic solvents, is suitable only for very fast reactions such as with alkenes. Silica gel will adsorb up to 4.7% by weight of ozone at −78°C. If an organic substrate is first adsorbed on the silica gel and then ozone added at −78°C, followed by warming to room temperature, oxidation of benzene rings and tertiary carbon–hydrogen bonds occurs conveniently and in good yield. Adamantane was converted to 1-adamantanol in 81–84% yield (Eq. 9.20) [65]. This "dry ozonation" [66] occurs without solvent, thus the name, but the small amounts of water that often contaminate silica gel may play an important role.

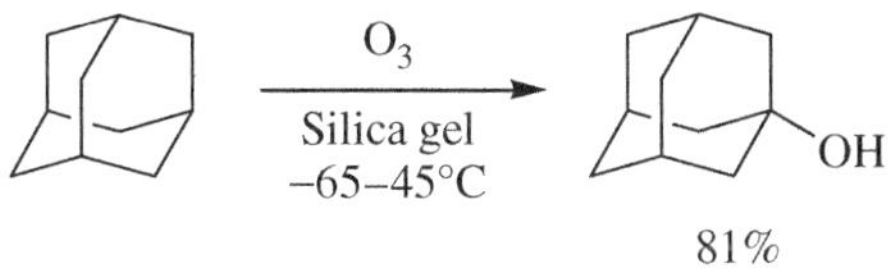

(9.20)

Alcohols are efficiently oxidized to aldehydes with *o*-iodosobenzoic acid (IBX), but IBX is insoluble and reactions are often performed under heterogeneous conditions. Binding IBX to linear polystyrene, a soluble polymer (Eq. 9.21), allows for the oxidation to occur under homogeneous conditions. With this reagent, benzyl alcohol is oxidized to benzaldehyde in dichloromethane in 100% yield [67].

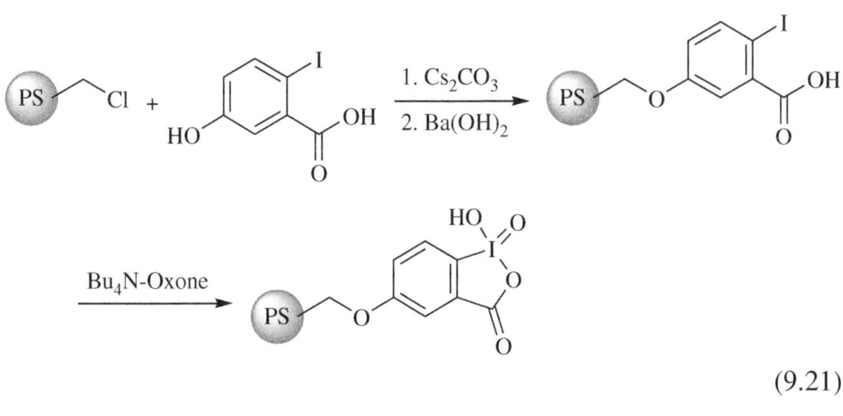

(9.21)

Linear polystyrene was used as a solid support in the synthesis of prostaglandin E_2 methyl ester (Eq. 9.22) [68]. Methanol precipitates the polymer-bound product at the end of the sequence, allowing excess reagents and by-products to be removed. The polymer and protecting group are removed with 48% hydrofluoric acid (HF) in 6 hours to give the product in 37% yield in eight steps as shown. This sequence allowed for the generation of a library of compounds by varying the two side chains attached to the polymer-bound cyclopentenone.

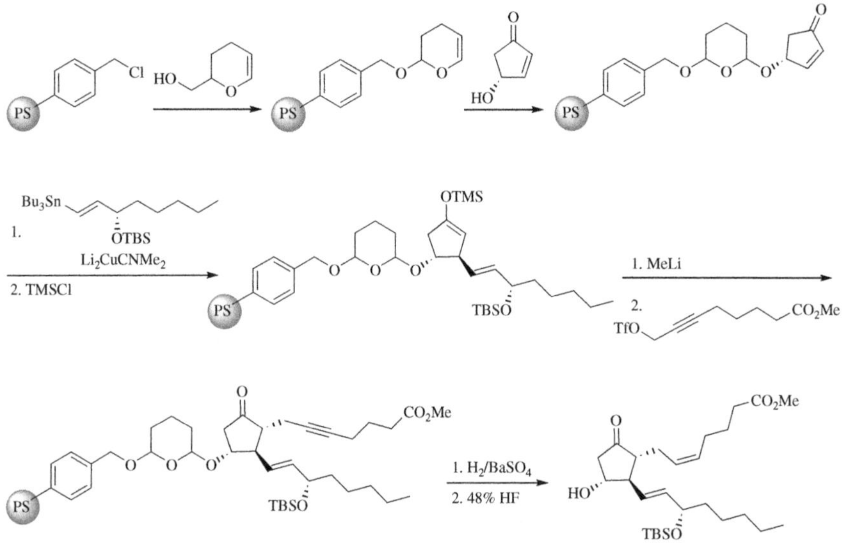

(9.22)

Ionic liquids have also been used as a soluble support for catalysts, reagents, and reactants [69]. As shown in Equation 9.23, a hypervalent iodine reagent is bound to an ionic liquid and used in the tosylation reaction shown [70]. The iodobenzene salt can be recovered in 95% yield and reused.

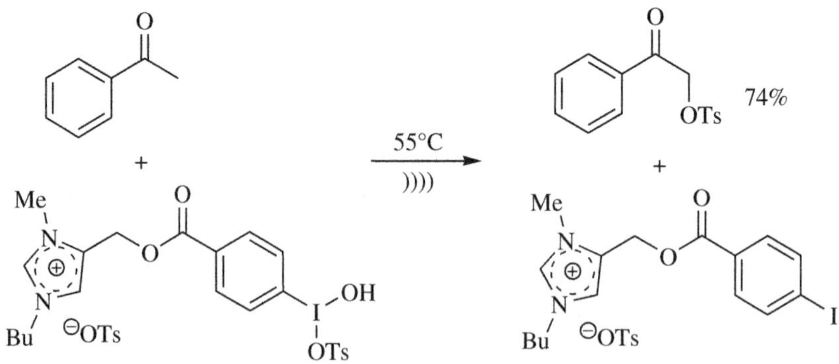

(9.23)

9.8 USING UNFAVORABLE EQUILIBRIA

Most of the techniques covered thus far are designed to improve contact among reactants where reactions would otherwise be too slow (kinetically limited). Some reactions are fast in forward and reverse directions, leading to small conversions at equilibrium (thermodynamically limited). Many of these may be pushed toward high conversions by a phase change. If one of the products of an unfavorable equilibrium is removed from the reaction solution by evaporation or crystallization while the reaction is underway, nearly all the starting materials may be converted. The equilibrium is continually reestablished based on remaining concentrations. Distillation is used to remove a lower-boiling product, as in transketalization, transesterification, preparation of higher anhydrides from acetic anhydride (Eq. 6.14), and many others. Azeotropic distillation of water from esterifications or enamine syntheses is common.

Ordinarily, sulfonate groups may be displaced from primary carbons by bromide ions. The reverse can be done with an exchange reaction driven by the escape of the volatile by-product bromomethane, bp 4°C (Eq. 9.24) [71].

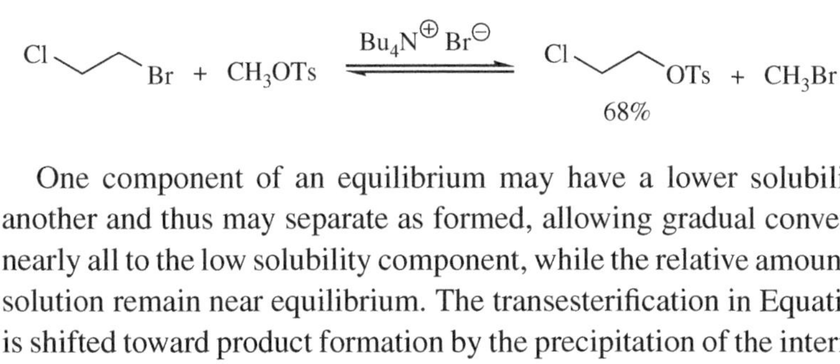

(9.24)

One component of an equilibrium may have a lower solubility than another and thus may separate as formed, allowing gradual conversion of nearly all to the low solubility component, while the relative amounts in the solution remain near equilibrium. The transesterification in Equation 9.25 is shifted toward product formation by the precipitation of the intermediate lithium salt, resulting in a 95% yield of the product [72].

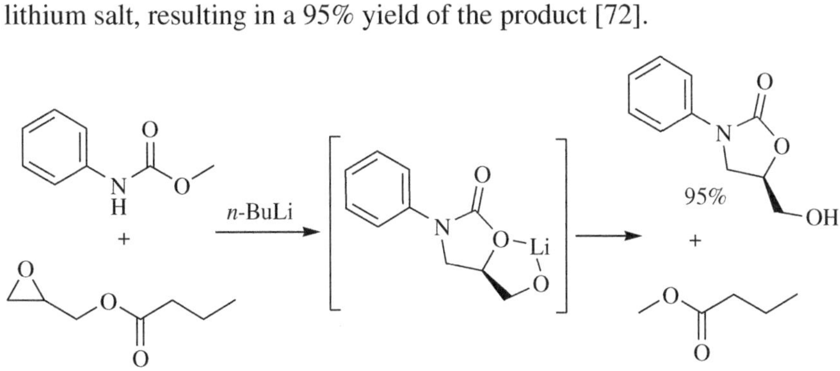

(9.25)

Equilibria among stereoisomers have been manipulated using solubility differences. Meso and racemic dibromoglutaric acids are present in equal amounts when equilibrated in the presence of HBr (Eq. 9.26). The meso compound is less soluble in ethanol and will crystallize from the equilibrium solution when cooled to 5–10°C. Filtering three crops of crystals, concentrating the solution after each, converted 82% of the mixture to the meso isomer [73]. In other examples, an enantiopure compound was converted to an enantiopure epimer by equilibration where the product solidified [74].

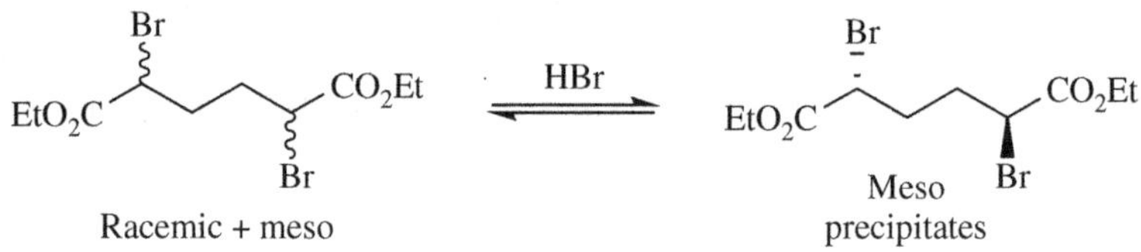

(9.26)

Finally, a racemic amine was converted to one pure enantiomer by equilibrating the enantiomers in the presence of (+)-10-camphorsulfonic acid (CSA), which preferentially crystallized the salt of the *S* enantiomer (Eq. 9.27) [75]. A catalytic amount of an aromatic aldehyde gave an imine that could racemize in solution in isopropyl acetate-acetonitrile, while the CSA salt of the *S* isomer crystallized, eventually driving the equilibrium to produce the *S*-isomer in 91% yield.

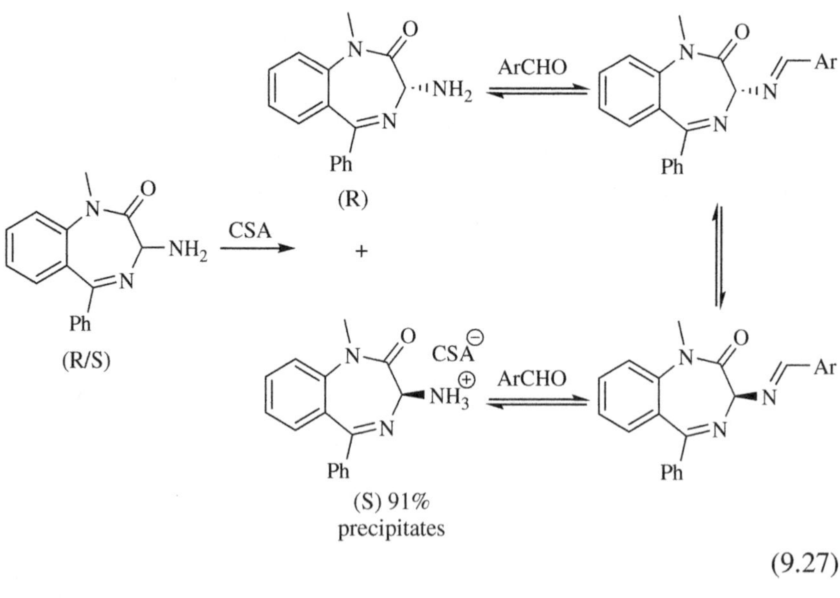

(9.27)

9.9 GREEN CHEMISTRY

Once yield was the primary factor that determined the success of a reaction, but increasingly, concerns about toxicity, sustainability, environmental impact and related legislation, are incorporated into the design of an experiment. Twelve principles of green chemistry, which call for safer reactants and procedures, less chemical and energy waste, avoiding the use of environmentally persistent materials and unnecessary reactions, and the use of renewable feedstocks, were published in 1998, and continue as guiding principles today [76].

Several metrics are available to quantify the "greenness" of an experiment [77]. The Environmental or E-factor is one of the most flexible, particularly in industrial applications, where a facility can measure quantities of material leaving as waste and leaving as a products. It calculates the ratio of total waste to total product (Eq. 9.28) [78]. While the type of waste, and not only the amount, is important, this simple calculation allows for an estimation of environmental impact. E-factors for typical bulk chemical production range from 1 to 5, while pharmaceutical industry processes produce *E*-factors ranging from 25 to over 100. Bulk chemical production generates more waste overall, but to remain profitable, it must resolve waste disposal issues. Pharmaceutical processes are more focused on manufacturing and quality, and their higher profit margins with lower overall volumes make waste disposal less of a concern, at least financially.

$$\mathrm{E-factor} = \frac{\text{total amount of waste}}{\text{total amount of desired product}} = \frac{(\text{mass of raw materials}) - (\text{mass of product})}{(\text{mass of product})} \tag{9.28}$$

In the E-factor calculation, all types of waste carry the same weight, but some are more hazardous or environmentally detrimental than others. A Q-factor has been proposed as a multiplier for the E-factor to account for this variation [79]. Q-values are somewhat arbitrary, and several methods have been proposed to quantify the value that take into account threshold limit value (TLV), safety, health, and environment (SHE) hazards as well as persistence, bioaccumulation, and ecotoxicity [77].

Another metric for estimating environmental impact is to calculate atom economy, which is determined by dividing the molecular weight of the desired product by the total of the molecular weights of all the products and by-products of the reaction, weighting as appropriate for a balanced equation (Eq. 9.29, where n is the molar coefficients from the balanced equation) [80]. This method does not require experimentation and can be applied during the planning stages of a synthetic route.

$$\%\text{ atom economy} = \frac{\text{molecular weight of desired product}}{\sum \text{molecular weight of each product} \times n} \times 100 \tag{9.29}$$

EcoScale is a semi-quantitative metric to rank green synthetic procedures, which takes into account yield, cost, safety, technical setup, temperature, time, and purification. Rankings are calculated by starting at 100 and subtracting points for negative influences [81].

Solvents are perhaps the largest source of waste in chemical reactions, so proper solvent selection can greatly benefit the greenness of an organic procedure. Increasingly, organic reaction conditions are being developed that use water or even no solvent. Environmental, health, and safety concerns must be weighed against the role that solvent plays in a reaction. The pharmaceutical industry has largely been a driver of the production of green solvent guides. CHEM21 is a European consortium that promotes and supports sustainable processes in chemistry and biology. Their solvent guide assigns scores based on Globally Harmonized System of Classification and Labeling of Chemicals (GHS) health statements, among other factors [82]. The Green Chemistry Institute of the American Chemical Society hosts a solvent selection tool that was originally developed by AstraZeneca [83]. A reagent guide is also available [84].

RESOURCES

1. The American Chemical Society Green Chemistry Institute provides resources for educators and students, and for industry and research. https://www.acs.org/green-chemistry-sustainability.html.
2. The Environmental Protection Agency also maintains web-based resources. https://www.epa.gov/greenchemistry/basics-green-chemistry.

3. Beyond Benign maintains substantial web-based resources to support green chemistry practice and education at all levels, including K-12, higher education, and professional development. https://www.beyondbenign.org/.

PROBLEMS

Answers are provided in Appendix C for questions marked with an asterisk.

9.1* The Grignard reagents from 2-(2-bromoethyl)-1,3-dioxolane and 2-(3-chloropropyl)-1,3-dioxolane are thermally unstable and decompose during their preparation under ordinary conditions [85, 86]. What can be done to overcome this problem? [87].

9.2* Ozone was passed into a solution of 26 g of *cis*-decalin in CCl_4 at 0°C for 147 hours. This gave 7 g of decahydronaphth-9-ol [88]. What could be done to improve the yield and shorten the reaction time?

9.3* Heating 1-phenyl-1-benzoylcyclopropane with excess ethyl bromoacetate and 20-mesh zinc at reflux for 17 hours gave a 47% yield of the Reformatsky product along with 55% recovered ketone [89]. What could be done to improve this reaction?

9.4* The oxidation of alkynes to 1,2-diketones with potassium permanganate is a well-known reaction, but the early examples gave carboxylate salts that are soluble in aqueous permanganate. What conditions would you choose to oxidize 1-phenyl-1-pentyne to 1-phenyl-1,2-pentanedione? [90]

9.5* Flow reactors are used to allow for the continuous mixing of reagents during a reaction in a flowing stream rather than the traditional flask [91]. While designs will vary, many such reactors have narrow bore tubing through which the reactants flow. The narrow tubing helps biphasic reactions but can present a problem if precipitates are formed. Can you suggest a solution? [92]

9.6* Two syntheses of 1,3,5-triacetylbenzene are shown below. Compare these reactions in terms of green chemistry principles. Balance the equations and calculate the percent atom economy.

Reaction A: [93]

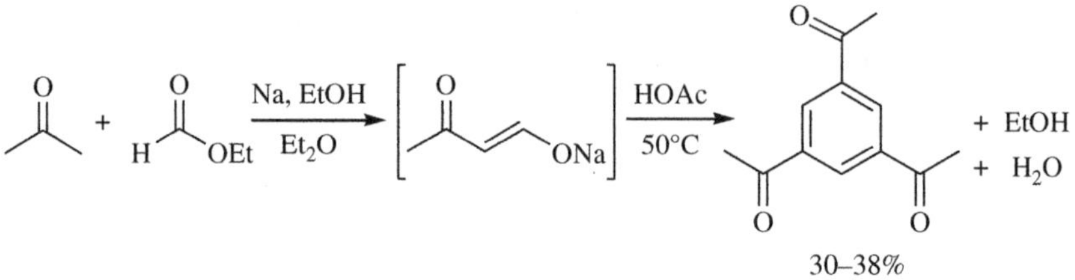

Reaction B: [94]

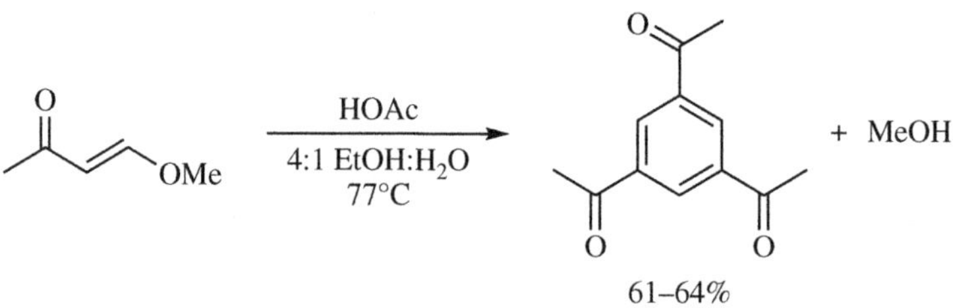

9.7* The *aza*-Baylis–Hillman reaction is especially atom efficient. An example is shown in Equation 9.30. Workup to isolate the product includes several extractions, using 2-propanol (10 mL), methanol (25 mL), H_2SO_4 (1 M, 10 mL, d = 10.6 g/mL), CH_2Cl (3 × 30 mL), $NaHCO_3$ (50 mL, d = 1.10 g/mL), NaCl (saturated, 50 mL, d = 1.2 g/mL), NaOH (1 M, 50 mL, 1.04 g/mL), and $MgSO_4$ (5 g) [95].

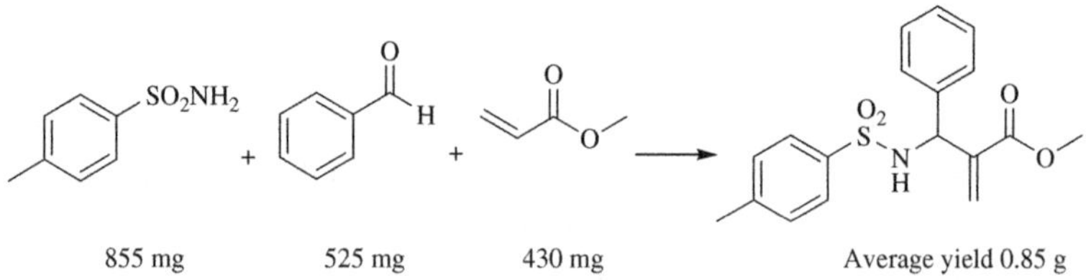

(9.30)

Calculate atom economy and the E-factor for this reaction. What could you do to improve the E-factor for this reaction? What components are the biggest negative influences on the E-factor?

9.8* Cyanohydrins are important synthetic intermediates, and investigations continue to improve sustainability. Trimethylsilyl cyanide (TMSCN) as a cyanating agent serves to form the trimethylsilyl (TMS)-protected cyanohydrin, which is more stable. Batch reactions using polystyrene-supported triphenylphosphine as a heterogeneous catalyst for the reaction resulted in an E-factor of 1.72. Under flow conditions, this reaction gave an E-factor of 0.16 [96]. A new heterogeneous catalyst, POLITAG-L-F, is able to activate TMSCN, making the reaction more efficient. Combining this catalyst with flow conditions resulted in an E-factor of 0.04 [97]. Explain the significance of these results.

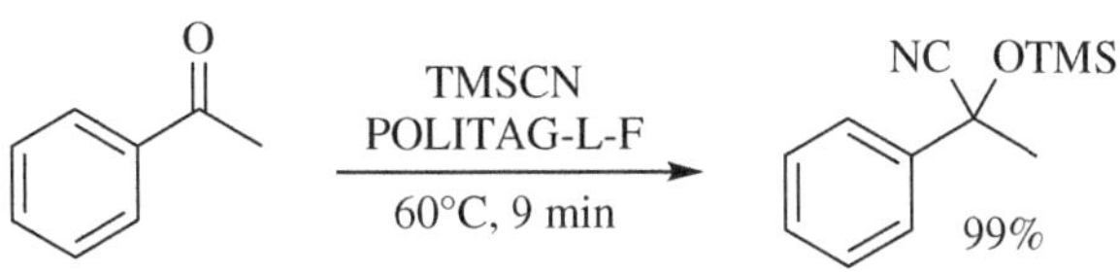

9.9* Coumarins include medically important compounds like Warfarin and Celecoxib. In order to access these structures, Pramanik and Mukherjee designed this reaction to synthesize 4-hydroxy-3-(5-(4-methoxyphenyl)-3-(4-methylbenzyl)-1-phenyl-1H-pyrazol-4-yl-2H-chromen-2-one. The reagents are stirred at room temperature while the product precipitates. After an hour, no further precipitation occurs, and the product is collected by filtration. The ethanol solvent is recycled and reused [98].

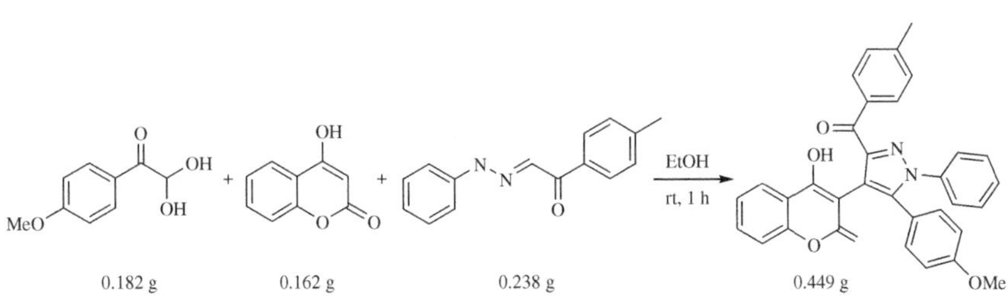

Calculate the atom efficiency and E-factor for this reaction. What factors allowed the authors to achieve such a green reaction?

REFERENCES

1. Pirrung, M. C. *The Synthetic Organic Chemist's Companion*; Wiley Interscience: Hoboken, 2007. Gordon, A. J.; Ford, R. A. *The Chemist's Companion: A Handbook of Practical Data, Techniques, and References*; Wiley Interscience: Hoboken, 1973
2. Partridge, J.; Chadha, N. K.; Uskovic, M. R. *Org. Synth.* **1985**, *63*, 44–56.
3. Scholz, K.-H.; Heine, H.-G.; Hartmann, W. *Org. Synth.* **1984**, *62*, 149–157.
4. Cantillo, D.; Sheibani, H.; Kappe, C. O. *J. Org. Chem.* **2012**, *77*, 2463–2473 McNab, H. *Aldrichimica Acta* **2004**, *37*, 19–26; Wiersum, U. E. *Aldrichimica Acta* **1984**, *17*, 31–40.
5. Kreilein, M. M.; Eppich, J. C.; Paquette, L. A. *Org. Synth.* **2005**, *82*, 99–107.
6. Parsons, B. A.; Dragojlovic, V. *J. Chem. Educ.* **2011**, *88*, 1553–1557.
7. Porcheddu, A.; Colacino, E.; Cravotto, G.; Delogu, F.; De Luca, L. *Beilstein J. Org. Chem.* **2017**, *13*, 2049–2055.
8. Leung, S. H.; Angel, S. A. *J. Chem. Educ.* **2004**, *81*, 1492–1493.
9. Whitney, S. E.; Rickborn, B. *J. Org. Chem.* **1988**, *53*, 5595–5596.
10. Cui, C.-X.; Li, H.; Yang, X.-J.; Yang, J.; Li, X.-Q. *Org. Lett.* **2013**, *15*, 5944–5947.
11. Belletire, J. L.; Fry, D. F. *J. Org. Chem.* **1987**, *52*, 2549–2555.
12. Kappe, C. O. *Angew. Chem. Int. Ed.* **2004**, *43*, 6250–6284.
13. Glasnov, T. N.; Kappe, C. O. *Org. Synth.* **2009**, *86*, 252–261.
14. Benito-Lopez, F.; Egberink, R. J. M.; Reinhoudt, D. N.; Verboom, W. *Tetrahedron* **2008**, *64*, 10023–10040.
15. Schettino, V.; Bini, R. *Chem. Soc. Rev.* **2007**, *36*, 869–880.
16. Matsumoto, K.; Acheson, R. M., Eds. *Organic Synthesis at High Pressures*; John Wiley & Sons, Inc: New York, 1991.
17. Dauben, W. G.; Lam, J. Y. L.; Guo, Z. R. *J. Org. Chem.* **1996**, *61*, 4816–4819.
18. Hayashi, Y.; Okado, K.; Ashimine, I.; Shoji, M. *Tetrahedron Lett.* **2002**, *43*, 8683–8686.
19. Yamamoto, Y.; Furuta, T.; Matsuo, J.; Kurata, T. *J. Org. Chem.* **1991**, *56*, 5737–5738.
20. Reichardt, C.; Welton, T. *Solvents and Solvent Effects in Organic Chemistry*, 4th ed.; Wiley-VCH Verlag & Co: Weinheim, Germany, 2011. Reichardt, C. *Org. Process Res. Dev.* **2007**, *11*, 105–113
21. Pearson, R. G.; Sobel, H.; Songstad, J. *J. Am. Chem. Soc.* **1968**, *90*, 319–326.

22. Smiley, R. A.; Arnold, C. *J. Org. Chem.* **1960**, *25*, 257–258.
23. Friedman, L.; Schechter, H. *J. Org. Chem.* **1960**, *25*, 877–879.
24. Fuchs, R.; Mahendran, K. *J. Org. Chem.* **1971**, *36*, 730–731.
25. Hallett, J. P.; Welton, T. *Chem. Rev.* **2011**, *111*, 3508–3576.
26. Ionic Liquids Database – ILThermo (v.2.0). *NIST Standard Reference Database 147.* http://ilthermo.boulder.nist.gov/ILThermo/ (accessed December 31, 2024).
27. Creary, X.; Willis, E. D. *Org. Synth.* **2005**, *82*, 166–169.
28. Chiappe, C.; Capraro, D.; Conte, V.; Pieraccini, D. *Org. Lett.* **2001**, *3*, 1061–1063.
29. Reichardt, C. *Chem. Rev.* **1994**, *94*, 2319–2358 Machado, V. G.; Machado, C. *J. Chem. Educ.* **2001**, *78*, 649–651.
30. Dehmlow, E. V.; Dehmlow, S. S. *Phase Transfer Catalysis*, 3rd, Revised and Enlarged ed.; VCH: Weinheim, Germany, 1993.
31. Halpern, M. E., Ed. *Phase-Transfer Catalysis*, ACS Symposium Series *659*; American Chemical Society: Washington, DC, 1997.
32. Starks, C. M., Ed. *Phase-Transfer Catalysis*, ACS Symposium Series *326*; American Chemical Society: Washington, DC, 1987.
33. Starks, C. M. *J. Am. Chem. Soc.* **1971**, *93*, 195–199.
34. Orsini, F.; Pelizzoni, F. *J. Org. Chem.* **1980**, *45*, 4726–4727.
35. Crandall, J. K.; Pradat, C. *J. Org. Chem.* **1985**, *50*, 1327–1329.
36. Reed, S. M.; Hutchison, J. E. *J. Chem. Educ.* **2000**, *77*, 1627–1629.
37. Loupy, A.; Pedoussaut, M.; Sansoulet, J. *J. Org. Chem.* **1986**, *51*, 740–742.
38. Hashimoto, T.; Maruka, K. *Chem. Rev.* **2007**, *107*, 5656–5682.
39. Shirakawa, S.; Yamamoto, K.; Liu, K.; Maruoka, K. *Org. Synth.* **2013**, *90*, 112–120.
40. Pedersen, C. J. *Org. Synth.* **1972**, *52*, 66–74.
41. Pedersen, C. J. *J. Am. Chem. Soc.* **1967**, *89*, 7017–7036.
42. Kim, D. W.; Song, C. E.; Chi, D. Y. *J. Org. Chem.* **2003**, *68*, 4281–4285.
43. Raber, D. J.; Guida, W. C. *J. Org. Chem.* **1976**, *41*, 690–696.
44. Sala, T.; Sargent, M. V. *J. Chem. Soc. Chem. Commun.* **1978**, 253–254 Vincent, J. B.; Chang, H. R.; Folting, K.; Huffman, J. C.; Christou, G.; Hendrickson, D. N. *J. Am. Chem. Soc.* **1987**, *109*, 5703–5711.
45. Zhu, L.; Wehmeyer, R. M.; Rieke, R. D. *J. Org. Chem.* **1991**, *56*, 1445–1453.
46. Rieke, R. D.; Hanson, M. V. *Tetrahedron* **1997**, *53*, 1925–1956.
47. Rieke, R. D. *Acc. Chem. Res.* **1977**, *10*, 301–306.

48. Rieke, R. D.; Bales, S. E.; Hudnall, P. M.; Burns, T. P.; Poindexter, G. S. *Org. Synth.* **1979**, *59*, 85–94.
49. Salaun, J.; Marguerite, J. *Org. Synth.* **1985**, *63*, 147–153.
50. Plutschack, M. B.; Pieber, B.; Gilmore, K.; Seeberger, P. H. *Chem. Rev.* **2017**, *117*, 11796–11893.
51. Leduc, A. B.; Jamison, T. F. *Org. Process Res. Dev.* **2012**, *16*, 1082–1089.
52. Lindley, J.; Mason, T. J. *Chem. Soc. Rev.* **1987**, *16*, 275–311.
53. Suslick, K. S. *Science* **1990**, *247*, 1439–1445 Suslick, K. S. *Sci. Am.* **1989**, *260*, 80–86.
54. Boudjouk, P. *J. Chem. Ed.* **1986**, *63*, 427–429.
55. Naruta, Y.; Nishigaichi, Y.; Mayurama, K. *Org. Synth.* **1993**, *71*, 118–124.
56. Han, B.-H.; Boudjouk, P. *J. Org. Chem.* **1982**, *47*, 5030–5032.
57. Suslick, K. S.; Casadonte, D. J. *J. Am. Chem. Soc.* **1987**, *109*, 3459–3461.
58. Lin, H.; Yang, M.; Huang, P.; Cao, W. *Molecules* **2003**, *8*, 608–613.
59. Shukla, G.; Verma, R. K.; Verma, G. K.; Singh, M. S. *Tetrahedron Lett.* **2011**, *52*, 7195–7198.
60. Drewry, D. H.; Coe, D. M.; Poon, S. *Med. Res. Rev.* **1999**, *19*, 97–148.
61. Onitsuka, S.; Jin, Y. Z.; Shaikh, A. C.; Furuno, H.; Inanaga, J. *Molecules* **2012**, *17*, 11469–11483.
62. Toy, P. H.; Janda, K. D. *Acc. Chem. Res.* **2000**, *33*, 546–554.
63. Dickerson, T. J.; Reed, N. N.; Janda, K. D. *Chem. Rev.* **2002**, *102*, 3325–3344.
64. Bergbreiter, D. E. *Chem. Rev.* **2002**, *102*, 3345–3384.
65. Cohen, Z.; Varkony, H.; Keinan, E.; Mazur, Y. *Org. Synth.* **1979**, *59*, 176–182.
66. Rubin, M. B. *Bull. Hist. Chem.* **2008**, *33*, 68–75.
67. Reed, N. N.; Delgado, M.; Hereford, K.; Clapham, B.; Janda, K. D. *Bioorg. Med. Chem. Lett.* **2002**, *12*, 2047–2049.
68. Chen, S.; Janda, K. D. *J. Am. Chem. Soc.* **1997**, *119*, 8724–8725.
69. Miao, W.; Chan, T. H. *Acc. Chem. Res.* **2006**, *39*, 897–908.
70. Handy, S. T.; Okello, M. *J. Org. Chem.* **2005**, *70*, 2874–2877.
71. Hahn, R. C.; Tompkins, J. *J. Org. Chem.* **1988**, *53*, 5783–5785.
72. Manninen, P. R.; Brickner, S. J. *Org. Synth.* **2005**, *81*, 112–120.
73. Watson, H. A. Jr.; O'Neill, B. T. *J. Org. Chem.* **1990**, *55*, 2950–2952.
74. Brands, K. M. J.; Davies, A. J. *Chem. Rev.* **2006**, *106*, 2711–2733.

75. Reider, P. J.; Davis, P.; Hughes, D. L.; Grabowski, E. J. J. *J. Org. Chem.* **1987**, *52*, 955–957.
76. Anastas, P. T.; Warner, J. C. *Green Chemistry: Theory and Practice*; Oxford University Press: Oxford, 1998.
77. Sheldon, R. A. *ACS Sustainable Chem. Eng.* **2018**, *6*, 32–48; Zhang, X.; Li, C.; Fu, C.; Zhang, S. *Ind. Eng. Chem. Res.* **2008**, *47*, 1085–1094, and references therein.
78. Sheldon, R. A. *Pure Appl. Chem.* **2000**, *72*, 1233–1246.
79. Sheldon, R. A. *ChemTech* **1994**, *24*, 38–47.
80. Trost, B. M. *Acc. Chem. Res.* **2002**, *35*, 695–705.
81. Van Aken, K.; Strekowski, L.; Patiny, L. *Beilstein J. Org. Chem.* **2006**, *2*, 3. doi: 10.1186/1860-5397-2-3 (accessed December 20, 2024).
82. Prat, D.; Wells, A.; Hayler, J.; Sneddon, H.; McElroy, C. R.; Abou-Shehada, S.; Dunn, P. J. *Green Chem.* **2016**, *18*, 288–296.
83. American Chemical Society Green Chemistry Institute. *Solvent Selection Tool*. https://www.acs.org/green-chemistry-sustainability/research-innovation/research-tools/solvent-tool.html (accessed December 31, 2024); ACS GCI Pharmaceutical Roundtable. Solvent Selection Guide Version 2.0, Issued March 21, 2011, https://www.acs.org/content/dam/acsorg/greenchemistry/industriainnovation/roundtable/acs-gci-pr-solvent-selection-guide.pdf (accessed December 31, 2024).
84. American Chemical Society Green Chemistry Institute. *Reagent Guides*. https://reagents.acsgcipr.org/reagent-guides/ (accessed December 20, 2024).
85. Forbes, C. P.; Wenteler, G. L.; Wiechers, A. J. *Chem. Soc. Perkin* **1977**, *1*, 2353–2355.
86. Eaton, P. E.; Mueller, R. H.; Carlson, G. R.; Cullison, D. A.; Cooper, G. F.; Chou, T. C.; Krebs, E.-P. *J. Am. Chem. Soc.* **1977**, *99*, 2751–2767.
87. Bal, S. A.; Marfat, A.; Helquist, P. *J. Org. Chem.* **1982**, *47*, 5045–5050.
88. Durland, J. R.; Adkins, H. *J. Am. Chem. Soc.* **1939**, *61*, 429–433.
89. Bennett, J. G.; Bunce, S. C. *J. Org. Chem.* **1960**, *25*, 73–79.
90. Srinivasan, N. S.; Lee, D. G. *J. Org. Chem.* **1979**, *44*, 1574.
91. McQuade, D. T.; Seeberger, P. H. *J. Org. Chem.* **2013**, *78*, 6384–6389.
92. Sedelmeier, J.; Ley, S. V.; Baxendale, I. R.; Baumann, M. *Org. Lett.* **2010**, *12*, 3618–3621.
93. Frank, R. L.; Varland, R. H. *Org. Synth.* **1947**, *27*, 91–93.

94. Alaimo, P. J.; Marshall, A.-L.; Andrews, D. M.; Langenhan, J. M. *Org. Synth.* **2010**, *87*, 192–200.

95. Gomez-Biagi, R. F.; Dicks, A. P. *J. Chem. Educ.* **2015**, *92*, 1938–1942.

96. Strappaveccia, G.; Lanari, D.; Gelman, D.; Pizzo, F.; Rosati, O.; Curini, M.; Vaccaro, L. *Green Chem.* **2013**, *15*, 199–204.

97. Ferlin, F.; Valentini, F.; Brufani, G.; Lanari, D. *ACS Sustainable Chem. Eng.* **2021**, *9*, 5740–5749.

98. Mukherjee, S.; Pramanik, A. *ACS Sustainable Chem. Eng.* **2020**, *8*, 403–414.

10

SURVEY OF ORGANIC SPECTROSCOPY

Most organic compounds are white solids or colorless liquids of rather similar appearance, and unknown samples present a puzzle for identification. Physical measurements such as melting point, boiling point, and refractive index are useful for matching against lists of values for limited numbers of known compounds. Actual structural information is readily obtained by means of various spectroscopic methods (or ultimately by X-ray crystallography). Here, we will briefly outline several techniques and then focus on the most heavily used technique, nuclear magnetic resonance (NMR) spectroscopy.

10.1 ELECTROMAGNETIC RADIATION

Organic molecules can absorb electromagnetic radiation, resulting in measurable changes that can provide a great deal of structural information [1]. Electromagnetic radiation can be characterized by either frequency, ν, or wavelength, λ, and these are related by the speed of light, c, as shown in

Understanding Organic Chemistry: Bridging Fundamental and Advanced Courses, Fourth Edition. Ann M. Fabirkiewicz and John C. Stowell.

Companion website: www.wiley.com/go/fabirkiewiczorganic4e

TABLE 10.1 Electromagnetic Radiation and Absorption Spectroscopy

Region	Wavelength (nm)	Frequency (Hz)	Energy (kJ/mol)	Molecular Transitions
X-rays	0.1	3×10^{18}	1.2×10^{6}	Ionization
	10	3×10^{16}	1.2×10^{4}	
Ultraviolet (UV)	10	3×10^{16}	1.2×10^{4}	Electronic transitions
	200	$\mathbf{1.5 \times 10^{15}}$	**600**	
	400	$\mathbf{7.4 \times 10^{14}}$	**300**	
Visible (vis)	400	7.4×10^{14}	300	Electronic transitions, colored compounds
	700	4.3×10^{14}	168	
Infrared (IR)	700	4.3×10^{14}	168	Bond stretching and bending
	2500	$\mathbf{1.2 \times 10^{14}}$	**48**	
	15,000	$\mathbf{2 \times 10^{13}}$	**7.8**	
	1,000,000	3×10^{11}	0.12	
Radiofrequency (NMR)	1,000,000	3×10^{11}	0.12	Nuclear spin flipping
	1 m	$\mathbf{3 \times 10^{8}}$	$\mathbf{1.2 \times 10^{-4}}$	
	5 m	$\mathbf{6 \times 10^{7}}$	$\mathbf{2.4 \times 10^{-5}}$	
	100 km	3×10^{3}	1.2×10^{-9}	

Numbers in boldface represent typical instrument ranges.

Equation 10.1. Energy is related to frequency and wavelength by Plank's constant, h, as shown in Equation 10.2. These relationships are summarized in Table 10.1.

$$\lambda = \frac{c}{\nu}, \quad c = 3 \times 10^{8}\ \text{m/s} \tag{10.1}$$

$$\Delta E = h\nu = \frac{hc}{\lambda}, \quad h = 6.626 \times 10^{-34}\ \text{J} \cdot \text{s} \tag{10.2}$$

The absorption of particular wavelengths (or frequencies) contributes specific energy to the molecule and the resulting changes can be characterized and used to identify functionality in the molecule. Correlation charts that list this information are included within each section of this chapter.

10.2 ULTRAVIOLET SPECTROSCOPY

Ultraviolet–visible (UV–vis) spectroscopy gives information on the extent, shape, and substituents of π-conjugation in molecules [2–4]. It is a measure of the energy gaps between the electronic ground and excited states. Intensity of the peaks is most often used to quantitate changes in concentration, and this technique is used to track the progress of a reaction, rather than to identify structural features in molecules.

10.2.1 Origin of the Signals

Ultraviolet–visible spectroscopy records absorbances in the range from 200 to 700 nm, with the visible range starting at about 400 nm. Absorption of light at these wavelengths excites an electron from the highest occupied molecular orbital (HOMO) to the lowest unoccupied molecular orbital (LUMO).

As an example, ethene absorbs light at 165 nm, just outside the measurable UV range. Using Equation 10.2, we see that this absorption corresponds to an energy of 725 kJ/mol (Eq. 10.3) and occurs as a result of an electron from a π orbital transferring to a π* orbital, called a π–π* transition (Fig. 10.1). Conjugation of the π system is necessary to lower the energy of the electronic transitions sufficiently to be recorded by instruments in common use. Butadiene, for example, shows an absorbance band at λ_{max} 217, ε_{max} 21,000.

$$\Delta E = \frac{(6.626 \times 10^{-34}\ \mathrm{J \cdot s})\ (3 \times 10^{8}\ \mathrm{m/s})\ (6.02 \times 10^{23}/\mathrm{mol})}{1.65 \times 10^{-7}\ \mathrm{m}}$$
$$= 7.25 \times 10^{5}\ \mathrm{J/mol} \qquad (10.3)$$

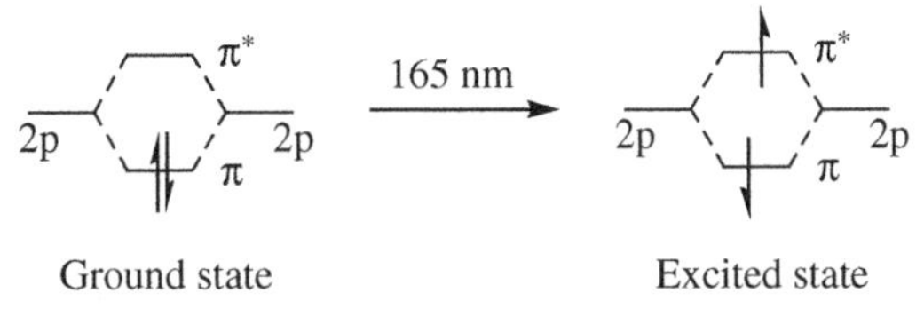

FIGURE 10.1 The π–π* electronic transition for ethylene.

10.2.2 Interpretation

A UV–vis spectrum plots intensity versus wavelength, recording absorbances directly. The wavelengths at which absorbances occur provide information about the energy of the electronic transitions. The intensity of the peak is proportional to the amount of light absorbed and is an intrinsic property of the sample. Thus, intense peaks indicate electronic transitions that occur more readily. According to the Beer–Lambert law, absorbance and concentration (in moles per liter) are related by pathlength, b, in cm, and molar absorption coefficient [5], ε, in units of cm^2/mol (Eq. 10.4). In general, allowed processes have ε values greater than 1000 and forbidden transitions have ε values less than 100.

$$A = \varepsilon bc \tag{10.4}$$

A correlation chart listing representative molecules and the transitions and typical wavelengths of maximum absorbance is provided in Table 10.2. UV–vis spectra are subject to significant solvent effects, and thus the ranges given are approximate. More extensive correlation charts are available [6].

UV–vis spectra are usually obtained from a dilute solution of the sample and thus require solvents that do not absorb within the measured wavelength range. Common solvents that are transparent in the UV range include water, acetonitrile, and hexane. Methanol and ethanol are transparent above 205 nm, and chloroform is transparent above 245 nm. Quartz sample cuvettes are transparent above 200 nm, and these are normally used. Glass is transparent over 300 nm, which limits its utility.

10.2.3 Visible Spectroscopy

Compounds with extended π conjugation have lower energies for the transitions discussed in the previous sections. These compounds are colored and have absorbances in the visible range of the spectrum. Table 10.3 correlates color with the approximate absorbed wavelength. RGB values correlated to wavelength are available [7].

TABLE 10.2 Ultraviolet Absorbances for Representative Molecules

Molecule	Transition	λ_{max} (nm)	ε_{max}
$H_2C{=}CH_2$	π–π*	165	15,000
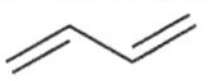	π–π*	217	21,000
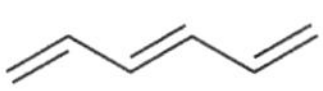	π–π*	253 263 274	~50,000 52,500 ~50,000
	π–π*	256	8000
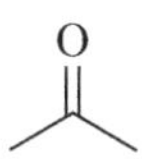	π–π* *n*-π*	188 279	900 15
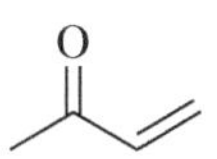	π–π* *n*-π*	212 320	7080 21
	π–π*	180 200 255	60,000 8000 215
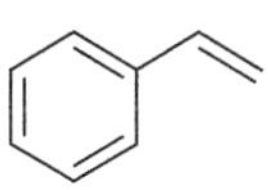	π–π*	244 282	12,000 450
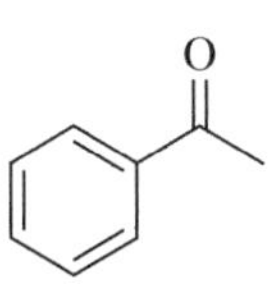	π–π* π–π* *n*-π*	240 278 319	13,000 1110 50
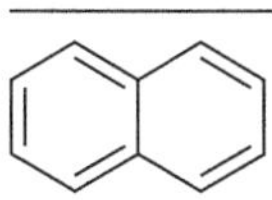	π–π*	221 286 312	133,000 9300 289

Adapted from Silverstein et al. [3].

TABLE 10.3 Visible Absorption Correlations

Wavelength Absorbed (nm)	400–450	450–490	490–570	590–610	620–630	640–690
Color absorbed	Violet	Blue	Green	Yellow	Orange	Red
Color observed	Yellow	Orange	Red	Violet	Blue	Green

10.3 INFRARED SPECTROSCOPY

Infrared (IR) spectroscopy is useful for determining the presence and identity of functional groups. These spectra measure the frequency of bending and stretching of bonds where the bond dipole changes with the movement. The stretching vibrations of double and triple bonds in alkenes and alkynes involve small bond dipole changes and give weak or no IR absorptions. For these cases, laser Raman spectroscopy gives strong, informative signals [8, 9].

10.3.1 Origin of the Signals

The IR region of the electromagnetic spectrum covers the wavelength range from 700 nm to 1 m. Within this range, wavelengths from 2.5 to 15 μm, considered the mid to long wavelength range, are absorbed by typical organic molecules.

Absorption of electromagnetic radiation in this range of wavelengths causes the stretching and bending of chemical bonds. If you can imagine a bond as a spring, the various vibrational modes can be defined as shown in Figure 10.2, using the methylene unit as an example.

10.3.2 Interpretation

The IR spectrum is a plot of percent transmittance versus wavenumber. Wavenumber, given in units of reciprocal centimeters, or cm^{-1}, is the reciprocal of wavelength. Plotting spectra in these units has the benefit of expanding the region of the spectrum most useful for analysis, from 4000 to about 1400 cm^{-1} (2500–7140 nm). The remaining portion of the spectrum, called the fingerprint region, represents a pattern unique to each molecule, and thus can be used for absolute identification. Spectral matching software is especially useful in this role.

Each chemical bond stretches or bends at a specific wavelength, and thus it is possible to use IR spectroscopy to readily identify the functional

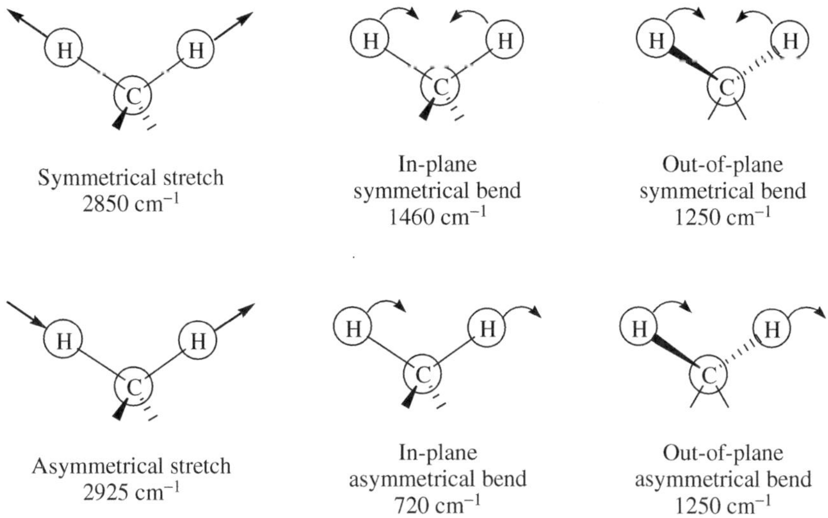

FIGURE 10.2 Stretching and bending modes for $-CH_2-$.

TABLE 10.4 Infrared Correlation Chart for Common Organic Functional Groups

Wavenumber (cm^{-1})	Bond/Motion	Functional Group
3400–3300	–OH stretch	Alcohols
3400–3300	–NH stretch	Amines
3300	≡CH stretch	Alkynes
3100–3000	=CH stretch	Alkenes, aromatics
3000–2800	–CH stretch	Aliphatics
2850–2750	–CH stretch	Aldehydes
2250	C≡N stretch	Nitriles
2150	C≡C stretch	Alkynes
1800–1600	C=O stretch	Acids and derivatives, aldehydes, ketones
1650–1500	C=C stretch	Alkenes, aromatics
1600–1500	–NH bend	Amines, amides
1500 and 1300	$-NO_2$ stretch	Symmetric and asymmetric
1200–1150	(O=)C–O stretch	Esters
1050–1000	C–O stretch	Alcohols, esters

groups within a molecule. The abbreviated correlation chart given in Table 10.4 lists ranges for the major organic functional groups. Expanded correlation charts are found on the Internet and in many organic texts.

10.4 MASS SPECTROMETRY

Spectroscopy involves the absorption and transmission of electromagnetic radiation. In mass spectrometry, a molecule is ionized with a high-energy electron beam. The resulting ionic fragments are sorted by mass and their abundance and masses are measured. Mass spectrometry is useful for determining molecular weights, what elements are present, and, by analysis of fragmentation patterns, some structural information [10].

10.4.1 Origin of the Signals

Mass spectra are generated by passing a vaporized sample through a high-energy electron beam, which bumps a single electron from a molecule. The radical cation formed decomposes as the molecule rearranges to accommodate the loss of the electron. These fragments are passed through a mass analyzer to detect fragments of increasing mass-to-charge ratio.

While it is possible to remove any electron from the molecule in the ionization process, those most likely to be lost are those most loosely held, and thus the electron is more often lost from a non-bonded pair or a π bond. When only one electron is lost, and no decomposition takes place, the mass-to-charge ratio for that species (the molecular ion) corresponds to the molecular weight of the molecule. When only σ electrons are present in the molecule, bond cleavage occurs with ionization and it can be difficult to determine the molecular weight of these molecules. Mass spectrometers of sufficient resolution to measure four to five places beyond the decimal can provide the molecular formula.

Fragmentation of the initially formed radical cation occurs to increase its stability. Rearrangements occur and result in the loss of a small stable molecule, such as carbon dioxide or water, when possible. Bond cleavages often occur at a branch point in the molecule to result in the loss of a radical and detection of the most substituted carbocation. Only charged particles are detected. A likely decomposition pathway is shown for ethylbenzene (Eq. 10.5). Significant peaks are observed in the mass spectrum at *m/z* of 106, 91, and 65.

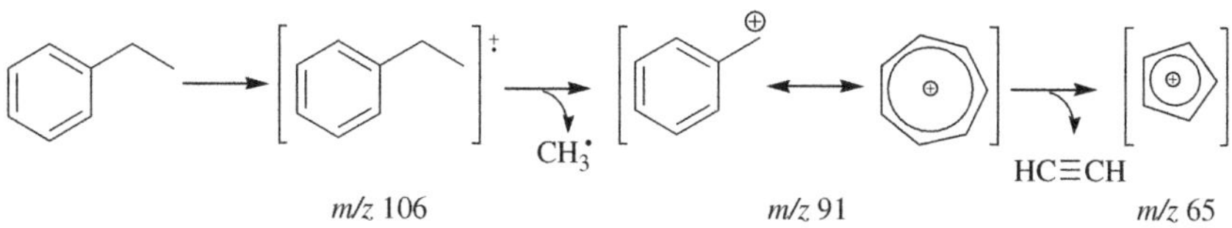

(10.5)

Rearrangements often result in bond cleavage at benzylic, allylic, and branch points, and are facilitated by carbonyl and hydroxyl groups. A decomposition pathway for 2-methyl-3-hexanone is shown as an example in Eq. 10.6. Significant peaks in the mass spectrum are present at *m/z* of 114, 71, and 43.

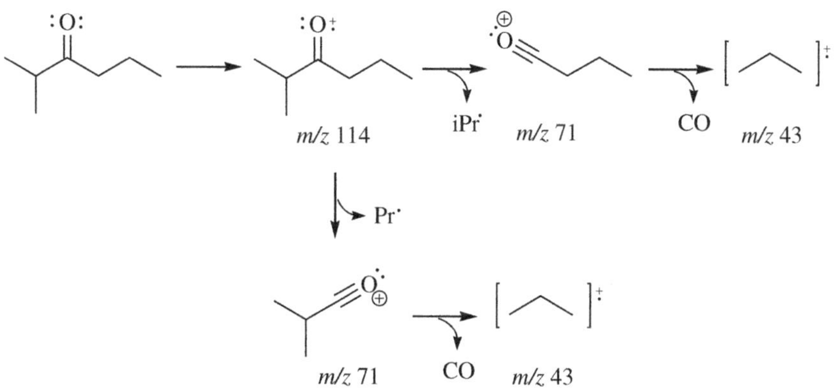

(10.6)

10.4.2 Interpretation

The mass spectrum is a plot of intensity versus mass-to-charge ratio, or *m/z*. Usually, a single electron is lost in the ionization process, so the charge is +1, and the peaks correspond to the mass of the fragments. Intensity corresponds with abundance; thus, the more intense peaks represent those fragments that are present in higher concentration, from decomposition pathways that are more likely.

Table 10.5 lists common fragments detected, and Table 10.6 lists common fragments lost. More extensive tables have been published [8, 10, 11].

TABLE 10.5 Commonly Detected Ions

m/z	Possible Fragment	*m/z*	Possible Fragment
16	O	43	$CH_3C{\equiv}O$ (+14 for homologs)
18	H_2O (contamination, alcohols)	45	CO_2H
26	$C{\equiv}N$	65	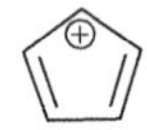
29	C_2H_5 (+14 for homologs), CHO	77	
32	O_2 (air)	92	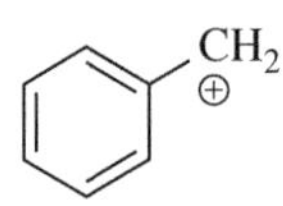

TABLE 10.6 Common Fragments or Molecules Lost

m/z	Possible Fragment	*m/z*	Possible Fragment
15	CH_3	32	CH_3OH
18	H_2O	35, 37	^{35}Cl, ^{37}Cl
19	F	43	$CH_3C{\equiv}O$
28	$C{\equiv}O$	47	CH_3S
29	C_2H_5 (+14 for homologs)	79, 81	^{79}Br, ^{81}Br
31	CH_3O	127	I

Mass spectrometry is particularly useful in the detection of isotopes. Due to the relatively high percentages of the heavy isotopes, the halogens chlorine and bromine are readily identified by the distinctive pattern of peaks in the molecular ion region of their mass spectra (Fig. 10.3).

10.5 NMR SPECTROSCOPY

NMR spectroscopy gives the connections that make up a chemical structure by showing neighboring relationships. Elemental nuclei that have an odd mass number and/or an odd atomic number have a magnetic moment, and many of these can be observed in an NMR spectrometer. Those nuclei that have the most practical value for organic chemists are ^1H, ^{13}C, ^{19}F, and ^{31}P.

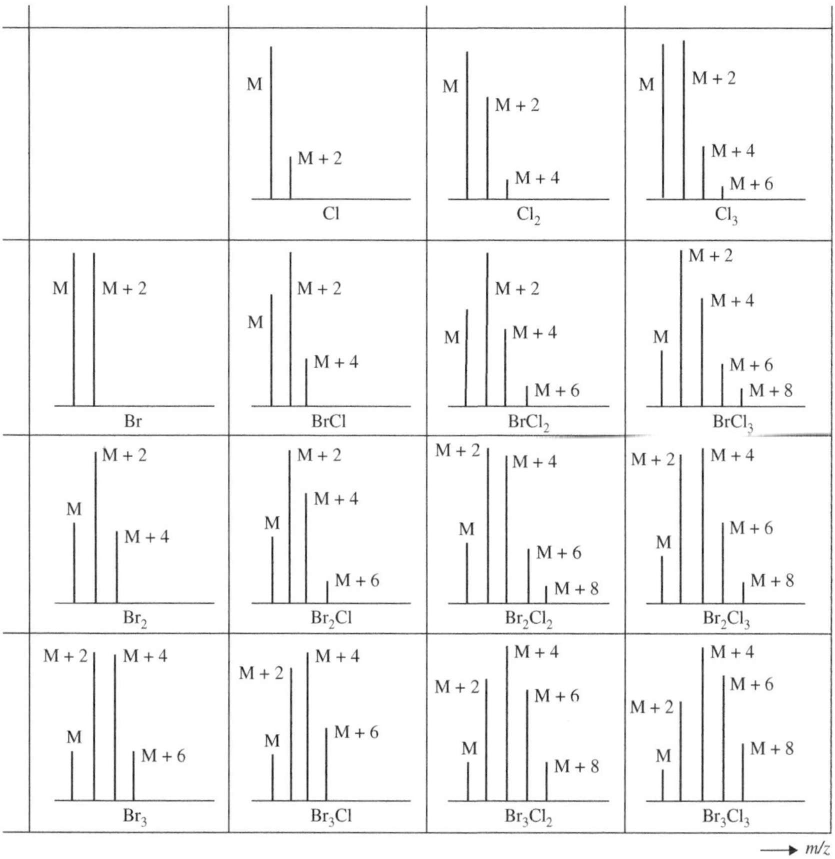

FIGURE 10.3 Peaks in the molecular ion region for compounds containing chlorine and bromine. Silverstein and Webster [12]/John Wiley & Sons.

10.5.1 Origin of the Signals

Modern NMR spectrometers have features that give high-resolution spectra with high sensitivity in a matter of seconds. Instruments in common use have magnetic field strengths ranging from 1.4 T in older continuous-wave instruments to 14 T for instruments with superconducting magnets. They are referred to by the resonance frequency of the ^1H nucleus, thus a 7 T magnet is referred to as a 300-MHz instrument.

Protons are spinning charged particles, and as such have a magnetic dipole (Fig. 10.4a). In the presence of an external magnetic field, B_0, these

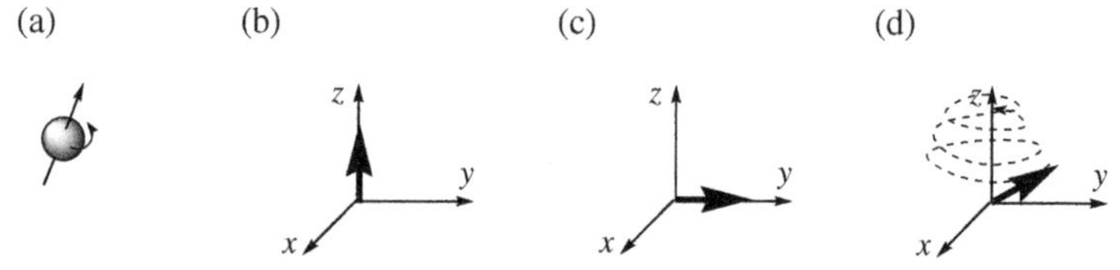

FIGURE 10.4 The effect of a 90° pulse on the net magnetization vector of a proton. B_0 lies along the positive z-axis. (a) A proton as a spinning particle, (b) the net magnetization vector, (c) a 90° pulse, and (d) relaxation.

magnetic dipoles will either align or oppose B_0. The net magnetization vector will be aligned with B_0 (along the z-axis) because there is a slight excess population in the lower energy state (Fig. 10.4b). While the excess population in the lower energy state is small, it is sufficient for NMR spectroscopy. The population distribution is given by the Boltzmann distribution and varies directly with B_0 and indirectly with temperature; thus, it is possible to increase the net magnetization vector, increasing the strength of the signal, by increasing B_0 or by decreasing temperature.

Protons spin with a characteristic frequency called the Larmor frequency and this too is related to B_0. At a field of 7.0462 T, ^{1}H nuclei precess at 300.00 MHz and ^{13}C nuclei precess at 75.430 MHz. Instruments of higher fields provide higher resolution of the frequencies for individual nuclei, as will be discussed further in Section 10.5.2.3.

An intense pulse of radio frequency energy (300 MHz for a ^{1}H spectrum) is applied from an oscillator to the sample, lasting typically 14 μs, and of sufficient power to cause 90° of precession around the x-axis toward the x, y plane. This moves the net magnetization vector from alignment along the z-axis to a position in the x, y plane (Fig. 10.4c). The net magnetization goes around the z-axis at 300 MHz inducing a 300-MHz alternating current in the coil on the x-axis, which is connected to the receiver. This current dies out over a few seconds as the nuclei gradually fan out around the z-axis and the net magnetization also equilibrates back out of the x, y plane toward the z-axis, a process called *relaxation*. With relaxation, the excess in the lower energy state is restored, and the precessional motion becomes incoherent (Fig. 10.4d).

Data collection occurs once the pulse ceases, while the nuclei return to their initial state. Multiple pulses allow data to be compiled, thus smaller samples or less sensitive nuclei, like ^{13}C, can be used. Initially, data is collected over time, and thus a decaying sine wave or free induction decay (FID) is obtained (Fig. 10.5a). The FID contains the overlapping resonance frequencies of each different proton in the sample, and Fourier transform (FT) allows for the extraction of these frequencies, resulting in the frequency-domain spectrum we expect (Fig. 10.5b).

The spectrum is a plot of intensity versus frequency and is conventionally plotted, with precessional frequency decreasing toward the right along the *x*-axis. The scale is in units of parts per million (ppm) difference from a standard, usually tetramethylsilane (TMS) with the signals to the left of TMS (less shielded nuclei) given positive numbers and those to the right of TMS (more shielded) given negative. The frequency differences from TMS are converted to ppm by multiplying by 1 ppm/300 Hz in the case of a 300 MHz spectrum. A well-tuned spectrometer can resolve proton signals differing by less than 0.3 Hz in 300 MHz.

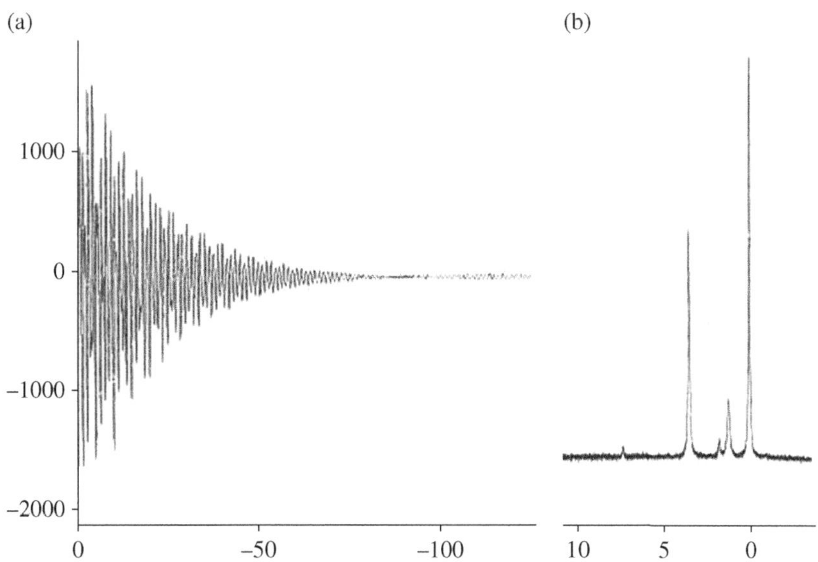

FIGURE 10.5 FID and NMR spectrum for methanol. (a) FID for methanol and (b) 60 MHz ^{1}H NMR spectrum for methanol. The signal at 0 is TMS.

This is a highly simplified description of the whole process. For more thorough descriptions, see the references [8, 13–15].

10.5.2 Interpretation of Proton NMR Spectra

Figure 10.6 illustrates the 400-MHz ^1H spectrum of jasmone. All the signals may be assigned to particular hydrogens in the structure using the chemical-shift values, the splitting multiplicity, and comparison with spectra of similar compounds.

In the introductory course, you learned to use four basic kinds of information from ^1H NMR spectra: number of signals, chemical-shift values, integrated signal area, and splitting patterns. We will now delve further into some of these, particularly the splitting patterns. A brief correlation chart is provided in Table 10.7, and a list of residual peaks for common solvents is shown in Table 10.8. More extensive listings for both can be found in the references.

10.5.2.1 Coupling Constants and Connectivity When two signals are coupled with each other, there will be an identical coupling constant as part of

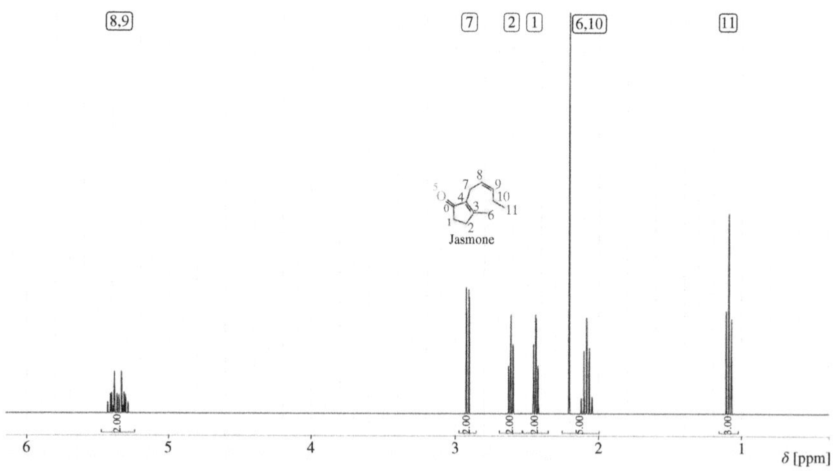

FIGURE 10.6 The 400-MHz ^1H NMR spectrum of jasmone, simulated with NMRium NMR simulator. Adapted from Patiny et al. [16].

TABLE 10.7 ^{1}H Correlation Chart for Common Organic Functional Groups

Peak Position (ppm)	H	Functional Group
10–12	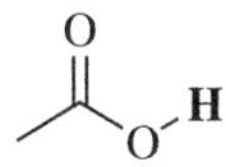	Carboxylic acids
9–11	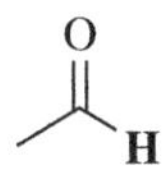	Aldehydes
8–11	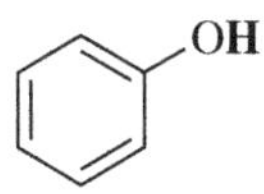	Phenols
6–8	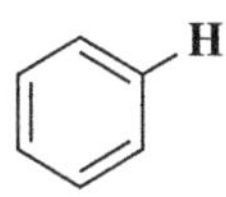	Aromatics
5–7	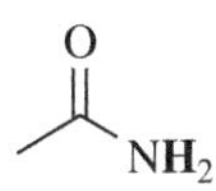	Amides
4–6	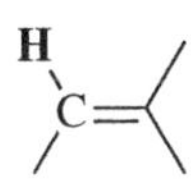	Alkenes
3–4	\\CH-O/	Alcohols, ethers
3–7	\\CH-X/	Alkyl halides RCl > RBr > RI
2–3	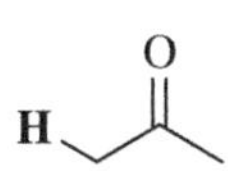	α protons
	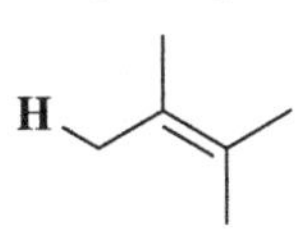	Allylics
	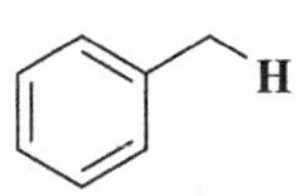	Benzylics
	–C≡C–**H**	Alkynes
0.5–2	–**CH**$_2$–R	Alkanes 3° > 2° > 1°
1–4	R**OH**	Alcohols
1–3	R**NH**$_2$	Amines

TABLE 10.8 Residual H NMR Peaks for Common Solvents

Position (ppm)	Structure	Name
7.26	$CDCl_3$	d-Chloroform
7.16	C_6D_6	d^6-Benzene
4.79	D_2O	Water
3.31	CD_3OD	d^4-Methanol
2.05	CD_3COCD_3	d^6-Acetone

the multiplicity of each signal. Coupling constants can often be measured to two or three significant figures (to about 0.1 Hz), which allows matching values in various multiplets to discover coupled neighbors. An equivalent set of n neighbors will provide $n + 1$ subpeaks in a signal, all equally spaced, and with area distribution following the binomial distribution.

For example, three equivalent neighbors will split a signal into a quartet of peaks of relative area 1:3:3:1. Nonequivalent neighbors will split with different coupling constants and give up to 2^n subpeaks, where n is the number of neighboring protons. For example, three all-different neighbors will give a doublet of double doublets (ddd), which will appear as eight peaks all with the same areas (although these may be difficult to resolve clearly). The distance between these signals is the coupling constant, J.

The size of the magnetic influence a proton receives from a neighboring proton depends on the distance between them, the intervening bonds, and their angular relationship. The coupling constant is a measure of this effect. It is specified in units of hertz (Hz), which do not vary with different applied field strengths. Typical values of coupling constants for various neighboring relationships are given in Table 10.9.

Splitting patterns may be analyzed with a tree diagram [17] as depicted in Figure 10.7 using the simulated spectrum [16] of *N*,*N*-dimethyl-2,4-dibromoaniline. Analysis of the coupling constants allows for unambiguous assignment of each hydrogen in the molecule using the values provided in Table 10.9. Coupling constants are determined by calculating the distance between each peak in a multiplet in ppm and multiplying by the field strength of the instrument in Hz/ppm. Hoye has developed a method that allows for determination of coupling constants

TABLE 10.9 Coupling Constants for Neighboring Hydrogens in Hertz

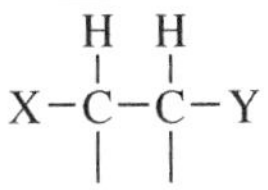	6–8	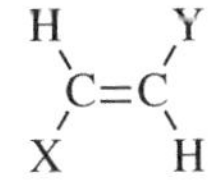	14–18	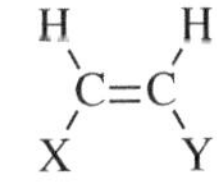	7–12
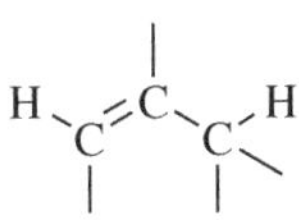	0.6–2	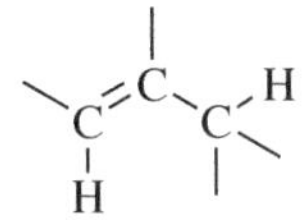	0–1.3	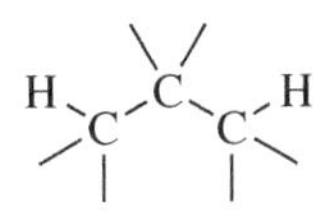	0
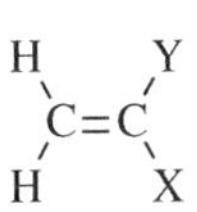	1	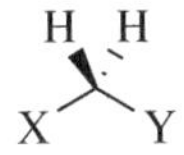	1–3	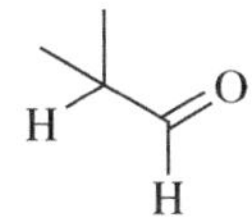	3
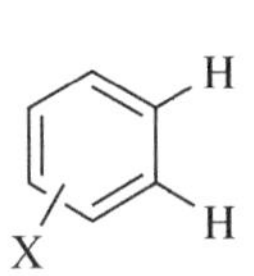	8	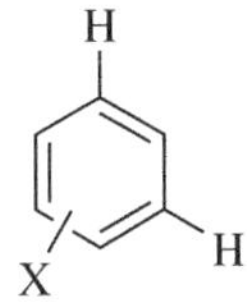	1–3	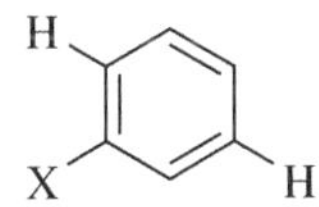	0–1
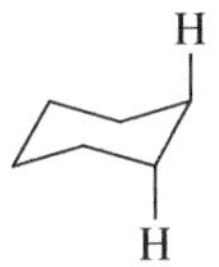	9		3	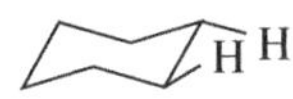	3

for complicated first-order systems [18]. The multiplicities and coupling constants for *N*,*N*-dimethyl-2,4-dibromoaniline are given in Table 10.10.

10.5.2.2 Non-first-Order Signals When we see a simple triplet, we conclude that the hydrogen(s) giving that signal has (have) two equivalent neighboring hydrogens. When we see a doublet of doublets (dd), we conclude that the hydrogen(s) giving that signal has (have) two nonequivalent neighbors. This is called *first-order analysis* and is possible when the chemical-shift difference, $\Delta\nu$, between neighboring hydrogens is much larger than the J value. Often this is not the case. If $\Delta\nu$ is less than 10 times the J value, the signals are distorted from simple first-order expectations. Frequency is proportional to the strength of the magnetic field, so while the chemical shift and J values remain the same regardless of the instrument, $\Delta\nu$ increases with increasing field strength.

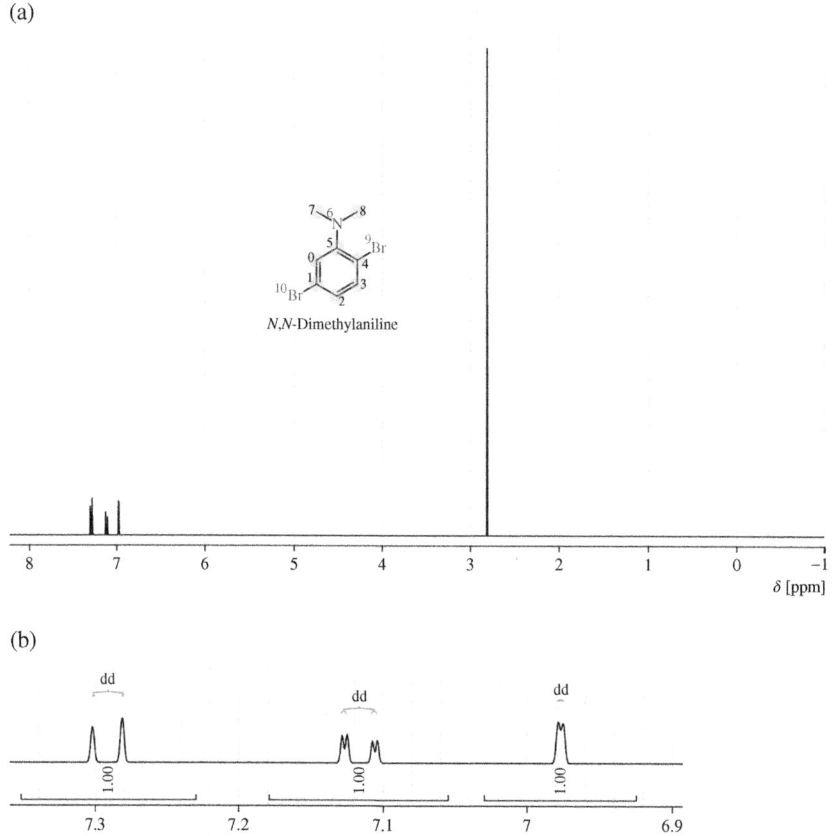

FIGURE 10.7 (a) 400 MHz NMR spectrum of *N,N*-dimethyl-2,4-dibromoaniline, simulated with NMRium NMR simulator [16]. (b) Tree diagram for the aromatic region, expanded to show detail.

TABLE 10.10 Multiplicities in the ^{1}H NMR Spectrum of *N,N*-Dimethyl-2,4-Dibromoaniline

Chemical Shift (ppm)	Integration	Multiplicity	*J* Values (Hz)			Assignment
7.29	1	dd	1.4		0.5	$-H_3$
7.12	1	dd	1.4	8.5		$-H_2$
6.80	1	dd		8.5	0.5	$-H_0$
2.8	6	s				$-H_{7,8}$

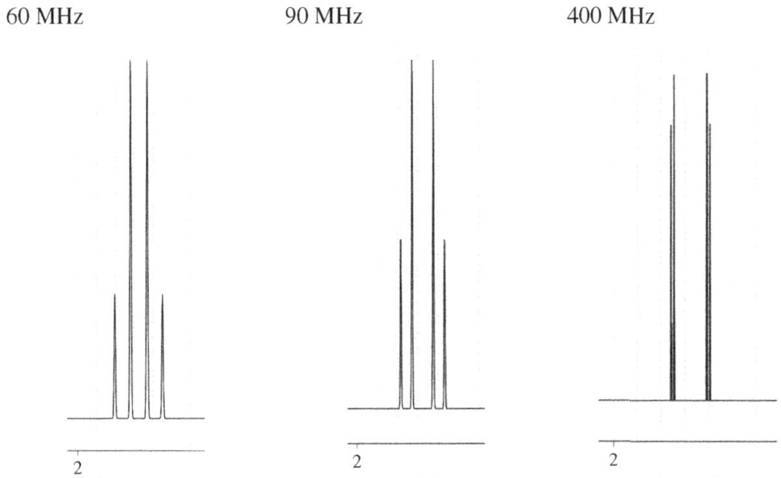

FIGURE 10.8 An AB spin system demonstrating the effect of increasing magnetic field strength, simulated with NMRium NMR simulator. Adapted from Patiny et al. [16].

Figure 10.8 shows a single hydrogen with a single neighbor, termed an AB system, with signals at 1.2 and 1.5 ppm. In the 60-MHz spectrum, the doublets are difficult to recognize as such, but as the magnetic field strength and thus frequency increase, the signals become more obviously a pair of doublets. One or both of these signals may be resplit by other neighbors, but the distorted doublet aspect of it should be discernable, with a corresponding J value, even if it is a dd, dt, or dq. Figure 10.9 shows simulated AB, A_2B, and ABX systems.

No attempt is made here to give the theoretical reasoning or predictive methods for these non-first-order patterns [18–20], but you should be able to *recognize* them as you analyze spectra and know what structural meaning they have.

10.5.2.3 Spectra at Higher Magnetic Fields Equation 10.7 shows the direct proportionality between the magnetic field strength and the oscillator frequency for proton resonance, where ν is the frequency in Hz and B_0 is the magnetic field strength in tesla (T). In the early days, ^{1}H NMR spectra

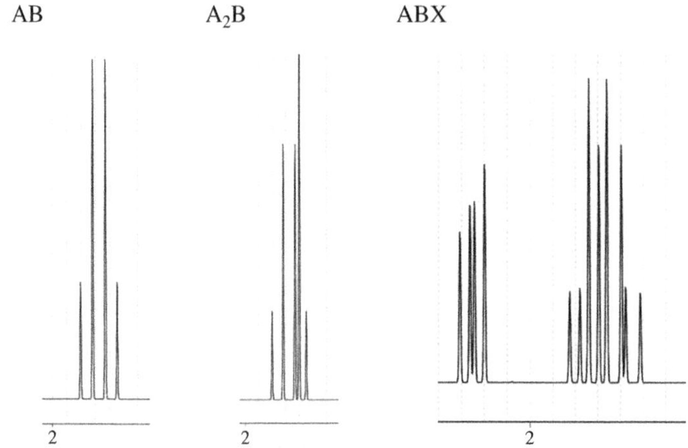

FIGURE 10.9 AB, A_2B, and ABX spin systems, simulated with NMRium NMR simulator. Adapted from Patiny et al. [16].

were usually run at 60 MHz in a magnetic field of 1.4092 T, whereas now 300 MHz or higher is common. What is gained at higher magnetic fields? Doubling the magnetic field of the instrument doubles the frequency difference between the signals and the spectrum is more spread out. If we simply stretch a 60-MHz spectrum out horizontally, we will not see more peaks or eliminate overlaps; however, in high-field spectra, the overlaps are lessened. The width of a multiplet is determined by the magnetic field effects of nearby nuclei, the size of which is fixed and independent of the instrument's magnetic field. Therefore, as the spectrum spreads out with higher magnetic fields, the individual multiplets do not widen but come away from each other, diminishing their overlaps and facilitating interpretation.

$$\nu = 4.2576 \times 10^7 B_0 \quad (10.7)$$

As the magnetic field strength is increased, $\Delta\nu$ increases proportionally. Chemical-shift differences have the same ppm values regardless of the applied magnetic field, and the J values remain the same in hertz as we go from one instrument to another. Therefore, on the ppm scale, the separation of peaks within a multiplet shrinks to half as we go to an instrument with twice the magnetic field strength. Thus, with increasing magnetic field, the multiplets appear to narrow but stay centered on the same ppm values (Fig. 10.8). The signals become more nearly first order, and interpretation is simplified. An example of this effect is shown in Figure 10.10. Spectacular detail has been resolved in spectra of complex molecules at high fields.

10.5.2.4 Spin Decoupling In some spectra, it may not be possible to distinguish coupling constants, or several J values may be essentially the same, and conclusions about neighboring relationships may not be possible. *Double-resonance* techniques will allow for differentiation of these signals. In spin decoupling, a second oscillator is set to continually irradiate at the frequency of one of the coupled protons. The rapid spin inversions that result for those nuclei will cancel out their magnetic influence on neighboring protons. When the spectrum is obtained, the splitting of neighbors normally caused by the nuclei that are under continuous irradiation will disappear, thus identifying those neighbors. This is a through-bond technique and allows for determination of atoms that are chemically bonded.

10.5.2.5 The Nuclear Overhauser Effect Through-space enhancement is also possible using a double-resonance technique. Neighboring relationships may be traced by relaxation transfer, which is the transfer of polarization to other nearby nuclei. If an oscillator is set at one spectral signal to disturb the equilibrium populations, the relaxation of these populations involves simultaneous inversions of other nuclei nearby. This results in more nuclei in the lower energy state and thus stronger signals from those neighbors during the irradiation of the first signal. This enhancement, called the *nuclear Overhauser effect* (NOE), is a through-space effect determined by the distance to the neighbor, and not by intervening bonds. It falls off steeply with distance, proportional to $1/r^6$. In practice, a spectrum is collected, and then a second spectrum is collected during irradiation of a specific signal. The first spectrum is subtracted from the second to reduce all signals to zero except those that were enhanced by the irradiation. This is called an *NOE difference spectrum*. These enhanced signals indicate proximity to the hydrogen whose signal was irradiated.

The stereochemistry at the carbon carrying the OH group in compound **10.1** was determined with an NOE difference spectrum [21]. Irradiation

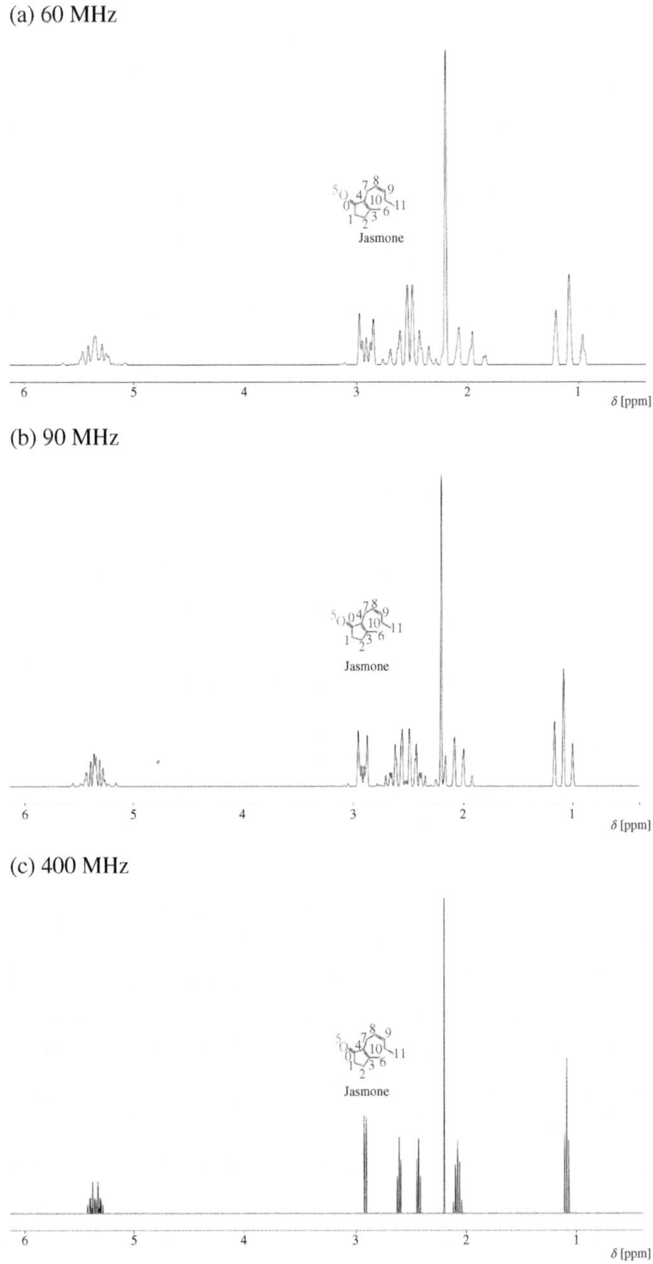

FIGURE 10.10 NMR spectra for jasmone at (a) 60 MHz, (b) 90 MHz, and (c) 400 MHz, simulated with NMRium NMR simulator. Adapted from Patiny et al. [16].

at the signal for the hydrogen on the alcohol carbon gave enhancement to a hydrogen on the one-carbon bridge and the nearest hydrogen on the aromatic ring. Thus, the OH must be nearer the two-carbon bridge and the compound is the endo isomer.

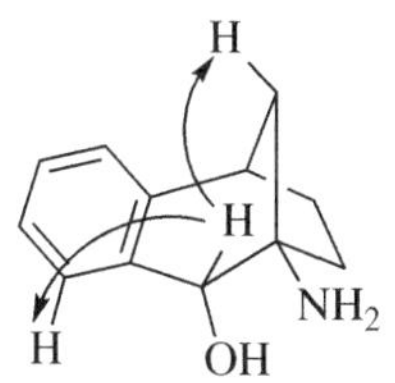

10.1

10.5.2.6 Stereochemical Effects Diastereomers are chemically different materials with different physical properties, and they give different spectra. The differences may be small but are often sufficient for measuring the percentage of each isomer in mixtures by integration. Diastereotopic groups reside in diastereomeric environments in molecules (Section 3.8) and thus give separate signals, and each can split the signal of the other if they are close together. When a signal appears more complex than it ought to on simple considerations and is also nearly symmetric, consider diastereotopic groups. If they are geminal hydrogens, they should split each other with considerable distortion.

Enantiomeric compounds give identical spectra in ordinary solvents, but diastereomeric complexes may form in the presence of a single enantiomer of a chiral complexing agent. Therefore, enantiomers may give separate measurable signals as discussed in Section 3.3. Enantiotopic groups or atoms in a molecule will also give separate signals in chiral media. For example, a chiral praseodymium reagent (Compound **3.14** with Pr in place of Eu) added to a solution of benzyl alcohol in CCl_4 gave two doublets, one for each enantiotopic hydrogen ($J = 13.0\,\text{Hz}$, $\Delta\nu = 0.13\,\text{ppm}$) [22]. Since it is a shift reagent, it also moved the signals at 6.63 ppm to the right. It gave two singlets for the enantiotopic methyl groups in dimethylsulfoxide and also distinguished the methyls in 2-propanol.

In the spectrum of jasmone (Fig. 10.6), note the complexity of the alkene region from 5.3 to 5.5 ppm. The result is the complex pattern

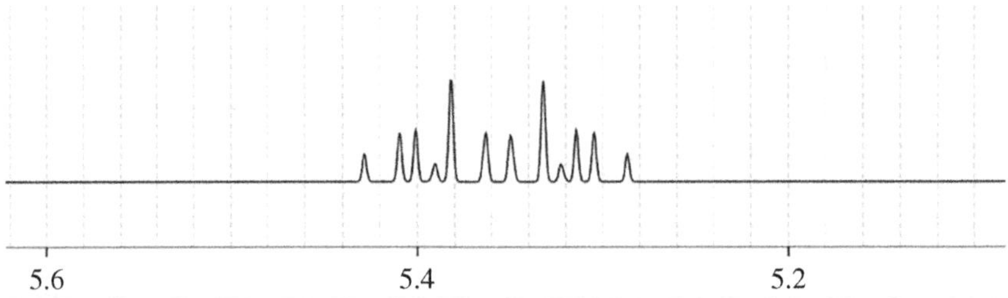

FIGURE 10.11 NMR spectrum of jasmone, expanded to show detail, simulated with NMRium NMR simulator. Adapted from Patiny et al. [16].

of lines seen expanded in the simulated spectrum in Figure 10.11. Assignment of individual hydrogens may be difficult from a simple proton NMR spectrum but may be determined using two-dimensional techniques (Section 10.5.2.7).

10.5.2.7 Two-Dimensional Proton Correlation Spectroscopy A two-dimensional coupling correlation ^{1}H NMR spectrum shows all the proton coupling relationships in a molecule in a single, although longer, experiment. This is more efficient than spin decoupling, which probes one relationship at a time and may require many experiments.

Two-dimensional correlations are brought out with multipulse irradiations [15, 23–25]. For a COrrelation SpectroscopY (COSY) experiment, a 90° pulse is followed by a period of precession (t_1) and then a second (mixing) 90° pulse transfers magnetization including frequencies between neighbors among the precessing sets of nuclei. This is followed by the acquisition of the FID. The sequence is repeated many times with incremental increases in t_1 to give a stack of spectra after Fourier transformation. Precessional differences during t_1 cause a cyclic rise and fall of signals in the series of spectra. A set of Fourier transformations through the stack of spectra at increments in the spectral frequency $F1$ produces a second frequency scale $F2$. A two-dimensional plot of $F1$ against $F2$ gives signals that are represented by contour lines indicating the intensity of the signal. On these plots, there are signals that have both frequencies of those pairs of protons that are coupled. Although this process is lengthy and complicated,

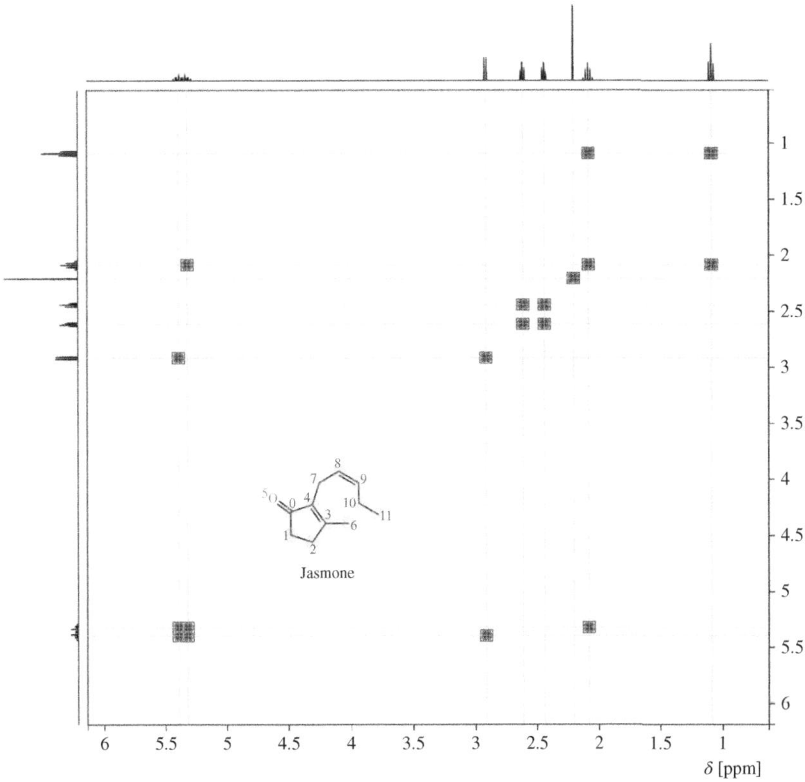

FIGURE 10.12 The 400-MHz two-dimensional $^1H-^1H$ shift correlation spectrum (COSY) of jasmone, simulated with NMRium NMR simulator. Adapted from Patiny et al. [16].

COSY spectra are readily available from automated spectrometers, and the simplicity of interpretation makes them very useful. How these plots are obtained will not be covered here, but interpretation is presented.

In the COSY spectrum of jasmone (Fig. 10.12), all the neighboring relationships may be traced by locating the two chemical-shift values correlated by the off-diagonal signals. The diagonal in this figure runs from the upper right to the lower left. The signals on the diagonal are not useful because they correlate a signal in $F1$ with itself in $F2$; however, those off the diagonal (duplicated on each side of the diagonal) give the neighboring relationships. The peaks at 5.4 and 5.38 ppm are readily assigned to the alkene at C8 and C9. The off-diagonal element identifies the peak at 2.9 ppm as the hydrogens at C7 and the peak at 2.1 ppm as the hydrogens at C10. The off-diagonal element at (1.09, 2.08) correlates the hydrogens on C10 and C11. In this way, peak assignments are made unambiguously for the entire molecule. COSY spectra are very useful for interpreting structures for complex molecules as illustrated in many recent journal articles [26].

Long-range COSY spectra may be obtained by allowing a delay for further precession before the mixing pulse and before the acquisition each time. This allows coupled pairs correlated by small values of J to give appreciable off-diagonal signals. In this way, connectivity can be traced over three or four bonds. For good examples, see the work of Cho and Harvey [27].

For complex molecules that give unresolved ^{1}H spectra, cross-sections of COSY spectra may be taken, giving a set of one-dimensional spectra. In these, signals that overlap in the conventional ^{1}H spectrum are often separated into different cross-sections where they can be seen clearly for their multiplet structure. An example is given in the work of Portlock et al. [28].

10.6 CARBON NMR SPECTRA

The abundant ^{12}C isotope has no nuclear magnetic moment, but the ^{13}C isotope, with 1.1% natural abundance, does [8, 29]. The signal from these carbons, however, is only $\frac{1}{64}$th the intensity from the same number of hydrogens. These two factors cause a carbon spectrum to be about $\frac{1}{6000}$th the intensity of a hydrogen spectrum of the same sample. This necessitates the use of pulsed irradiation and FT methods and the summation of many repeated spectra (Section 10.5.1). Typically, the spectra are obtained at 2–10 s intervals and, for a routine 5–20-mg sample, less than 15 minutes is required to accumulate a good summation.

10.6.1 General Characteristics

10.6.1.1 Chemical-Shift Range The signals for carbon occur 2.98×10^6 ppm to the right of those for protons. At 7.0462 T, the frequency for carbon resonance is 75.430 MHz, and for hydrogen, it is 300.00 MHz. The chemical-shift values are generally recorded as ppm from the carbons of TMS and extend over a range of about 230 ppm. Since this is more than 10 times the range found for hydrogens, obscuring overlaps are much less of a problem. The sp^3 hybrid carbons bonded only to hydrogens and other carbons generally occur in the 6–50 ppm range. Those carrying electronegative atoms give signals 10–60 ppm greater. The sp^2 hybrid carbons of alkenes and aromatic rings occur in the 100–165 ppm range and those in carbonyl groups are at 155–220 ppm. The sp hybrid carbons in alkynes are at 65–85 ppm. It is possible to calculate carbon NMR chemical shifts with reasonable accuracy [30].

10.6.1.2 Number of Signals Each structurally different carbon in a molecule gives a separate signal. Identical carbons will, of course, coincide. A *tert*-butyl group gives two signals, one for the three equivalent CH_3 carbons and one for the central carbon. *p*-Nitroanisole gives five signals, two of which represent two carbons each.

10.6.1.3 Splitting of Signals The low natural abundance of ^{13}C makes it highly improbable that two would be side by side in a molecule; therefore, $^{13}C-^{13}C$ splitting is not observed in routine spectra. Protons split the signals of the carbon to which they are bonded by 100–240 Hz. Protons on an adjacent or the next farther carbon give only 4–6-Hz splitting. This coupling greatly complicates ^{13}C spectra as these multiplets are likely to overlap and the large coupling constants make it difficult to know which peaks belong together as multiplets. Routinely, broad irradiation (spin decoupling) of all the protons is used to reduce the ^{13}C spectrum to all singlets, as depicted in Figure 10.13. This provides an advantage in increased sensitivity as well because a signal that is concentrated entirely in a sharp singlet stands taller, giving a far greater signal/noise ratio.

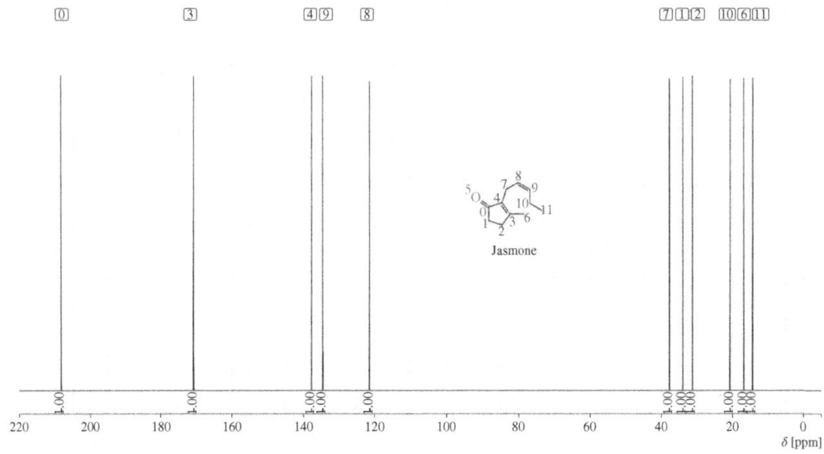

FIGURE 10.13 400 MHz ^{13}C NMR spectrum of jasmone, simulated with NMRium NMR simulator. Adapted from Patiny et al. [16].

10.6.1.4 Area of Signals During the accumulation of spectra at short intervals, the excited ^{13}C nuclei relax toward equilibrium distribution after each pulse. Each carbon in a structure does this at a different rate and those with no hydrogen atoms bonded to them are particularly slow. Those less completely relaxed give less signal in the succeeding pulse, which leads to accumulated signals of varying intensities where those from carbons bearing no hydrogen are weak. Thus, the signal areas are not quite proportional to the number of carbons they represent in the structure. If desired, long intervals between pulses may be used to obtain signal areas nearly proportional to the ratio of carbons represented, but this may require excessive instrument use time.

The broad irradiation of the proton resonances for spin decoupling also causes relaxation transfer (NOE) to the carbons to which they are attached. This causes an increase in the proportion of ^{13}C nuclei in the lower-energy state beyond that in thermal equilibrium, leading to a beneficial signal enhancement of up to threefold. This varies according to the nearness of hydrogens. Those carbons bearing no hydrogens give relatively weak signals.

Both the variation in the relaxation time and the NOE result in signal areas that are not proportional to the number of atoms represented; therefore, integrals from ^{13}C spectra are not often used. Nevertheless, qualitatively a larger signal may represent twice as many carbons as a smaller signal, and very small signals are often assignable to carbons that bear no hydrogens.

The instruments maintain a constant calibration by using a deuterium signal to lock the field-to-frequency ratio. Therefore, deuterated solvents are necessary. Common solvents and their peak positions are listed in Table 10.12.

10.6.2 Interpretation of ^{13}C NMR Spectra

The various characteristics of carbon NMR spectra have been discussed in the preceding section. An abbreviated correlation chart is provided in Table 10.11, and more detailed ones are available in the general references and online [31]. Carbon NMR peaks for common solvents are provided in Table 10.12. A good collection of spectra is available online as well [32].

The carbons of benzene rings may be assigned by starting with the value for benzene (128.5 ppm) and adding the parameters in Table 10.13.

TABLE 10.11 ^{13}C Correlation Chart for Common Organic Functional Groups

Peak Position (ppm)	C	Functional Group
200–220	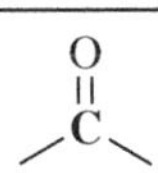	Ketones
180–200	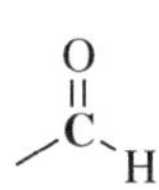	Aldehydes
160–180	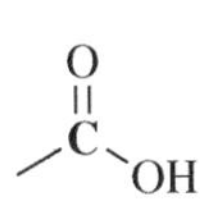 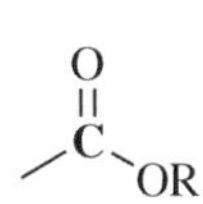 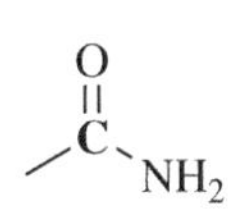	Acids, esters, amides
120–140	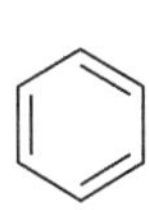	Aromatics
120–140	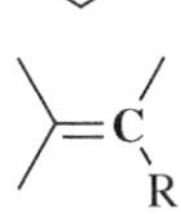	Substituted alkenes

(*Continued*)

TABLE 10.11 (*Continued*)

Peak Position (ppm)	C	Functional Group
110–140	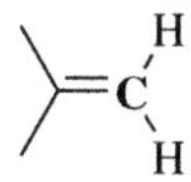	Unsubstituted alkenes
110–120	R–C≡N	Nitriles
60–80	–C≡C–	Alkynes
60–80	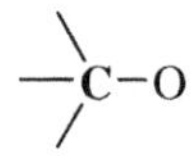	Alcohols, ethers 3° > 2° > 1°
20–40	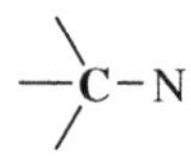	Amines 3° > 2° > 1°
40–75	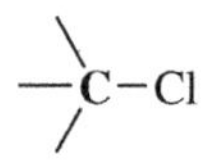	Alkyl chlorides
30–50	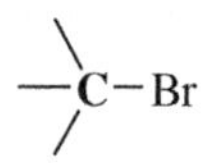	Alkyl bromides
0–40	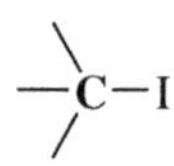	Alkyl iodides
0–20		Alkanes

TABLE 10.12 Carbon NMR Peaks for Common Solvents

Position (ppm)	Structure	Name
206, 29.8	CD_3COCD_3	d^6-Acetone
128	C_6D_6	d^6-Benzene
77.0	$CDCl_3$	d-Chloroform
40.5	CD_3OD	d^4-Methanol
39.5	CD_3SOCD_3	d^6-DMSO

TABLE 10.13 Chemical-Shift Changes at each Ring Carbon on Replacement of a Hydrogen with Group X [8, 30]

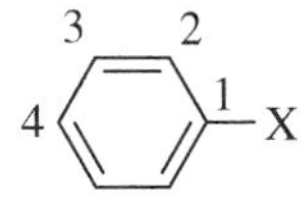

X	1	2	3	4
$-CH_3$	9.3	0.8	0	−2.9
$-CH_2CH_3$	15.6	−0.4	0	−2.6
$-CH(CH_3)_2$	20.1	−2.0	0	−2.5
$-CO_2CH_3$	2.1	1.1	0.1	4.5
$-COCH_3$	9.1	0.1	0	4.2
$-OCH_3$	31.4	−14.4	1.0	−7.7
$-OCOCH_3$	23.0	−6.4	1.3	−2.3
$-NHCOCH_3$	11.1	−9.9	0.2	−5.6
$-NH_2$	18.0	−13.3	0.9	−9.8
$-NO_2$	20.0	−4.8	0.9	5.8
−Cl	6.2	0.4	1.3	−1.9
−Br	−5.5	3.4	1.7	−1.6
−OH	26.9	−12.7	1.4	−7.3
−CN	−15.4	3.6	0.6	3.9
−CHO	8.2	1.2	0.6	5.8

As you might expect, electron-donating, ortho-, para-directing groups resonance-shield the ortho and para carbons, and the effects of any substituent are small at the meta position. Where more than one substituent is on the ring, the parameters are additive, as shown for 4-nitroanisole, **10.2**. In the proton-decoupled ^{13}C spectrum, carbons 1 and 4 are apparent because they are quite small owing to incomplete relaxation and the lack of NOE.

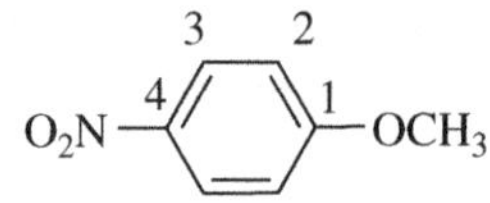

10.2

	Increments	Calculated	Measured
C-1	128.5 + 31.4 + 5.8	= 165.7	164.7
C-2	128.5 − 14.4 + 0.9	= 115.0	114.0
C-3	128.5 + 1.0 − 4.8	= 124.7	125.7
C-4	128.5 − 7.7 + 20.0	= 140.8	141.5
OCH_3			55.9

10.7 CORRELATION OF ^{1}H AND ^{13}C NMR SPECTRA

A great deal of information is lost when carbon spectra are decoupled, though the simpler spectra that result are worth the trade-off. Some of this information can be retrieved through correlation techniques, which plot the carbon spectrum on one axis and the proton spectrum along the other. Heteronuclear coupling correlation (HETCOR) is measured in the carbon spectrum and is a less sensitive and now less common technique. Heteronuclear multiple-quantum coherence (HMQC) provides similar spectra but is proton-detected, allowing greater sensitivity. Heteronuclear single-quantum coherence (HSQC) spectra are particularly useful in analyzing large molecules, such as proteins. Heteronuclear multiple-bond correlation (HMBC) allows correlation over more than one bond connection.

Varying the delay between pulses gives a stack of spectra, which are Fourier-transformed to give a two-dimensional spectrum. Contoured peaks correlate the signal for a proton with the signal for the carbon to which it is attached. The HSQC spectrum for jasmone is shown in Figure 10.14. Unambiguous assignments in the alkyl region can be made by correlating the proton nuclear magnetic resonance (PMR) spectrum on the x-axis with the carbon nuclear magnetic resonance (CMR) spectrum on the y-axis. The singlet at 2.2 ppm is assigned to C6 based on splitting and correlated to the peak at 17 ppm in the carbon spectrum. The triplet at 1.1 ppm and the multiplet at 2.1 ppm are assigned to the ethyl group at

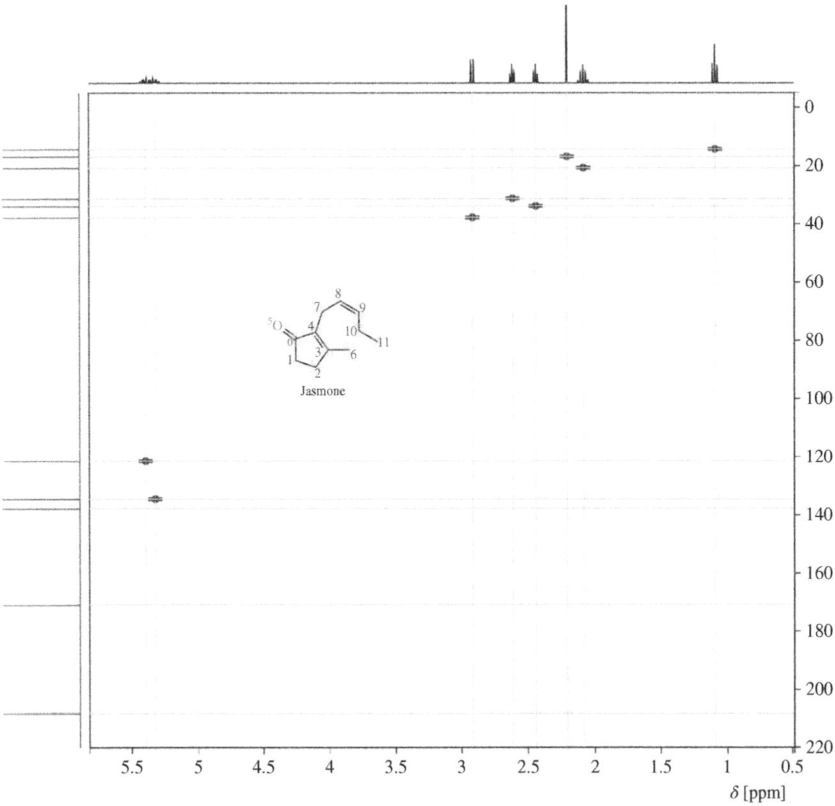

FIGURE 10.14 400 MHz ^{1}H–^{13}C HSQC spectrum of jasmone, simulated with NMRium NMR simulator. Adapted from Patiny et al. [16].

C11 and C10, respectively, based on their splitting, and these correlate to the peaks at 14 and 20 ppm, respectively, in the carbon spectrum.

As with COSY spectra, appropriate delays can be introduced in the sequence to give long-range correlations. This can be optimized at about 145 Hz as in the HMQC and HSQC experiments, giving one bond couplings, or it can be optimized at about 8 Hz to obtain the longer-range couplings in the HMBC experiment. A good example of signal assignments on the basis of long- and short-range HETCOR spectra can be seen in the work of Lankin et al. [33].

A variety of spectroscopy techniques exist, each with its own advantages and disadvantages in particular applications [34].

RESOURCES

1. ChemCalc [35] allows for calculation of exact molecular mass and predicts the molecular ion region of the mass spectrum: http://www.chemcalc.org/.
2. The National Institute of Standards and Technology (NIST) maintains an extensive database of spectral and thermodynamic data. https://webbook.nist.gov/.
3. NMRium [16, 17] allows for simulation of spin systems, and the prediction of ^1H and ^{13}C NMR spectra as well as COSY and HSQC spectra (it allows user entry of chemical structures). https://www.nmrium.org/.
4. NMRium also has a series of exercises for practice. https://www.nmrium.org/teaching#exercises.
5. *Understanding NMR Spectroscopy*, Lectures by James Keeler, is available for personal use and describes more advanced theory. https://www-keeler.ch.cam.ac.uk/lectures/.
6. NMR acronyms can be confusing. A guide, along with brief descriptions of the experiments. is available. https://www.acdlabs.com/blog/nmr-acronyms-decoded/.
7. Orange NMR is an NMR simulator available for Apple® devices available at the Apple app store.

PROBLEMS

Questions marked with an asterisk are answered in Appendix C.

10.1* It is possible to approximate λ_{max} and ε_{max} for conjugated alkenes using the Fieser–Kuhn equations (Eqs. 10.8 and 10.9), where M is the number of substituents on the double bonds, n is the number of double bonds, R_{endo} is the number of endocyclic double bonds, and R_{exo} is the number of exocyclic double bonds.

$$\lambda_{max} = 114 + 5M + n(48.0 - 1.7n) - 16.5R_{endo} - 10R_{exo} \quad (10.8)$$

$$\varepsilon_{max} = 1.74 \times 10^4\, n \quad (10.9)$$

Calculate λ_{max} and ε_{max} for zeaxanthin, **10.3**. What color is zeaxanthin? Explain.

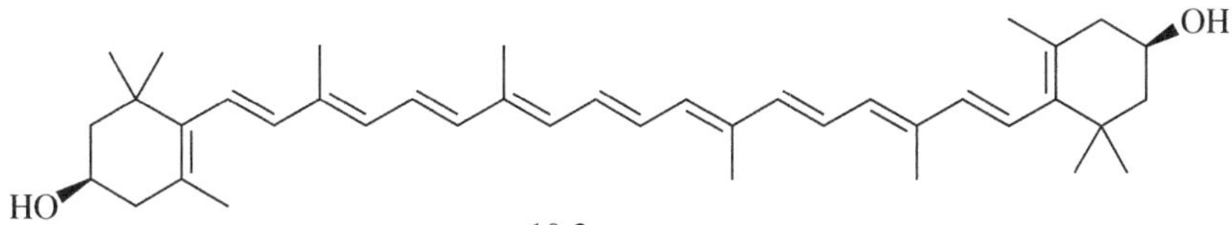

10.3

10.2* Alcohol dehydrogenase is an enzyme that catalyzes the conversion of ethanol to acetaldehyde in yeast (Eq. 10.10). Its activity can be monitored by following the increase in absorbance at 340 nm due to the formation of NADH. NAD+ does not absorb at this wavelength. Given the absorbance data below, calculate the activity of the enzyme in units per milliliter. One unit of alcohol dehydrogenase is defined as the amount needed to reduce 1 μmole of NAD+ per minute. The molar absorption coefficient, ε_{340}, is 6220/M/cm.

(10.10)

Time (min)	A_{340}
0	0.000
0.5	0.107
1.0	0.209
1.5	0.305
2.0	0.417
2.5	0.530
3.0	0.626

10.3* The vibrational frequency of an IR absorption can be calculated using the equation for a simple harmonic oscillator (Eq. 10.11)

$$\tilde{\nu} = \frac{1}{2\pi c}\sqrt{f\frac{M_x + M_y}{M_x M_y}} \qquad (10.11)$$

where c is the speed of light, f is the force constant for the bond, and M_x and M_y are the masses of the two atoms. The force constant for the carbon–oxygen double bond is approximately 12×10^5 dyne/cm. Calculate the approximate frequency of the carbonyl bond. Does your calculation match the value in the IR correlation chart?

10.4* The force constant of a bond is influenced by a number of factors including the steric, electronic, and chemical environments. Conjugation weakens the double bond character, which lowers the frequency of the observed stretch. Match the molecules below to these carbonyl stretching frequencies: 1666, 1686, 1720, 1722, 1731/cm.

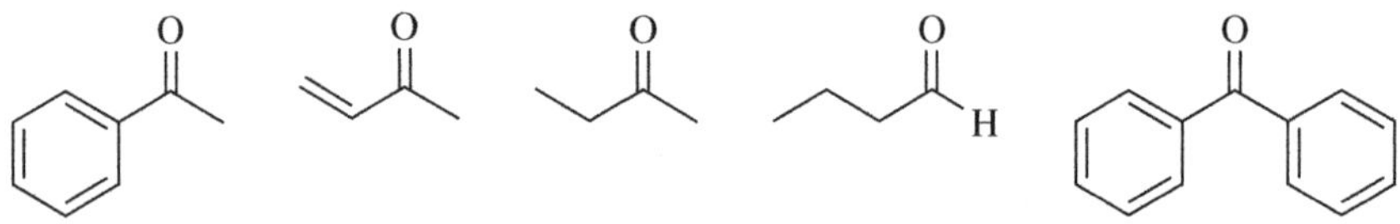

10.5* The mass spectra for isomers A and B are given below. Assign structures to both molecules.

Compound A		Compound B	
m/z	Relative Intensity	*m/z*	Relative Intensity
51	9.3	65	7.2
77	13.7	77	2.7
79	11.9	78	5.9
103	7.4	91	100
105	100	92	10.8
106	9.0	105	3.8
120	26.6	120	25.9

10.6* Assign a likely structure to Compound C, whose mass spectrum is provided below. The molecular weight is 201.9 g/mole.

m/z	Relative Intensity	*m/z*	Relative Intensity
27	20.9	123	49.6
39	29.3	200	18.3
41	100	202	35.4
121	51.9	204	16.5

10.7* Using Table 10.11 in the text, calculate the positions of the peaks in the aromatic region of the carbon NMR spectrum for vanillin, **10.4**. Find the spectrum online and compare it to the spectrum you calculated.

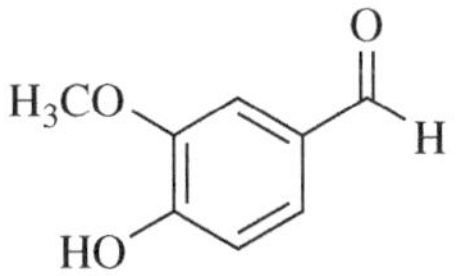

10.4

10.8* The ^{13}C NMR of 4-methyl-2-pentanol shows six signals at 22.4, 23.1, 23.9, 24.8, 48.7, and 65.8 ppm. Why should we not expect five signals as in 2-methylpentane?

10.9* A compound of molecular formula $C_8H_9ClO_2$ has peaks at 55.8, 56.8, 112.9, 113.4, 116.2, 123.1, 149.5, and 153.9 in its proton-decoupled ^{13}C NMR spectrum. Give the structure of the compound and assign as many signals as possible to the carbons in the structure.

10.10* The 90-MHz ^1H NMR spectrum of a certain compound showed an overlapping doublet and triplet, the individual maxima of which appeared at 1.254, 1.332, 1.357, 1.410, and 1.451 ppm. Using the J values, assign each peak to either the doublet or the triplet, then calculate the position of each. Calculate the separation between the peaks in each multiplet in hertz and in parts per million at both 90 and 300 MHz. Plot the peaks at both 90 and 300 MHz using the scale 1 cm = 0.01 ppm.

10.11* Figure 10.15 gives the ^1H, ^{13}C, COSY, and HSQC spectra for a compound with molecular formula $C_{10}H_{16}O$. Assign a structure to the molecule. Assign all the peaks in the ^1H and ^{13}C spectra, using COSY and HSQC to resolve ambiguities.

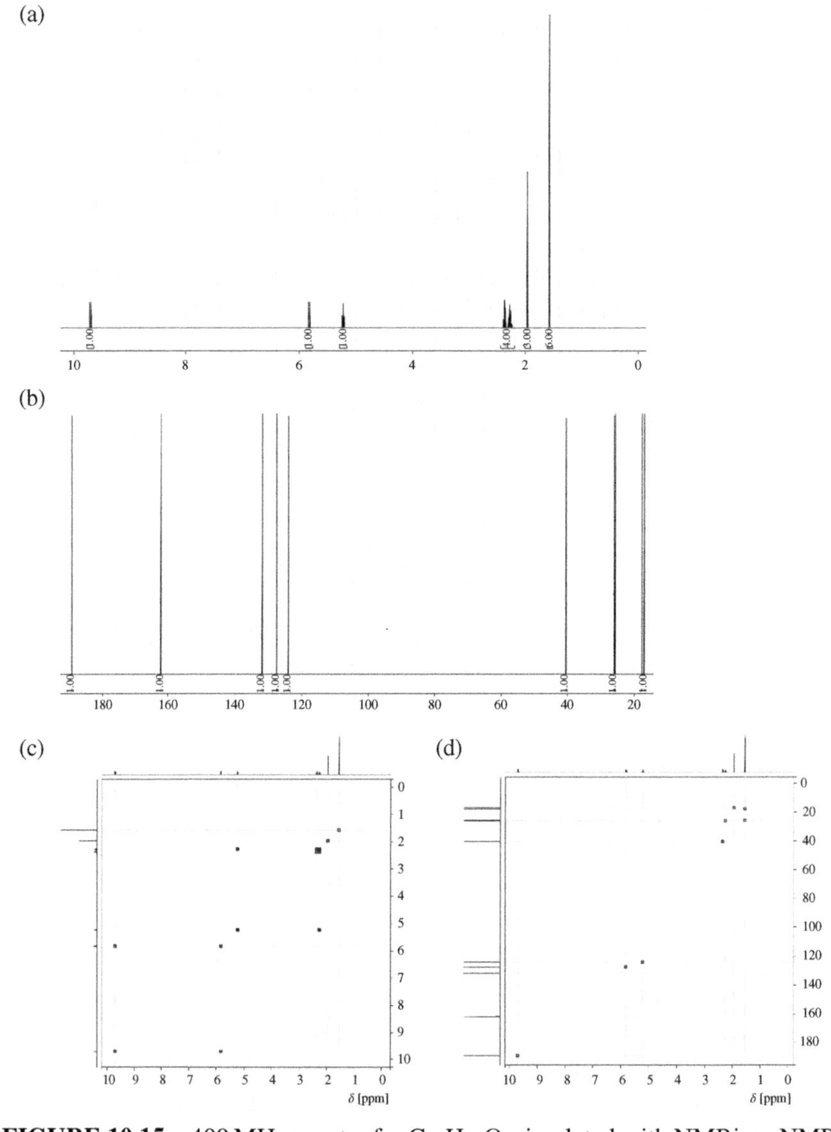

FIGURE 10.15 400 MHz spectra for $C_{10}H_{16}O$, simulated with NMRium NMR simulator. Adapted from Patiny et al. [16]: (a) ^{1}H NMR, (b) ^{13}C NMR, (c) COSY, and (d) HSQC.

10.12* Figure 10.16 gives the ^{1}H, ^{13}C, COSY, and HSQC spectra for a compound with molecular formula $C_{10}H_{12}O_3$. Assign a structure to the molecule. Assign all the peaks in the ^{1}H and ^{13}C spectra, using COSY and HSQC to resolve ambiguities.

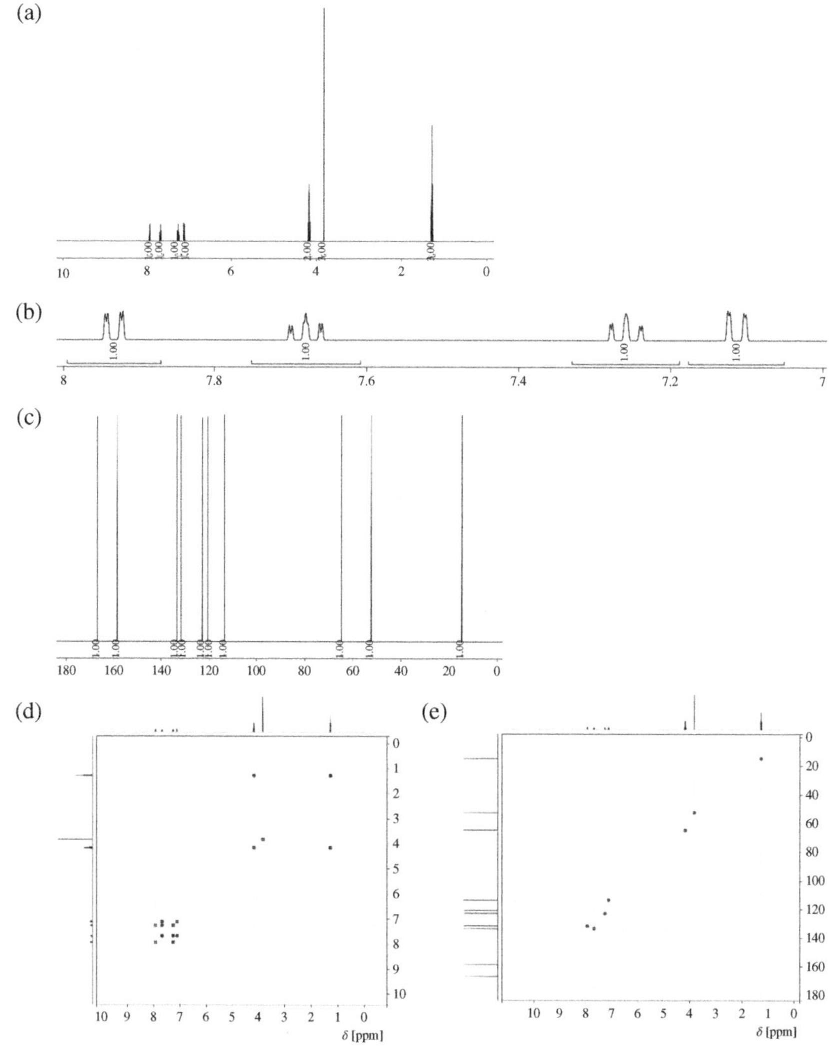

FIGURE 10.16 400 MHz spectra for $C_{10}H_{12}O_3$, simulated with NMRium NMR simulator. Adapted from Patiny et al. [16]: (a) ^{1}H NMR, (b) expanded aromatic region, (c) ^{13}C NMR, (d) COSY, and (e) HSQC.

10.13* Figure 10.17 gives the ^1H, ^{13}C, COSY, and HSQC spectra for a compound with molecular formula $C_{10}H_{16}O_2$. Assign a structure to the molecule. Assign all the peaks in the ^1H and ^{13}C spectra, using COSY and HSQC to resolve ambiguities.

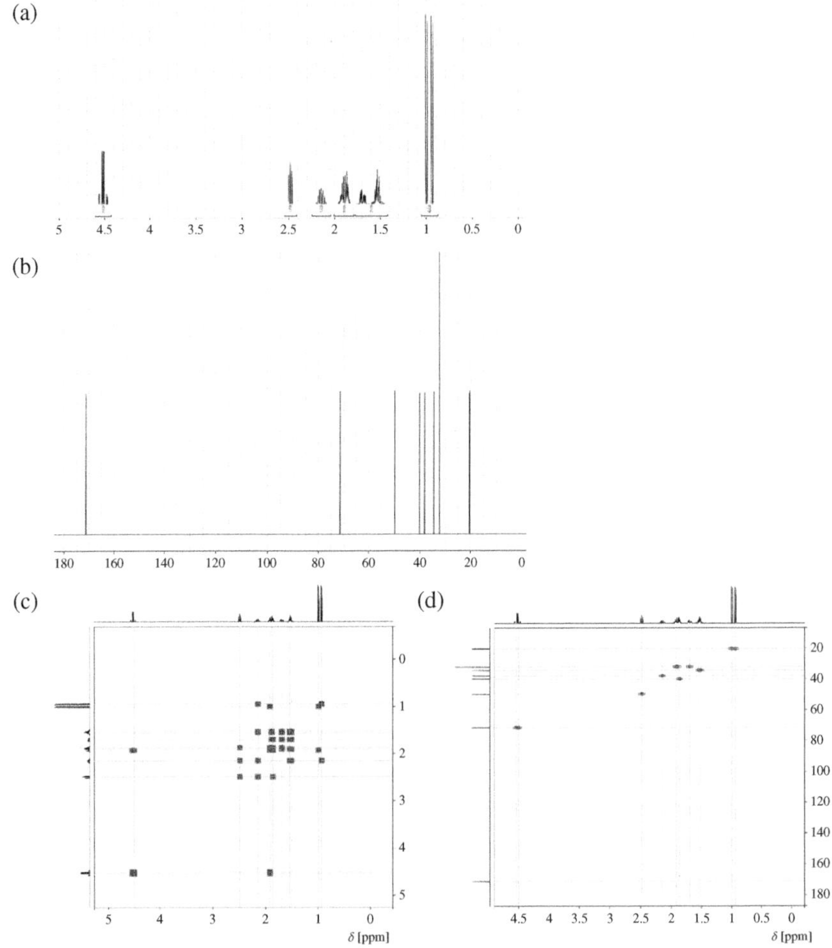

FIGURE 10.17 400 MHz spectra for $C_{10}H_{16}O_2$, simulated with NMRium NMR simulator. Adapted from Patiny et al. [16]: (a) ^1H NMR, (b) ^{13}C NMR, (c) COSY, and (d) HSQC.

REFERENCES

1. Harwood, L. M.; Claridge, T. D. W. *Introduction to Organic Spectroscopy*; Oxford University Press: Oxford, 1997.
2. Pavia, D. L.; Lampman, G. M.; Kriz, G. S.; Vyvyan, J. R. *Introduction to Spectroscopy*, 5th ed.; Cengage Learning: Stanford, CT, 2015, pp. 577–613.
3. Silverstein, R. M.; Bassler, G. C.; Morrill, T. C. *Spectrometric Identification of Organic Compounds*, 5th ed.; John Wiley & Sons, Inc: New York, 1991, pp. 289–315.
4. Agilent. *The Basics of UV–Vis Spectrophotometry*. https://www.agilent.com/cs/library/primers/public/primer-uv-vis-basics-5980-1397en-agilent.pdf (accessed December 23, 2024).
5. This is the preferred terminology. *IUPAC Compendium of Chemical Terminology (The Gold Book)*. http://goldbook.iupac.org/ (accessed December 23, 2024)
6. Bruno, T. J.; Svoronos, P. D. N. *CRC Handbook of Fundamental Spectroscopic Correlation Charts*; CRC Press: Boca Raton, 2006.
7. Wolfram Demonstrations Project. *Colors of the Visible Spectrum*. https://demonstrations.wolfram.com/ColorsOfTheVisibleSpectrum/ (accessed December 23, 2024).
8. Silverstein, R. M.; Webster, F. X.; Kiemle, D. J. *Spectrometric Identification of Organic Compounds*, 7th ed.; John Wiley & Sons, Inc: New York, 2005.
9. For an excellent example of the value of laser Raman spectroscopy, see Smith, L. M.; Smith, R. G.; Loehr, T. M.; Daves, G. D. Jr.; Daterman, G. E.; Wohleb, R. H. *J. Org. Chem.* **1978**, *43*, 2361–2366.
10. McLafferty, F. W. *Interpretation of Mass Spectra*, 4th ed.; University Science Books: Sausalito, 1993.
11. McLafferty, F. W.; Venkataraghavan, R. *Mass Spectral Correlations*, Advances in Chemistry Series *40*, 2nd ed.; ed.; American Chemical Society: Washington, DC, 1982.
12. Silverstein, R. M.; Webster, F. X. *Spectrometric Identification of Organic Compounds*, 6th ed.; ed.; John Wiley & Sons, Inc: New York, 1998, p. 35.
13. Williams, K. R.; King, R. W. *J. Chem. Ed.* **1989**, *66*, A213–A219 A243–A248; **1990**, *67*, A93–A99; **1990**, *67*, A100-A105; **1990**, *67*, A125-A137.
14. Macomber, R. S. *J. Chem. Ed.* **1985**, *62*, 213–214.

15. Hornack, J. P. *The Basics of NMR.* https://www.cis.rit.edu/htbooks/nmr/ (accessed 23 December 2024).
16. Patiny, L.; Musallam, H.; Bolaños, A.; Zasso, M.; Wist, J.; Karayilan, M.; Ziegler, E.; Liermann, J. C.; Schlörer, N. E. *Beilstein J. Org. Chem.* **2024**, *20*, 25–31 https://www.nmrium.org/nmrium (accessed December 23, 2024).
17. Mann, B. *J. Chem. Ed.* **1995**, *72*, 614–615.
18. Hoye, T. R.; Zhao, H. *J. Org. Chem.* **2002**, *67*, 4014–4016.
19. Hoye, T. R.; Hanson, P. R.; Vyvyan, J. R. *J. Org. Chem.* **1994**, *59*, 4096–4103.
20. Ault, A. *J. Chem. Educ.* **1970**, *47*, 812–818.
21. Grunewald, G. L.; Ye, Q. *J. Org. Chem.* **1988**, *53*, 4021–4026.
22. Fraser, R. R.; Petit, M. A.; Miskow, M. *J. Am. Chem. Soc.* **1972**, *94*, 3253–3254.
23. Morris, G. A. *Magn. Reson. Chem.* **1986**, *24*, 371–403.
24. Kessler, H.; Gehrke, M.; Griesinger, C. *Angew. Chem., Int. Ed.* **1988**, *27*, 490–536.
25. Williams, K. R.; King, R. W. *J. Chem. Ed.* **1990**, *67*, A125–A137.
26. As an example, see:Ma, S.-G.; Gao, R.-M.; Li, Y.-H.; Jiang, J.-D.; Gong, N.-B.; Li, L.; Lü, Y.; Tang, W.-Z.; Liu, Y.-B.; Qu, J.; Lü, H.-N.; Li, Y.; Yu, S.-S. *Org. Lett.* **2013**, *15*, 4450–4453.
27. Cho, B. P.; Harvey, R. G. *J. Org. Chem.* **1987**, *52*, 5679–5684.
28. Portlock, D. E.; Lubey, G. S.; Borah, B. *J. Org. Chem.* **1989**, *54*, 2327–2330.
29. Simpson, J. H. *Organic Structure Determination Using 2-D NMR Spectroscopy: A Problem-Based Approach*, 2nd ed.; Elsevier: Amsterdam, 2012.
30. Brown, D. W. *J. Chem. Educ.* **1985**, *62*, 209–212.
31. Reich, H. C-13 Chemical Shifts. https://organicchemistrydata.org/hansreich/resources/nmr/?index=nmr_index%2F13C_shift (accessed February 1, 2025).
32. National Institute of Advanced Industrial Science and Technology (AIST). Spectral Database for Organic Compounds SDBS. https://sdbs.db.aist.go.jp/ (accessed February 1, 2025).
33. Shieh, H.-L.; Cordell, G. A.; Lankin, D. C.; Lotter, H. *J. Org. Chem.* **1990**, *55*, 5139–5145.
34. Soulsby, D. Modern NMR experiments: applications in the undergraduate curriculum. In Soulsby, D.; Anna, L. J.; Wallner, A. S.; Eds., *NMR Spectroscopy in the Undergraduate Curriculum*, ACS Symposium Series 1128; American Chemical Society: Washington, DC, 2013, pp. 7–41; Reynolds, W. F.; Enriquez, R. G. *J. Nat. Prod.* **2002**, *65*, 221–244.
35. Patiny, L.; Borel, A. *J. Chem. Inf. Model.* **2013**, *53*, 1223–1228.

APPENDIX A

COMMON CHEMICAL ABBREVIATIONS

[α]	Specific rotation in degrees
9-BBN	9-Borabicyclononane
α	Observed optical rotation in degrees
abs	Absorbance
ABT	1-Aminobenzotriazole
Ad	Adamantyl
AIBN	Azobisisobutyronitrile
anh	Anhydrous
aq	Aqueous
atm	Atmosphere(s)
BDMS	Bromodimethylsulfonium bromide
BINAL-H	2,2′-Dihydroxy-1,1′-binaphthyl-lithium aluminum hydride
BINAP	2,2′-Bis(diphenylphosphino)-1,1′-binaphthyl
[bmim][BF_4]	1-*n*-Butyl-3-methylimidazolium tetrafluoroborate
[bmim][PF_6]	1-*n*-Butyl-3-methylimidazolium hexafluorophosphate
Bn	Benzyl

Understanding Organic Chemistry: Bridging Fundamental and Advanced Courses,
Fourth Edition. Ann M. Fabirkiewicz and John C. Stowell.

Companion website: www.wiley.com/go/fabirkiewiczorganic4e

Boc	*t*-Butyloxycarbonyl
BOM	Benzyloxymethyl
BPO	Benzoyl peroxide
Bu	Butyl
Bz	Benzoyl
CA	Chemical Abstracts
CAS	Chemical Abstracts Service
cat	Catalyst
CBz	Carboxybenzyl
CSA	Camphorsulfonic acid
CuAAC	Copper-catalyzed azide-alkyne cycloaddition
DABCO	1,4-Diazabicylo[2.2.2]octane
DAST	Diethylaminosulfur trifluoride
DBA	Dibutylamine
DBN	1,5-Diazabicyclo[4.3.0]non-5-ene
DBU	1,7-Diazabicyclo[5.4.0]undec-7-ene
DCE	Dichloroethane
DCM	Dichloromethane
DDQ	Dichlorodicyanobenzoquinone
de	Diastereomeric excess
DEAE	Diethylaminoethanol
DEPT	Distortionless enhancement by polarization transfer
DIBAL	Diisobutylaluminum hydride
DIPT	Diisopropyl tartrate
DMAP	4-(*N*,*N*-dimethylamino)pyridine
DME	1,2-Dimethoxyethane
DMF	Dimethylformamide
DMP	Dess–Martin periodinane
DMSO	Dimethylsulfoxide
dppf	1,1′-Bis(diphenylphosphino)ferrocene
DPPH	*O*-(diphenylphosphinyl)hydroxylamine
dr	Diastereomeric ratio
DTBMP	2,6-Di-tert-butyl-4-methylpyridine
dtbbpy	4,4′-Bis(tert-butyl)-2,2′-bipyridine
ee	Enantiomeric excess
EEAC	Electric eel acetylcholinesterase
er	Enantiomeric ratio
Et	Ethyl
FID	Free induction decay

FT	Fourier transform
F-TEDA	*N*-Fluoro-*N′*-(chloromethyl)triethylenediamine bis(tetrafluoroborate)
GHS	Globally Harmonized System of Classification and Labelling of Chemicals
h	Hour(s)
HDMS	Hexamethyldisilazane
HMBC	Heteronuclear multiple-bond coherence
HMPA	Hexamethylphosphoramide
HMQC	Heteronuclear multiple-quantum coherence
HOMO	Highest occupied molecular orbital
HSOMO	Highest singly occupied molecular orbital
HSQC	Heteronuclear single-quantum coherence
Hz	Hertz
IBX	*o*-Iodoxybenzoic acid
i-Pr	Isopropyl
LAH	Lithium aluminum hydride
LCAO	Linear combination of atomic orbitals
LDA	Lithium diisopropylamide
LICA	Lithium isopropylcyclohexylamide
LUMO	Lowest unoccupied molecular orbital
m/z	Mass-to-charge ratio
Me	Methyl
min	Minute(s)
MOM	Methoxymethyl
Ms	Methanesulfonyl, mesyl
NaDT	Sodium decatungstate
NBS	*N*-bromosuccinimide
NIS	*N*-iodosuccinimide
NIST	National Institute of Science and Technology
NMP	*N*-methylpyrrolidone
NMR	Nuclear magnetic resonance
NOE	Nuclear Overhauser effect
OBO	4-methyl-2,6,7-trioxa-bicyclo[2.2.2]octan-1-yl
Oxone	Potassium peroxymonosulfate, $KHSO_5 + 0.5\ KHSO_4 + 0.5\ K_2SO_4$
PAMO-P3	Phenylacetone monooxygenase
PBSF	Perfluoro-1-butanesulfonyl fluoride
PCC	Pyridinium chlorochromate

PDC	Pyridinium dichromate
Ph	Phenyl
Piv	Pivaloyl (dimethylpropanoyl)
PMHS	Polymethylhydrosiloxane
PPA	Polyphosphoric acid
PPE	Polyphosphate ester
ppm	Parts per million
PPTS	Pyridinium *p*-toluenesulfonate
PTC	Phase-transfer catalyst/catalysis
py	Pyridine
rac	Racemic
Red-Al	Sodium bis(2-methoxyethoxy)aluminum hydride
rt	Room temperature
SAMP	(*S*)-1-amino-2-methoxymethylpyrrolidine
SDS	Safety data sheet
SPhos	2-(2′,6′-dimethoxybiphenyl)dicyclohexylphosphine
STAB-H	Sodium triacetoxyborohydride
TBAF	Tetrabutylammonium fluoride
TBATB	Tetrabutylammonium tribromide
TBDPS	*t*-Butyldiphenylsilyl
TBS	*t*-Butyldimethylsilyl
TEA	Triethylamine
TEMPO	2,2,6,6-Tetramethylpiperidine 1-oxyl
Tf	Trifluoromethanesulfonate, triflate
TFA	Trifluoroacetyl
TFAA	Trifluoroacetic anhydride
TFAE	2,2,2-Trifluoro-1-(9-anthryl)ethanol
THF	Tetrahydrofuran
THP	Tetrahydropyranyl
THPTA	Tris-hydroxypropyltriazolylmethylamine
TK	Transketolase
TMBS	Trimethylbromosilane
TMDS	1,1,3,3-Tetramethyldisiloxane
TMEDA	*N*,*N*,*N*,*N*-Tetramethyl-1,2-ethylenediamine
TMP	Trimethylolpropane
TMS	Trimethylsilyl, tetramethylsilane (in NMR)
Ts	Toluenesulfonyl, tosyl

RESOURCES

1. Chemical Abstracts Services, a division of the American Chemical Society. *CAS Standard Abbreviations and Acronyms*. https://www.cas.org/training/documentation/references/cas-standard-abbreviations (accessed January 6, 2025).
2. Daub, G. H.; Leon, A. A.; Silverman, I. R.; Daub, G. W.; Walker, S. B. *Aldrichchim. Acta* 1984, *17*, 9–23.
3. American Chemical Society. *ACS Style Quick Guide*. https://pubs.acs.org/doi/full/10.1021/acsguide.40303 (accessed January 6, 2025).
4. Vogel, T.; Cardinal, S.; Butkovich, N.; Wrublewski, D.; Baysinger, G. References. Section 4.3. In *ACS Guide to Scholarly Communication*, Banik, G. M.; Baysingerk, G.; Kamat, P. V.; Pienta, N. J., Eds.; ACS Publications, 2020, DOI: 10.1021/acsguide.

APPENDIX B

COMMON PROTECTING GROUPS

Understanding Organic Chemistry: Bridging Fundamental and Advanced Courses,
Fourth Edition. Ann M. Fabirkiewicz and John C. Stowell.

Companion website: www.wiley.com/go/fabirkiewiczorganic4e

name (Abbreviation)	Formation/removal	Stability
Alcohols		
Methoxymethyl (MOM)	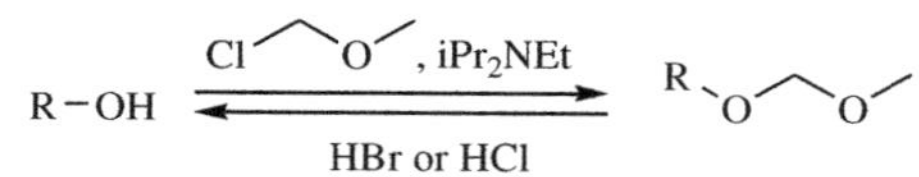	Base, mild acid
Benzyloxymethyl (BOM)	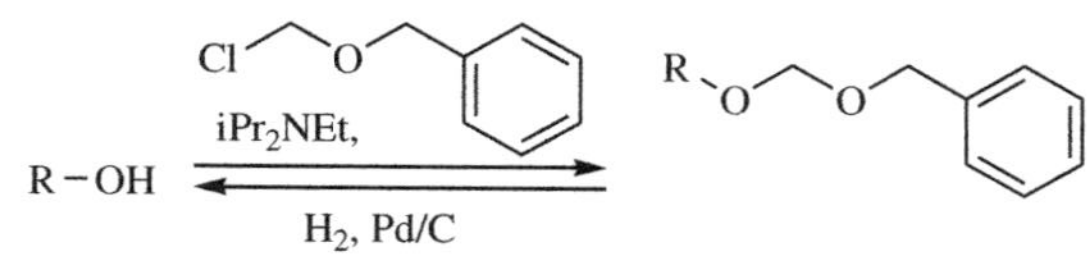	Base, acid
Tetrahydropyranyl (THP)	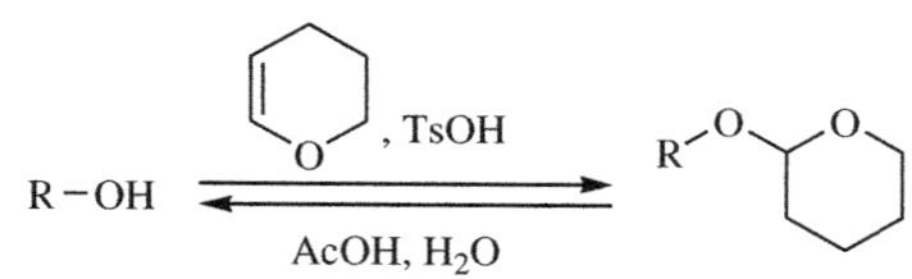	Base
Trimethylsilyl (TMS)	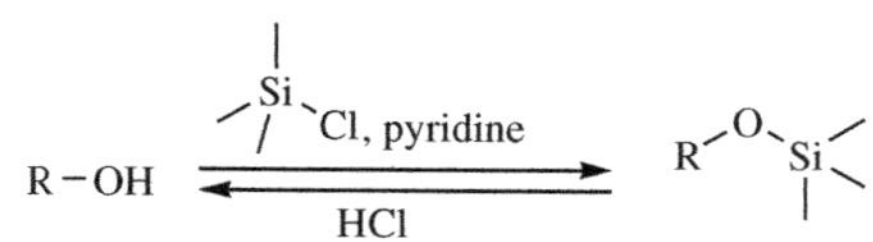	Base
t-Butyldimethylsilyl (TBDMS or TBS) *t*-Butyldiphenylsilyl (TBDPS)	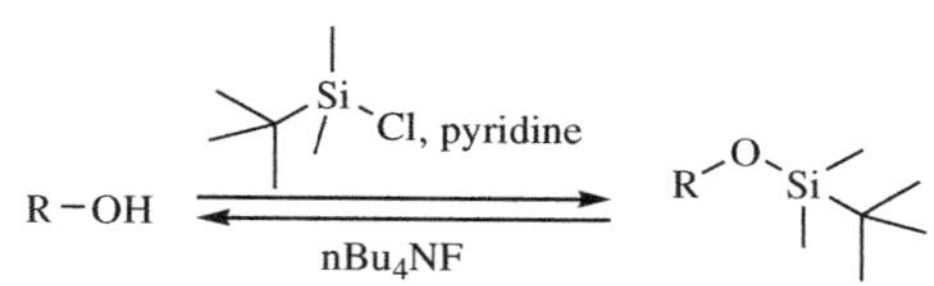	Base, mild acid, can be selective for 1° alcohols

Aldehydes/ketones		
Dimethyl acetal	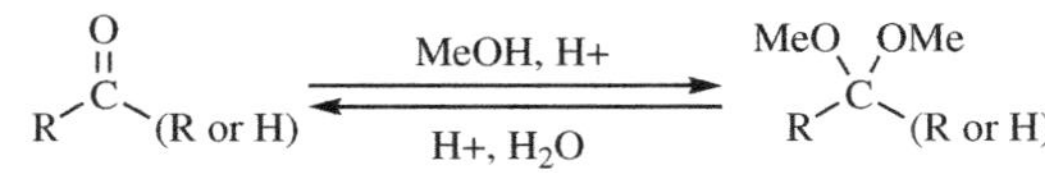	Base
	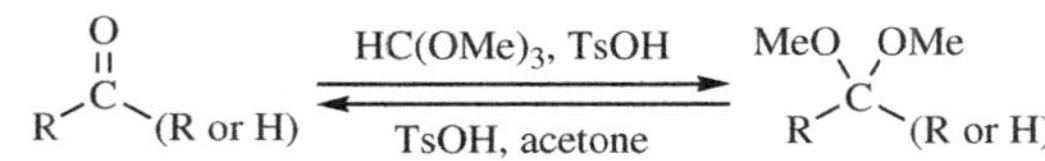	
Ethylene glycol acetal	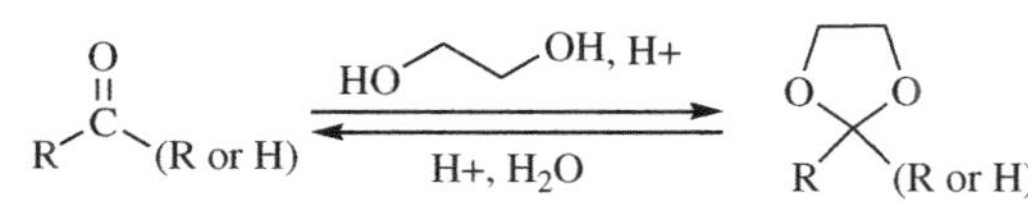	Base
1,3-Dithiane acetal	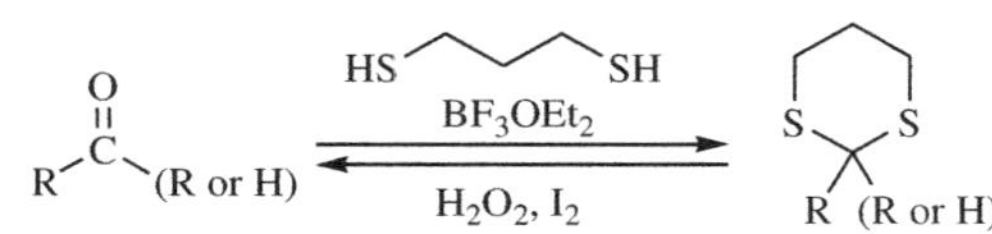	More often used as a masked carbonyl
Amines		
Benzyloxycarbamate (CBz)	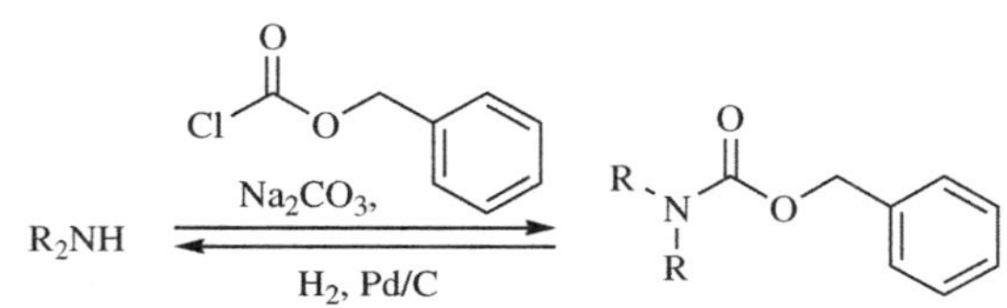	Base, oxidation

(*Continued*)

name (Abbreviation)	Formation/removal	Stability
t-Butoxycarbonyl (Boc)	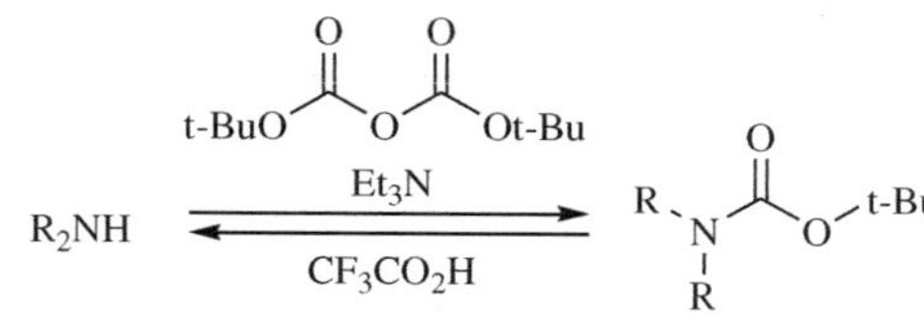	Base, mild acid, oxidation
Benzyl (Bn)	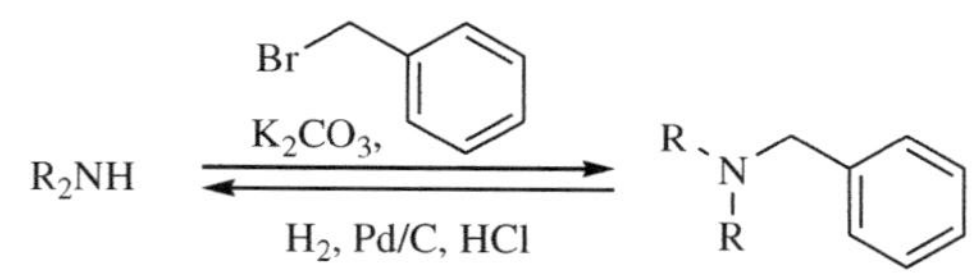	Base, nucleophiles
Carboxylic acids		
Benzyl ester	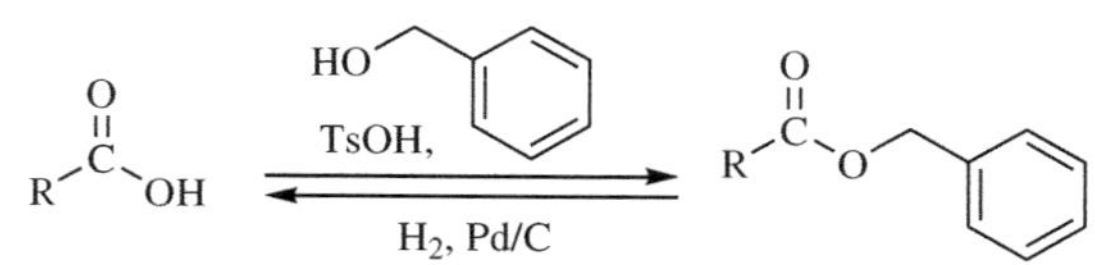	Base, mild acid
4-Methyl-2,6,7-trioxa-bicyclo [2.2.2]octan-1-yl (OBO)	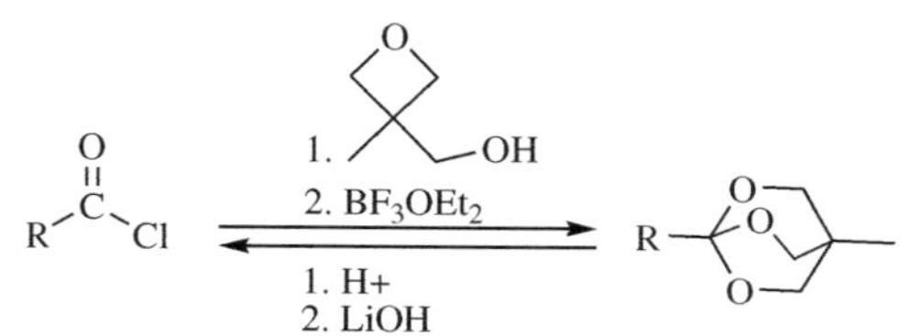	Base

RESOURCES

1. Beaudoin, D.; Murphy, P. The Organic Synthesis Archive. https://synarchive.com/protecting-group.
2. Mueller, R. Organic Chemistry Portal. Protecting Groups. https://www.organic-chemistry.org/protectivegroups/.
3. Wuts, P. G. M.; Greene, T. W. *Greene's Protective Groups in Organic Synthesis*, 4th ed.; John Wiley and Sons, Inc: Hoboken, New Jersey, 2007.

APPENDIX C

ANSWERS TO SELECTED PROBLEMS

Chapter 1

1.1 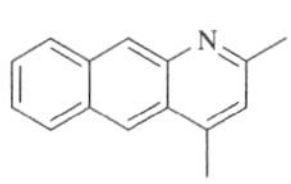

1.2 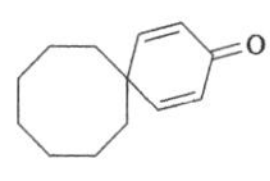

1.3 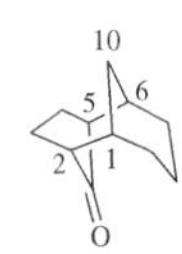

1.4 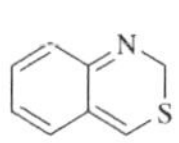

1.5 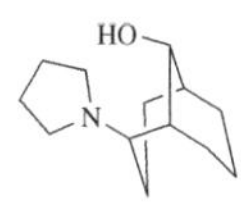

1.6 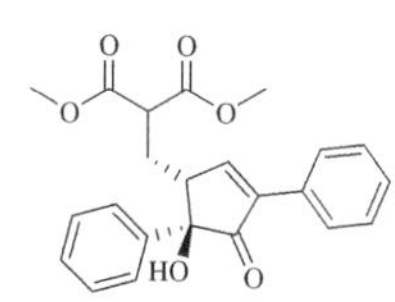

+ enantiomer

1.7 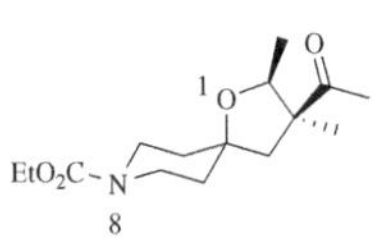

1.8 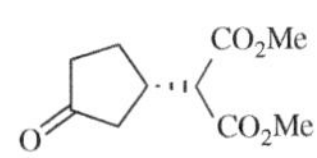

1.9 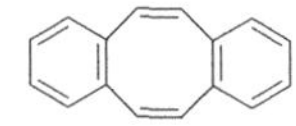

1.10 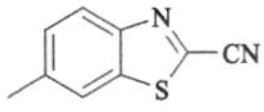

Chapter 2

2.1 A structure search in Organic Syntheses finds the procedure in Jalalian, N.; Olofsson, B. *Org. Synth.* **2013**, *90*, 1. The compound

Understanding Organic Chemistry: Bridging Fundamental and Advanced Courses, Fourth Edition. Ann M. Fabirkiewicz and John C. Stowell.

Companion website: www.wiley.com/go/fabirkiewiczorganic4e

is 1-[Hydroxy(tosyloxy)iodo]-3-trifluoromethylbenzene, or Koser's reagent. It is a mild oxidant, and references given in the Org. Syn. paper include the following three references: (a) Zhdankin, V. V.; Stang, P. J. *Chem. Rev.* **2008**, *108*, 5299–5358; (b) Merritt, E. A.; Olofsson, B. *Synthesis* **2011**, 517–538; and (c) Koser, G. F. *Aldrichimica Acta* **2001**, *34*, 89–102.

2.3 The structure of crizotinib is shown below. The IUPAC name is (*R*)-3-[1-(2,6-dichloro-3-fluorophenyl)ethoxy]-5-[1-(piperidin-4-yl)-1*H*-pyrazol-4-yl]pyridin-2-amine. PubChem indicates that it is a tyrosine kinase inhibitor, and the links in the use and manufacturing section indicate its specific use as a drug for locally advanced or metastatic non-small cell lung cancer.

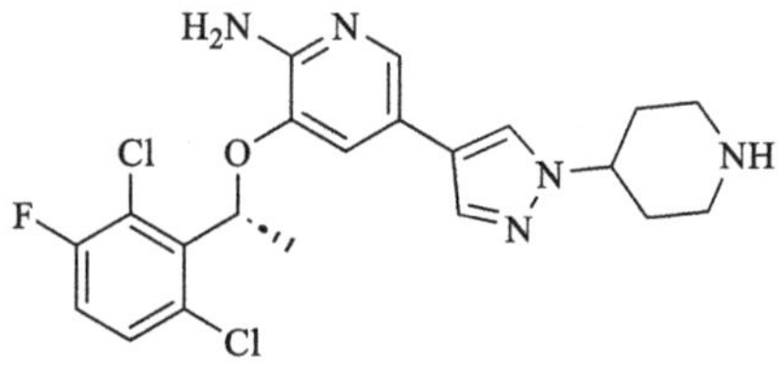

2.5 A structure search in PubChem finds 6,7,8,9-tetrahydro-6,10-methano-6*H*-pyrazino (2,3-h) benzazepine, or Chantix, a smoking-cessation drug. The patent is held by Pfizer, and patent numbers are 6890927, 6410550, and 7265119. The structure of Chantix is given below:

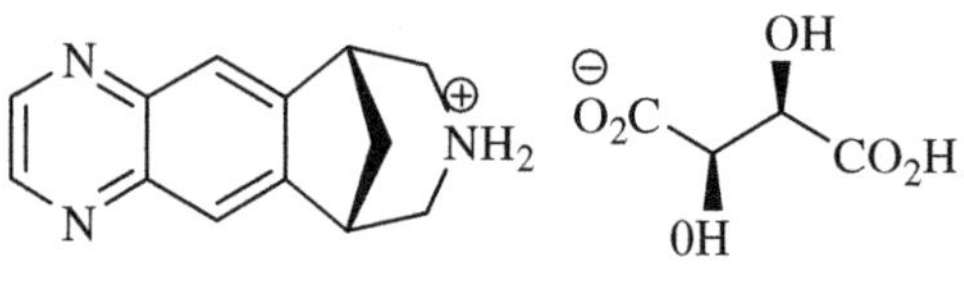

2.6 A PubChem search of the alcohol (replace R- with H-) finds a list of compounds. The closest matches to the given structure are lovastatin (Mevacor), simvastatin (Zocor), and nicostatin (Advicor), a niacin–lovastatin mixture.

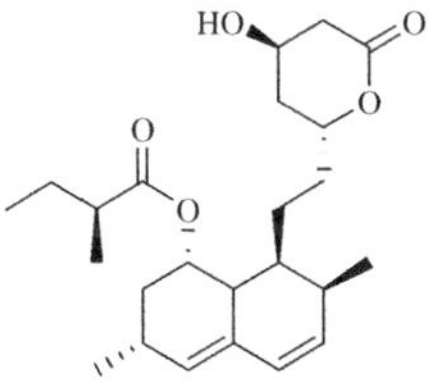

lovastatin

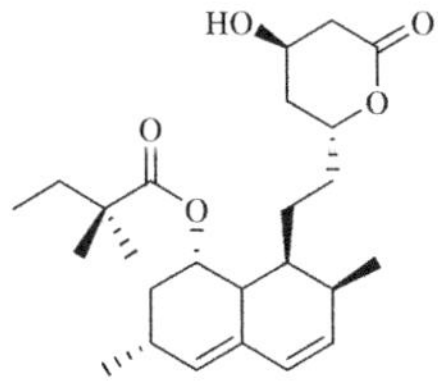

simvastatin

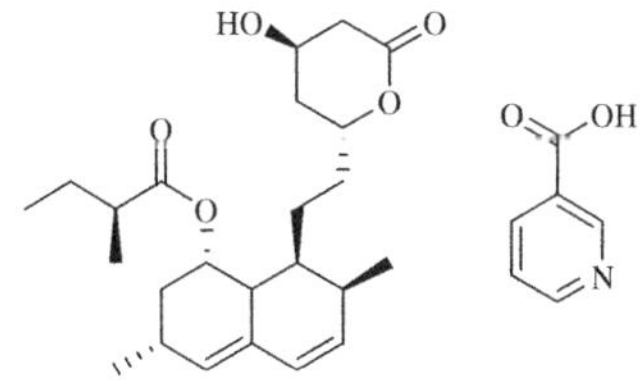

nicostatin

Statins are a class of drugs that lower serum cholesterol. A number of papers are listed for these drugs at the PubChem site. Two papers that describe studies of lovastatin (1) and simvastatin (2) analogs are found with a search of these terms at pubs.acs.org. Those references are: (1) Lee, T. J.; Holtz, W. J.; Smith, R. L.; Alberts, A. W.; Gilfillan, J. L. *J. Med. Chem.* **1991**, *34*, 2474–2477; (2) Hartman, G. D.; Halczenko, W.; Duggan, M. E.; Imagire, J. S.; Smith, R. L.; Pitzenberger, S. M.; Fitzpatrick, S. L.; Alberts, A. W.; Bostedor, R. *J. Med. Chem.* **1992**, *35*, 3813–3821.

2.11 A PubChem search for β-conhydrine produces the CAS number 495-20-5 and its IUPAC name, (1*R*)-1-[(2*S*)-piperidin-2-yl]propan-1-ol. From Wikipedia, we find that it is isolated from poison hemlock. PubChem lists toxicity of 400 μg/kg as an LD_{low} in guinea pigs.

2.12 BDQ is bedaquiline, used in the treatment of drug-resistant tuberculosis. Its specific rotation is −166.98 at 20°C with a concentration of 0.505 in DMF, according to a PubChem search. GHS symbols indicate acute toxicity, health hazards, and environmental hazards.

2.13 A search for (*S*)-binol at pubs.acs.org produces many references. Brunel, J. M. *Chem. Rev.* **2005**, *105*, 857–897, a review article, lists the reaction by which the binol enantiomers can be prepared and purified, as well as many possible uses in organic reactions. The chemical name is 1,1′-bi-2-naphthol and the structure is

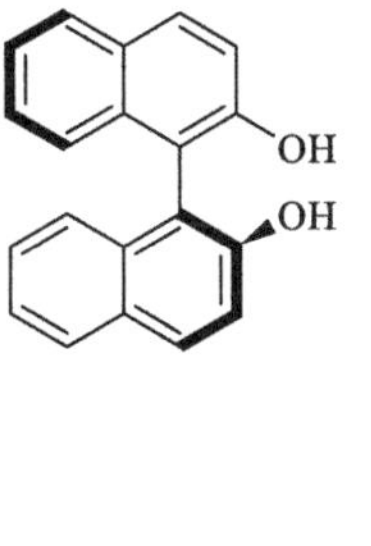

Chapter 3

3.1 The first step is a nucleophilic attack on the sulfonic acid halide and will not change the configuration at the chiral carbon:

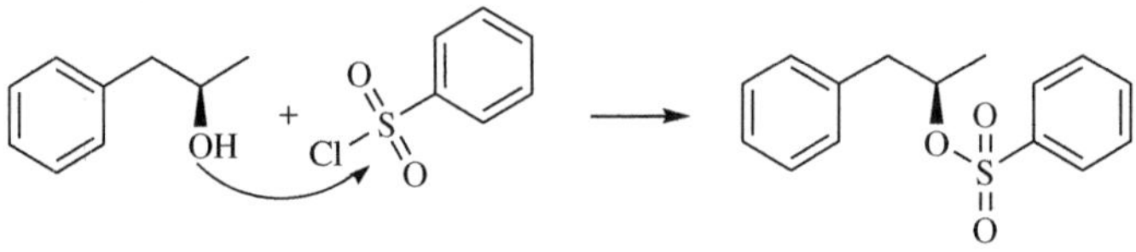

The second step is an S_N2 reaction which will invert the configuration at the chiral carbon:

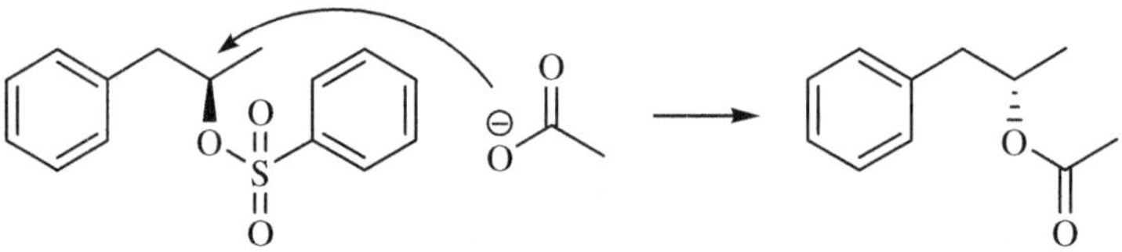

The final reaction is hydrolysis of the ester which will not change the configuration at the chiral carbon:

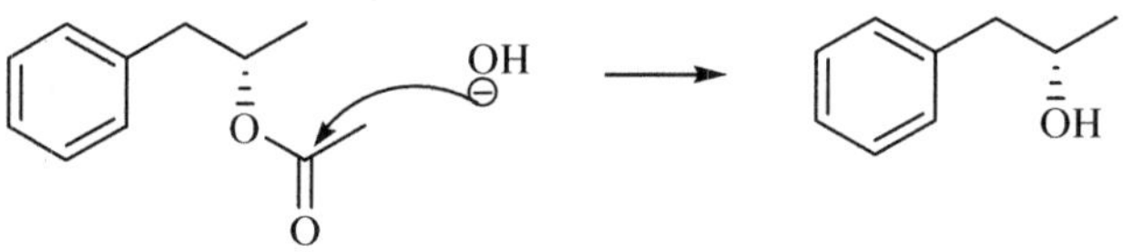

3.4 Pig liver esterase was used to catalyze the reaction, resulting in 83% yield and 80% ee before recrystallization.

3.5 Both esters can be synthesized by the sequence below, starting from either D- or L-malic acid. PPTS is pyridinium *p*-toluenesulfonate.

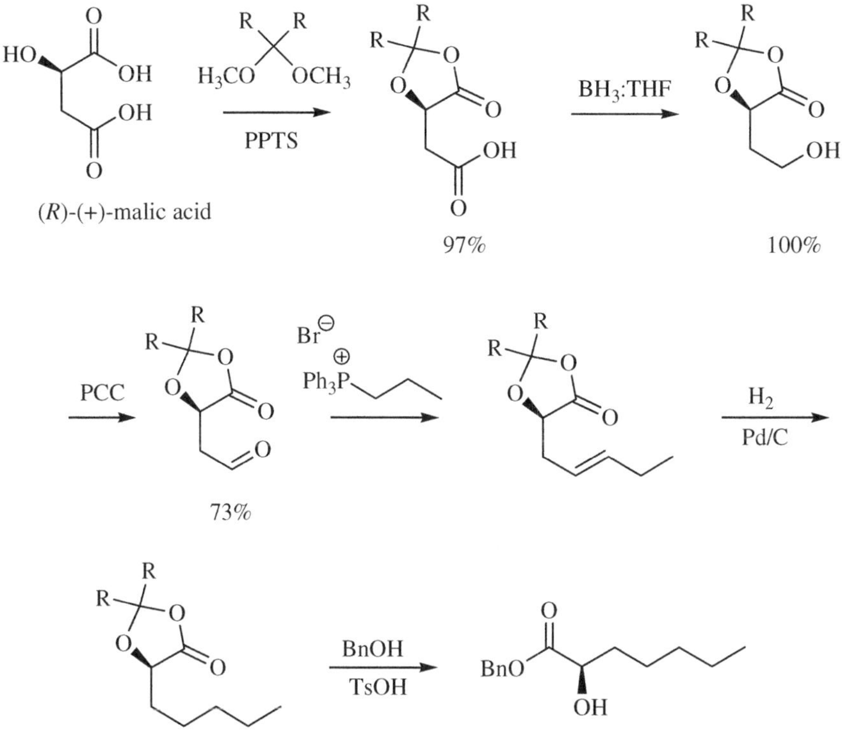

(*R*)-(+)-benzyl 2-hydroxyheptanoate
30%, 3 steps

3.10 The chiral lithium salt is allowed to equilibrate with N-Boc-2-lithiopiperidine at −45°C, and through carbanion inversion, one stereoisomer predominates. The reaction mixture is cooled to freeze the equilibrium and propanal added to quench the equilibrium. The chiral dilithium anion shown in the text, and its enantiomer, shows excellent selectivity.

3.11 H_a and H_b, H_c and H_d are diastereotopic, H_a and H_c, H_b and H_d are enantiotopic. C-2 and C-6 are enantiotopic.

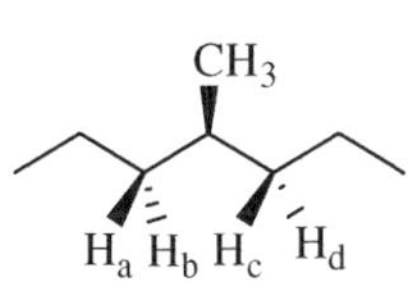

3.12 The two methoxy groups are diastereotopic.

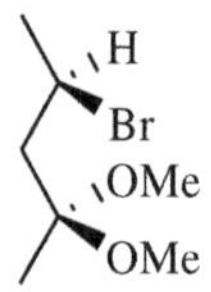

3.14 The hydride reduction results in a trans double bond which, following the Claisen rearrangement, results in the (*R*)-ester.

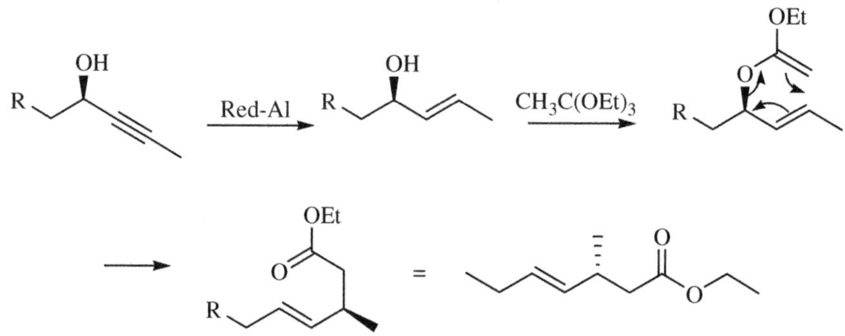

To obtain the (*S*)-ester, reduce the alkyne to the cis-alkene using catalytic hydrogenation with Lindlar's catalyst.

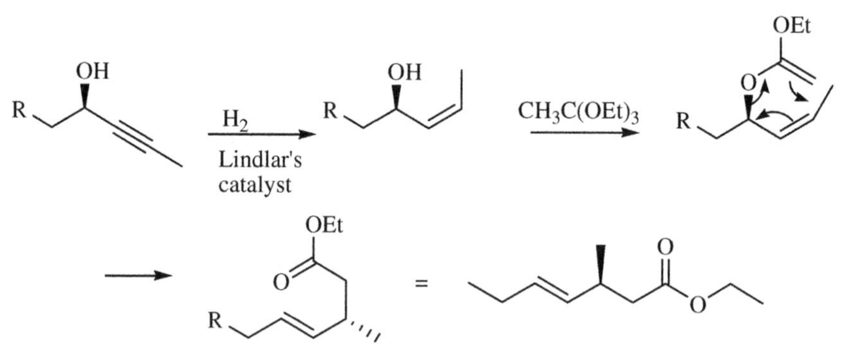

3.19 The enamine intermediate makes the stereochemistry of the methyl group irrelevant.

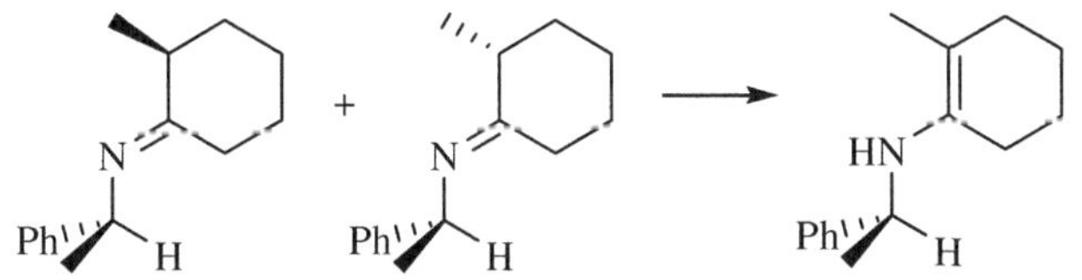

A six-membered ring intermediate was proposed to explain the stereochemistry of the Michael addition. Chiral induction occurs via coordination between N and the carbonyl carbon.

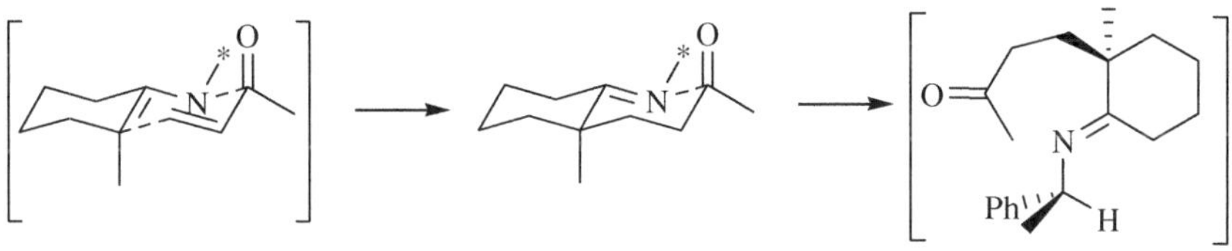

3.20 The first structure, on the left, has the (*S*)– configuration at the chiral carbon of the amine component. The remaining chiral carbons in that molecule are both (*S*)–.

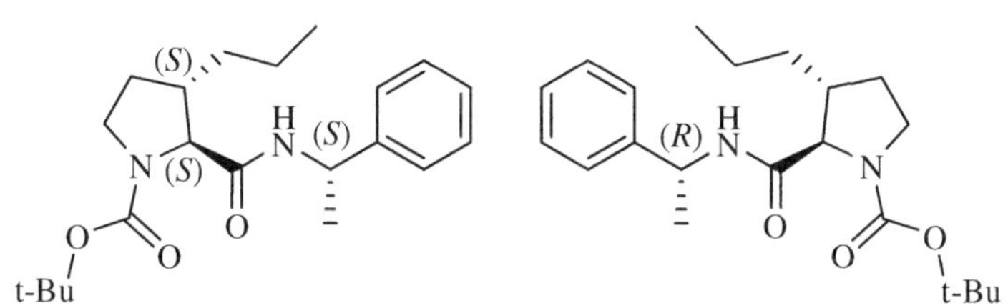

3.22 Using a chiral amine to form the chiral amide as the deprotonating agent results in a chiral intermediate lithium salt. Chirality transfer allows the selectivity observed in these reactions. The two open-access references describe a series of experiments to determine the importance of temperature and additional lithium salts to this reaction.

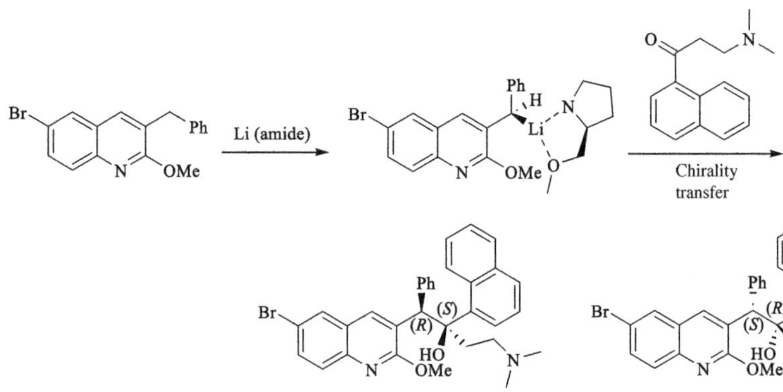

Chapter 4

4.1 The kinetic isotope effect of 6.4 implies that a bond to hydrogen is broken in the rate-determining step. The initial proton is likely abstracted from protonated 1,4-diazabicyclo[2.2.2]octane (DABCO), as it is most acidic. The isotopic scrambling observed in the crossover experiment allows for the source of the second proton to be something other than protonated DABCO. The authors suggest residual water or the starting material as the source of that proton.

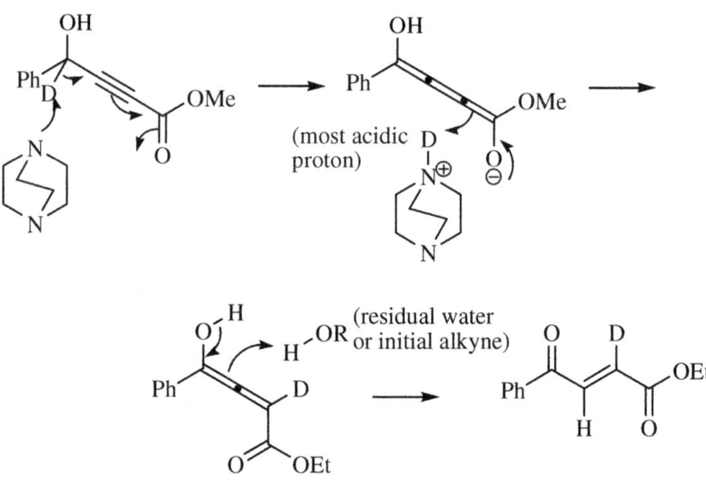

4.2 Solvent insensitivity supports a concerted mechanism, implying there is no polarity in the transition state. That the *E* isomer in the product double bond is highly supported in all substrates is consistent with a concerted mechanism and with a transition state in which the E isomer is sterically favored. The six-membered ring transition state supports these data.

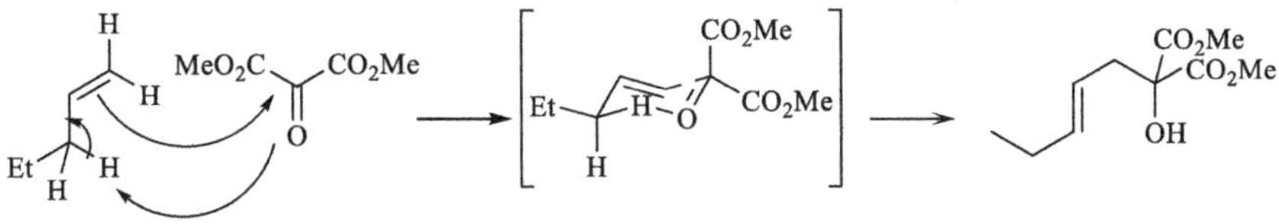

4.6 (a) The reaction is pseudo first order in the starting material, because a large excess of hydroxide is used in the experiments.

(b) Note that doubling the concentration of NaOH doubles the rate, and thus the reaction is first order in NaOH.

(c) The overall rate expression is:

$$\text{rate} = k[\text{A}][\text{OH}^-]$$

(d) The rate constant is calculated at 0.05 M NaOH:

$$k = \frac{k_{obs}}{[\text{B}]} = \frac{8.67 \times 10^{-2}\ \text{sec}^{-1}}{0.050\frac{\text{mol}}{\text{L}}} = 1.73\ \text{mol L}^{-1}\ \text{sec}^{-1}$$

4.7 The Eyring plots for k_1 and k_2 are as follows:

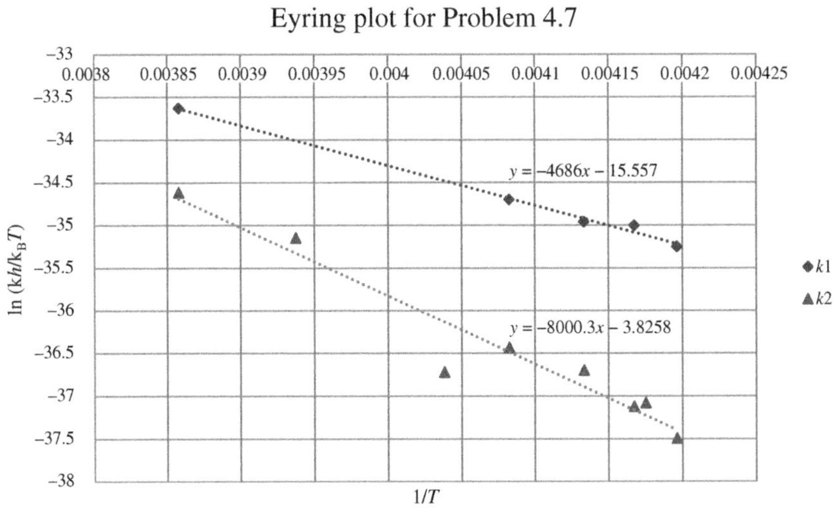

The slope of the line is $-\Delta H^{\ddagger}/R$ and the intercept is $\Delta S^{\ddagger}/R$, therefore:

For k_1:

$$\Delta H^{\ddagger} = -(-4686) \times 8.314 = 38.9\ \text{kJ/mol}$$

$$\Delta S^{\ddagger} = (-15.557) \times 8.314 = -129\ \text{J/mol-K}$$

The $\Delta S^{\ddagger}$ value indicates an ordered system in the transition state, consistent with the attack on the carbonyl shown.

For k_2:

$$\Delta H^{\ddagger} = -(-8000) \times 8.314 = 66.5\ \text{kJ/mol}$$

$$\Delta S^{\ddagger} = (-3.826) \times 8.314 = -31.9\ \text{J/mol-K}$$

The enthalpy of activation is consistent with multiple bonds broken in the transition state. The entropy value supports the formation of several products from one intermediate.

4.8 The graph shows a linear relationship between time and the reciprocal of [ICl], indicating second-order kinetics for that reagent, as shown in Eq. 4.12. Because the concentration of $C_6H_{12}ICl$ is much higher, the data from the graph can be used to calculate k_{pseudo} substituting values from the graph into Eq. 4.12.

$$\frac{1}{[A]_t} - \frac{1}{[A]_0} = kt$$

Run 1: $476 - 227 = k_{\text{pseudo}}(1000) \Rightarrow k_{\text{pseudo}} = 0.249$

Run 2: $748 - 625 = k_{\text{pseudo}}(1000) \Rightarrow k_{\text{pseudo}} = 0.123$

Note that the rate doubles with a doubling of the concentration of $C_6H_{12}ICl$, making the reaction first order in this reagent.

The rate expression is: $\dfrac{d[\text{ICl}]}{dt} = k[\text{ICl}]^2[C_6H_{12}\text{ICl}]$

The rate constant is calculated as:

$$k_{\text{pseudo}} = k[C_6H_{12}\text{ICl}] \Rightarrow 0.249 = 0.0349k \Rightarrow$$

$$k = 7.13\ \text{L}^2\text{mol}^{-2}\text{sec}^{-1}$$

4.10 Order in 2-buten-1-ol: +1. It was used to measure pseudo-first-order rates.

Order in $PdCl_4^{-2}$: +1. Compare runs 3 and 4, in which doubling the concentration doubles the rate.

Order in H^+: −1. Compare runs 8 and 10, in which doubling the concentration halves the rate.

Order in Cl^-: −2. Compare runs 6 and 7. Using Eq. 4.18 gives an order of −2.32, which rounds to −2. Calculating the values obtained by various run comparisons gives values ranging from −1.44 to −2.32, with an average of −1.73.

$$\log \frac{\text{rate}_1}{\text{rate}_2} = n \log \frac{[A]_1}{[A]_2}$$

$$\log \frac{1.5}{0.77} = n \log \frac{0.9}{1.2} \Rightarrow n = -2.32$$

$$\text{Rate expression: } \frac{d[\text{2-buten-1-ol}]}{dt} = \frac{k\left[PdCl_4^{-2}\right][\text{2-buten-1-ol}]}{[H^+][Cl^-]^2}$$

To calculate the actual rate constant, substitute the expression for k_{obs} into the rate expression above and use values from any run in the table (run 1 shown as an example).

$$-\frac{d[\text{2-buten-1-ol}]}{dt} = k_{obs}[\text{2-buten-1-ol}]$$

$$k_{obs} = \frac{k\left[PdCl_4^{-2}\right]}{[H^+][Cl^-]^2} \Rightarrow 0.52 \times 10^{-5} = \frac{k(0.005)}{(0.02)(0.6)^2} \Rightarrow$$

$$k = 7.49 \times 10^{-5}\ \text{mole}^2/\text{L}^2/\text{sec}$$

The average value of all the runs is 8.68×10^{-5} mole^2/L^2/sec, with a range of 6.53×10^{-5} to 1.04×10^{-4}.

4.12 In Eq. 4.49, all the alkenes would have shown an isotope effect because a bond to hydrogen is broken during the one step of the reaction. In Eq. 4.50, the first step, forming the aziridine oxide, is the rate-determining step. In the second step, for the second and third alkenes, there is a choice for the oxygen to abstract an H or a

D from the methyl group, and an isotope effect is expected. For the first alkene, there is no H versus D option, and no isotope effect is observed. The second mechanism fits the data presented.

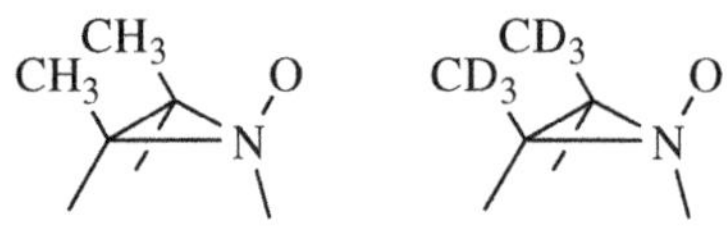

These intermediates would arise from the first alkene and the reaction would not show an isotope effect because there would be no H/D choice in the abstraction reaction.

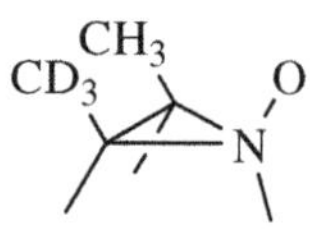

This intermediate would arise from the second or third alkene and its reaction would show an isotope effect because O could abstract either H or D.

4.15 The stereochemistry of the dichloro compound was proven using the reaction below (Eq. 4.5 in the text). An S_N2 reaction with sodium iodide followed by anti-elimination can result in the trans double bond (confirmed by spectral comparison) only for the dichloro compound shown.

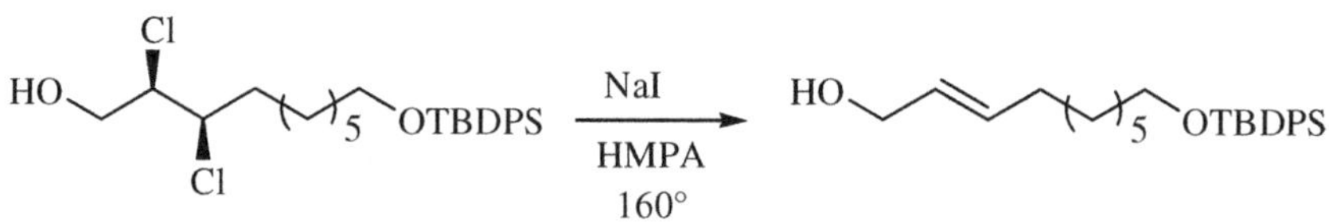

4.16

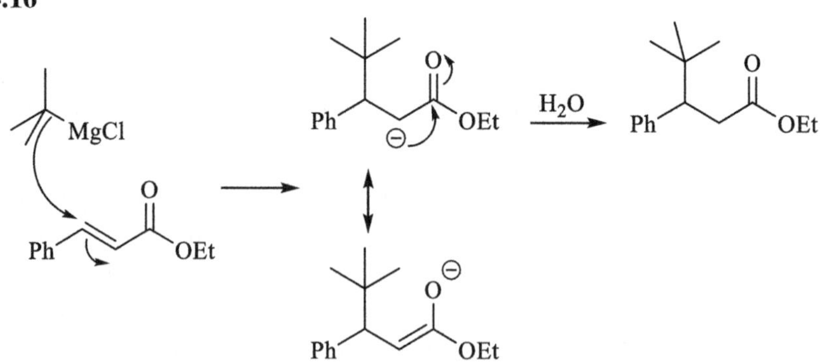

4.17

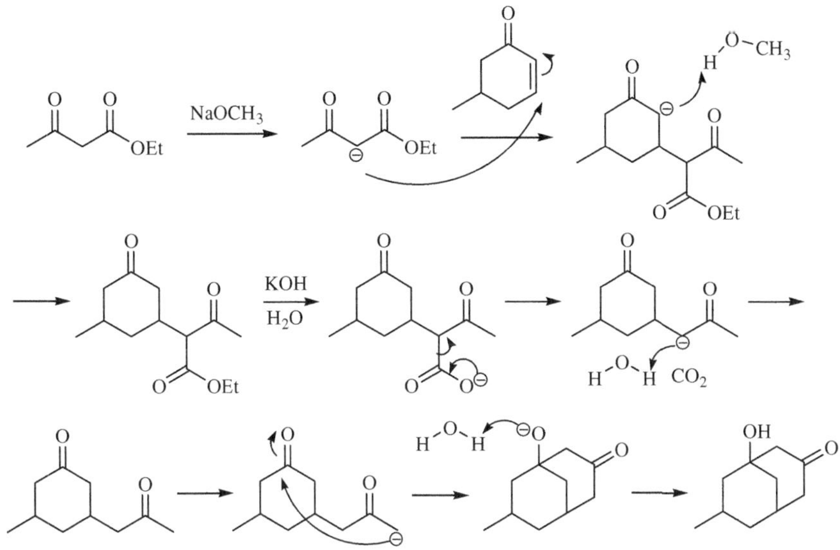

4.19

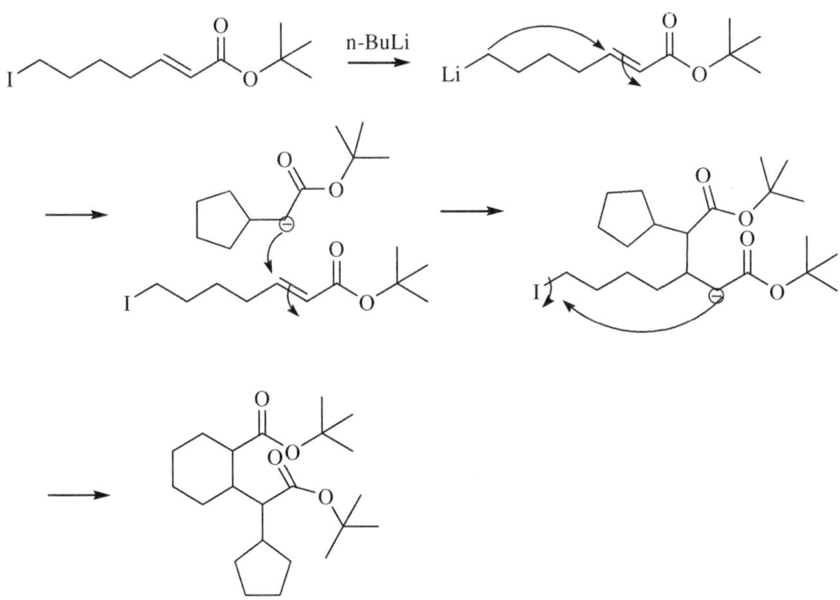

4.20

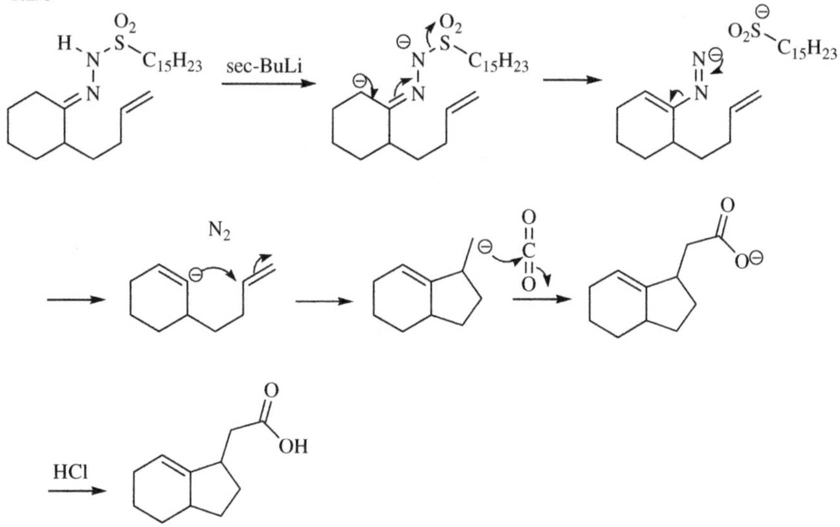

4.23

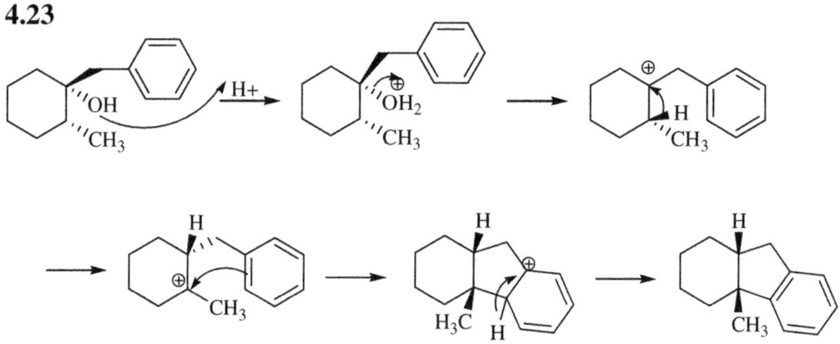

4.26

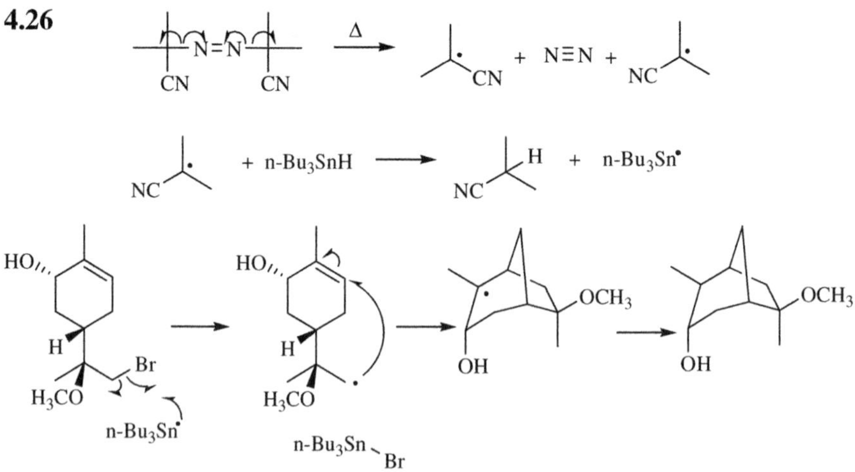

4.27

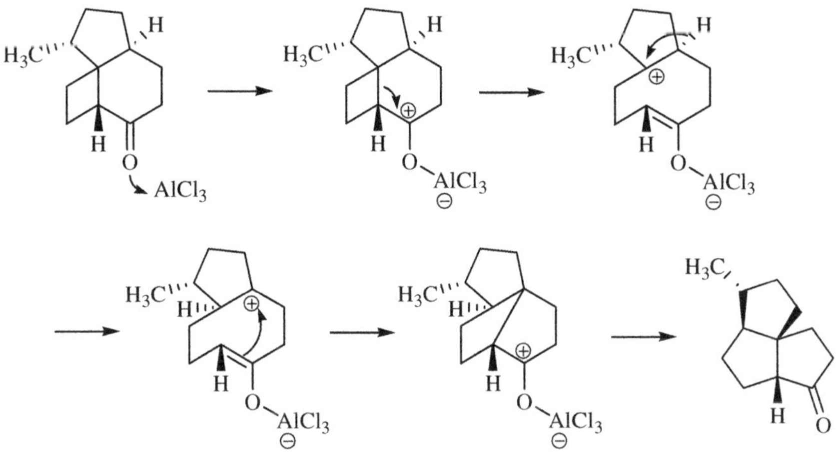

4.30 This mechanism is an ortho-ester Claisen rearrangement.

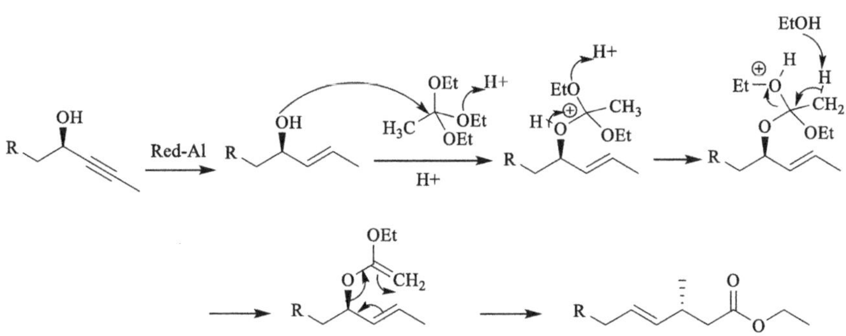

4.31

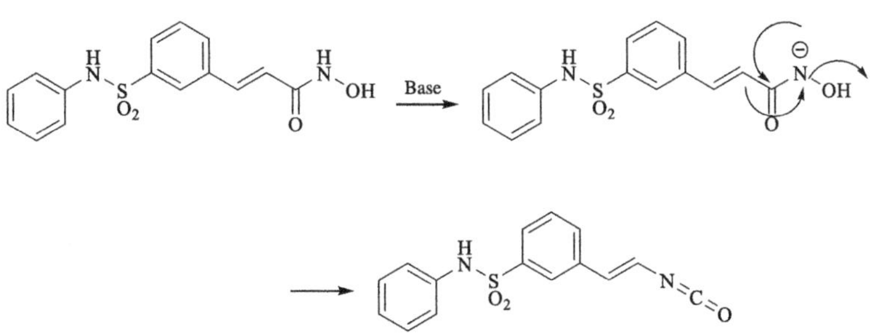

Chapter 5

5.1a [$_\pi 4_s + _\pi 2_s$] cycloaddition

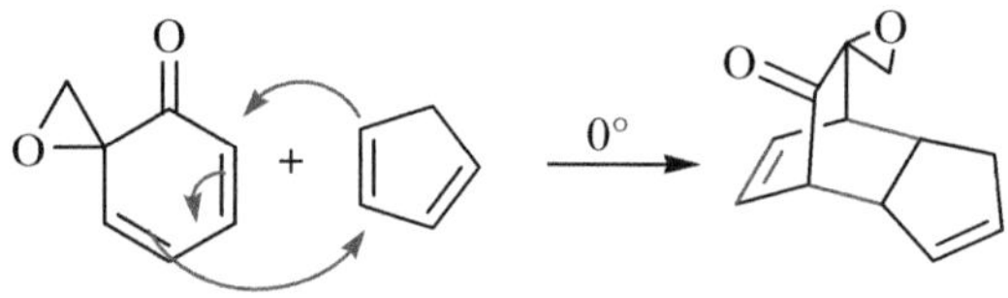

5.1b [$_\pi 6_s + _\pi 4_s$] cycloaddition

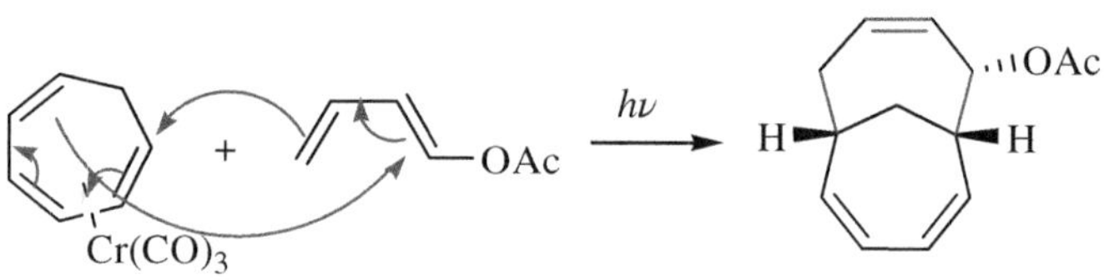

5.1c The first step is a cyclization, followed by a [1,5] sigmatropic shift, then a 6e− electrocyclic ring opening.

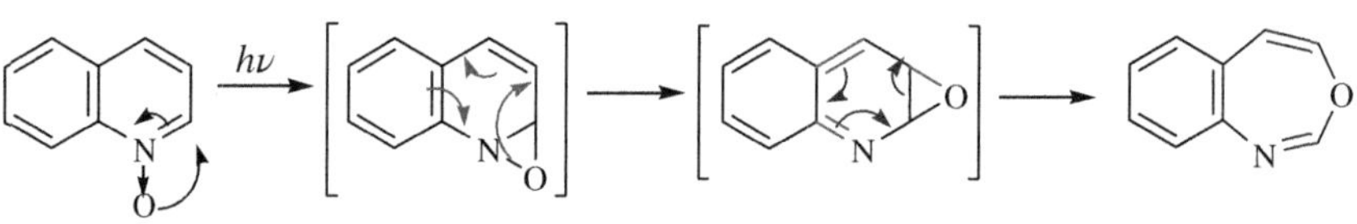

5.1d [$_\pi 4_s + _\pi 2_s$] cycloaddition

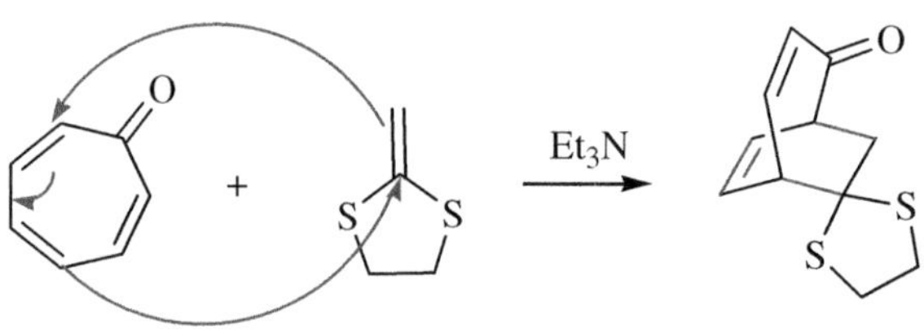

5.1e [3,3] sigmatropic shift

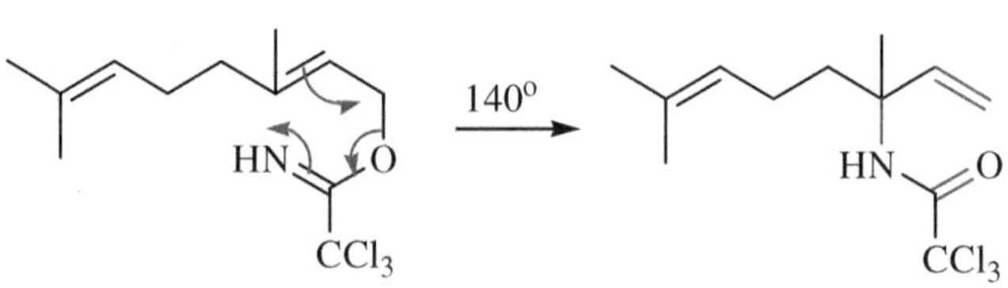

5.1f Ene reaction

5.2 For the cation,

Ψ_1 and Ψ_2 are occupied. The electron density at C-1 is the sum of the probabilities at C-1 times the number of electrons in each molecular orbital: $2\psi_1{}^2 + 2\psi_2{}^2 = 0.667$. Subtract this from 1 (the charge on a neutral atom) to calculate the charge on C-1 = 1 − 0.667 = +0.333. Note that the resonance structures show a positive charge distributed over three atoms, C-1, C-3, and C-5.

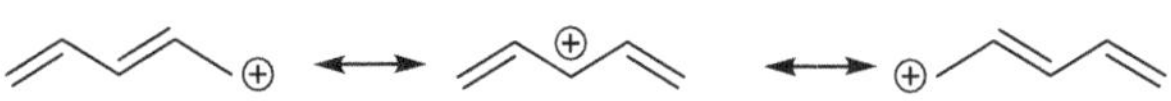

For the anion,

Ψ_1, Ψ_2, and Ψ_3 are occupied. The electron density at C-1 is the sum of the probabilities at C-1 times the number of electrons in each molecular orbital: $2\psi_1{}^2 + 2\psi_2{}^2 + 2\psi_3{}^2 = 1.333$. Subtract this from 1 (the charge on a neutral atom) to calculate the charge on C-1 = 1 − 1.333 = −0.333. Note that the resonance structures show a negative charge distributed over three atoms, C-1, C-3, and C-5.

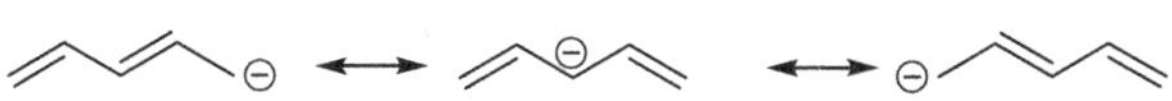

5.4 The thermally allowed processes can use 2, 4, or 6 electrons in the tropone, and 2 or 4 electrons in the cyclopentadiene.

Product a: 6 + 4 addition, s + s, thermally allowed.

Product b: 4 + 4 addition, s + s, thermally forbidden.

Product c: 4 + 2 addition, s + s, thermally allowed.

Product d: 6 + 2 addition, s + s, thermally forbidden.

Product e: 4 + 2 addition, s + s, thermally allowed.

Product f: 4 + 4 addition, s + s, thermally forbidden.

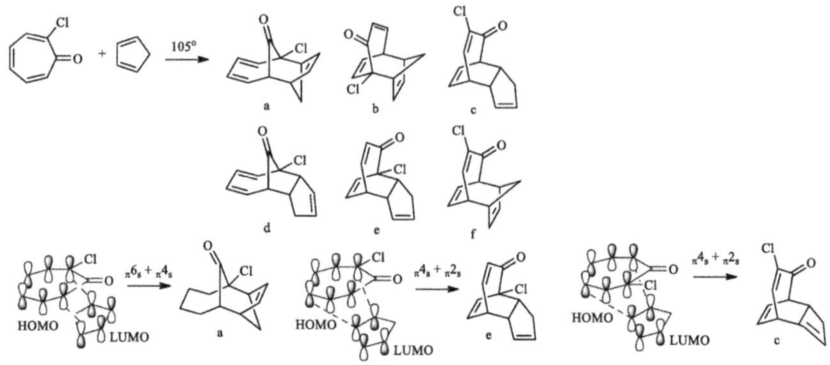

5.7 The reaction is a thermal 12 electron ring closure which occurs with conrotatory motion, leaving the Hs trans to one another.

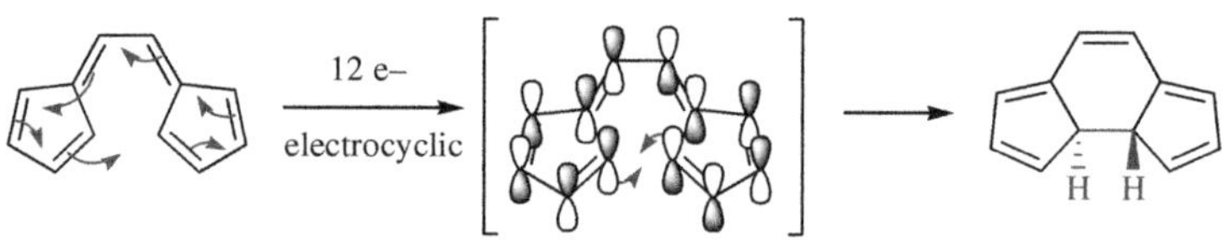

5.9 There are two reactions: a [3,3] sigmatropic shift (Cope) followed by a second [3,3] sigmatropic shift (Claisen).

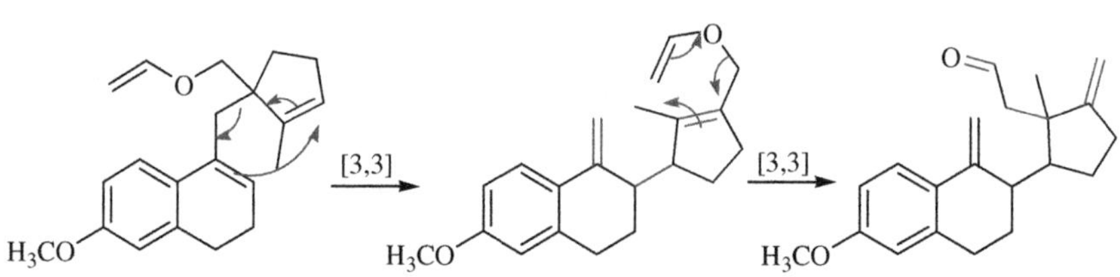

5.13 The reaction is a $[_{\pi}6_s + {_{\pi}4_s}]$ cycloaddition. The exo transition state explains the stereochemistry and is well-known in these systems (Garst, M. E.; Roberts, V. A.; Houk, K. N.; Rondan, N. G. *J. Am. Chem. Soc.* **1984**, *106*, 3882–3884).

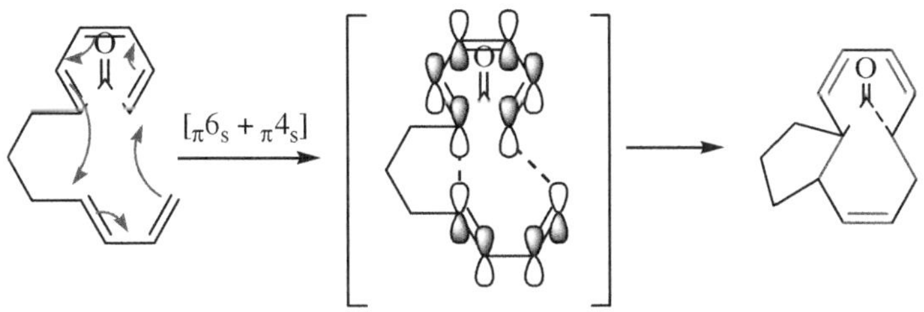

5.14 Catalytic hydrogenation results in a cis double bond, with the first structure preferred for steric reasons.

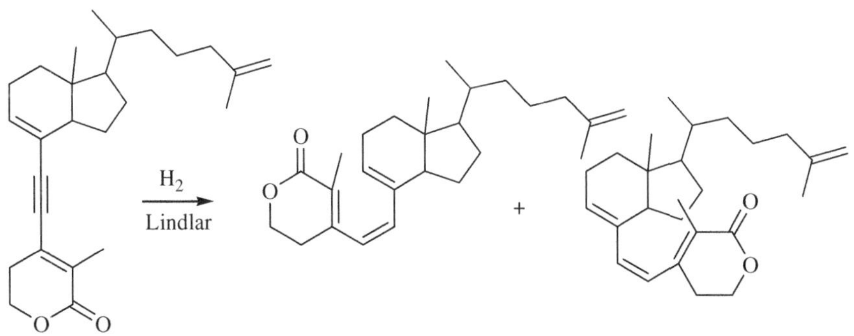

Reduction by Dibal opens the lactone.

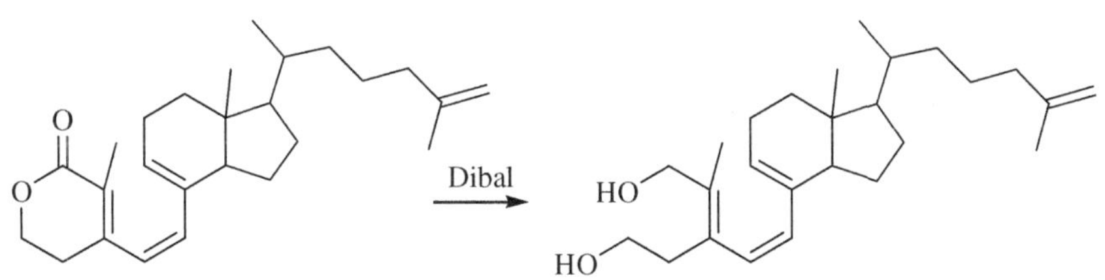

The final [1,7] antarafacial sigmatropic shift can only occur from the first of the two cis compounds.

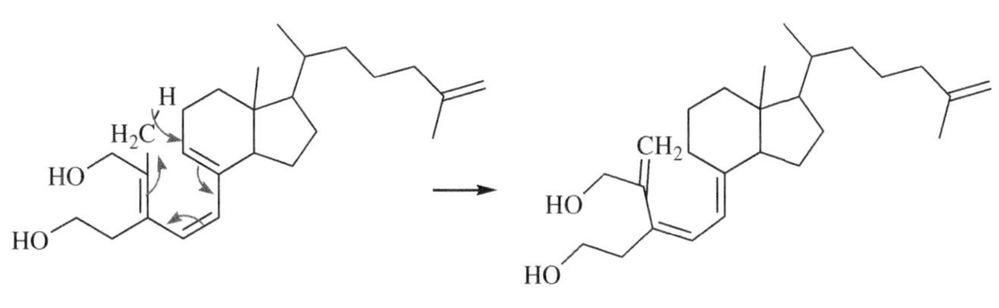

The orbitals for the [1,7] antarafacial shift are:

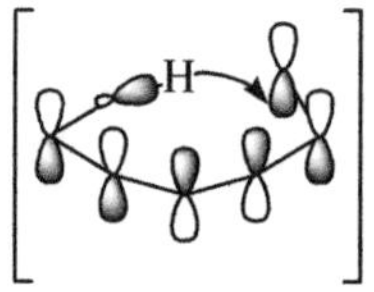

5.17 A [3,3] sigmatropic shift (Cope rearrangement) leads to a product that has five vinylic hydrogens.

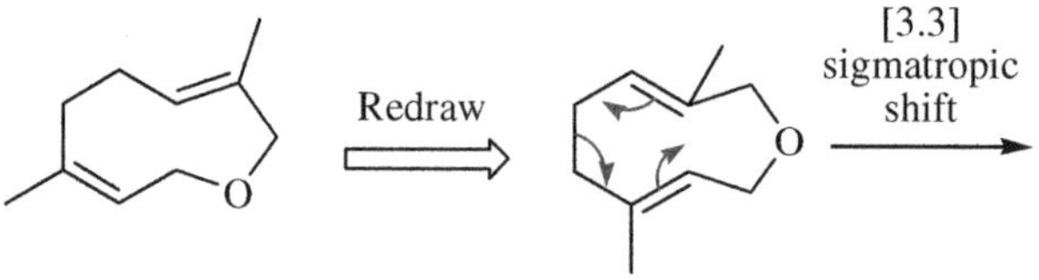

5.19 The endo product predominates in *in vitro* experiments, likely due to secondary orbital interactions between the developing π bond and the benzene ring. *in vivo*, the enzyme catalyzes a preferentially exo process.

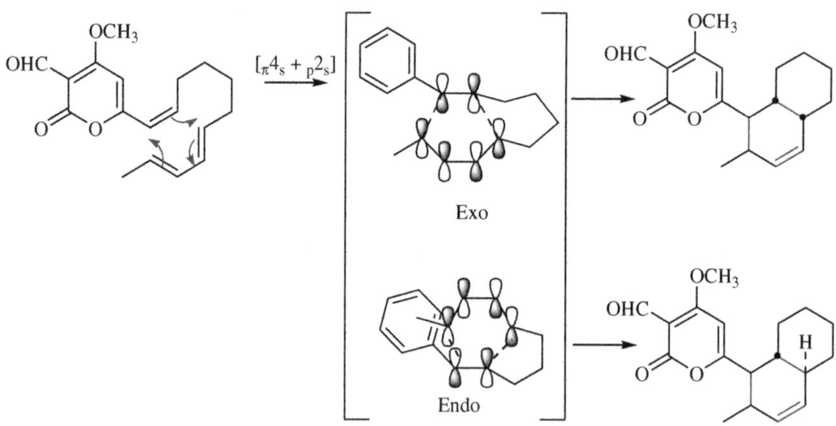

5.22

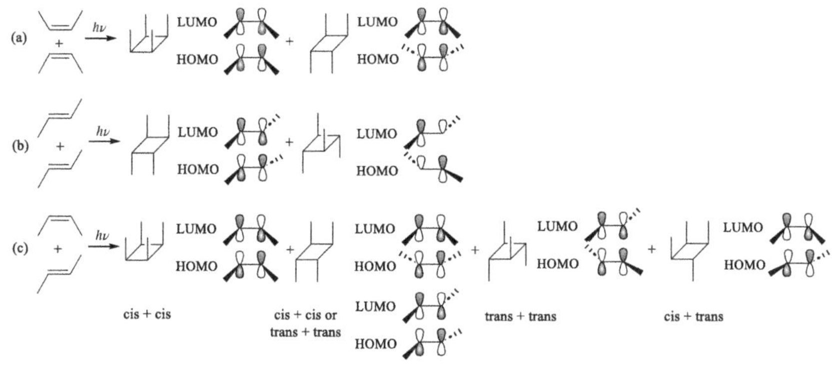

5.23 This is an example of an enzymatic ene reaction. The large difference in k_{cat} for the H- versus D-substituted isochorismate indicates that a bond to H is broken in a rate-determining step (primary isotope effect). That the deuterium is transferred to pyruvate quantitatively indicates a concerted reaction.

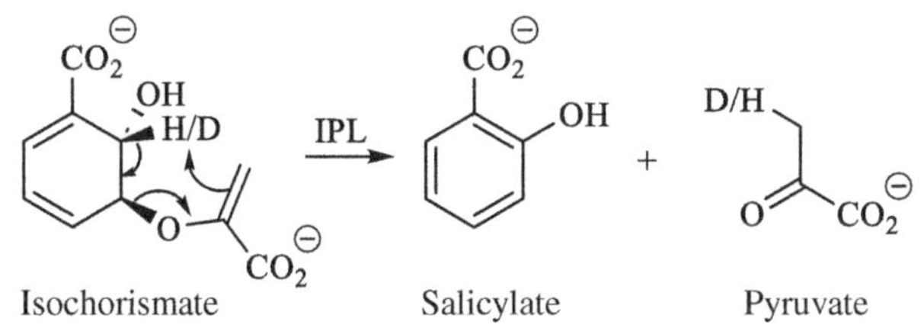

5.24 Product A is the result of a concerted ene reaction at the less substituted double bond, with the six-membered ring intermediate controlling the stereochemistry of the reaction. Product B is likely the result of a stepwise reaction that involves a complex between tin and the carbonyl group. The complex increases the polarity of the carbonyl group and thus the rate of the reaction.

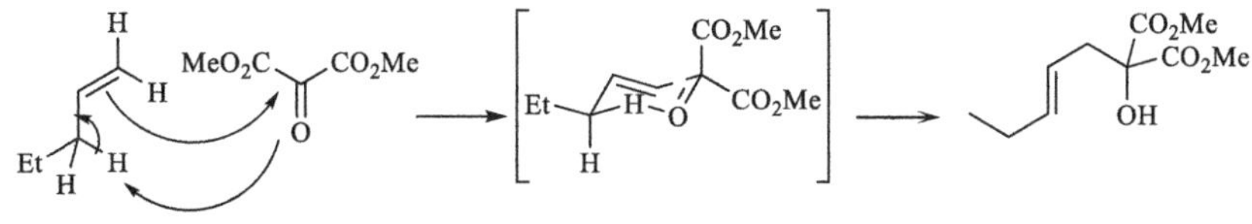

Chapter 6

6.1 The first step hydrolyzes the acetal and forms the oxime. The second step dehydrates the oxime to give the nitrile and forms the ester.

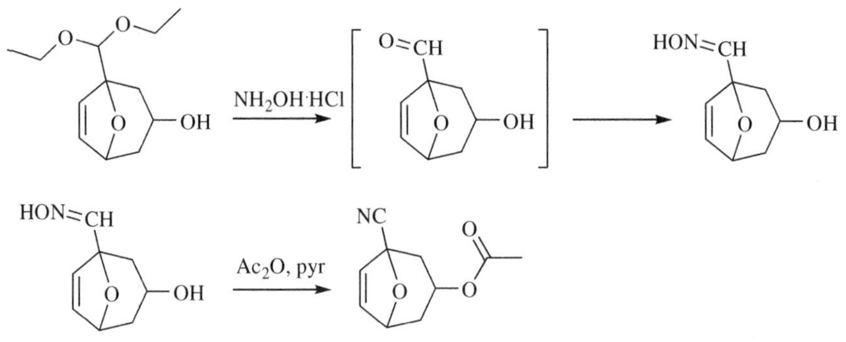

6.5 Ozonolysis followed by oxidation of the ozonide forms the diacid. Thionyl chloride forms the diacid chloride, which gives the diamide on treatment with dimethylamine. The diamide is reduced with lithium aluminum hydride (LAH).

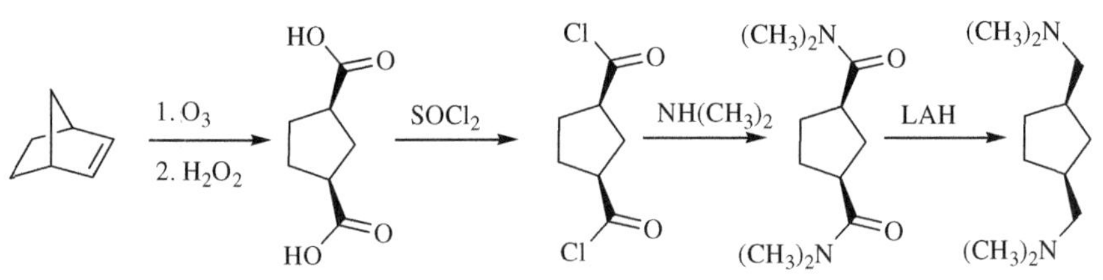

6.6 The hydroxylamine is formed first, then reduced with LAH.

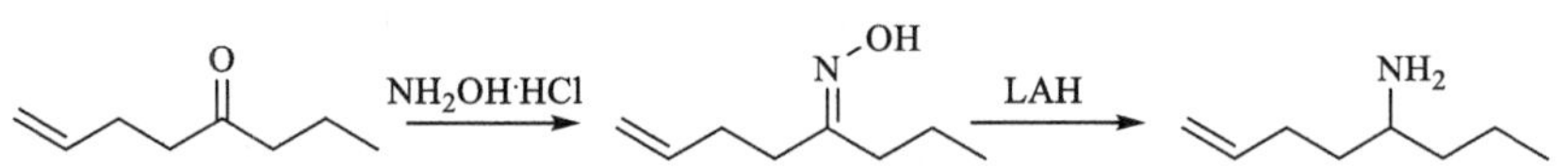

6.9 The 1,3-dioxolane is formed in two steps: first epoxidation followed by condensation with 3-pentanone.

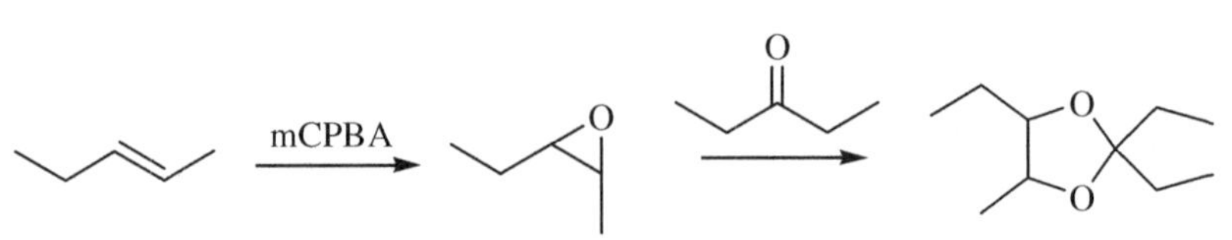

6.10 LDA forms the enolate which reacts with phenylselenium bromide to form the α-substituted ketone. Oxidation to the selenoxide is followed by elimination.

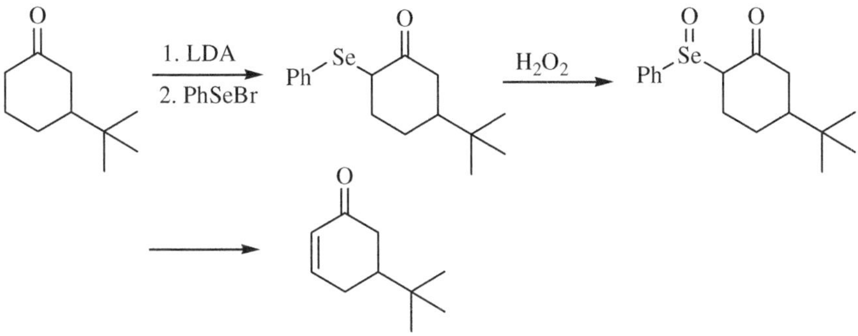

6.15 The imino ester forms first, and heat allows the orthoester to form.

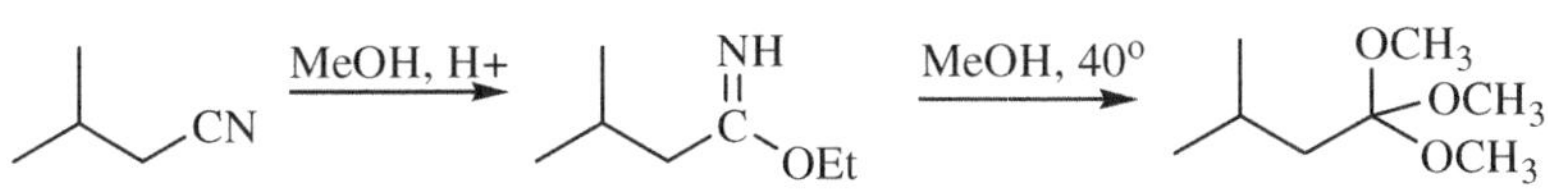

6.17 Dess–Martin oxidation of the alcohol is followed by base-catalyzed elimination with 1,8-diazabicyclo[5.4.0]undec-7-ene (DBU).

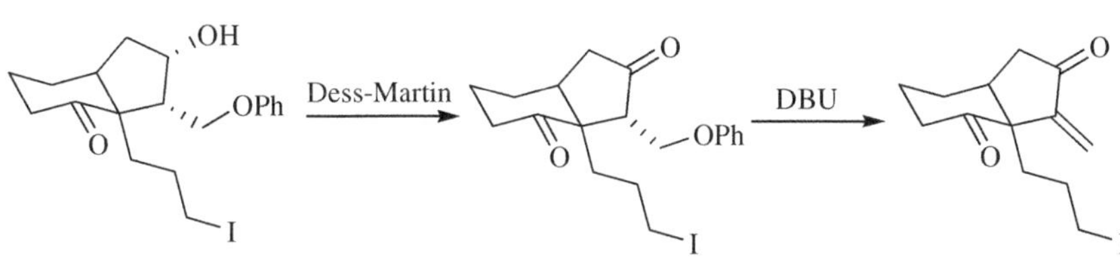

6.18 The first step forms the tosylhydrazone. Two equivalents of *n*-butyl lithium, in the Bamford–Stevens reaction, result in elimination to form the alkene.

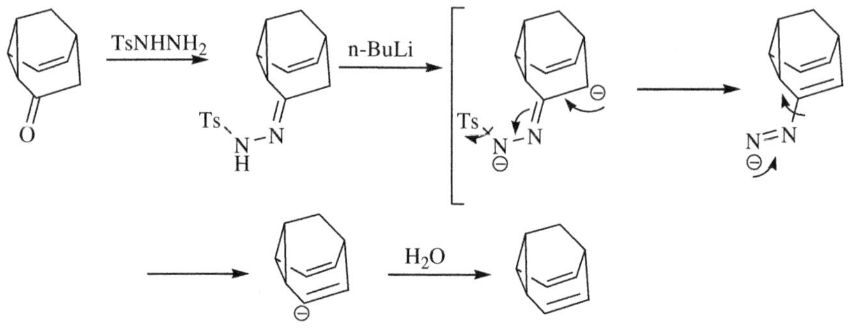

6.22 The aldehyde is reduced with sodium borohydride, protecting it during subsequent oxidation of the alkene to form the epoxide. The aldehyde is then regenerated by Dess–Martin periodinane.

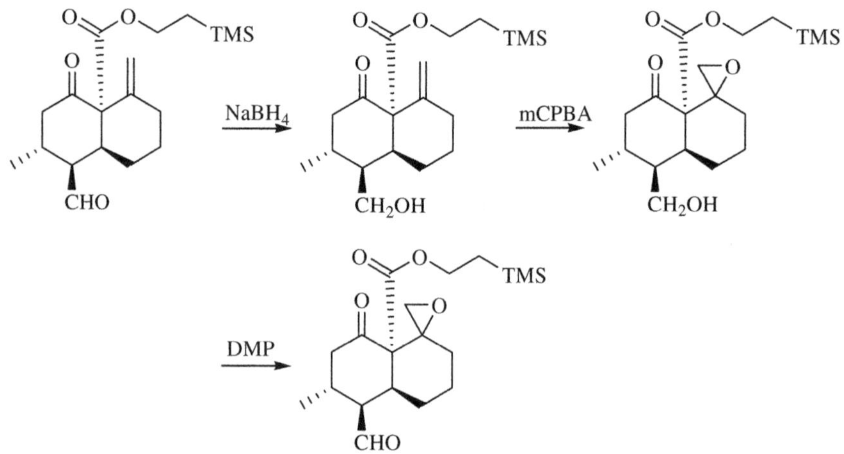

6.25 The anhydride is formed, and the alcohol is acetylated in the first step. The anhydride is opened by transesterification. Borane:THF is used to selectively reduce the acid in the presence of the ester. The lactone is formed under basic catalysis and the ester hydrolyzed under acidic conditions. The alcohol is fluorinated, and methyl lithium attacks the carbonyl to provide the diol. The monoacetate is formed with acetic anhydride, and the tertiary alcohol is protected. Reduction of the ester forms the primary alcohol, which is tosylated and then treated with sodium iodide to generate the product.

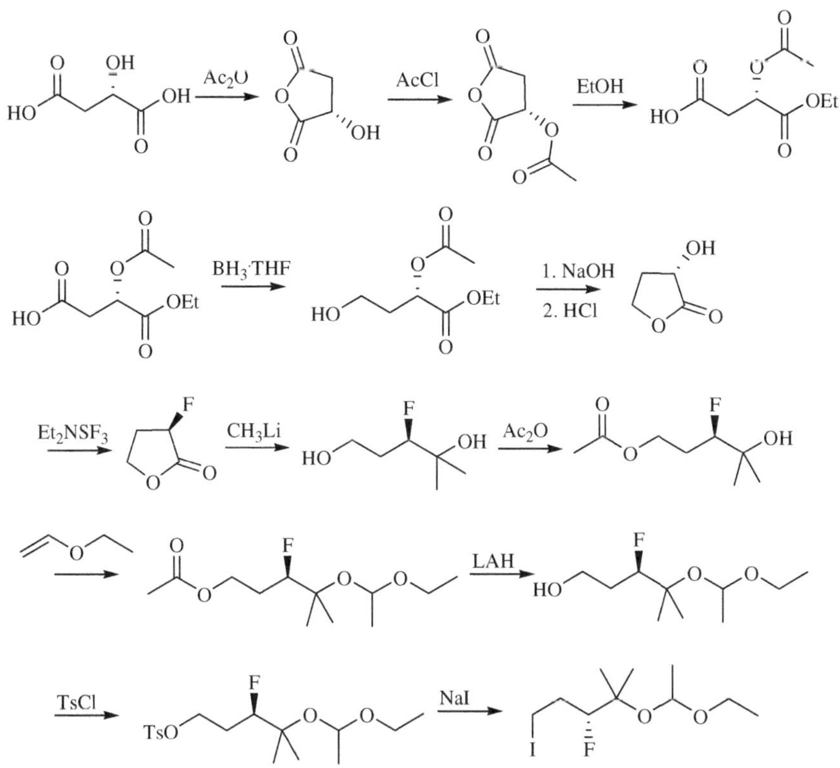

6.28 A Baeyer–Villiger oxidation is followed by opening the lactone and brominating the resultant alcohol.

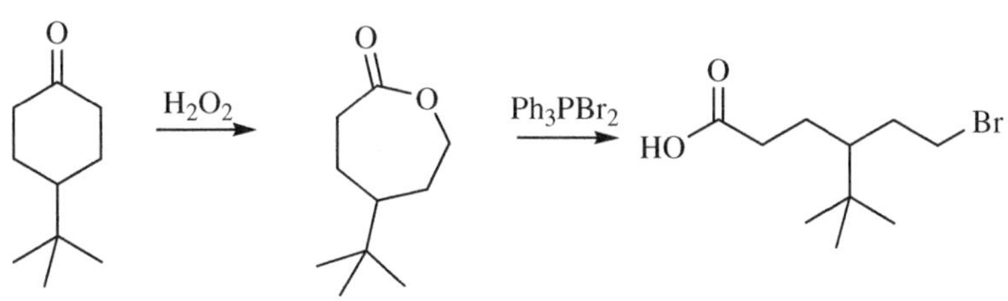

6.29 The methyl groups are first oxidized to acids, then the anhydride is formed.

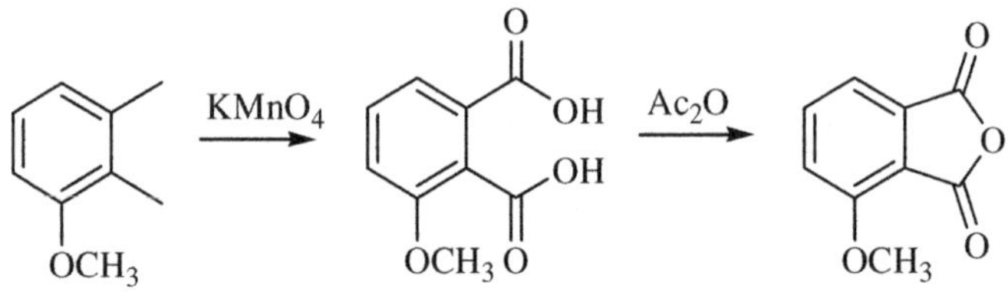

6.32 Reduction of the ester is followed by Grignard formation of the secondary alcohol, which is oxidized using Jones reagent. The acetal is formed, and ozonolysis of the alkene gives the aldehyde.

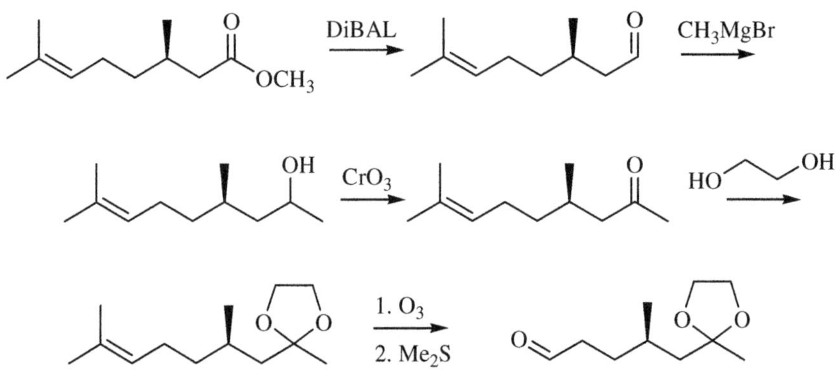

6.34 The acid is reduced to the alcohol, then reoxidized under mild conditions to form the aldehyde.

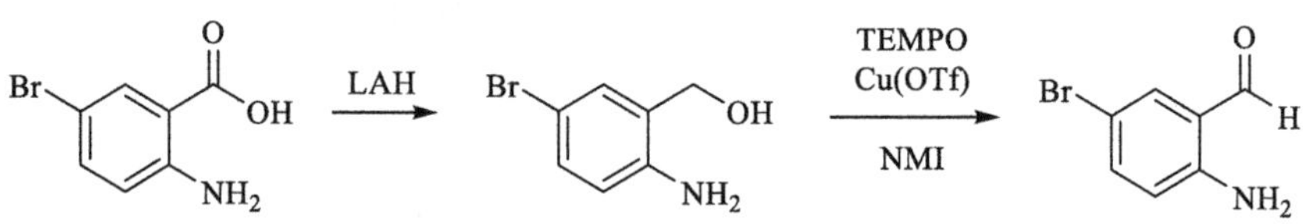

6.36 DIBAL reduces the ester to the alcohol, which is then brominated and reduced.

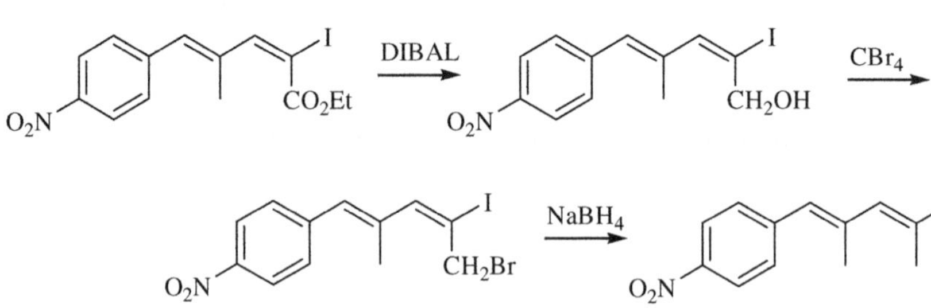

Chapter 7

7.1 Friedel–Crafts acylation of the benzene ring closes six-membered ring.

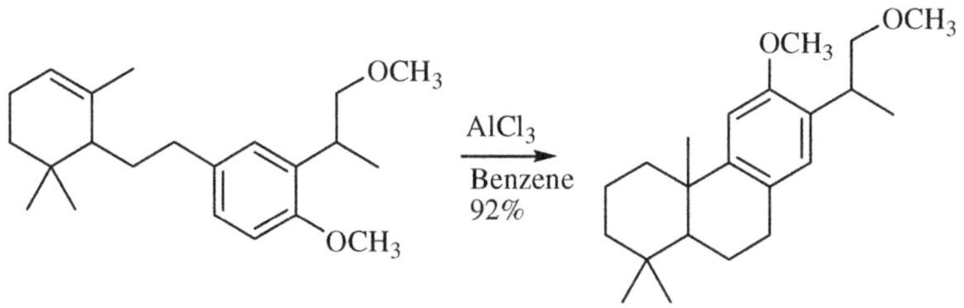

7.2 The acid is converted first to the acid chloride and then to the amide. *s*-BuLi removes a proton from the aromatic ring and then an attack on DMF attaches the aldehyde group. Sodium borohydride reduces the aldehyde to an alcohol, setting up the esterification.

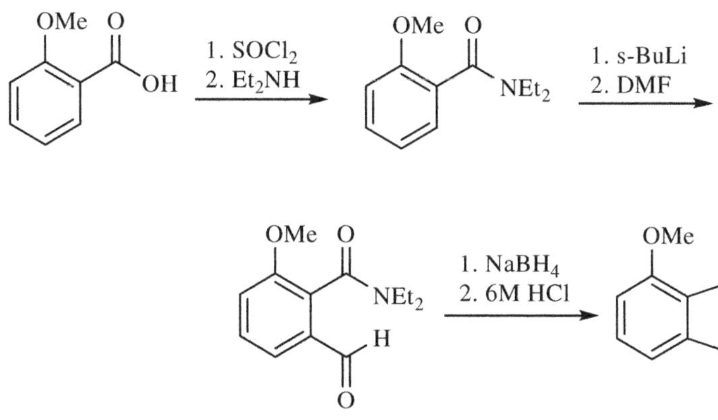

7.4 Vinyl ketones tend to polymerize in base, so the TMS ether is used as the nucleophile in the Michael addition.

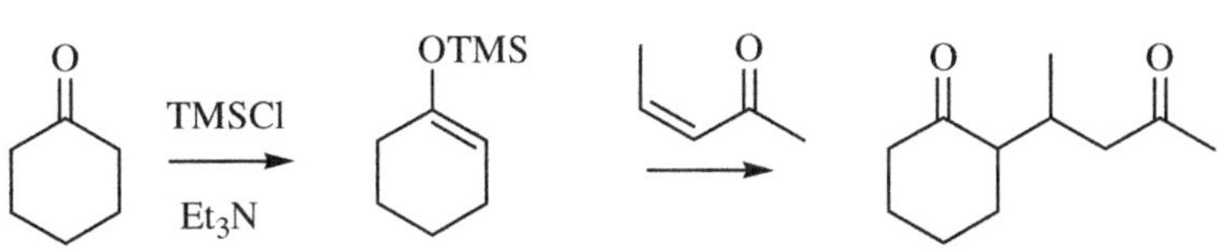

7.7 This is a photochemical ring closure.

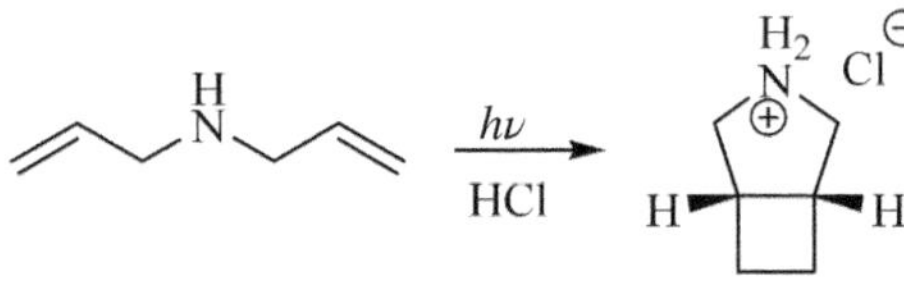

7.9 LDA is used to generate the dianion of butanoic acid which attacks the epoxide on the least hindered side. The five-membered ring closes to form the lactone.

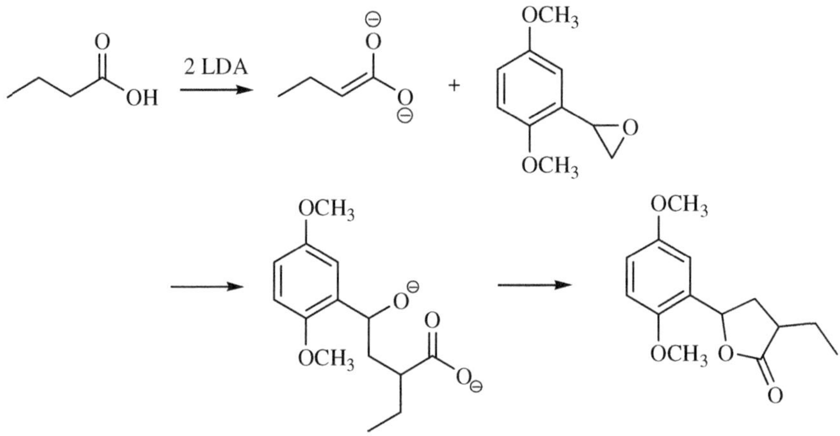

7.10 Dibromocarbene gives the three-membered ring, followed by coupling of the dialkylcuprate to give the dimethyl compound. Hydrolysis of the ester is followed by oxidation.

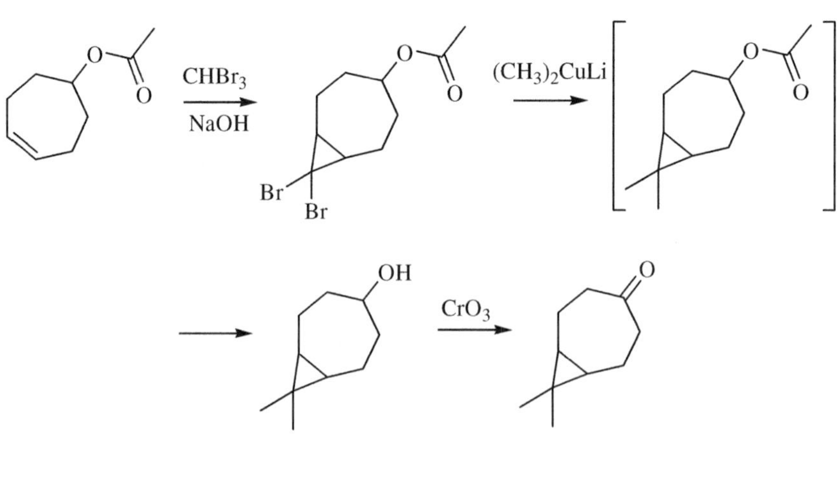

7.17 The alcohol is mesylated and then iodinated. The metal halogen exchange proceeds best with primary alkyl iodides. Cyclization followed by nucleophilic attack and hydrolysis of the nitrile gives the product.

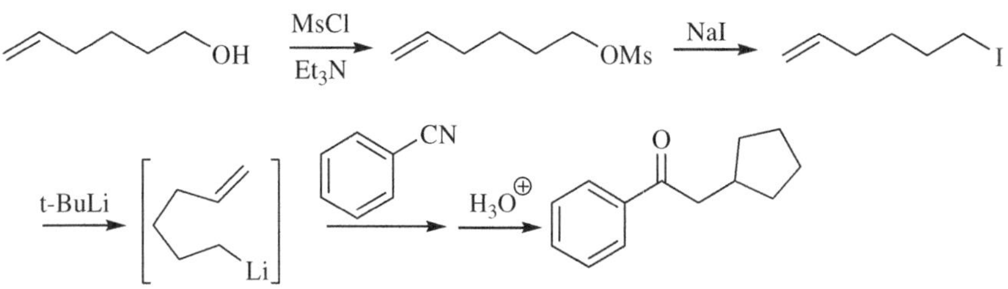

7.18 Alkylation of the enolate is followed by reduction of the ester. The alcohol is tosylated and then treated with sodium cyanide to form the nitrile. Ozonolysis gives the aldehyde.

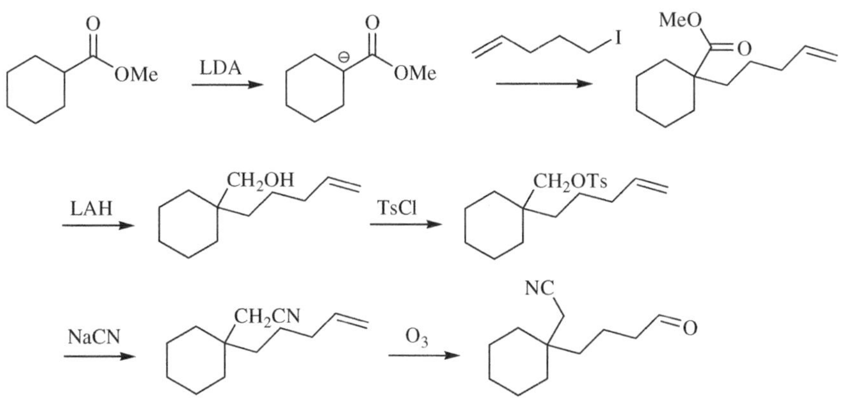

7.23 The silyl enol ether provides the dienophile in a Diels–Alder reaction.

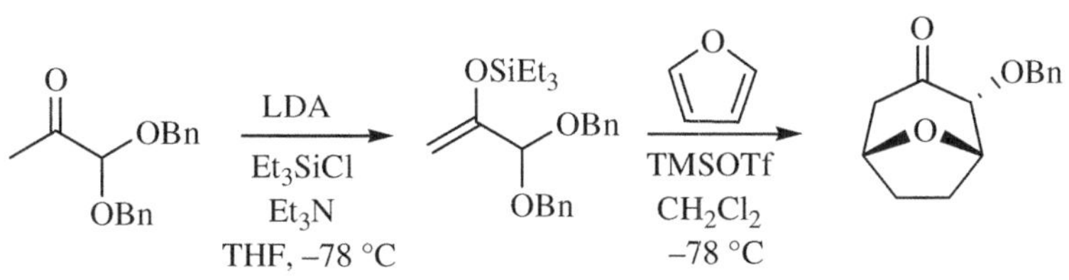

7.25 The initial Diels–Alder adduct is aromatized by enolization. The hydroxyl groups are then methylated.

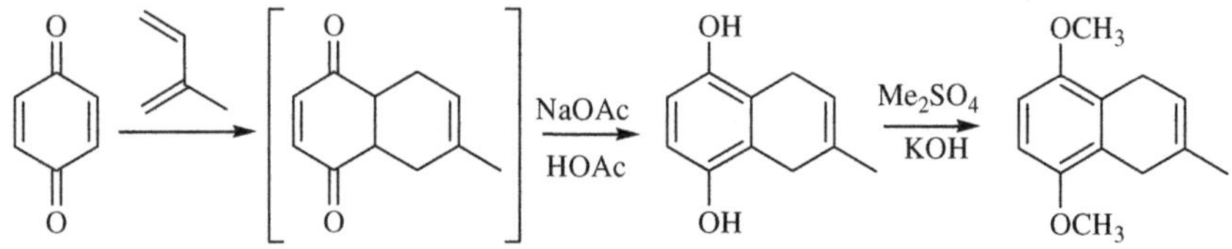

7.29 Acid-catalyzed cyclization is accomplished with concentrated sulfuric acid. Hydrogenation of the double bond gives the product.

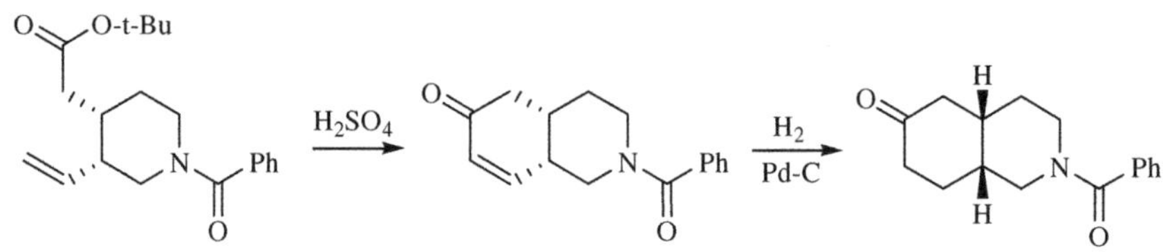

7.34 Alkylation and dehydration are followed by hydroboration to give the expected anti-Markovnikov alcohol. Swern oxidation provides the ketone.

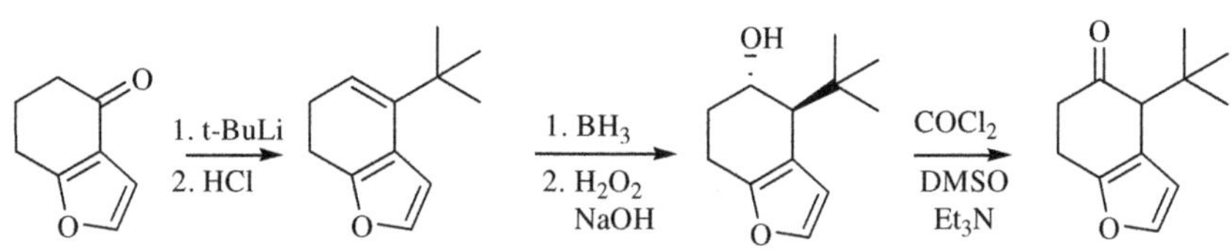

Chapter 8

8.1 Addition of HBr to 3-buten-2-one was followed by the protection of the carbonyl as the acetal. The Grignard reaction with propanal followed by acetylation and deprotection provided the product.

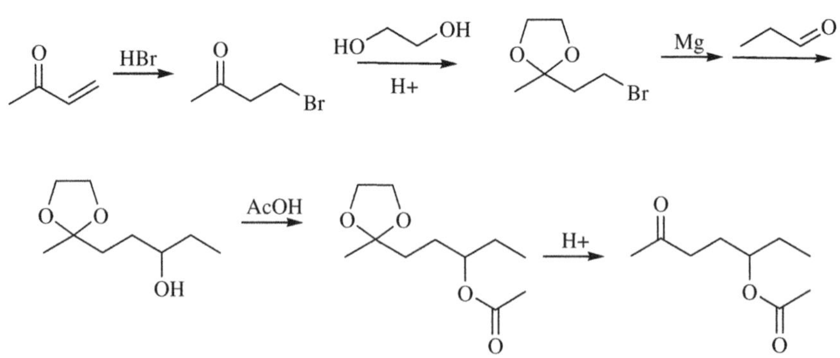

8.2 Dehydration was followed by alkylation of the alkyne. Hydrogenation gave the *cis*-alkene.

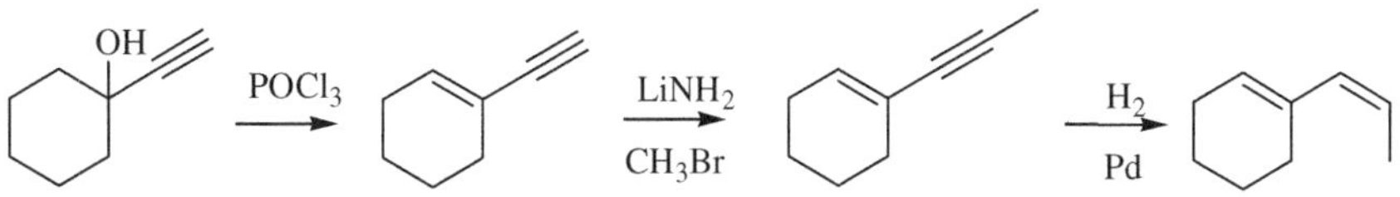

8.7 Addition to cyclohexenone was followed by an oxidative rearrangement, which sets up the Diels–Alder reaction.

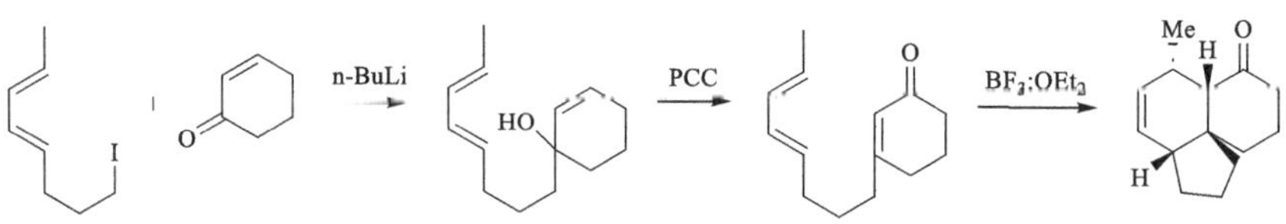

8.10 Cyclohexyne is formed, which then undergoes a Diels–Alder reaction with diphenylisobenzofuran.

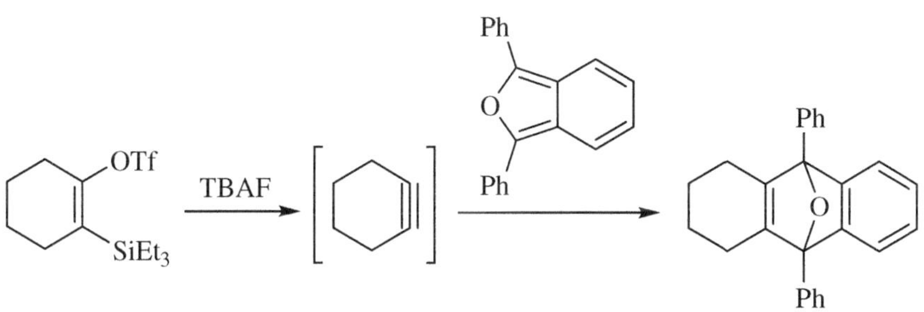

8.12 Dialkylation of ethyl acetoacetate formed the cyclopentene, and decarboxylation provided the product.

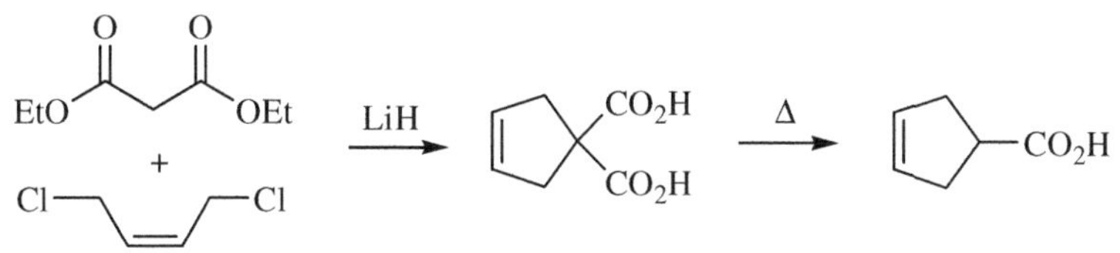

8.14 The imine is formed in the first step, then 1-ethyl-3-(3-dimethylaminopropyl)carbodiimide hydrochloride (EDC) is used to form the ester, in place of the more common *N,N′*-dicyclohexylcarbodiimide (DCC), because it avoids the need for the lengthy purification DCC requires.

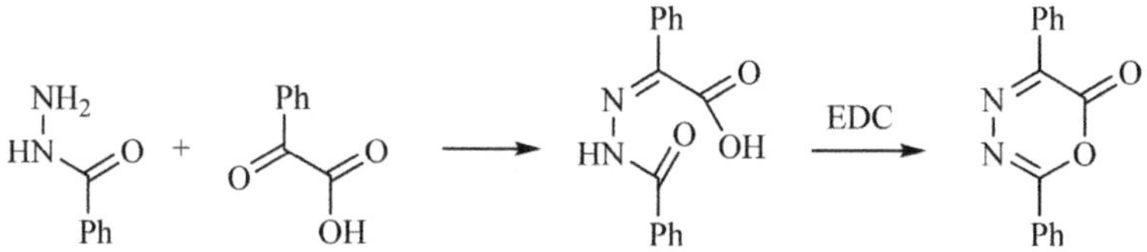

8.19 The amide is formed using methyl malonyl chloride, and then a Lacey–Dieckmann cyclization closes up the ring. A decarboxylation leads to the tetramic acid product. This is an interesting class of compounds, and the student is encouraged to consult the reference.

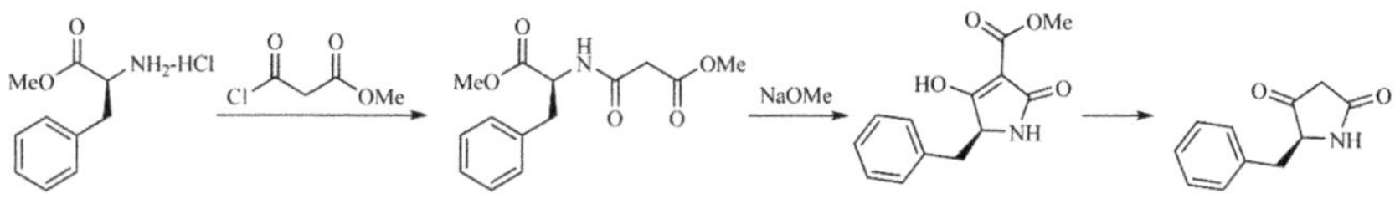

8.27 The δ,γ-unsaturation suggested a Claisen rearrangement.

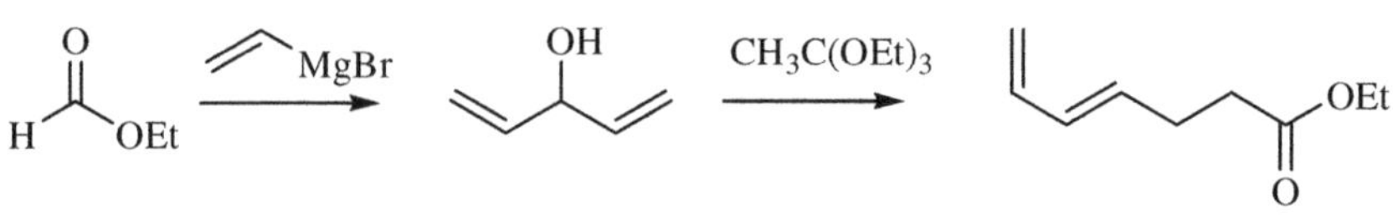

8.31 The cyanoenolate was alkylated twice and then the nitrile was reduced.

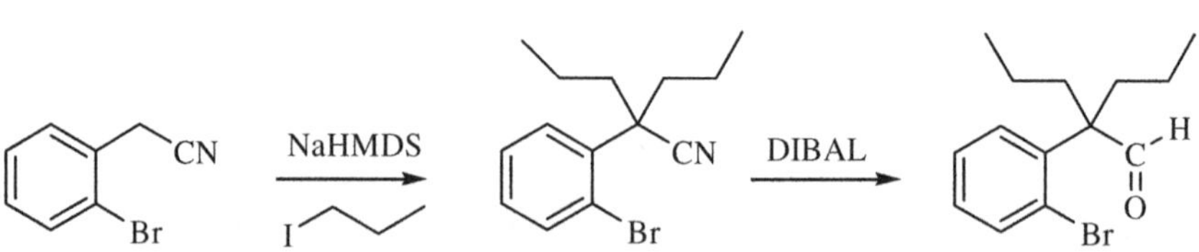

8.35 Benzyne is formed with the addition of *n*-BuLi, which undergoes a Diels–Alder reaction. The cycloadduct ring opens to form the para-benzoquinone.

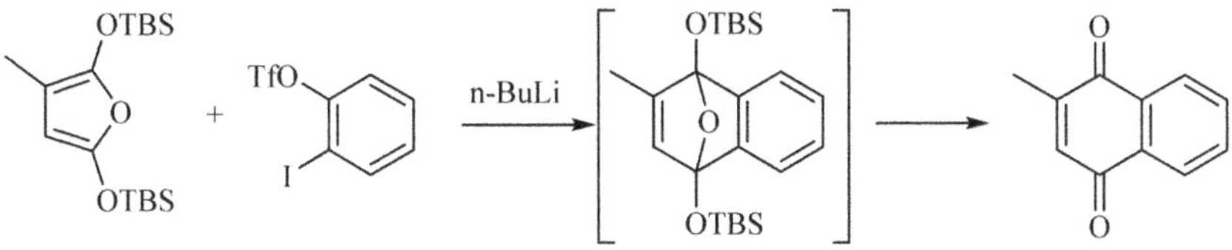

8.36 An ene reaction sets up the intramolecular Diels–Alder reaction.

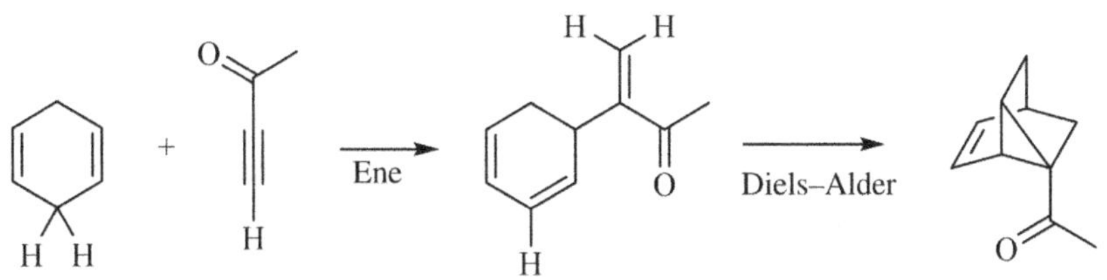

8.37 Benzylic halogenation is performed under UV radiation. Triphenylphosphine forms the ylide, which undergoes a Wittig reaction to form the new double bond.

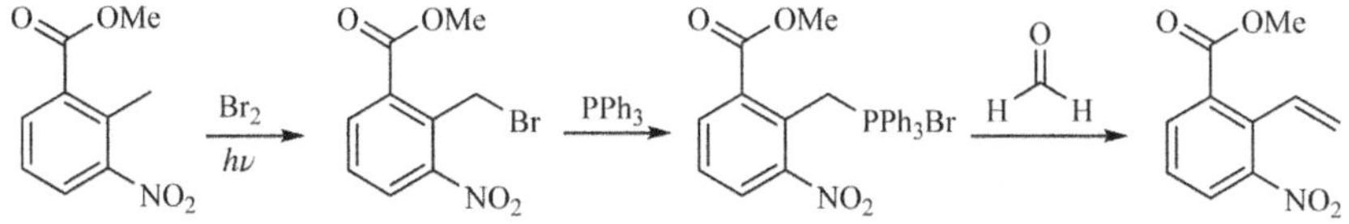

Chapter 9

9.1 The authors in the 1982 paper (Ref. 87: Bal, S. A.; Marfat, A.; Helquist, P. *J. Org. Chem.* **1982,** *47*, 5045–5050) used Rieke magnesium, prepared using anhydrous magnesium chloride and potassium metal (Ref. 48: Rieke, R. D.; Bales, S. E.; Hudnall, P. M.; Burns, T. P.; Poindexter, G. S. *Org. Synth.* **1979**, *59*, 85–94). Ultrasonication may also have worked.

9.2 The dry ozonation technique (Ref. 65: Cohen, Z.; Varkony, H.; Keinan, E.; Mazur, Y. *Org. Synth.* **1979,** *59*, 176–182) would adsorb the hydrocarbon on silica gel followed by treatment with ozone.

9.3 Ultrasonic irradiation using zinc dust and iodine may improve the yield (Ref. 56: Han, B.-H.; Boudjouk, P. *J. Org. Chem.* **1982,** *47*, 5030–5032). Activated Rieke zinc, formed by the reduction of zinc chloride with potassium, may also work.

9.4 The 1,2-diketone is likely an intermediate in the formation of the carboxylic acid (Lee, D. G.; Lee, E. J.; Chandler, W. D. *J. Org. Chem.* **1985**, *50*, 4306–4309), which is formed when the Mn^{3+} ion is present to continue oxidation to the acid. Under neutral conditions, using acetone and water as co-solvents (Ref. 90: Srinivasan, N. S.; Lee, D. G. *J. Org. Chem.* **1979**, *44,* 1574), the diketone is the major product. Phase-transfer conditions would also work.

9.5 Sonication of the tubing in which the mixing occurred allowed for dispersal of the precipitate.

9.6 The references provide detailed experimental procedures for both reactions shown. With the information provided in the question, however, the advantages of Reaction B are clear. The higher yield is an advantage, the procedure is simpler and requires less hazardous reagents, and the waste products are less hazardous. The authors of Ref. 95 provide a listing of seven other syntheses of the product and while some occur in higher yield, none are as simple and cost-effective as Reaction B. If students check the references, they will find that Reaction A uses substantially more of several solvents.

Percent atom efficiency:

Reaction A:

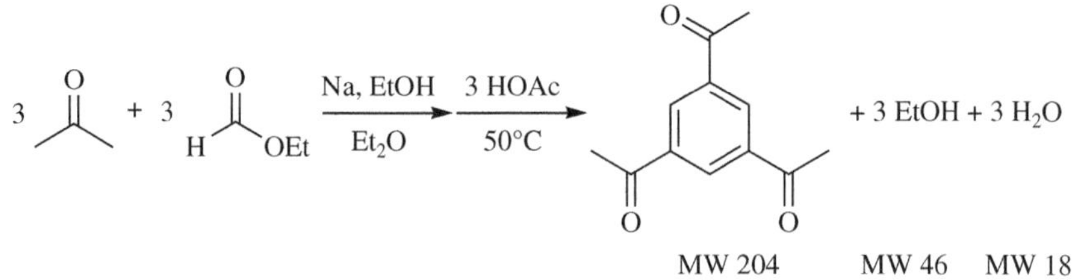

$$\%AE = 204/(204 + 3 \times 46 + 3 \times 18) = 52\%$$

Reaction B:

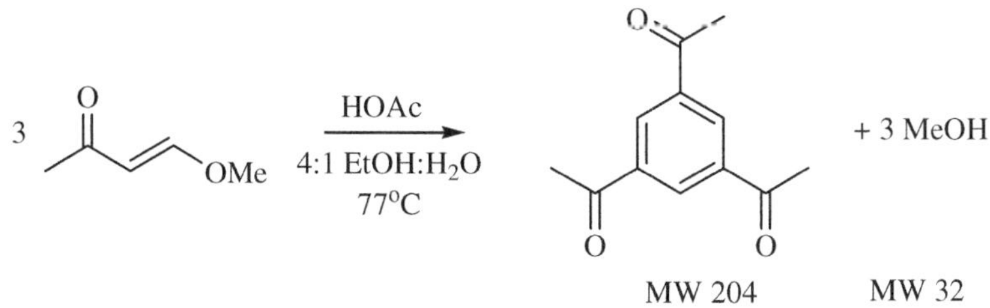

$$\%AE = 100 \times 204/(204 + 3 \times 32) = 68\%$$

9.7 Judging by atom economy alone, the reaction looks pretty good, but the workup has a significant negative effect on the greenness of this reaction.

$$\text{Atom economy} = [(345.41)/(171.22 + 106.12 + 86.04)]^{*}\ 100$$
$$= 95\%$$

$$\text{E-factor (based on average student yield)} = 382\ g/0.865\ g = 442$$

To calculate the weight of waste, multiply volume by density to get the weight of each component, then add them up, and subtract off the weight of the product. The number 382 g might vary a little depending on rounding and such.

The biggest contributor to the large E-factor is workup, and within workup, the dichloromethane and aqueous extractions contribute the most. Try to substitute a less dense, less hazardous solvent for dichloromethane, and look for a purification technique that requires fewer steps to bring the E-factor down. Dichloromethane is worth 31% of the waste. Substituting ethyl acetate would bring that down to 21%.

9.8 Under batch conditions, a higher volume of solvent contributes to the higher E-factor. The heterogeneous catalyst simplifies purification steps because of the ease with which it can be removed. Use of flow conditions helps minimize unnecessary solvent use. The POLITAG catalyst increases the efficiency of TMSCN by activating it (transferring a fluoride atom to the silicon). The E-factor drops in each subsequent experiment as the reaction is made more efficient.

9.9

$$\text{Atom efficiency} = [528/(182 + 162 + 238)] * 100 = 91\%$$

$$\text{E-factor} = [(0.182 + 0.162 + 0.238) - 0.449]/0.449 = 0.29$$

The E-factor is so low because there is no workup step. Products precipitate and the ethanol used in rinsing is recycled. This multicomponent reaction is atom-efficient because the reagents are all integrated into the product.

Chapter 10

10.1 Zeaxanthin has a calculated λ_{max} of 453 nm and ε_{max} of 1.9×10^{-5}/M/cm (0.019 cm^2/mol) and observed λ_{max} of 452 nm and ε_{max} of 1.4×10^{-5}/M/cm (0.014 cm^2/mol).[1,2]

Table 10.3 predicts that compounds that absorb at 452 nm, at the border between purple and blue, will appear yellow-orange. Zeaxanthin is a carotenoid, so this color range is expected.[3]

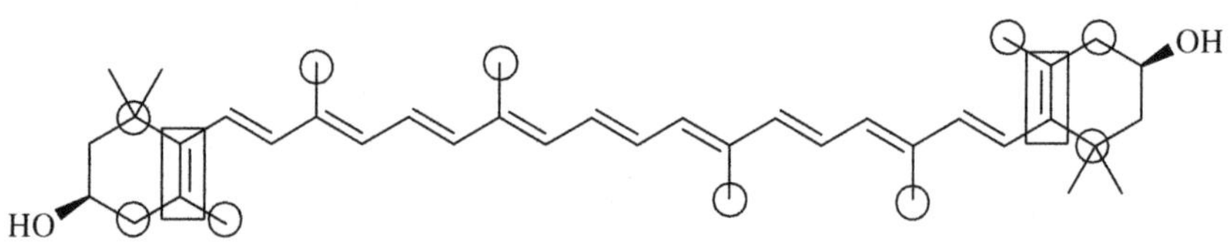

(substitutions are circled, endo double bonds are boxed)

$$\lambda_{max} = 114 + 5M + n(48.0 - 1.7n) - 16.5R_{endo} - 10R_{exo}$$

$$\lambda_{max} = 114 + 5(10) + 11(48.0 - 1.7(11)) - 16.5(2) - 10(0)$$

$$= 453 \text{ nm}$$

$$\varepsilon_{max} = 1.74 \times 10^4 n = 1.74 \times 10^4(11) = 1.9 \times 10^5$$

[1]Sajilata, M. G.; Singhal, R. S.; Kamat, M. Y. *Compr. Rev. Food Sci. Food Saf.* **2008**, *7*, 29–49.
[2]Widomska, J.; Kostecka-Gugała, A.; Latowski, D.; Gruszecki, W. I.; Strzałka, K. *Biophys. Chem.* **2009**, *140*, 108–114.
[3]Yarnell, A. *Chem. Eng. News* **2005**, *83*, 10.

10.2 Plotting absorbance versus time gives a straight line with a slope of 0.209/min. Convert this ΔA/min to Δc/min using the molar absorption coefficient and the Beer–Lambert law. Then, calculate the activity of the enzyme in units per milliliter.

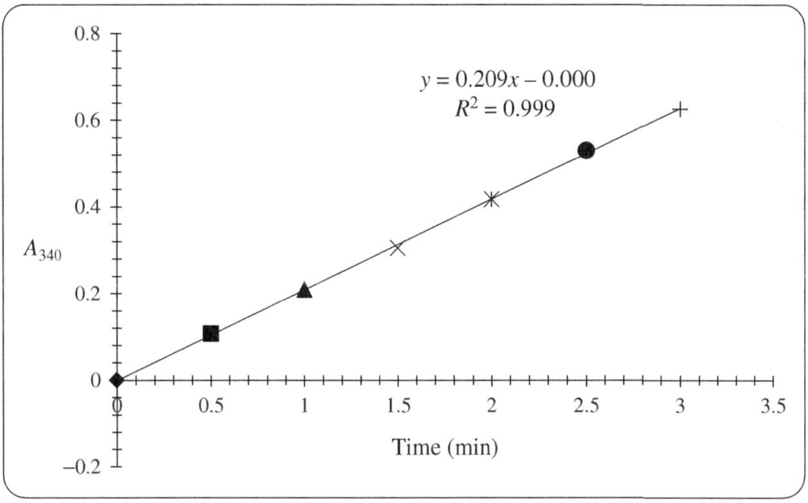

$$\frac{\Delta c}{\min} = \left(\frac{\Delta A}{\varepsilon b}\right)\frac{0.209/\min}{\left(\frac{6220}{\text{M}-\text{cm}}\right)(1\ \text{cm})} = 3.36 \times 10^{-5}\ \text{M NADH}/\min$$

$$\text{activity} = \left(\frac{3.3 \times 10^{-5}\ \text{M NADH}}{\min}\right)\left(\frac{1 \times 10^{6}\ \mu\text{mole}}{\text{mole}}\right) \times \left(\frac{\text{L}}{1000\ \text{mL}}\right) = \frac{0.0336\ \text{units}}{\text{mL}}$$

10.3

$$\tilde{\nu} = \frac{1}{2\pi c}\sqrt{f\frac{M_x + M_y}{M_x M_y}}$$

$$M_C = 1.99 \times 10^{-23}\ \text{g}, M_O = 2.65 \times 10^{-23}\ \text{g}$$

$$\tilde{\nu} = \frac{1}{2\pi(3 \times 10^{10}\ \text{cm/sec})}\sqrt{(12 \times 10^{5})\frac{1.99 \times 10^{-23} + 2.65 \times 10^{-23}}{(1.99 \times 10^{-23})(2.65 \times 10^{-23})}}$$

$$= 1729\ \text{cm}^{-1}$$

which is where the carbonyl stretch shows up.

10.4

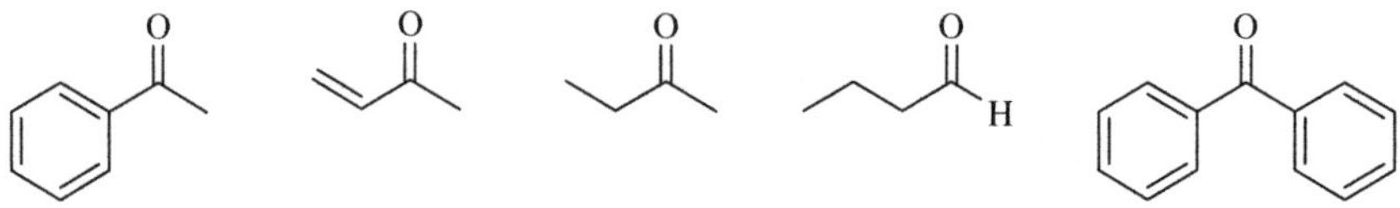

1686 cm^{-1} 1720 cm^{-1} 1722 cm^{-1} 1666 cm^{-1} 1731 cm^{-1}

10.5 Compound A is isopropylbenzene and compound B is propylbenzene.

10.6 Compound C is 1,2-dibromopropane.

10.7

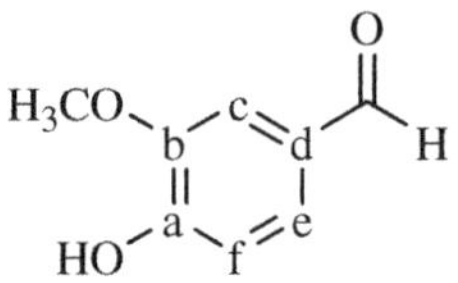

Peak	Base	–OH	–OCH$_3$	–CHO	Cal'd	Obs'd[1]
a	128.5	+26.9	−14.4	+5.8	146.8	152.18
b	128.5	−12.7	+31.4	+0.6	147.8	147.50
c	128.5	+1.4	−14.4	+1.2	116.7	109.34
d	128.5	−7.3	+1.0	+8.2	130.4	129.77
e	128.5	+1.4	−7.7	+1.2	123.4	127.49
f	128.5	−12.7	+1.0	+0.6	117.4	114.75

10.8 The two methyl groups on carbon 4 are diastereotopic.

10.9

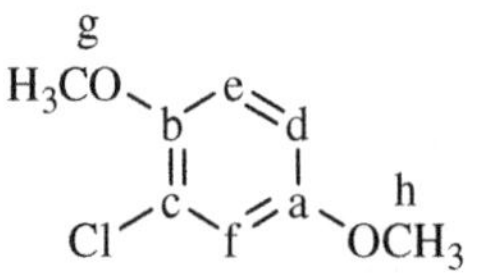

55.8 (h)	56.8 (g)	112.9 (f)
113.4 (e)	116.2 (d)	123.1 (c)
149.5 (b)	153.9 (a)	

[1]Values obtained from Spectral Database of Organic Compounds (SDBS). http://sdbs.riodb.aist.go.jp/sdbs/cgi-bin/cre_index.cgi

55.8 (h), 56.8 (g), 112.9 (f), 113.4 (e), 116.2 (d), 123.1 (c), 149.5 (b), and 153.9 (a).

10.10 Calculating the distances between the peaks allows for the assignment of the peaks at 1.254, 1.357, and 1.451 to the triplet. The doublet is then the peaks at 1.332 and 1.410. The peak position for the triplet is 1.357 (the center peak of the triplet) and the doublet is 1.371 (the average position of the two doublet peaks.

	At 90 MHz	At 300 MHz
$J_{triplet}$	$(1.451 - 1.357) \times 90 = 8.46$ Hz	8.46 Hz/300 MHz = 0.028 ppm
$J_{doublet}$	$(1.410 - 1.332) \times 90 = 7.02$ Hz	7.02 Hz/300 MHz = 0.023 ppm

Peak positions

At 90 MHz	At 300 MHz
1.254	1.357 − 0.028 = 1.329
1.332	1.371 − 0.023 = 1.348
1.357	1.357
1.410	1.371 + 0.023 = 1.394
1.451	1.357 + 0.028 = 1.385

A scaled plot of the peaks shows an ambiguous separation of the doublet and triplet at 90 MHz, with a clearer differentiation at 300 MHz.

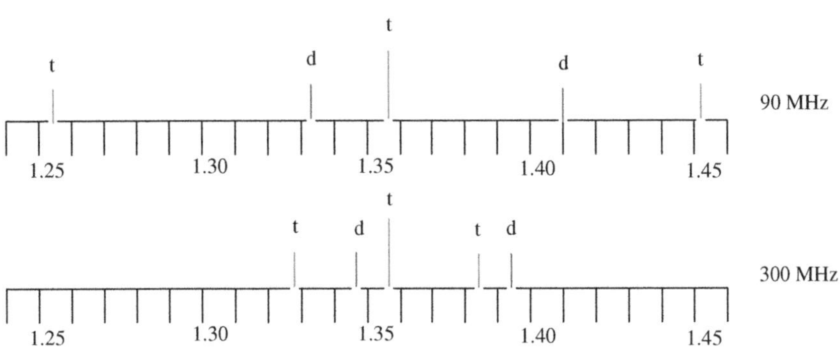

10.11 Peak assignments in the PMR and CMR are shown in the figures below.

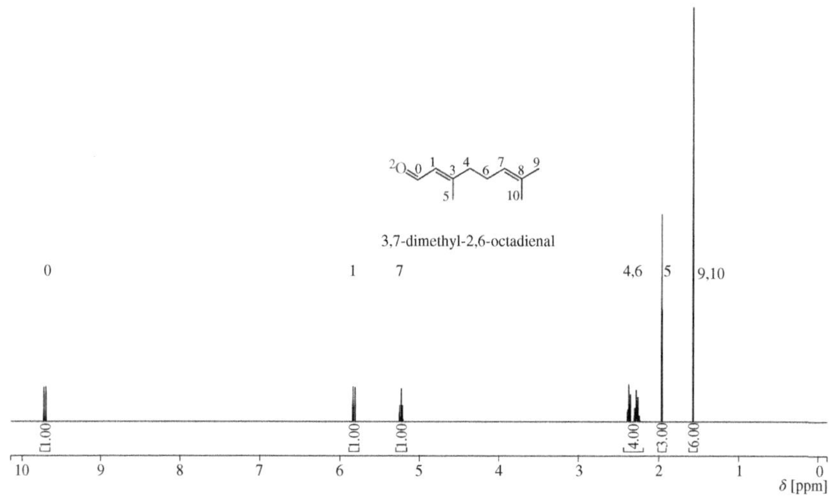

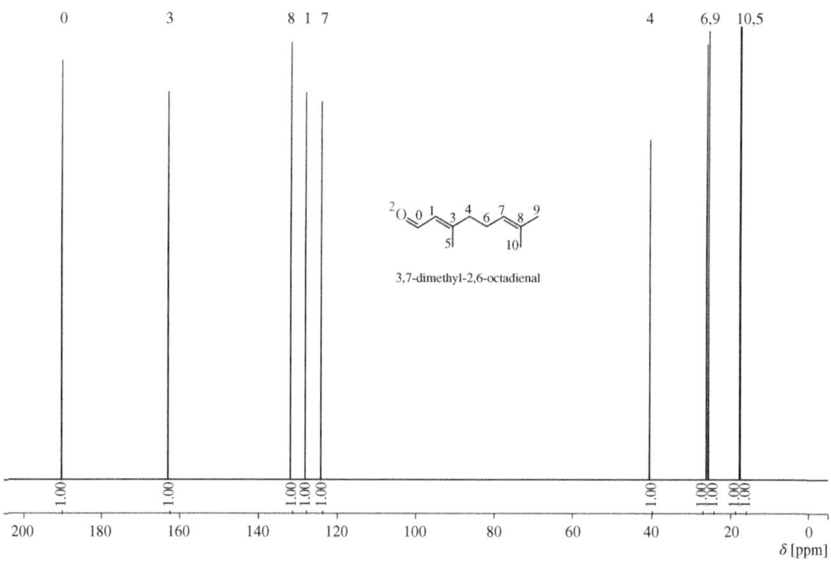

10.12 Peak assignments in the PMR and CMR are shown in the figures below. Analysis of the aromatic region is also shown.

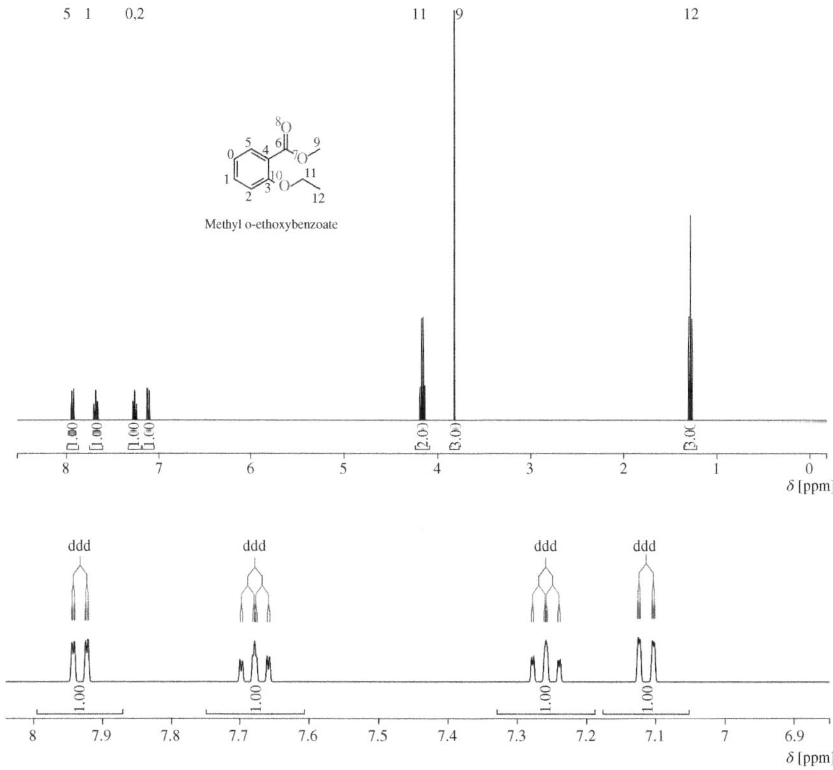

Peak Assignment	Peak Position	Multiplicity	*J* Value (Hz)	*J* Assignment
5	7.93	ddd	8.07, 1.43, 0.46	*J*(5-0) = 8.07 ortho *J*(5-1) = 1.43 meta *J*(5-2) = 0.46 para *J*(0-2) = 1.24 ortho *J*(1-2) = 8.33 para
1	7.68	ddd	8.33, 7.35, 1.43	
0	7.26	ddd	8.07, 7.35, 1.24	
2	7.11	ddd	8.33, 1.24, 0.46	

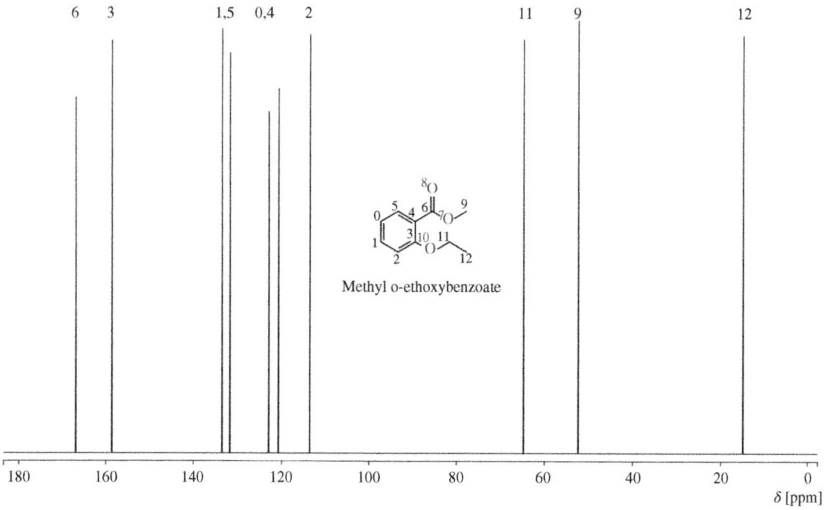

10.13 Peak assignments in the PMR and CMR are shown in the figures below.

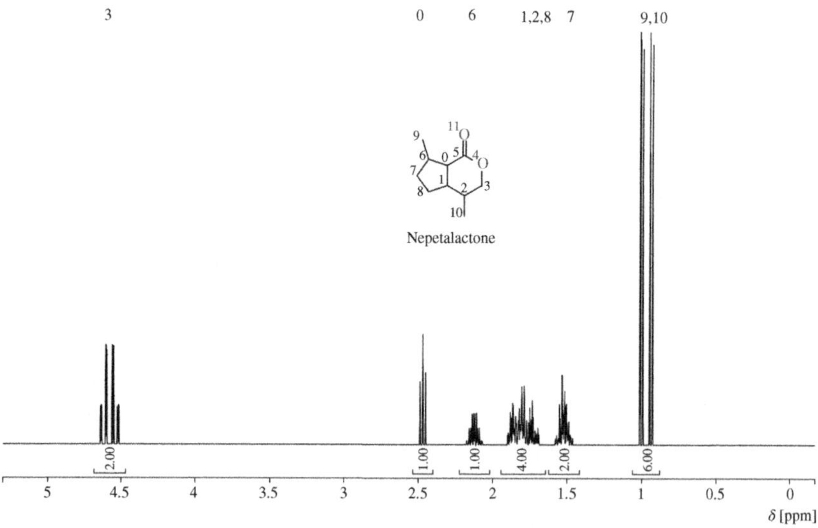

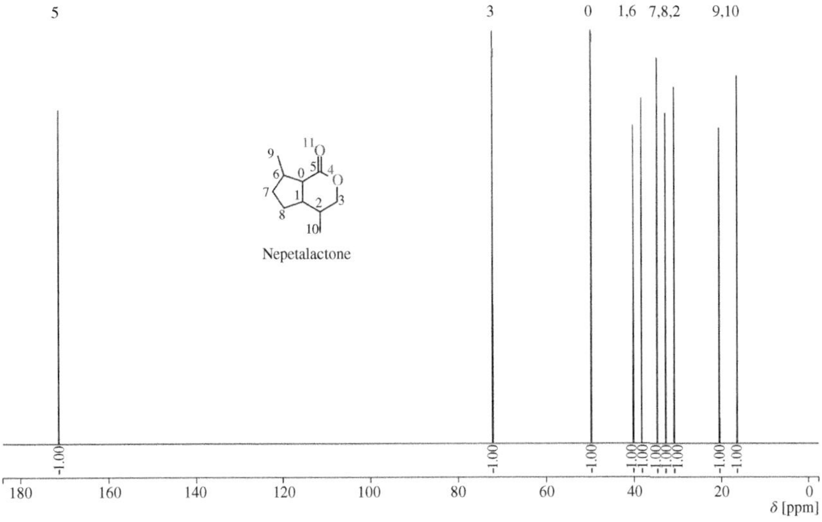
5
3
0
1,6
7,8,2
9,10
11
9
6
0
5
4
O
O
7
1
2
3
8
10
Nepetalactone
-1.00
-1.00
-1.00
-1.00
-1.00
-1.00
-1.00
-1.00
-1.00
-1.00
180
160
140
120
100
80
60
40
20
0
δ [ppm]

INDEX

Note: *Italic* page numbers refer to *figures* and **Bold** page numbers reference to **tables**.

Understanding Organic Chemistry: Bridging Fundamental and Advanced Courses, Fourth Edition. Ann M. Fabirkiewicz and John C. Stowell.

Companion website: www.wiley.com/go/fabirkiewiczorganic4e

Printed and bound by CPI Group (UK) Ltd, Croydon, CR0 4YY
17/05/2026
14880776-0001